AF753748

MODELING SOLID-STATE PRECIPITATION

MODELING SOLID-STATE PRECIPITATION

ERNST KOZESCHNIK

MOMENTUM PRESS

MOMENTUM PRESS, LLC, NEW YORK

Modeling Solid-State Precipitation

Copyright © Momentum Press®, LLC, 2013.

All rights reserved. No part of this publication may be reproduced, stored in a retrieval system, or transmitted in any form or by any means—electronic, mechanical, photocopy, recording, or any other—except for brief quotations, not to exceed 400 words, without the prior permission of the publisher.

First published by Momentum Press®, LLC
222 East 46th Street, New York, NY 10017
www.momentumpress.net

ISBN-13: 978-1-60650-062-0 (paperback)
ISBN-10: 1-60650-062-7 (paperback)
ISBN-13: 978-1-60650-064-4 (e-book)
ISBN-10: 1-60650-064-3 (e-book)

DOI: 10.5643/9781606500644

Cover design by Dagmar Fischer
Interior design by Exeter Premedia Services Private Ltd.,
Chennai, India

10 9 8 7 6 5 4 3 2 1

Printed in the United States of America

Contents

List of Symbols

Symbol	Meaning	Unit	Chapter
a	Atomic distance	m	2
a_0	Attachment frequency of monomers to unit area of nucleus surface	s^{-1}	2
a_i	Activity of component	dim.less	1
b	Burger's vector	m	2
b_n	a_0/q_n	dim.less	2
A	Helmholtz energy	J	1
A_n	Surface area of particle/droplet of size n	m^2	2
B	Diffusional mobility	$m^4/(Js)$	7
c	Concentration	$\#/m^3$	2
c_{eq}	Equilibrium concentration	$\#/m^3$	2
c_B^α	Concentration of element B in phase α	mol/m^3	3
$[[c_i^{\alpha\beta}]]$	Concentration jump of element i across phase boundary, $c_i^{\beta\alpha} - c_i^{\alpha\beta}$	mol/m^3	3
c_{Mi}	Mean concentration of element i inside the grain	mol/m^3	6
c_{Pi}	Mean concentration of element i in the precipitate	mol/m^3	6
c_{Mi}^{c}	Concentration of element i in the grain center	mol/m^3	6
c_{Mi}^{eq}	Equilibrium concentration of element i in the matrix/grain boundary	mol/m^3	6
c_{Mi}^{grain}	Initial concentration of element i in the matrix	mol/m^3	6
d	Density	kg/m^3	2
d	Interface thickness	m	4
d_{chem}^{β}	Molar chemical driving force for precipitation of β	J/mol	1
d_{chem}^{Va}	Molar (chemical) vacancy super saturation	J/mol	2
$d_F^{(m)}$	Total molar driving force	J/mol	1, 2, 3
d_F^{D}	Driving force dissipated by long-range diffusion	J/mol	3

d_F^{IF}	Driving force dissipated by interfacial friction	J/mol	3
D_B^{α}	Diffusion coefficient for element B in phase α	m^2/s	3
D_{gi}	Tracer diffusion coefficient of element i in the grain boundary	m^2/s	6
D_{bi}	Tracer diffusion coefficient of element i in bulk	m^2/s	6
d_g	Grain size (diameter)	m	7
E^i	Elastic modulus of phase i	Pa	1
E_i	Total (bond) energy of component i	J	1
F	Rayleigh dissipative function	J/s	3
f_i	Fractional composition of component i	dim.less	3
G	Gibbs energy	J	1
G	Shear modulus	Pa	2
G^*	Critical nucleation energy	J	2
ΔG_{Va}^{f}	Gibbs energy of vacancy formation	J	1
Δg_{Va}^{f}	Molar Gibbs energy of vacancy formation	J/mol	7
Δg	Molar Gibbs energy of mixing	J/mol	1
ΔG	Gibbs energy of mixing	J	1
ΔG_{nucl}	Volume free energy change	J	2
ΔG_{surf}	Energy required to create unit area of precipitate/matrix interface	J/m^2	2
ΔG_{vol}	Specific volume free energy change on nucleus formation	J/m^3	2
$\Delta G_{vol}^{chem,va}$	Chemical component of specific volume free energy change related to vacancy annihilation/generation	J/m^3	2
ΔG_{vol}^{chem}	Chemical component of specific volume free energy change	J/m^3	2
ΔG_{vol}^{el}	Elastic component of specific volume free energy change	J/m^3	1, 2
g	Molar Gibbs energy	J/mol	1
g_{EX}	Excess molar Gibbs energy	J/mol	1
g_{curv}	Molar curvature-induced Gibbs energy	J/mol	1
g_i^0	Molar Gibbs energy of pure component i	J/mol	1
g_{IS}	Molar Gibbs energy of ideal solution	J/mol	1
g_{MM}	Molar Gibbs energy of mechanical mixture	J/mol	1
g_{RS}	Molar Gibbs energy of regular solution	J/mol	1
H	Enthalpy	J	1
h	Molar Enthalpy	J/mol	1

H_{mig}	Migration enthalpy	J	7
ΔH^{f}_{Va}	Enthalpy of vacancy formation	J	1
Δh^{f}_{Va}	Molar vacancy formation enthalpy	J/mol	2, 7
Δh_{sol}	Molar enthalpy of solution/formation	J/mol	4
H_{J}	Dislocation jog density	m^{-3}	7
J	Diffusional flux	$mol/(m^2s)$	7
J	Transient nucleation rate	$\#/(m^3s)$	2
J_s	Steady-state nucleation rate	$\#/(m^3s)$	2
J^{α}_{B}	Diffusion flux of element B in phase α	$mol/(m^2s)$	3
$J^{\alpha\beta}_{B}$	Diffusion flux of element B into the phase boundary between phases α and β	$mol/(m^2s)$	3
j_{ki}	Flux of component i in the grain boundary around precipitate k	$mol/(m^2s)$	6
J_{ki}	Flux of component i perpendicular to the grain boundary corresponding to precipitate k	$mol/(m^2s)$	6
K	Kinetic constant in nucleation rate expression	$\#/(m^3s)$	2
k_B	Boltzmann constant	J/K	1
k_{co}	Coefficient in the continuums coalescence model	#	6
k_{sc}	Number of size classes in the NKW model	#	5
L_{ij}^{n}	Interaction energy in Redlich–Kister polynomials	J/mol	1
L_B	Characteristic diffusion length for element B	m	3
L	Lagrange function	J	3
m	Number of precipitates in the system	#	3
m_g	Number of precipitates at the grain boundary of a single grain	#	6
M	Mass	kg	2
M^{IF}	Interface mobility	$m^4/(Js)$	2
n	Number of monomers in cluster	#	2
$\dot{n}$	Growth velocity of clusters	#/s	2
n^*	Number of monomers in critical cluster	#	2
n_S	Number of atoms per unit interface area	$\#/m^2$	4
N	Number of potential nucleation sites	#	2
N_i	Number of moles of component i	#	1
N_n	Number of clusters of size n	#	2
N_A	Avogadro number	#/mol	1
O_n	Embryo of size n	dim.less	2
P	Pressure	Pa	1

P_{curv}	Curvature-induced pressure	Pa	1
p_{curv}	Molar curvature-induced pressure	Pa/mol	1
P_{ij}	Probability of finding a bond between component *i* and *j*	dim.less	1
Q_D	Activation energy for diffusion	J	7
q_D	Molar activation energy for diffusion	J/mol	7
q_n	Detachment frequency of monomers from cluster of size *n*	s^{-1}	2
q_i	Independent characteristic variable(in TEP model)	*variable*	3
r, ρ	Radius of particle/droplet	m	2, 3
r^*	Critical (nucleation) radius	m	2, 3
$\bar{r}$	Mean radius of precipitates	m	3
$\bar{\rho}$	Normalized radius of precipitates ($r/\bar{r}$)	dim.less	3
R	Universal gas constant	J/(molK)	1
R, R_i	Total/partial resistance	s	2
t	time	s	2
T	Temperature	K	1
T_{crit}	Critical temperature in regular solution model	K	1
S	Entropy	J/K	1, 2, 3
s	Molar entropy	J/(molK)	1
s_{IS}	Molar entropy of ideal solution	J/(molK)	1
s_{RS}	Molar entropy of regular solution	J/(molK)	1
$S_{cl,mix}$	Configurational mixing entropy of clusters	J/K	2
ΔS_{Va}^{f}	Entropy of vacancy formation	J	1
Δs_{Va}^{f}	Molar entropy of vacancy formation	J/mol	1
$\dot{S}^*$	Internal entropy production	J/(Ks)	3
u	Drift velocity (diffusion)	m/s	7
U	Internal energy	J	1
U_{Va}^{f}	Internal energy of vacancy formation	J	1
v	Velocity of the phase boundary	m/s	3
v^{α}	Molar volume of a phase α	m^3/mol	1
v_S	Partial molar volume of substitutional elements	m^3/mol	3
Δv	Volumetricmisfit	dim.less	1
V	Volume	m^3	1
V^{ext}	Extended volume (KJMA)	m^3	5
V^{tot}	Total volume (KJMA)	m^3	5

$V_{\mathrm{Va}}^{\mathrm{f}}$	Vacancy formation volume	m^3	1
W	Work	J	2
W^*	Work to form critical nucleus	J	2
W_n	Work to form nucleus with n monomers	J	2
X	Mole fraction of component	dim.less	1
X_{B}^{α}	Mole fraction of element B in phase α	dim.less	3
$X_{\mathrm{B}}^{\beta\alpha}$	Mole fraction of element B on the β-side of the $\alpha\beta$-interface (phase boundary)	dim.less	3
$X_{\mathrm{B}}^{\alpha\beta}$	Mole fraction of element B on the α-side of the $\alpha\beta$-interface (phase boundary)	dim.less	3
y_{Va}	Site fraction of structural vacancies on substitutional sublattice	dim.less	7
$y_{\mathrm{Va}}^{\mathrm{eq}}$	Equilibrium site fraction of structural vacancies	dim.less	7
Z	Zeldovich factor	dim.less	2
Z	Number of atoms in unit cell	#	4
Z_n	Number of clusters of size n	#	2
Z'_n	Total surface area of clusters of size n	m^2	2
z_{L}	Coordination number/number of bonds per atom	#	1, 4
$z_{\mathrm{L,eff}}$	Effective number of bonds per atom	#	4
z_{S}	Number of broken bonds per interface atom	#	4
$z_{\mathrm{S,eff}}$	Effective number of broken bonds per interface atom	#	4
β	Attachment frequency	s^{-1}	2
β^*	Attachment frequency of monomers to critical cluster	s^{-1}	2
Δ	Region $2k_{\mathrm{B}}$T around the critical size n^* where thermal fluctuations govern the cluster size evolution	#	2
δ	Thickness of the grain boundary region	m	6
ε^*	Linear misfit	dim.less	1
ε^*	Effective atomic interaction energy	J	4
ε_{ij}	Bond energy between components i and j	J	1
ϕ_{B}	Thermodynamic factor	dim.less	7
Φ	General potential	m^2	2
γ	Specific interfacial energy	J/m^2	1
$\gamma_{\alpha\alpha}$	Specific grain boundary energy between two grains α	J/m^2	2
$\gamma_{\alpha\beta}$	Specific interfacial energy between phases α and β	J/m^2	2
Λ	Radius of precipitate diffusion zone at the grain boundary	m	6
Γ	Dislocation line energy	J/m	2

μ_i	Chemical potential of component *i*	J/mol	1
ν_P	Poisson's ratio	dim.less	1
Ω_{MC}	Micro-canonical density of states	dim.less	1
Ω	Atomic volume	m^3	2
$\bar{\Omega}$	Volume corresponding to one mol of lattice sites	m^3	7
ρ	Radius of particle / precipitate	m	1
σ	Surface tension	J/m^2	2
τ	Incubation time	s	2
ω_{ij}	Regular solution parameter for system *ij*	s	2
ζ^{ext}	Extended volume fraction (KJMA)	dim.less	5

List of Figures

List of Tables

PREFACE

Precipitation is a phenomenon where a high number of relatively small particles of a new phase form within a comparably large volume of a parent phase. The new phase is distinguished from the parent phase by crystal structure, chemical composition, local ordering, and so forth, or any combination of these. The evolution of precipitates commonly occurs in the three stages: nucleation, growth, and coarsening, with more or less overlap in the temporal sequence.

The newly formed precipitates can be located on random positions within undisturbed bulk volume or along lattice heterogeneities, such as dislocations, grain boundaries, or inclusions. The crystal lattice of the precipitates can be coherent with the lattice of the parent phase, or it can be semicoherent or incoherent. The shape of the precipitates can be spherical, cuboidal, octo-cuboidal, plate-like, elongated, dendritic and more, depending on the degree of anisotropy of the mechanical, chemical, and thermodynamic properties of matrix and precipitate.

Precipitation processes can occur in sequences, where kinetically favorable (often metastable) phases form temporarily, to be replaced later on by kinetically less favorable, but thermodynamically more stable, phases. For instance, in Al-alloys, these sequences are well known and of great technological importance. At low temperatures, the precipitation process often starts with clusters of only a few atoms or monolayer Guinier–Preston zones, then continues with metastable precursor phases until, finally, the thermodynamically stable precipitates sustain. During tempering of martensitic steel, carbon atoms initially assemble in the form of irregular carbon clouds, the form metastable epsilon-carbide without involving the movement of substitutional atoms, until precipitates that incorporate slow-diffusing substitutional atoms also occur at higher heat treatment temperatures and/or longer heat treatment times.

Deformation can substantially alter the precipitation characteristics of a material. On one hand, newly introduced dislocations provide fast short-circuit diffusion paths, which accelerate diffusion-controlled processes. The interaction of dislocations leads to the generation of deformation-induced vacancies, which can substantially increase the vacancy concentration over the level given by the equilibrium thermal vacancy concentration. This vacancy excess leads to enhanced diffusion and, thus, faster precipitation kinetics. Moreover, the dislocations act as heterogeneous nucleation sites and, thus, affect the precipitate nucleation kinetics. After the deformation and/or quenching process, the excess vacancies annihilate again at dislocation jogs, grain boundaries, Frank loop, or incoherent phase boundaries or, eventually, become trapped at solute atoms.

These introductory notes emphasize an important aspect of modeling and simulation of precipitation processes: Solid-state precipitation involves a variety of physical phenomena that interact strongly with each other. Often, these interactions occur on different length scales, thus making it difficult to capture the relevant processes and interactions within a single "unified" model. Many order-of-magnitude differences, for instance, in the kinetics of diffusion between interstitial and substitutional elements, as well as interface mobility, substantially challenge the numerical algorithms implemented in software treating precipitation processes. Researches are confronted with the necessity of having to adapt the model complexity to a suitable degree of abstraction such that the solution to the precipitation problem can be obtained with reasonable computational effort and within the desired amount of time.

The interested reader will find many of these aspects reflected in the different modeling approaches that are dealt with in this book. Several of these models operate on different length scales and, even on the same length scale, incorporate a different degree of detail. The variety of models stretches from straightforward analytical solutions that can simply be typed into the pocket calculator, to model formulations that can only be solved numerically on high-level computers. Furthermore, we should keep in mind that several research fields that contribute to precipitation modeling, such as alloy thermodynamics, interface properties, diffusional transport, nucleation theory and so forth, are research areas by themselves, involving their own scientific communities.

In the past decades, modeling and simulation of solid-state precipitation has attracted increasing attention in academia and industry due to their high potential in designing optimal mechanical-technological properties of advanced structural materials as well as in improving production routes with the goal of increased productivity and decreased costs for expensive alloying elements.

Unfortunately, modeling of precipitation is not a trivial task, particularly in the early stages. The mathematical expression for the nucleation rate, that is the rate at which new precipitates are created per unit volume and time, contains two essential quantities that dominate the nucleation behavior of a precipitation system: The *driving force* for precipitation and the *interfacial energy*. Whereas the first quantity is nowadays readily available for many alloy systems of interest through the powerful method of Computational Thermodynamics, the interfacial energy is most often treated as an unknown fitting parameter. Several scientists and researchers continuously try to reduce the problem of modeling the behavior of precipitation systems on a rigorous physical basis to the much simpler problem of *fitting* their simulation to experimental results. In precipitation modeling, this can most easily be done by "turning a single screw": The interfacial energy.

Indeed, the interfacial energy has the power to control precipitation through its enormous impact on the nucleation behavior. Only minor changes in this quantity are commonly required to entirely change the precipitation behavior of the system. Unfortunately, the more powerful any such unknown fitting quantity acts in controlling the system behavior, the more we lose insight into the physical behavior of the system if we treat it as an unknown parameter.

To clarify this aspect, I recall a fundamental statement made in 1962 by Richard Wesley Hamming in the preface of his book, *Numerical Methods for Scientists and Engineers*:

The purpose of computing is insight, not numbers

With this short declaration, Hamming condenses the philosophy of *physical modeling* of scientific and engineering processes in a most beautiful way. When modeling precipitation, we shall *not* be focused on merely fitting our experimental data. By this, we lose insight into the behavior of the system. Instead, we shall strive for *physical* models describing fundamentally whatever processes occur in our precipitation system. Only this will carry us along our way towards the final goal of establishing a computational framework for *predictive* (precipitation) modeling.

Precipitation is a complex phenomenon, particularly in multicomponent systems. In this book, the state-of-the-art in precipitation modeling is outlined, although no claim can be made for completeness. The most relevant physical processes are discussed in terms of their impact on precipitation. The last chapters additionally contain some recommendations for how to set up precipitation simulations in practice and which software can be used for what class of problems. Unfortunately, not all aspects of precipitation modeling can be covered in sufficient depth. Therefore, the reader is sometimes referred to literature dealing with certain aspects in more detail. Some of the topics that I have not treated in this book will hopefully be part of a second edition.

I want to express my sincere gratitude to numerous people that have made this book possible: My friends, partners, and mentors that have many times inspired me to see things from other perspectives, the many partners from the industry that have contributed with their personal and financial support to projects dealing with various facets of precipitation, people from the scientific community that have helped me greatly with solutions to questions that I could not find myself, to my students and colleagues that have continuously encouraged me to pursue my visions; to the members of my group in Vienna who have motivated and supported me with their commitment and expertise and contributed to this book in many ways. I acknowledge the precious help of Weisheng Cao of CompuTherm, of Herng-Jeng Jou of QuesTek and Gerhard Inden from Max-Planck-Institute for Iron Research (MPIE) for their articles in the software chapter. I am grateful to Koen and Joel, who invited me to write this book and encouraged me to complete it. I also acknowledge the contribution of Austrian Railways (ÖBB), Austrian Airways (AUA), and Lufthansa (LH) in providing valuable office space throughout my many journeys, as well as *Wieden Bräu* and Pizzeria *Don Giovanni* in 1040 Vienna, where countless evenings were spent in stimulating atmosphere while working on the book. And most important: Sincere thanks go to my wife, Daniela, who takes care of the many things of real life.

Have fun with this book, and may it be of value to you.

Ernst Kozeschnik
(August 2012)

CHAPTER 1

Thermodynamic Basis of Phase Transformations

Precipitation is the phenomenon by which particles of a new phase form out of a supersaturated solid solution. Precipitation reactions commonly occur within a closed region in space. As we are usually able to define the precise spatial boundaries between the precipitation system and the surroundings, precipitation reactions can be considered within the framework of a *thermodynamic system*.

In a thermodynamic system, all processes are described by the laws of *thermodynamics*. The state of the system is characterized by a set of *thermodynamic parameters* associated with the system. Typical parameters of this kind are experimentally measurable macroscopic quantities related to the constitution of the system, such as temperature, T, pressure, P, volume, V, or chemical composition, N_i (or X_i), of the system or individual phases, with N_i being the number of moles of component i and X_i being the mole fraction defined as N_i/N. The set of parameters uniquely defining the constitution of the system is called *thermodynamic state*. In principle, there exist a large number of possible combinations of state variables defining a thermodynamic system. Some of these combinations offer an appropriate and useful framework for description of the processes occurring in the given thermodynamic system, some do not.

A most relevant quantity defining the energy of a thermodynamic system is the *free energy*, which denotes that part of the total energy of the system which is "free" to do thermodynamic work. The most often used *state functions* representing the free energy are the *Gibbs energy*, G and the *Helmholtz energy*, A. For the latter, the symbol F is also often found in literature. Whereas the Helmholtz energy is commonly used in the field of chemistry, the Gibbs energy represents a convenient framework for describing the state of a system in the course of solid-state phase transformations.

Since precipitation reactions, or phase transformations in a more general sense, are accompanied by changes in the free energy of the thermodynamic system, in this chapter, the basics of statistical and solution thermodynamics are reviewed and outlined as far as they are relevant to the modeling of precipitation phenomena. For more detailed information on general thermodynamics, the interested reader is referred to, for example, refs. [1–5]. Special reference, in the present context, may be given to the textbook by Hillert [6] where many issues related to the thermodynamics of solid-state phase transformations are covered in good depth and clarity.

1.1 The Gibbs Energy

The *internal energy* of a thermodynamic system is commonly denoted with the symbol U and it can be written as

$$U = TS - PV + \sum_i \mu_i N_i, \tag{1.1}$$

with T being absolute temperature, S being entropy, P and V representing pressure and volume, and N_i and μ_i being the number of moles of component i and the corresponding chemical potential, respectively. The SI unit of U is joule (J). The internal energy comprises all forms of energy, which are intrinsic to a thermodynamic system. These are, for example, of nuclear and atomic nature (energy of protons, electrons, neutrons), of chemical nature (bonds between molecules and atoms), or of mechanical nature (elastic and plastic deformation).

In many cases of interest, some of the state variables and state functions in equation (1.1) adopt constant values. Think, for instance, of systems under constant pressure (isobaric) or temperature (isothermal). In these cases, it is straightforward to lump together all constant energy contributions and take them into account in the form of an appropriate *reference state*.[1] Moreover, instead of investigating the total energy of a system, for practical purposes, it is often sufficient and convenient to consider only that amount of energy that is available for doing mechanical work and/or which can drive chemical reactions and phase transformations. This "reduced" energy is denoted as the *free energy* of a system.

It has already been indicated above that in condensed matter, a convenient choice of free energy is the Gibbs energy, G, defined as

$$G = U + PV - TS. \tag{1.2}$$

The SI unit of G is joule (J). After introducing the enthalpy, H, of the system with

$$H = U + PV, \tag{1.3}$$

the Gibbs energy can finally be written as

$$G = H - TS. \tag{1.4}$$

It is interesting to note that the internal energy, U, contains the term $-PV$ and H contains the term $+PV$. These terms cancel out in the definition of the Gibbs energy, G, which thus emphasizes that G explicitly excludes energy contributions from purely mechanical work. G is thus a convenient state function for the free energy of isobaric, isothermal systems—conditions, which typically apply to systems involving solid-state phase transformations.

With some algebra on the partial derivatives of the state functions and comparison of coefficients, it can be shown that the derivatives of the Gibbs energy with respect to the state parameters T, P, and N_i deliver the quantities $-S$, V, and μ_i with

[1] The use of reference states is common practice in thermodynamics as thermodynamic state functions, particularly the internal energy, are practically inaccessible to a quantitative measurement. It is simply not possible to determine all contributions to the total energy (nuclear, magnetic, chemical, physical, mechanical, etc.) and fix an absolute value for U by experiment.

$$\left(\frac{\partial G}{\partial T}\right)_{P,\mathbf{N}_i} = -S$$
$$\left(\frac{\partial G}{\partial P}\right)_{T,\mathbf{N}_i} = V$$
$$\left(\frac{\partial G}{\partial N_i}\right)_{T,P,\mathbf{N}_{j\neq i}} = \mu_i. \tag{1.5}$$

All partial derivatives are evaluated with the subscript state parameters held constant. The bold typeface represents vectors.

For application to precipitation processes, the third derivative, delivering the chemical potential μ_i of component i, is of particular importance. Accordingly, in isobaric and isothermal systems, and knowing the chemical potentials of all components, the Gibbs energy can be evaluated from the sum of potentials as

$$G = \sum_i \mu_i N_i. \tag{1.6}$$

There exist several other useful relations among the thermodynamic state functions. For these, however, the interested reader is referred to the corresponding general literature on thermodynamics (see introduction to this chapter).

1.2 Molar Gibbs Energy and Chemical Potentials

The Gibbs energy, G, is an *extensive* thermodynamic quantity, which means that the value of G depends on the size of the system. Other extensive quantities are, for instance, the volume V or entropy S. Typical *intensive* thermodynamic quantities are pressure P and temperature T, which are obviously independent of the system size.

Since precipitation systems are usually closed thermodynamic systems,[2] it appears beneficial to express all extensive thermodynamic quantities with respect to constant amount of matter. The extensive quantity "free energy," as defined in the previous section, thus transforms into a *specific* free energy, which can now be associated with, for example, one kilogram of matter or one mole of atoms. In the latter case, this quantity is denoted as the *molar Gibbs energy*, g, with the SI-unit (J/mol), and given as

$$g = \frac{G}{N}. \tag{1.7}$$

N is the total number of moles of atoms in the system. For the partial quantities N_i the following balanced equation holds:

$$\sum_i N_i = N. \tag{1.8}$$

[2] In a *closed* thermodynamic system, no mass flow between the system and its surrounding is allowed. Exchange of heat or work is possible, though.

In a molar framework, that is, in systems with a total of one mole of atoms (or molecules), the number of moles of component i is expressed in terms of the *mole fraction* variables X_i defined as

$$X_i = \frac{N_i}{N}. \tag{1.9}$$

For the sum of all mole fractions, we have

$$\sum_i X_i = 1. \tag{1.10}$$

On investigating the relations in equation set (1.5) after substituting $G = Ng$, we quickly find that the derivatives of g with respect to T and P now deliver molar quantities for entropy and volume, that is, the *molar entropy* $s = S/N$ and the *molar volume* $v = V/N$. Insertion of $G = Ng$ to the mass derivative of G delivers

$$\frac{\partial G}{\partial N_i} = \mu_i = \frac{\partial}{\partial N_i}(Ng) = g + \frac{\partial g}{\partial N_i}. \tag{1.11}$$

This equation gives the relation between the chemical potential μ_i and the molar Gibbs energy g. The derivative with respect to N_i can be further assessed after application of the chain rule with

$$\frac{\partial}{\partial N_i} = \sum_j \frac{\partial}{\partial X_j}\frac{\partial X_j}{\partial N_i}. \tag{1.12}$$

The mole fraction derivative can be evaluated on substituting $X_j = N_j/N$ with

$$\frac{\partial X_j}{\partial N_j} = \frac{\partial}{\partial N_j}\left(\frac{N_j}{N}\right) = \frac{N - N_j}{N^2} \quad \text{and} \quad \frac{\partial X_j}{\partial N_k} = \frac{\partial}{\partial N_k}\left(\frac{N_j}{N}\right) = -\frac{N_j}{N^2}. \tag{1.13}$$

Substituting equation set (1.13) into equation (1.11) finally delivers the relation between the chemical potential μ_i, the molar Gibbs energy g, and the mole fraction variables X_i with

$$\mu_i = g + \frac{\partial g}{\partial X_i} - \sum_j X_j \frac{\partial g}{\partial X_j}. \tag{1.14}$$

In analogy to the treatment in the previous section, for the relation between the molar Gibbs energy and the chemical potentials, we obtain

$$g = \sum_i \mu_i X_i. \tag{1.15}$$

1.3 Solution Thermodynamics

In the previous sections, general properties of thermodynamic functions have been investigated, irrespective of the detailed nature of the thermodynamic system. In this section, various models for the Gibbs energy of solid solutions and alloys, with different levels of sophistication, will be introduced and discussed. The first example will restrain to binary systems, which are thermodynamic systems with exactly two atomic species. They will be denoted as atoms A and atoms B subsequently. The concluding part of this section is concerned with a popular model capable of describing the Gibbs energy of "real" alloys (the so-called CALPHAD approach), taking into account all chemical interactions between atoms in multicomponent systems.

1.3.1 Mechanical Mixture and Ideal Solution

Consider two blocks of matter, one composed of pure atoms A and the other one composed of pure atoms B. In a thought experiment these two blocks are brought together and mixed. Interestingly, from the viewpoint of free energy, there exist two fundamentally different situations after merging the blocks, depending on whether the mixing is performed on the macro-/mesoscopic or the microscopic scale.

On a macroscopic scale, we might imagine that the two blocks are first disintegrated into small pieces (e.g., by crushing or grinding). In the next step, the pieces are mixed and finally compacted such that a single solid block with macroscopically observable and clearly separatable chunks of pure A and B is created. A possible result of this procedure is schematically shown in the top right corner of Figure 1.1, where a so-called *mechanical mixture* is sketched. The chunks of material A are clearly visible in a matrix of material B.

When assessing the free energy of a mechanical mixture, and using lowercase symbols for molar quantities in the following, we postulate that the molar free energies of the pure components, g_A^0 and g_B^0, are known. The free energy g_{MM} of the merged binary system is then straightforwardly given by the weighted sum of the two free energies as

$$g_{MM} = X_A g_A^0 + X_B g_B^0. \tag{1.16}$$

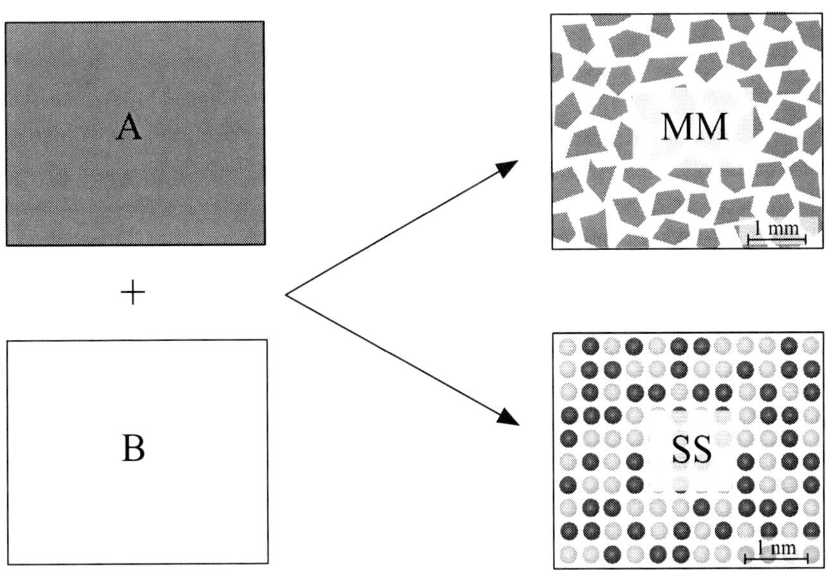

Figure 1.1. Two blocks of atoms A and B (left) are merged into either a mechanical mixture, MM (top right), or a solid solution, SS (bottom right).

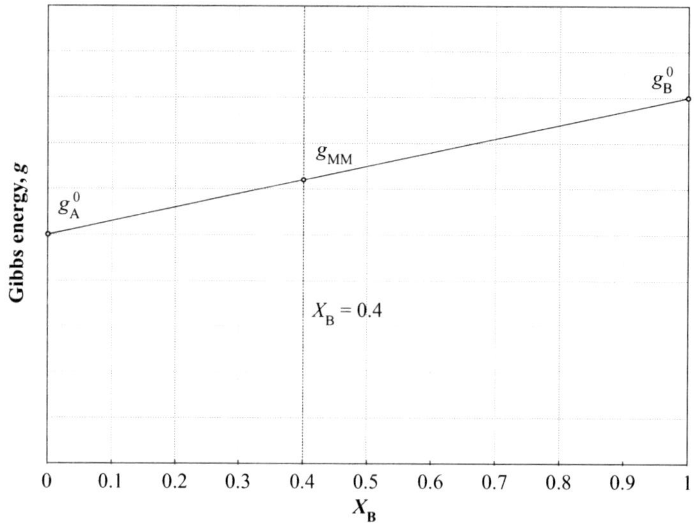

Figure 1.2. Molar Gibbs energy g_{MM} of a mechanical mixture of atoms A and B as the weighted sum of the free energies of the pure substances.

For general, multicomponent mechanical mixtures, we obtain

$$g_{MM} = \sum_i X_i g_i^0, \tag{1.17}$$

where the sum is taken over all components, *i*. Graphically, the free energy construction for a mechanical mixture is shown in Figure 1.2, where g_{MM} is obtained for a system containing 40% B-atoms. The construction follows along a straight line connecting the free energies of the pure components A and B.

In contrast to the simple weighted superposition of pure substance energies for a mechanical mixture, the free energy of a solid solution behaves entirely differently, the reason being that mixing of A and B now occurs on a significantly lower level, that is, in atomic dimensions. Whereas the comparably large chunks of A are practically immobile in the B matrix and we are confronted with a more or less static arrangement of regions A and B, individual atoms can exchange lattice position with, for example, vacancies (empty lattice sites) on the microscopic scale, and by this mechanism move through the volume of the material. The process by which atoms move through a material is known as *diffusion.*

If two blocks of matter are brought into close contact, on a microscopic scale, mixing of atoms occurs. From experience, we know that this process continues until complete mixing of the two substances is achieved. On a macroscopic scale, we arrive at a homogeneous solid solution as soon as the mixing process is complete. Think, for instance, of a drop of ink that is very carefully injected into a glass of water. When, thus, excluding convective transport of ink molecules, we will still observe that the ink entirely dissolves in the water after some time because the ink molecules are constantly relocated inside the liquid by diffusion.[3] From

[3] Note that diffusion in liquids does not occur on the basis of vacancy/atom exchange. Liquid atoms/molecules are not bound to specific locations on a crystal lattice.

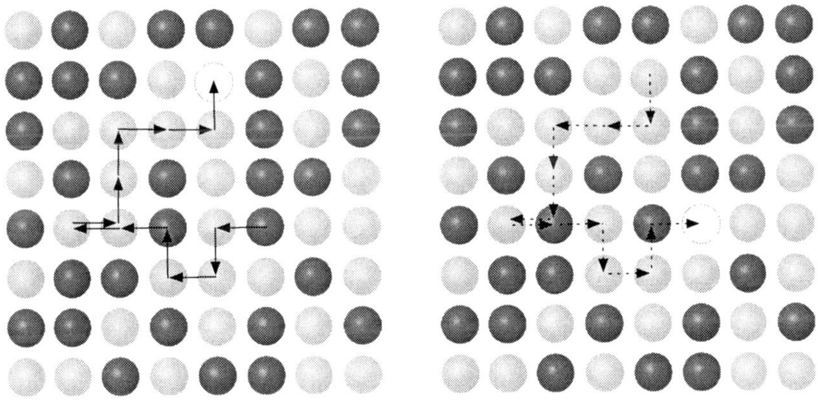

Figure 1.3. Random walk of a lattice vacancy and two possible arrows of time.

the second law of thermodynamics, we know that the driving force for this process is the production of entropy.

Consider a single vacancy on a crystal lattice moving through the crystal by random walk.[4] Each individual vacancy–atom exchange carries one of the lattice atoms over one atomic distance in random direction. The effect of this sequence of elementary transport processes is that the atoms of the solution are in permanent motion. When assessing these processes in more detail, we start on the microscopic scale and investigate the atom–vacancy exchange on a two-dimensional 8×8 lattice.

Figure 1.3 sketches a random walk of a lattice vacancy over 12 steps, 10 of which are effectively displacing lattice atoms.[5] When looking at an arbitrary configuration of atoms before and after a given number of vacancy steps, we make an interesting observation: An observer is not able to distinguish which of the two states, initial or final, has occurred earlier in time by simply looking at the atomic configurations. It is not possible for him to determine the *arrow of time* from only looking at the results of the two experiments. When considering an elementary atom–vacancy exchange, we find that the forward jump probability of the exchange process is identical to the backward jump probability, such that the initial state is established again. On a microscopic scale, we thus observe the *principle of time reversal.*

On a macroscopic scale, the situation is different. Figure 1.4 shows a sequence of snapshots related to the ink-in-water experiment. An observer, familiar with the experiment and looking at the three images, will immediately be able to put the images into the correct order. The initial state is the one with the drop of ink in the middle of the system. The final state is the one with the ink entirely dissolved and all concentration gradients in the glass leveled out. On a macroscopic scale, one would never expect that a well-defined drop of ink spontaneously forms out

[4] Random walk designates the process by which a vacancy travels ("walks") through a crystal without preferred direction. This walk is based on random site exchange with any of its nearest neighbors. Random walk implies that there exists no chemical or mechanical interaction between the different atoms as well as the atoms and the vacancy apart from the periodic lattice potential.

[5] The fact that not all vacancy–atom exchanges will effectively displace an atom is taken into account in a *correlation factor* in diffusion theory. Its theoretical value is constant for each type of crystal lattice and it is 0.727 for bcc and 0.781 for fcc.

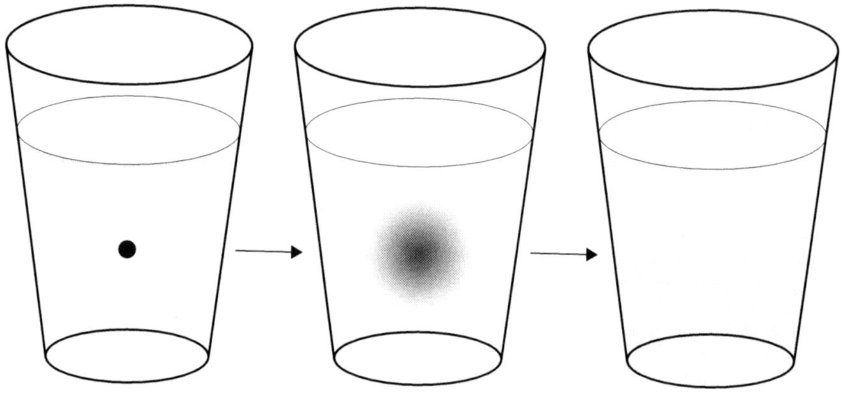

Figure 1.4. Ink droplet dissolving in water and the arrow of time.

of the homogeneous solution of the two substances.[6] On a macroscopic scale, we observe that diffusion is an *irreversible* process.

The Austrian physicist and philosopher Ludwig Boltzmann (1844–1906) has first addressed the irreversibility of random processes in large physical systems in his treatment of kinetic gas theory. He discovered that the probability of finding a system in a given constitution is related to the number of possible microstates, a relation that can be expressed in terms of the proportionality between the entropy of the system, S, and the natural logarithm of the microcanonical density of states, Ω_{MC}, as

$$S = k_B \ln \Omega_{MC}. \tag{1.18}$$

The proportionality constant k_B is known as the *Boltzmann constant*. It has a value of $k_B = 1.3806504 \times 10^{-23}$ J/K.

Equation (1.18) has fundamental character, and we will utilize it later also in the development of nucleation theory. At this stage, it opens the possibility of determining the driving force for diffusional processes.

An important and general aspect of equation (1.18) is that the density of states, Ω_{MC}, is an a priori undetermined quantity and it must be concretized with an appropriate model describing the process under consideration. In the context of solution thermodynamics, Ω_{MC} represents the number of possible microstates that a solid solution containing different atomic species can adopt.

In a binary system, Ω_{MC} represents the number of possibilities of how a set of A and B atoms can be arranged on a crystal lattice. In general, the corresponding permutation of atoms is given with

$$\Omega_{MC} = \frac{N!}{N_A!\, N_B!}. \tag{1.19}$$

[6] Of course, we cannot exclude that the ink molecules spontaneously form a droplet out of the homogeneous ink–water mixture at some instant in the future. However, this process is *very, very* unlikely. For additional discussion of this aspect, compare also Chapter 2 on nucleation.

The factorials in equation (1.19) can be approximated with Stirling's formula for $\ln N! = N \ln N + N + O(\log N)$. For equation (1.18), we then obtain

$$S = k_B \left(N \ln N - N_A \ln N_A - N_D \ln N_B \right) \tag{1.20}$$

With $X_A = N_A/N$ and $X_B = N_B/N$, we arrive at the final result for the entropy of a binary solution with

$$S = -k_B N \left(X_A \ln X_A + X_B \ln X_B \right). \tag{1.21}$$

For one mole of atoms, N is identical to Avogadros' number, $N_A = 6.02214179 \times 10^{23}$. With the identity $R = N_A k_B$, R being the *Universal Gas Constant* with a value of $R = 8.314472$ J/(mol·K), we finally obtain for the *molar configurational entropy*,[7] s_{IS}, of an ideal binary solution

$$s_{IS} = -R \left(X_A \ln X_A + X_B \ln X_B \right). \tag{1.22}$$

The term "ideal" is related to the premise that no chemical and/or mechanical interactions between the atoms exist and that the solution of A and B atoms has thus an entirely random character. The multicomponent expression for the entropy of an ideal solution reads

$$s_{IS} = -R \sum_i X_i \ln X_i . \tag{1.23}$$

The entropy function for a binary alloy, equation (1.22), is shown in Figure 1.5. The configurational entropy in the limit of pure substances is zero and it has a negative maximum at the 50% mixture. On adding the negative molar entropy expression, equation (1.22) multiplied with temperature to the molar Gibbs energy of the mechanical mixture, equations (1.16) and (1.17), the *molar Gibbs energy of an ideal solution* is obtained with

$$g_{IS} = X_A g_A^0 + X_B g_B^0 + RT \left(X_A \ln X_A + X_B \ln X_B \right) \tag{1.24}$$

or

$$g_{IS} = \sum_i X_i g_i^0 + RT \sum_i X_i \ln X_i . \tag{1.25}$$

Figure 1.6 shows a graphical representation of the Gibbs energy of an ideal binary solution, g_{IS}, for a chemical composition of $X_B = 0.4$. The contribution of the entropy of mixing to the free energy is clearly seen as the difference between the free energy of the ideal solution, g_{IS}, and the free energy of the mechanical mixture, g_{MM}. Since the entropy of mixing is always positive and the entropic part of the free energy enters as $-Ts_{IS}$, we conclude that the entropy of mixing always stabilizes a solid solution over phase separation. Moreover, the entropic contribution scales linearly with temperature and the influence of entropy on the state of the systems thus becomes increasingly prominent with increasing temperature.

The figure also shows the relation between the free energy and the chemical potentials, μ_A and μ_B, which are obtained as the intersection of the tangent to the free energy curve with the

[7] The configurational entropy is also often denoted as the *entropy of mixing*, as it represents the entropy that is produced when two or more pure substances are merged into a homogeneous solid solution.

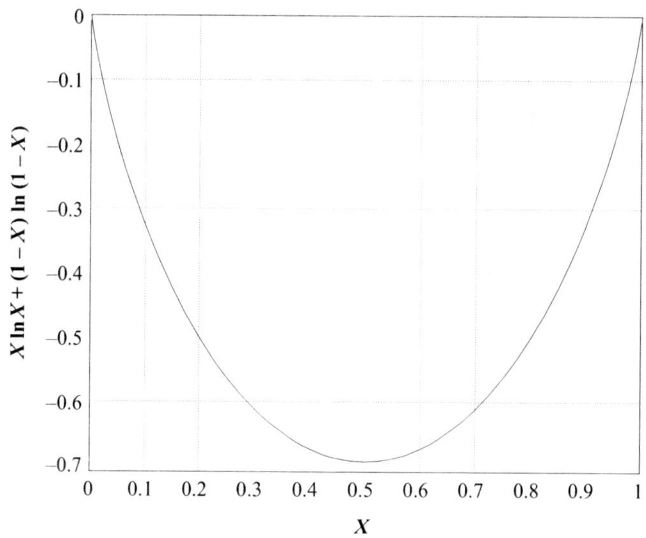

Figure 1.5. Shape of the function $X \ln X + (1 - X)\ln(1 - X)$ for a binary solution.

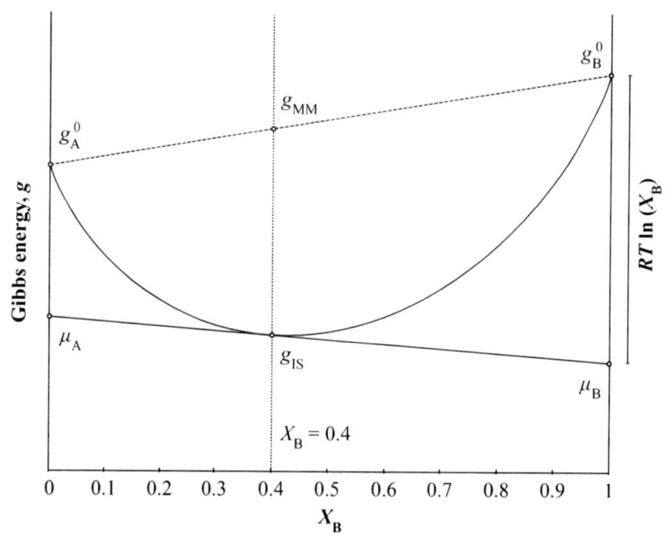

Figure 1.6. Molar Gibbs energy, g_{IS} of an ideal solution of atoms A and B.

vertical axes at pure A or B. This construction emphasizes the physical meaning of the chemical potential as the change in free energy of the system on addition or removal of an infinitesimal amount of A or B.[8] The chemical potential can also be interpreted as the *partial free energy* associated with an atomic species in the present chemical environment.

[8] Keep in mind that, in a binary molar framework, the addition of component A is accompanied by the removal of the same amount of B. In the general definition of the chemical potential, equation (1.5), only the target component is varied, whilst the amount of all other components is held constant.

The chemical potential of component i in an ideal solution can be evaluated by inserting equations (1.24) or (1.25) into the expression for the chemical potential in a molar framework, equation (1.14). For a binary ideal solution, we have

$$\begin{aligned}
\mu_i &= g_{\mathrm{IS}} + \frac{\partial g_{\mathrm{IS}}}{\partial X_i} - X_{\mathrm{A}} \frac{\partial g_{\mathrm{IS}}}{\partial X_{\mathrm{A}}} - X_{\mathrm{B}} \frac{\partial g_{\mathrm{IS}}}{\partial X_{\mathrm{B}}} \\
g_{\mathrm{IS}} &= X_{\mathrm{A}} g_{\mathrm{A}}^0 + X_{\mathrm{B}} g_{\mathrm{B}}^0 + RT\left(X_{\mathrm{A}} \ln X_{\mathrm{A}} + X_{\mathrm{B}} \ln X_{\mathrm{B}}\right) \\
\frac{\partial g_{\mathrm{IS}}}{\partial X_{\mathrm{A}}} &= g_{\mathrm{A}}^0 + RT\left(\ln X_{\mathrm{A}} + 1\right) \\
\frac{\partial g_{\mathrm{IS}}}{\partial X_{\mathrm{B}}} &= g_{\mathrm{B}}^0 + RT\left(\ln X_{\mathrm{B}} + 1\right).
\end{aligned} \qquad (1.26)$$

After some algebra, we obtain for the chemical potential of components A and B in an ideal solution

$$\begin{aligned}
\mu_{\mathrm{A}} &= g_{\mathrm{A}}^0 + RT \ln X_{\mathrm{A}} \\
\mu_{\mathrm{B}} &= g_{\mathrm{B}}^0 + RT \ln X_{\mathrm{B}},
\end{aligned} \qquad (1.27)$$

or for component i

$$\mu_i = g_i^0 + RT \ln X_i. \qquad (1.28)$$

The relation between the chemical potentials and the Gibbs energies of the pure components is emphasized in Figure 1.6, where the quantity $RT \ln X_{\mathrm{B}}$ is shown as the distance between g_{B}^0 and μ_{B} on the axis for $X_{\mathrm{B}} = 1$.

1.3.2 The Regular Solution

Real thermodynamic solutions rarely behave like an ideal system because there exist various types of chemical and/or mechanical interactions between the individual atomic species. In this chapter, a simple model, known as the *regular solution* model, is discussed, which takes into account attractive or repulsive forces between unlike atoms. Though its simplicity, the model is capable of illustrating basic features of solid solutions, thus making it an important thermodynamic prototype system. The fundamental properties of regular solutions are reflected in a very similar manner also in systems with more complex interactions.

Consider a binary solid solution with atoms A and B. In a first approximation, the energy of the solution can be calculated by summing up the chemical bond energies of all nearest-neighbor pairs, a concept that has been brought forward in the 1930s by Bragg and Williams [7] and which has later been used extensively in the development of thermodynamic and kinetic models.[9] Exemplarily, Figure 1.7 sketches nearest-neighbor interactions on a 2D primitive cubic lattice.

---

[9] A popular approach in this context is the lattice Monte Carlo method, where the local energy of a system is often determined as the sum of pair interactions between n-nearest neighbor atoms.

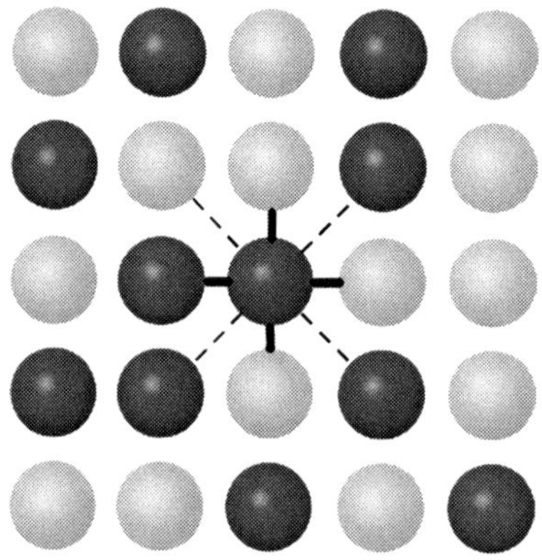

Figure 1.7. First-nearest (bold lines) and second-nearest (dashed lines) neighbor interactions on a 2D primitive cubic lattice.

With z_L being the coordination number (number of nearest-neighbor positions) and considering pure substances A and B first, the total bond energies E_A and E_B of each of the two substances is obtained by counting the number of bonds as

$$E_{AA} = \frac{1}{2} z_L N_A \cdot \varepsilon_{AA}$$
$$E_{BB} = \frac{1}{2} z_L N_B \cdot \varepsilon_{BB}. \quad (1.29)$$

ε_{AA} and ε_{BB} are the bond energies between the like atoms A and B, N_A and N_B are the number of atoms in the blocks of pure substances, and z_L is the coordination number (number of nearest neighbors). The factor ½ avoids counting the bonds twice.

When the two blocks of atoms A and B are merged into a homogeneous solid solution, the total bond energy of the solid solution is obtained by considering the probability of occurrence of certain bonds and summation over all combinations. Accordingly, the probability, P_{AB}, for an atom A being located next to an atom B is given by the fraction, X_B, of atoms B multiplied with the total number of atoms, N, as $P_{AB} = NX_B$. Analogously, the probability for an A atom being in a nearest-neighbor position to another A atom is $P_{AA} = NX_A$. As the total number of atoms A is NX_A, we have $z_L NX_A X_B$ bonds between A and B atoms. Consequently, for all bonds in the solid solution we obtain

$$E_{AA}^{SS} = \frac{1}{2} z_L N_A X_A \cdot \varepsilon_{AA}$$
$$E_{BB}^{SS} = \frac{1}{2} z_L N_B X_B \cdot \varepsilon_{BB}$$
$$E_{AB}^{SS} = z_L NX_A X_B \cdot \varepsilon_{AB}. \quad (1.30)$$

The difference in energies of the pure substances and the solid solution is then given by the difference between the two states as

$$\Delta H = E_{AA} + E_{BB} - E_{AA}^{SS} - E_{BB}^{SS} - E_{AB}^{SS}$$
$$= \frac{1}{2} z_L NX_A X_B \cdot (\varepsilon_{AA} + \varepsilon_{BB} - 2\varepsilon_{AB}). \quad (1.31)$$

The quantity ΔH is called the *enthalpy of mixing* or the *enthalpy of solution*. From practical chemical experiments, we find that dissipation of this quantity is experienced by the release of energy in the form of heat.

After substitution of

$$\omega_{\mathrm{AB}} = z_{\mathrm{L}}\left(\varepsilon_{\mathrm{AA}} + \varepsilon_{\mathrm{BB}} - 2\varepsilon_{\mathrm{AB}}\right) \tag{1.32}$$

and considering one mole of substance, we finally arrive at the *molar Gibbs energy* of a *regular solution* with

$$\begin{aligned} g_{\mathrm{RS}} &= g_{\mathrm{IS}} + \Delta h = g_{\mathrm{MM}} - Ts_{\mathrm{IS}} + \Delta h \\ &= X_{\mathrm{A}} g_{\mathrm{A}}^{0} + X_{\mathrm{B}} g_{\mathrm{B}}^{0} + RT\left(X_{\mathrm{A}} \ln X_{\mathrm{A}} + X_{\mathrm{B}} \ln X_{\mathrm{B}}\right) + \frac{1}{2} X_{\mathrm{A}} X_{\mathrm{B}} \cdot \omega_{\mathrm{AB}}. \end{aligned} \tag{1.33}$$

and

$$g_{\mathrm{RS}} = \sum_{i} X_i g_i^0 + RT \sum_{i} X_i \ln X_i + \frac{1}{2} \sum_{i} \sum_{j>i} X_i X_j \cdot \omega_{ij}. \tag{1.34}$$

The quantity ω_{AB} is commonly denoted as the *regular solution parameter*, as it determines the properties of the regular solution in excess of the properties of the ideal solution.

The investigation of the fundamental properties of a regular solution can be facilitated when considering only the excess energy that is associated with the mixing process, that is, using the energies of the pure substances as a reference state. For the *Gibbs energy of mixing*, Δg_{RS}, we thus obtain

$$\Delta g_{\mathrm{RS}} = g_{\mathrm{RS}} - g_{\mathrm{MM}} = RT\left(X_{\mathrm{A}} \ln X_{\mathrm{A}} + X_{\mathrm{B}} \ln X_{\mathrm{B}}\right) + \frac{1}{2} X_{\mathrm{A}} X_{\mathrm{B}} \cdot \omega_{\mathrm{AB}} \tag{1.35}$$

and

$$\Delta g_{\mathrm{RS}} = RT \sum_{i} X_i \ln X_i + \frac{1}{2} \sum_{i} \sum_{j>i} X_i X_j \cdot \omega_{ij}. \tag{1.36}$$

An analysis of the system behavior on the basis of equations (1.33) and (1.34), or analogously equations (1.35) and (1.36), delivers three cases depending on the value of ω:

- $\omega > 0$: The bonds between the like atoms AA and BB are stronger than the bonds between unlike atoms AB. In this case, bonds between unlike atoms are energetically unfavorable and the system tries to minimize the number of AB bonds by phase separation and formation of regions consisting of (predominately) A or B atoms. In concentrated alloys, one often observes spinodal structures (spinodal decomposition) with irregular interconnected areas of either A or B. Dilute systems are characterized by phase morphologies determined by the phenomenon of precipitation, where smaller local regions of the minority element form.
- $\omega = 0$: The enthalpy of mixing is zero and the regular solution reduces to an ideal solution. Mixing of A and B is purely random.
- $\omega < 0$: The bonds between unlike atoms are energetically favorable. Each A atom likes to be surrounded by as many B atoms as possible, and vice versa. Maximization of the number of AB bonds leads to short-range ordering, where the nearest-neighbor atoms arrange in a regular structure. In concentrated alloys, large short-range ordered regions can often be observed. In dilute systems, the characteristics strongly depend on the second-nearest

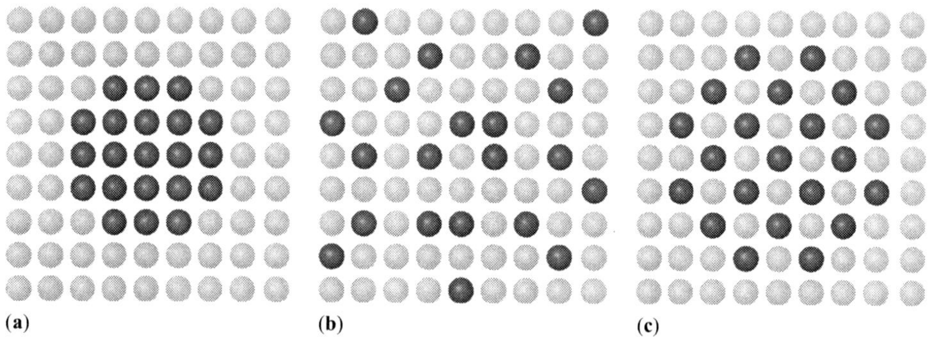

Figure 1.8. Three typical configurations of a binary regular solution as determined by the value of the regular solution parameter: (a) $\omega > 0$, phase separation, (b) $\omega = 0$, ideal (random) solution, (c) $\omega < 0$, short-range ordering.

neighbor interaction, which can be either attractive or repulsive for other atoms of the same kind. If the minority element prefers unlike atoms on the first-nearest neighbor positions and like atoms on the second-nearest neighbor sites, the system exhibits formation of local regions with short-range order. Otherwise, the minority element is dissolved in the solution with maximal average distances between the like atoms, thereby maximizing the number of unlike bonds in the first- and second-nearest neighbor shell. In this case, no short-range order is observed at all.

Examples for the three fundamental configurations of a regular solution in dependence of ω, and assuming $T = 0$ K,[10] are schematically shown in Figure 1.8. On the left image (a), all dark-colored atoms concentrate in a spherical precipitate, thus minimizing the number of AB bonds. In the middle image (b), no distinct trends toward any kind of ordering or phase separation can be identified. In the right image (c), the dark-colored atoms assemble into spherical form, however, this time maximizing the number of AB bonds.

Another important feature of regular solutions is exposed when investigating the impact of temperature on the system characteristics. According to equation (1.33), the Gibbs energy contains (*a*) contributions from the mechanical mixture, connecting the pure energies of A and B in the Gibbs energy—composition (*G-X*) diagram along a straight line, (*b*) the ideal entropy of mixing, and (*c*) a term depending on the regular solution parameter ω. It is interesting, now, to recognize again that the entropic contribution (*b*) to the Gibbs energy is always negative and linearly dependent on temperature, *T*. The shape of the entropy function has been shown in Figure 1.5. In contrast, the regular solution term is always positive if the regular solution parameter ω is positive, scaling with $X\,(1-X)$.

The shape of the functions contributing to the energy of a regular solution is shown in Figure 1.9 for a typical case of $\omega > 0$.[11] Interestingly, the entropy and enthalpy functions show considerable differences in the slopes of the curve close to the *x*-axis values of 0 and 1. Since

[10] This condition excludes the influence of entropy which always leads to a certain degree of disorder. The three configurations thus represent "ideal" situations with undisturbed arrangement of atoms for phase separation, random solution, and short-range ordering.

[11] Note that the enthalpy function in Figure 1.9 has been multiplied by an arbitrary weighting factor of 3 in order to emphasize some basic features of the two functions.

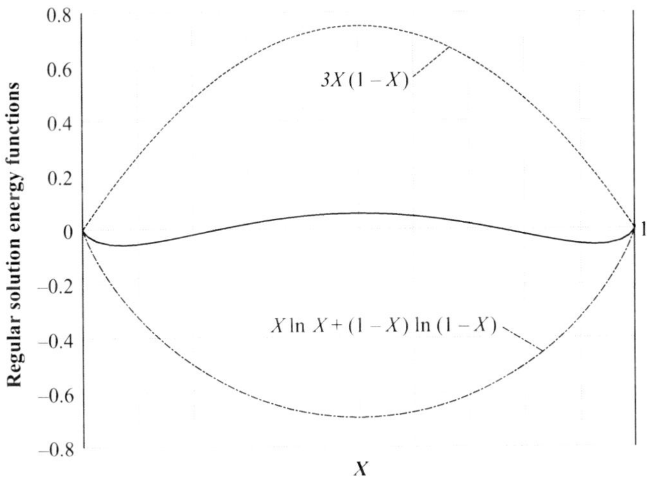

Figure 1.9. Gibbs energy of a regular solution as composed of two functions $X \ln X + (1-X)\ln(1-X)$ and $3X(1-X)$ describing the entropy and enthalpy of mixing.

the entropy function has a significantly higher negative slope close to these values compared to the enthalpy function,[12] the sum of both curves has also always a negative slope close to the x-axis limit. The regular solution model thus implies that the solution of at least a minimum amount of a minority element in the matrix of the majority element is not only an energetically favorable process, but also a thermodynamic requirement. According to the regular solution model, at any temperature $T > 0$, the separation into two phases will never lead to absolutely pure substances A and B, but there will always exist some equilibrium amount of component A in the B-rich substance and some B in the A-rich substance. How much of each component is dissolved in the other component depends on the values of the regular solution parameter, ω, and temperature, T, according to equations (1.35) and (1.36).

In many practical cases, the regular solution parameter, ω, is constant for a given thermodynamic system and temperature, T, remains to be an independent variable. This feature can be used to study the relation between the G-X (or ΔG-X) curves of a thermodynamic system with the associated phase diagram, which is shown in Figure 1.10. The upper part of this figure plots various ΔG-X curves for a regular solution between 300 and 800 K, with the Gibbs energies of the pure substances A and B taken as reference state.[13] The regular solution parameter is assumed to be $\omega = 11{,}500$ J/mol. The lower part of the figure presents the phase diagram, which is uniquely determined by the extremal points of the G-X curves as discussed subsequently.

The relation between the ΔG-X curves and the phase diagram can be understood when investigating the characteristics of the ΔG-X curves in more detail. When starting at the highest temperature of 800 K, we find that the Gibbs energy of mixing exhibits only a single minimum

[12] The derivative of the entropy function at the limits of 0 and 1 delivers a slope of minus infinity, whereas the derivative of the enthalpy function is plus and minus unity at these values.

[13] The values for g_A^0 and g_B^0 are different in most practical systems and they are functions of temperature. However, it can be shown that using the Gibbs energy of mixing, instead of the absolute value of the Gibbs energy, will not change the basic features of a regular solution.

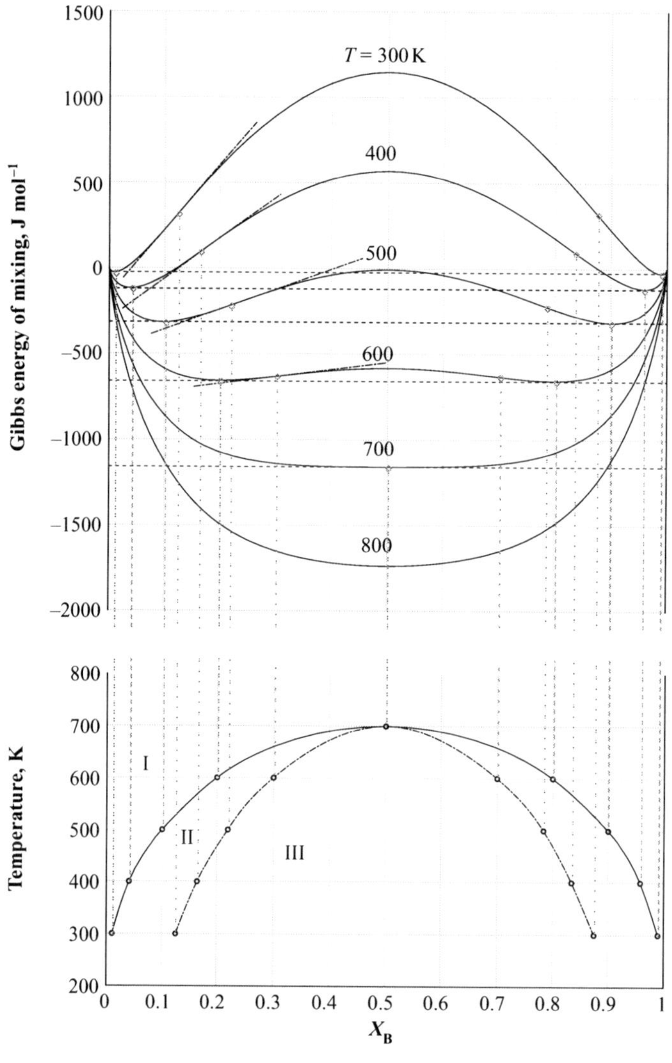

Figure 1.10. Relation between the *G-X* curves of a regular solution and the associated phase diagram.

exactly at the intermediate composition for A and B. This observation is quickly reasoned based on the dominant influence of the entropic part of the Gibbs energy in equations (1.35) and (1.36), which scales linearly with T and leads to an increased tendency for atomic mixing with increasing temperature.

With decreasing temperature, the influence of the entropy of mixing becomes weaker compared to the enthalpy of mixing, which is assumed to be independent of temperature in the regular solution model. At some specific temperature, denoted as the *critical temperature*, T_{crit}, the regular solution exhibits an inherent instability characterized by the occurrence of two inflection points and w-shaped *G-X* and ΔG-*X* curves.[14] At any temperature below T_{crit}, the

[14] The transformation of the *G-X* curve to the ΔG-*X* curve does not change the properties of the system and both representations can be used interchangeably.

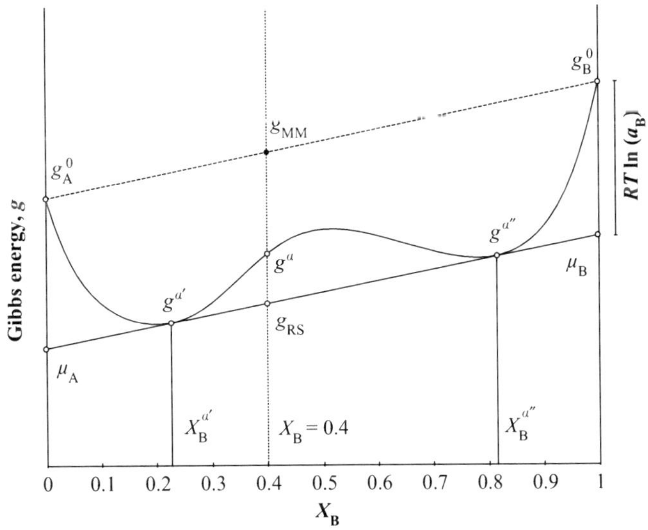

Figure 1.11. Gibbs energy, g_{RS}, of a regular solution of atoms A and B below the critical temperature.

lowest Gibbs energy is obtained with phase separation and two coexisting phases with different chemical composition in thermodynamic equilibrium.

Figure 1.11 shows a typical *G-X* curve for a system with phase separation. Let us denote the homogeneous mixture of A and B atoms by the phase name α.[15] The Gibbs energy of the homogeneous mixture of the two components is indicated with the symbol, g^{α}. From the tangent construction, it becomes evident that the total free energy of the system can be minimized if the phase α decomposes into a mixture of the two phases, α' and α''. The Gibbs energy of the mixture of the two phases can be read along the common tangent to the *G-X* curve with a value of g_{RS}. The Gibbs energy associated with the two individual phases is given at the intersections of the tangent with the *G-X* curve at the compositions of $X_B^{\alpha'}$ and $X_B^{\alpha''}$ with values designated as $g^{\alpha'}$ and $g^{\alpha''}$. It is evident that this construction uniquely defines the minimum of the total Gibbs energy and, thus, the equilibrium state of the thermodynamic system.

In analogy to the construction in Figure 1.6, the chemical potentials are read at the intersections of the common tangent with the vertical axes at values of 0 and 1. The tangent construction thus shows that the chemical potentials of the two phases, α' and α'', are identical in equilibrium. Thermodynamic equilibrium can, therefore, not only be described by a minimum of the Gibbs energy, but also by the identity of the chemical potentials in all phases participating in the equilibrium. In Section 1.4, we will observe a third condition that is related to the thermodynamic driving force for phase separation. All of these conditions can be utilized to evaluate thermodynamic equilibrium from the knowledge of the Gibbs energy functions of the individual phases.

The tangent construction discussed in Figure 1.11 can be performed for the *G-X* and ΔG-*X* curves at various temperatures and the observed equilibrium compositions of the stable phases

[15] A "phase" is defined as a region in space with homogeneous chemical composition, crystal structure, and other chemical and physical properties.

can be used to construct the phase diagram of the system by plotting them against temperature. The corresponding diagram is shown in the bottom plot of Figure 1.10. The equilibrium compositions form a phase boundary that separates the homogeneous solid solution from the region where phase separation is thermodynamically favorable.

Figure 1.11 also shows a region that is bounded by the inflection points of the ΔG-X curves, as indicated by the tangents in the left half of the ΔG-X curves in the top diagram. The physical relevance of the inflection points is graphically sketched in Figure 1.12 where these points separate two regions of the G-X curve with different physical response to local chemical composition fluctuations. Consider an A-B system that separates into the two equilibrium phases, α' and α''. The nominal B-content is X_1 and the system is in a state of homogeneous solid solution initially. The composition X_1 is located between the equilibrium composition of α' and the inflection point S_1. The secant line drawn above X_1 sketches the change in energy generated by a local chemical composition fluctuation around X_1. The partial Gibbs energies of the regions enriched and depleted in B are shifted along the G-X curve. The total Gibbs energy change associated with this fluctuation is read at the line connecting the two points.

The figure shows that, in the region between $X_B^{\alpha'}$ and X_1, any small fluctuation in local chemical composition is associated with an increase in the Gibbs energy of the system. Small enough perturbations are unstable, therefore, and the system tends to fall back into the homogeneous solution state again. Only if the amplitude of the perturbation is large enough, that is, the fluctuation forms a concentrated B-rich region, the total energy will decrease on further increasing the amplitude of the perturbation. However, in order to reach the state where the fluctuation becomes stable, a certain energy barrier must be overcome. Only in the presence of sufficiently "strong" fluctuations, the solution is able to decompose and exhibit phase separation.

This situation is entirely different in the region inside the two inflection points. Consider a nominal composition of X_2 now. According to the secant construction in Figure 1.12, any perturbation of local chemistry immediately decreases the Gibbs energy and the perturbation will

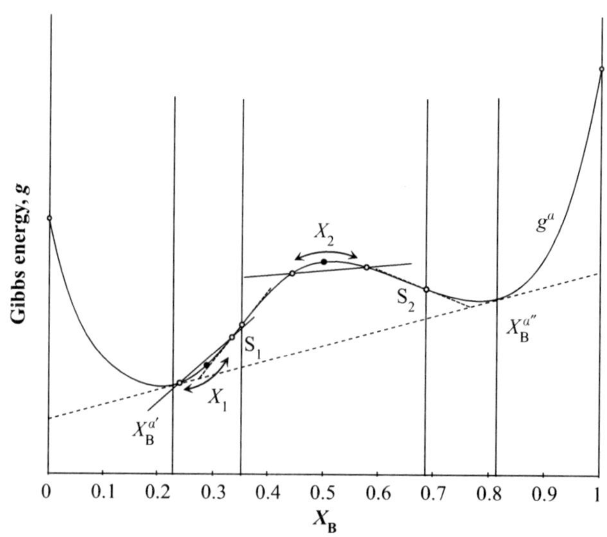

Figure 1.12. Inflection points separating two regions of the G-X curve with different behavior with respect to the system response to local compositional perturbations.

thus be amplified. The system is capable of exhibiting *spontaneous phase separation* without the necessity of overcoming any kind of energy barrier. The line in the phase diagram connecting the inflection points for various temperatures is called the *spinodal*.

According to the phase diagram shown in Figure 1.10, the lines for the phase boundary and the spinodal define three distinct regions:

I. This is the region where a homogeneous solid solution of A and B atoms is thermodynamically stable.
II. In the region between the phase boundary and the spinodal line, phase separation is thermodynamically favorable. However, decomposition of an initially homogeneous solid solution must be facilitated by large enough perturbations in local chemical composition. For phase separation, an energy barrier must be overcome.
III. In region III, phase separation is thermodynamically favorable *and* the system is capable of amplifying any chemical composition fluctuation. Inside the spinodal, the homogeneous solution of A and B atoms can spontaneously decompose into the thermodynamically stable configuration with two separate phases.[16]

Finally, expressions are provided for the evaluation of the critical points in the regular solution model. These can be found from the curve discussion and evaluation of derivatives of the Gibbs energy as defined in equations (1.33) and (1.34), or the Gibbs energy of mixing as given in equations (1.35) and (1.36). With $X_B = 1 - X_A$ and for equation (1.35), the Gibbs energy of mixing and the derivatives are

$$\Delta g_{RS} = RT\left(X_A \ln X_A + (1-X_A)\ln(1-X_A)\right) + \frac{1}{2} X_A (1-X_A)\cdot\omega_{AB}$$

$$\frac{\partial \Delta g_{RS}}{\partial X_A} = RT\left(\ln(X_A) - \ln(1-X_A)\right) + \frac{1}{2}(1-2X_A)\cdot\omega_{AB}$$

$$\frac{\partial^2 \Delta g_{RS}}{\partial X_A^2} = RT\left(\frac{1}{X_A} + \frac{1}{1-X_A}\right) - \omega_{AB}. \tag{1.37}$$

After setting the second derivative of the Gibbs energy of mixing to zero at a composition of $X_B = X_A = 0.5$, the critical temperature T_{crit} is obtained with

$$T_{crit} = \frac{\omega_{AB}}{4R}. \tag{1.38}$$

The chemical potential of component *i* in a regular solution can be evaluated after inserting equations (1.33) or (1.34) into the expression for the chemical potential in a molar framework, equation (1.14). For a binary regular solution, we have

[16] The position of the spinodal line can be shifted by elastic stress caused by differences in the partial volumes of the atomic species involved in the phase separation process. This issue is discussed in Section 1.5.2.

$$\mu_i = g_{\mathrm{RS}} + \frac{\partial g_{\mathrm{RS}}}{\partial X_i} - X_{\mathrm{A}} \frac{\partial g_{\mathrm{RS}}}{\partial X_{\mathrm{A}}} - X_{\mathrm{B}} \frac{\partial g_{\mathrm{RS}}}{\partial X_{\mathrm{B}}}$$

$$g_{\mathrm{RS}} = X_{\mathrm{A}} g_{\mathrm{A}}^0 + X_{\mathrm{B}} g_{\mathrm{B}}^0 + RT\left(X_{\mathrm{A}} \ln X_{\mathrm{A}} + X_{\mathrm{B}} \ln X_{\mathrm{B}}\right) + \frac{1}{2} X_{\mathrm{A}} X_{\mathrm{B}} \cdot \omega_{\mathrm{AB}}$$

$$\frac{\partial g_{\mathrm{RS}}}{\partial X_{\mathrm{A}}} = g_{\mathrm{A}}^0 + RT\left(\ln X_{\mathrm{A}} + 1\right) + \frac{1}{2} X_{\mathrm{B}} \cdot \omega_{\mathrm{AB}}$$

$$\frac{\partial g_{\mathrm{RS}}}{\partial X_{\mathrm{B}}} = g_{\mathrm{B}}^0 + RT\left(\ln X_{\mathrm{B}} + 1\right) + \frac{1}{2} X_{\mathrm{A}} \cdot \omega_{\mathrm{AB}}. \tag{1.39}$$

For the chemical potential of components A and B in a binary regular solution, we finally obtain

$$\mu_{\mathrm{A}} = g_{\mathrm{A}}^0 + RT \ln X_{\mathrm{A}} + \frac{1}{2} X_{\mathrm{B}}^2 \cdot \omega_{\mathrm{AB}} = g_{\mathrm{A}}^0 + RT \ln a_{\mathrm{A}}$$

$$\mu_{\mathrm{B}} = g_{\mathrm{B}}^0 + RT \ln X_{\mathrm{B}} + \frac{1}{2} X_{\mathrm{A}}^2 \cdot \omega_{\mathrm{AB}} = g_{\mathrm{B}}^0 + RT \ln a_{\mathrm{B}}. \tag{1.40}$$

In equation (1.40), the *activity* a_i has been introduced. This quantity is traditionally used to describe the thermodynamic behavior of general nonideal solutions. In Figure 1.11, the quantity $RT\ln a_{\mathrm{B}}$ is shown as the distance between g_{B}^0 and μ_{B} on the axis for $X_{\mathrm{B}} = 1$.

1.3.3 General Solutions—The CALPHAD Approach

In the previous section, we have introduced the *chemical potential*, a_i, as a convenient quantity to describe the excess Gibbs energy of a regular solution. The concept of activities can as well be applied to general solution, by simply replacing the mole fraction variables, X_i, by the corresponding activities, a_i, as done, for instance, in equation (1.40). Activities of elements in thermodynamic equilibrium can be measured with suitable experimental techniques, and activity data are available for numerous technologically relevant alloy systems. In this section, the focus is shifted onto a thermodynamic model, which nowadays represents a powerful and standardized framework for numerical evaluation of multicomponent multiphase thermodynamic phase equilibria: the CALPHAD method.[17]

"Real" thermodynamic systems are typically characterized by complex chemical interactions, which cannot be described by either an ideal or a regular solution. However, these two simple models already capture a substantial amount of the physical behavior of a thermodynamic system. For this reason, the ideal solution is often taken as a reference system for general solid solutions, whereas the regular solution is a convenient and simple candidate model for studying the fundamental properties of solid solutions. These objectives have been dealt with in the previous sections. In this section, the concepts are developed further for "real" systems where more complex chemical interactions between atoms of the solution can also be taken into account.

[17] CALPHAD stands for CALculation of PHAse Diagrams. The theoretical concept behind the method has been developed in the late 1960s and early 1970s. The CALPHAD approach provides the conceptual basis for practically all advanced software for computational thermodynamics that is nowadays available.

The Gibbs energy of a *general solution*, g, can be defined in terms of the Gibbs energy of the ideal solution, g_{IS}, and the *excess* Gibbs energy, g_{EX}, as

$$g = g_{\mathrm{IS}} + g_{\mathrm{EX}}. \tag{1.41}$$

The excess Gibbs energy describes that part of the total free energy, which characterizes the thermodynamic system in excess of the ideal solution. Excess Gibbs energy contributions typically come from complex chemical interactions, magnetism, or short-range ordering.

With equation (1.25) and some of the excess Gibbs energy terms explicitly split up into separate contributions, the Gibbs energy of a general solution can be expressed as

$$g = \sum_i X_i g_i^0 + RT \sum_i X_i \ln X_i + g_{\mathrm{EX}}^{\mathrm{RK}} + g_{\mathrm{EX}}^{\mathrm{magn}} + g_{\mathrm{EX}}^{\mathrm{SRO}} + \cdots. \tag{1.42}$$

The first two terms of the Gibbs energy expression in equation (1.42) represent the contributions from the ideal solution model. The following terms are some of the typical CALPHAD energy contributions, with $g_{\mathrm{EX}}^{\mathrm{RK}}$ being the excess energy contributions from nonideal chemical interactions, $g_{\mathrm{EX}}^{\mathrm{magn}}$ being the contribution from magnetism, and $g_{\mathrm{EX}}^{\mathrm{SRO}}$ being the contribution from short-range ordering.

The first of the three excess energy terms, $g_{\mathrm{EX}}^{\mathrm{RK}}$, is denoted after a model of Redlich and Kister [8], who suggested a series of polynomials for the mathematical representation of general functions (such as the excess Gibbs energy) with

$$g_{\mathrm{EX}}^{\mathrm{RK}} = \sum_{i,j} X_i X_j \left(X_i - X_j\right)^n \cdot L_{ij}^n. \tag{1.43}$$

The functions of mole fraction variables, X, in the sum are known as *Redlich–Kister polynomials*. They have shown to be convenient and general functions to describe even complex interactions in thermodynamic systems. The weighting parameters, L_{ij}^n, are to be determined for each system under consideration and they can be conveniently stored in thermodynamic databases. They are general functions of temperature and pressure. The exponent, n, in equation (1.43) is typically a number between 0 and 3, which is commonly considered to be sufficient for describing even complex thermodynamic interactions.

Figure 1.13 shows a series of Redlich–Kister polynomials for different values of the exponent n. Note that even numbers of the exponent represent symmetric shapes of the polynomials, whereas the odd numbers lead to unsymmetrical shapes.

From the initial CALPHAD-type models, the currently used framework has experienced significant further development in the last decades. Advanced models are nowadays based on the compound energy model that accounts for phases where groups of atoms are located on separate sublattices[18] or where multiple sublattices represent stoichiometric constraints stemming

[18] Think, for instance, of metals with substitutional and interstitial elements, such as Fe-C-based alloys. In steel, the Fe atoms share a common sublattice with other alloying elements, such as Mn, Si, Cr, etc., with the light C and N atoms occupying the interstitial positions of the cubic face centered or cubic body centered sublattice of the Fe. The elements located on one sublattice are assumed to interchange position only with atoms located on the same sublattice.

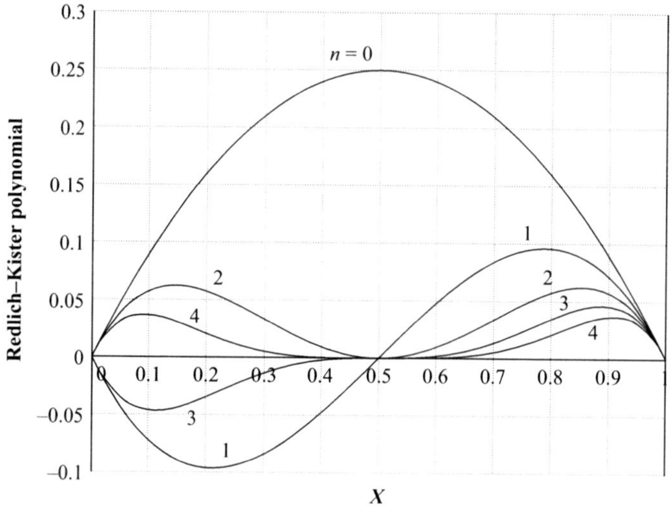

Figure 1.13. Redlich–Kister polynomials, equation (1.43), for different values of the exponent n and $X_j = 1 - X_i$.

from complex crystal structures. Ferromagnetic and paramagnetic behavior is accounted for in special formalisms and short-range order can be modeled in various different approaches. Even a short description of these models is way beyond the scope of this book. The interested reader is therefore referred to the comprehensive textbooks on computational thermodynamics and the CALPHAD method, for example, refs. [9,10]. A short example of a thermodynamic database is outlined in ref. [11], together with a brief description of the mathematical framework for numerical evaluation of multicomponent thermodynamic phase equilibria based on the compound energy formalism and CALPHAD-type databases.

1.4 Multiphase Systems and Driving Force for Precipitation

In the preceding Section 1.3.2 on the regular solution model, it was demonstrated that below the critical temperature, T_{crit}, the introduction of chemical interactions in the form of a regular solution parameter ω can lead to the decomposition of a homogeneous solid solution α into two phases α' and α''. In the following, this situation is slightly modified by assuming that a single homogeneous parent phase α decomposes into two phases α' and β. This situation applies, for instance, to the case where a *precipitate phase* β forms out of a supersaturated *matrix phase* α. The crystal structure of the precipitate β is not necessarily of the same type of the parent phase. We further assume that the precipitate is a B-rich phase and the matrix is a homogeneous solid solution of A and B atoms.

A typical Gibbs energy—composition G-X diagram is presented in Figure 1.14, showing the tangent construction for the two-phase $\alpha' + \beta$ equilibrium (solid line) as well as the initial condition of a supersaturated solution of the single-phase α (dashed lines). The nominal composition of the solid solution is denoted with X_B^α and the chemical potentials are read at the vertical y-axes with μ_A^α and μ_B^α.

In the initial, supersaturated solution, the thermodynamic system maintains an unstable state because the formation of some amount of β phase can lower the total Gibbs energy of the

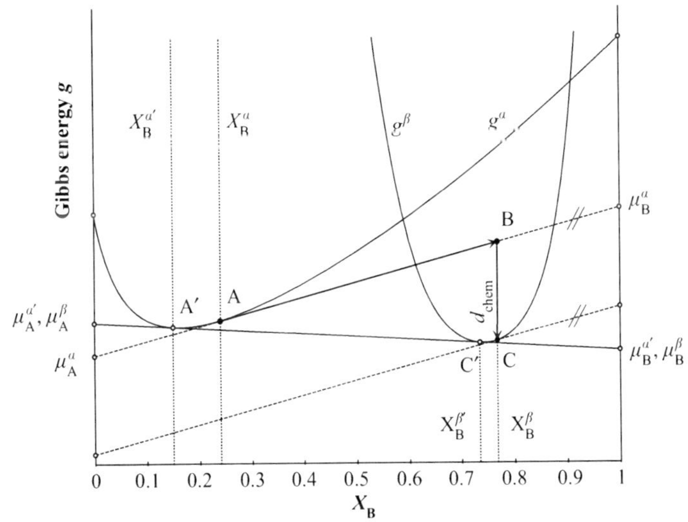

Figure 1.14. *G-X* diagram for a multiphase thermodynamic equilibrium sketching the procedure for evaluation of the chemical driving force d_{chem}.

system. Thus, in the supersaturated condition, a *driving force*,[19] d_{chem}, exists for precipitation of the β phase from α. The value of d_{chem} can be determined with the following procedure:

1. Take an infinitesimal amount of the supersaturated solid solution with a composition of X_B^{α}. Using the Gibbs energy representation in terms of chemical potentials, at point A, we have $g^A = X_A^{\alpha}\mu_A^{\alpha} + X_B^{\alpha}\mu_B^{\alpha}$.
2. Modify the chemical composition of this portion of α phase from X_B^{α} to X_B^{β}, equivalent to moving from point A to point B along the line connecting the chemical potentials μ_A and μ_B. At the end of this step, the infinitesimal amount of new phase has the thermodynamic characteristics of the α phase, but a composition corresponding to β. The Gibbs energy of this substance at point B is $g^B = X_A^{\beta}\mu_A^{\alpha} + X_B^{\beta}\mu_B^{\alpha}$. Due to the infinitesimal changes, the properties of the solution do not change during this process.
3. Change the crystal structure (if applicable) and thermophysical properties of the infinitesimal amount of α to that of the β phase. This step moves the Gibbs energy along the line from point B to point C. The distance between B and C corresponds to the molar chemical driving force d_{chem}, or, in other words, the Gibbs energy change on transforming one mole of substance α into β. The Gibbs energy of the final configuration at point C is $g^C = X_A^{\beta}\mu_A^{\beta} + X_B^{\beta}\mu_B^{\beta}$.

The procedure described by steps I to III above is not an exact representation of the integral Gibbs energy change during the phase transformation from the initial supersaturated to the final equilibrium state. Figure 1.14 instead suggests that the driving force calculation as depicted

[19] The subscript "chem" emphasized that the driving force for precipitation sketched in Figure 1.14 stems from chemical interactions only. Driving force contributions for phase transformations can also come from other sources, such as, for instance, capillarity, mechanical stresses, or plastic deformation gradients.

here is an extrapolation procedure, where a sufficiently small amount of α is transformed into β such that the global chemical environment remains unchanged. The process of evaluation of the properties of the solid solution is based on the current values of the chemical potentials of A and B.

If a certain finite amount of B-rich β is formed, the α phase becomes depleted in B and the composition of the matrix phase gradually evolves from point A in the figure toward point A′. Simultaneously, the chemical potentials evolve with the change in slope of the tangent to the Gibbs energy curve of α.

Arriving at the equilibrium state, the α phase has transformed into α' with a composition X_{B}^{α} having evolved to $X_{\mathrm{B}}^{\alpha'}$. Simultaneously, the composition of β has also evolved from X_{B}^{β} to $X_{\mathrm{B}}^{\beta'}$. While the final equilibrium state is given by the common tangent construction (as depicted already in Figure 1.11), each intermediate state is constructed in such a way that a maximum reduction of the Gibbs energy is achieved in each progressive step. This condition is satisfied if the tangents to the G-X curves of the two phases are maintained parallel to each other. In the course of the precipitation process, the chemical driving force gradually decreases. In the final equilibrium state, the driving force has decreased to zero.

Since the chemical driving force is obtained as the difference of the Gibbs energies at points B and C, we finally have

$$d_{\mathrm{chem}}^{\beta} = -\sum_i X_i^{\beta}\left(\mu_i^{\beta} - \mu_i^{\alpha}\right). \tag{1.44}$$

In some cases of practical computational thermodynamics, where the Gibbs energies of phases are described with the multiple-sublattice model of the CALPHAD approach, it can become difficult, or even impossible, to evaluate chemical potentials in some of the precipitation phases. In these cases, a modified expression can be utilized for evaluation of chemical driving forces with

$$d_{\mathrm{chem}}^{\beta} = \sum_i X_i^{\beta}\mu_i^{\alpha} - g^{\beta}, \tag{1.45}$$

where use has been made of the universal relationship between the chemical potentials and the Gibbs energy, equation (1.15).

On basis of these arguments, it is finally concluded that in thermodynamic equilibrium the total driving force,[20] $d_{\mathrm{F}}^{(m)}$, for all possible phases are either zero (the phase is participating in multiphase equilibrium with its phase fraction being zero) or negative (the phase is thermodynamically unstable), expressed as

$$d_{\mathrm{F}}^{(m)} \leq 0, \tag{1.46}$$

for all phases, m, of the thermodynamic system. Together with the condition of minimum Gibbs energy of the system and the equality of all chemical potentials in all phases with nonzero phase fraction (compare Section 1.3.2), this condition represents a third alternative necessary condition for thermodynamic equilibrium in multicomponent multiphase systems.

---

[20] In contrast to the chemical driving force, d_{chem}, the total driving force, $d_{\mathrm{F}}^{(m)}$, also contains contributions from capillarity forces, mechanical stress, etc. These additional contributions are in most cases decreasing the value of the total driving force compared to the maximum available chemical driving force.

1.5 Curvature and Elastic Stress

When investigating the free energy of thermodynamic systems, so far, we have only considered bulk chemical interactions between the atoms of the solid solution. The free energy of such a solution has been expressed as the weighted sum of the Gibbs energies of the pure substances, the ideal entropy of mixing, and the excess Gibbs energy, as given in equation (1.42). When dealing with the thermodynamics of solid-state precipitation, there are two additional phenomena that have to be taken into account to fully describe the total energy of the system.

1.5.1 The Gibbs–Thomson Equation

The first contribution to the chemical free energy comes from *capillarity*. Capillarity exerts pressure on curved surfaces due to anisotropies in atomic bonding. Since this pressure is inversely proportional to the radius of curvature, this effect is also denoted as *curvature-induced pressure*. Figure 1.15 presents a sketch of the forces acting on the atoms of the precipitate (dark color) located directly at the interface. Accordingly, nearest neighbor atoms of the same kind experience stronger binding forces compared to the unlike atoms at the surface.[21] As a result, a net force exists on all interface atoms pointing toward the center of the precipitate. If these contributions are integrated over the precipitate surface, a pressure is created that is acting on the precipitate. The value of this curvature-induced pressure, P_{curv}, is given with

$$P_{\text{curv}} = \frac{2\gamma}{\rho}, \tag{1.47}$$

where ρ is the radius of the spherical precipitate and γ is the *specific interfacial energy*. The units of γ are [J/m^2] or [N/m]. The first representation of units corresponds to an energy density along

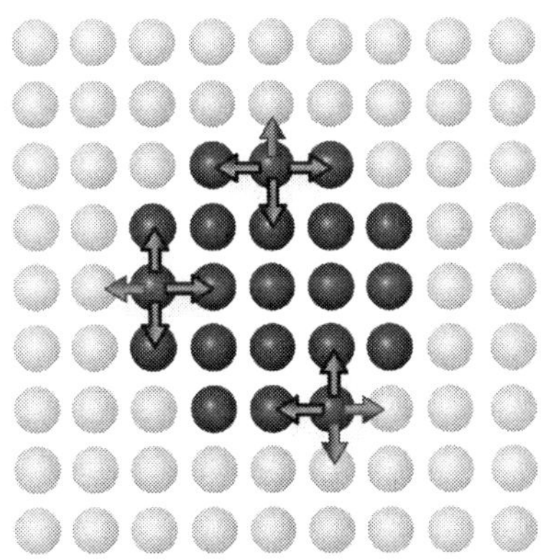

Figure 1.15. Sketch of atomic bonds across the phase boundary of a precipitate figuring the origin of interfacial pressure. Interface atoms have weaker bonds to the outside of the precipitate, thus introducing a net force pointing toward the inside of the sphere.

[21] This statement is easily reasoned on the basis that the precipitate would otherwise not be stable. If the bonds to the unlike atoms were equal or stronger, the precipitate atoms would prefer dissolving in the matrix to decrease the free energy of the system.

the precipitate surface, whereas the second representation is a *specific force*. According to equation (1.47), this energy contribution becomes relevant and is most prominent for small precipitates. The curvature-induced pressure term approaches infinity if ρ goes toward zero. In contrast, P_{curv} vanishes for $\rho \to \infty$, which is commonly assumed for all phases in equilibrium thermodynamics.

It is important to recognize that pressure on a precipitate modifies the thermodynamic properties of the system. Under the constraint of incompressibility of the precipitate phase, the molar free energy contribution, g_{curv}, from curvature-induced pressure becomes

$$g_{curv} = Pv^{\beta} = \frac{2\gamma}{\rho} v^{\beta}, \tag{1.48}$$

with v^{β} representing the molar volume of the precipitate. As all quantities entering g_{curv} have a positive sign, the energy contribution from curvature-induced pressure is also always positive.

Figure 1.16 shows a typical *G-X* diagram[22] for the Gibbs energy of a precipitate in the limit of $\rho \to \infty$ (solid line), as well as a diagram taking into account the curvature-induced pressure term (dashed line). From the figure, we recognize that capillarity effects shift the equilibrium composition of B atoms both in the matrix phase and in the precipitate. In thermodynamic equilibrium without capillarity, matrix and precipitate compositions are given at points A and C. With curvature-induced pressure, the B content in the matrix is shifted to higher values indicated at point A′. Simultaneously, the equilibrium content of B in the precipitate changes from point C to C′.[23] Although this effect might appear to be weak, we will see that it plays a

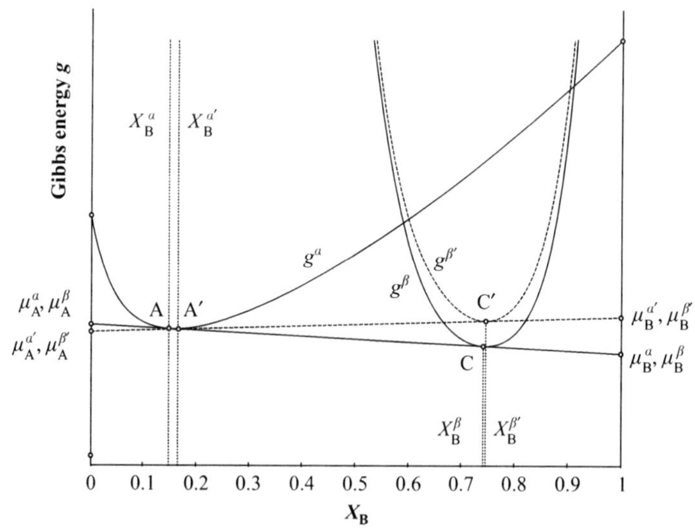

Figure 1.16. *G-X* diagram of a multiphase thermodynamic equilibrium taking into account the capillarity effect from curvature-induced pressure on small precipitates.

[22] A phase diagram that takes into account the influence of interfacial curvature is known as a *coherent phase diagram*.

[23] The effect of capillarity on the precipitate is usually weaker than the effect on the matrix composition. For this reason, the effect on the precipitate composition is often neglected.

major role in the evolution of precipitates in the later stages of precipitation, where coarsening (Ostwald ripening) occurs. This issue is dealt with in Section 3.6.

From Figure 1.16, the relation between the G-X curves and composition are found to be

$$\frac{\partial g^{\alpha}}{\partial X_{\mathrm{B}}}=\frac{g^{\beta}-g^{\alpha}}{X_{\mathrm{B}}^{\beta}-X_{\mathrm{B}}^{\alpha}} \quad \text{and} \quad \frac{\partial g^{\alpha'}}{\partial X_{\mathrm{B}}}=\frac{g^{\beta'}-g^{\alpha'}}{X_{\mathrm{B}}^{\beta'}-X_{\mathrm{B}}^{\alpha'}}. \tag{1.49}$$

The primed quantities in equation (1.49) correspond to a system with curvature-induced pressure. With the derivatives given at points A and A′, the curvature of the G-X curve in the vicinity of A can be approximated as

$$\frac{\partial^2 g^{\alpha}}{\partial(X_{\mathrm{B}})^2}=\frac{\dfrac{\partial g^{\alpha'}}{\partial X_{\mathrm{B}}}-\dfrac{\partial g^{\alpha}}{\partial X_{\mathrm{B}}}}{X_{\mathrm{B}}^{\alpha'}-X_{\mathrm{B}}^{\alpha}}. \tag{1.50}$$

On inserting equation (1.49) into equation (1.50) and assuming that the composition difference $X_{\mathrm{B}}^{\beta}-X_{\mathrm{B}}^{\alpha}$ is sufficiently large compared to $X_{\mathrm{B}}^{\alpha'}-X_{\mathrm{B}}^{\alpha}$, which allows us to utilize the approximation $X_{\mathrm{B}}^{\beta'}-X_{\mathrm{B}}^{\alpha'}\approx X_{\mathrm{B}}^{\beta}-X_{\mathrm{B}}^{\alpha}$, we obtain

$$\frac{\partial^2 g^{\alpha}}{\partial(X_{\mathrm{B}})^2}=\frac{\dfrac{g^{\beta'}-g^{\alpha'}}{X_{\mathrm{B}}^{\beta'}-X_{\mathrm{B}}^{\alpha'}}-\dfrac{g^{\beta}-g^{\alpha}}{X_{\mathrm{B}}^{\beta}-X_{\mathrm{B}}^{\alpha}}}{X_{\mathrm{B}}^{\alpha'}-X_{\mathrm{B}}^{\alpha}}\approx\frac{\dfrac{g^{\beta'}-g^{\beta}}{X_{\mathrm{B}}^{\beta}-X_{\mathrm{B}}^{\alpha}}}{X_{\mathrm{B}}^{\alpha'}-X_{\mathrm{B}}^{\alpha}}. \tag{1.51}$$

The difference in molar Gibbs energy of the precipitate phase $g^{\beta'}-g^{\beta}$ corresponds to the excess Gibbs energy from capillarity pressure, g_{curv}, given in equation (1.48) and we obtain for the change in equilibrium composition around the precipitate

$$X_{\mathrm{B}}^{\alpha'}-X_{\mathrm{B}}^{\alpha}\approx\left[\left(X_{\mathrm{B}}^{\beta}-X_{\mathrm{B}}^{\alpha}\right)\frac{\partial^2 g^{\alpha}}{\partial(X_{\mathrm{B}})^2}\right]^{-1}\cdot\frac{2\gamma}{\rho}v_{\mathrm{m}}^{\beta}. \tag{1.52}$$

In many cases, the second derivative of the Gibbs energy is not readily available. Equation (1.52) can be further simplified if we assume that the thermodynamic system behaves approximately like an ideal solution. Using the second derivatives of the ideal solution Gibbs energy (for instance, equation (1.37) with $\omega = 0$), the second derivative can be replaced with

$$\frac{\partial^2 g^{\alpha}}{\partial(X_{\mathrm{B}})^2}\approx RT\left(\frac{1}{X_{\mathrm{A}}}+\frac{1}{X_{\mathrm{B}}}\right)=\frac{RT}{X_{\mathrm{A}}X_{\mathrm{B}}}\approx\frac{RT}{X_{\mathrm{B}}}. \tag{1.53}$$

In the last step, we have made advantage of the relation $X_{\mathrm{A}}\approx 1$ in the dilute solution limit. For the shift in matrix composition, we finally obtain

$$X_{\mathrm{B}}^{\alpha'}-X_{\mathrm{B}}^{\alpha}\approx\frac{X_{\mathrm{B}}^{\alpha}}{X_{\mathrm{B}}^{\beta}-X_{\mathrm{B}}^{\alpha}}\cdot\frac{2\gamma v^{\beta}}{RT}\cdot\frac{1}{\rho}, \tag{1.54}$$

or

$$X_{\mathrm{B}}^{\alpha'}\approx X_{\mathrm{B}}^{\alpha}\left(1+\frac{2\gamma v^{\beta}}{\left(X_{\mathrm{B}}^{\beta}-X_{\mathrm{B}}^{\alpha}\right)RT}\cdot\frac{1}{\rho}\right). \tag{1.55}$$

Equation (1.55) is known as the linearized form of the *Gibbs–Thomson equation.* This relation represents a reasonable approximation of the composition shift for larger precipitates, that is, for lower values of the excess Gibbs energy coming from capillarity.

A more general treatment can be performed when considering a situation where the phase α is a dilute solution in equilibrium with the precipitate phase β (see, e.g., ref. [12]). In the absence of the effect of interface curvature, the Gibbs energy, G^α, of the matrix phase α can be written as[24]

$$G^\alpha = N_A^\alpha\left[g_A^\alpha + RT\ln\frac{N_A^\alpha}{N_A^\alpha + N_B^\alpha}\right] + N_B^\alpha\left[g_B^\alpha + RT\ln\frac{N_B^\alpha}{N_A^\alpha + N_B^\alpha}\right] + \omega_{AB}\frac{N_A^\alpha N_B^\alpha}{N_A^\alpha + N_B^\alpha}. \quad (1.56)$$

N_A^α and N_B^α are the number of moles of components A and B in α and ω_{AB} is the regular solution parameter. g_A^α and g_B^α are the molar Gibbs energies of pure A and B.

The precipitate phase β can be introduced into the system as a perfectly ordered phase. Neglecting configurational entropy, consequently, the Gibbs energy, G^β, of β is

$$G^\beta = \left(N_A^\beta + N_B^\beta\right) g_{A_xB_y}^\beta, \quad (1.57)$$

where $g_{A_xB_y}^\beta$ is the molar Gibbs energy of the β phase having the fixed stoichiometry A_xB_y.

If the two phases are in equilibrium, transferring a small amount of dn moles of atoms from the α phase of composition, $X_B^\alpha = 1 - X_A^\alpha$, to the β phase of composition, $X_B^\beta = 1 - X_A^\beta$, does not change the total energy of the system. This condition can be written as

$$dn\left(1 - X_B^\beta\right)\left.\frac{\partial G^\alpha}{\partial N_A^\alpha}\right|_{X_B^\alpha} + dnX_B^\beta\left.\frac{\partial G^\alpha}{\partial N_B^\alpha}\right|_{X_B^\alpha} = dn\frac{\partial G^\beta}{\partial N^\beta}, \quad (1.58)$$

with $N^\beta = N_A^\beta + N_B^\beta$. For a dilute regular solution, the Gibbs energy of the β phase is then observed with

$$g_{A_xB_y}^\beta = \left(1 - X_B^\beta\right)\left[g_A^\alpha + RT\ln\left(1 - X_B^\alpha\right)\right] + X_B^\beta\left[g_B^\alpha + RT\ln X_B^\alpha + \omega_{AB}\right]. \quad (1.59)$$

If the effect of interface curvature is taken into account via the interfacial energy, γ, and the precipitate radius, ρ, the Gibbs energy of the β phase reads

$$G^{\beta'} = N^\beta g_{A_xB_y}^\beta + \gamma A, \quad (1.60)$$

where A is the surface area of the precipitate. For a spherical precipitate, we have

$$\frac{4\pi}{3}\rho^3 = N^\beta v^\beta, \quad (1.61)$$

24 Compare also Section 1.3.2.

where v^β is the molar volume of phase β. The partial derivative of the Gibbs energy of β in the presence of interface curvature is then given by

$$\frac{\partial G^{\beta'}}{\partial N^\beta} = g^\beta_{\mathrm{A}_x\mathrm{B}_y} + \frac{8\pi\rho\gamma}{4\pi\rho^2/v^\beta} = g^\beta_{\mathrm{A}_x\mathrm{B}_y} + \frac{2\gamma\, v^\beta}{\rho}, \tag{1.62}$$

and the equilibrium condition in this system becomes

$$g^\beta_{\mathrm{A}_x\mathrm{B}_y} + \frac{2\gamma\, v^\beta}{\rho} = \left(1 - X^\beta_\mathrm{B}\right)\left[g^\alpha_\mathrm{A} + RT\ln\left(1 - X^{\alpha'}_\mathrm{B}\right)\right] + X^\beta_\mathrm{B}\left[g^\alpha_\mathrm{B} + RT\ln X^{\alpha'}_\mathrm{B} + \omega_\mathrm{AB}\right]. \tag{1.63}$$

The general form of the Gibbs–Thomson equation is finally obtained by substraction of the two equilibrium relations, equations (1.59) and (1.63), with

$$\frac{2\gamma\, v^\beta}{\rho RT} = \left(1 - X^\beta_\mathrm{B}\right)\ln\left(\frac{1 - X^{\alpha'}_\mathrm{B}}{1 - X^\alpha_\mathrm{B}}\right) + X^\beta_\mathrm{B}\ln\left(\frac{X^{\alpha'}_\mathrm{B}}{X^\alpha_\mathrm{B}}\right). \tag{1.64}$$

It can easily be shown that in a multicomponent system with elements A, B, C, … and a precipitate of composition $\mathrm{A}_x\mathrm{B}_y\mathrm{C}_z\ldots$, the equivalent equation reads

$$\frac{2\gamma\, v^\beta}{\rho RT}\left(x + y + z + \cdots\right) = x\ln\left(\frac{X^{\alpha'}_\mathrm{A}}{X^\alpha_\mathrm{A}}\right) + y\ln\left(\frac{X^{\alpha'}_\mathrm{B}}{X^\alpha_\mathrm{B}}\right) + z\ln\left(\frac{X^{\alpha'}_\mathrm{C}}{X^\alpha_\mathrm{C}}\right) + \cdots. \tag{1.65}$$

Unfortunately, the general form of the Gibbs–Thomson equation is not easy to solve and, for practical reasons, three approximations can be made. These are

1. $X^\beta_\mathrm{B} = 1$
   If the precipitate consists of mainly B atoms, this approximation leads to the well-known exponential form of the Gibbs–Thomson equation, which becomes

$$X^{\alpha'}_\mathrm{B} \approx X^\alpha_\mathrm{B} \cdot \exp\left(\frac{2\gamma v^\beta}{RT} \cdot \frac{1}{\rho}\right). \tag{1.66}$$

   The Gibbs–Thomson equation in this form gives a better approximation for small precipitates compared to the linearized version presented in equation (1.55). Moreover, it is the most often used form in literature, however, sometimes applied erroneously and neglecting the constraint of practically pure B precipitates.

2. $X^{\alpha'}_\mathrm{B} \approx X^\alpha_\mathrm{B}$
   Before this approximation can be applied, the general form of the Gibbs–Thomson equation is first reformulated as

$$\frac{2\gamma\, v^\beta}{\rho RT} = \left(1 - X^\beta_\mathrm{B}\right)\ln\left(1 + \frac{X^\alpha_\mathrm{B} - X^{\alpha'}_\mathrm{B}}{1 - X^\alpha_\mathrm{B}}\right) + X^\beta_\mathrm{B}\ln\left(\frac{X^\alpha_\mathrm{B} - X^{\alpha'}_\mathrm{B}}{X^\alpha_\mathrm{B}}\right). \tag{1.67}$$

Series expansion of the logarithmic terms then delivers

$$\frac{2\gamma v^{\beta}}{\rho RT} = \left(1 - X_{\mathrm{B}}^{\beta}\right)\frac{X_{\mathrm{B}}^{\alpha} - X_{\mathrm{B}}^{\alpha'}}{1 - X_{\mathrm{B}}^{\alpha}} + X_{\mathrm{B}}^{\beta}\,\frac{X_{\mathrm{B}}^{\alpha} - X_{\mathrm{B}}^{\alpha'}}{X_{\mathrm{B}}^{\alpha}} \tag{1.68}$$

and, finally,

$$X_{\mathrm{B}}^{\alpha'} \approx X_{\mathrm{B}}^{\alpha}\left(1 + \frac{2\gamma v^{\beta}}{RT}\frac{\left(1 - X_{\mathrm{B}}^{\alpha}\right)}{\left(X_{\mathrm{B}}^{\beta} - X_{\mathrm{B}}^{\alpha}\right)}\cdot\frac{1}{\rho}\right). \tag{1.69}$$

This form of the Gibbs–Thomson equation is very similar to the linearized version presented in equation (1.55).

3. $X_{\mathrm{B}}^{\alpha'} << 1$, $X_{\mathrm{B}}^{\alpha} << 1$
If the matrix concentration of B atoms at the precipitate surface is small, the first term in the general form of the Gibbs–Thomson equation (1.64) can be straightforwardly neglected in comparison to the second term. Problems might occur, however, if $X_{\mathrm{B}}^{\alpha'} \approx X_{\mathrm{B}}^{\alpha}$. The Gibbs–Thomson equation can then be written as

$$\frac{2\gamma v^{\beta}}{\rho RT} = -\left(1 - X_{\mathrm{B}}^{\beta}\right)\left(X_{\mathrm{B}}^{\alpha'} - X_{\mathrm{B}}^{\alpha}\right) + X_{\mathrm{B}}^{\beta}\cdot\frac{X_{\mathrm{B}}^{\alpha'} - X_{\mathrm{B}}^{\alpha}}{X_{\mathrm{B}}^{\alpha}}. \tag{1.70}$$

In this form, the first term is again small compared to the second one and might be neglected. The approximate Gibbs–Thomson equation for case 3 finally reads

$$X_{\mathrm{B}}^{\alpha'} \approx X_{\mathrm{B}}^{\alpha}\exp\left(\frac{2\gamma v^{\beta}}{RTX_{\mathrm{B}}^{\beta}}\cdot\frac{1}{\rho}\right). \tag{1.71}$$

The different approximations of the Gibbs–Thomson equation presented above are compared with each other in Figure 1.17 for the case of an almost pure B precipitate in a dilute α matrix. Figure 1.18 compares the equations for a precipitate with a B-fraction of only 0.2. The symbols denote the individual approximations, with "0" ··· Gibbs–Thomson equation without approximation, equation (1.64); "1" ··· approximation 1, equation (1.66); "2" ··· approximation 2, equation (1.69); "3" ··· approximation 3, equation (1.71); and "lin" ··· linearized version, equation (1.55).

In the first case, Figure 1.17, all variants of the Gibbs–Thomson equation give reasonable forecast. Approximations "2" and "3" are practically undistinguishable from the full version of the Gibbs–Thomson equation "0". This observation is expected because the precipitate is almost pure B and the matrix is a dilute solution. The differences between the predictions summarized in Figure 1.18 for the case of a precipitate with only 20% B atoms are substantial for approximation "1" and the linearized version. The approximations "2" and "3" are still rather close to the Gibbs–Thomson equation without approximations and can be used with fair accuracy also for the case of $X_{\mathrm{B}}^{\beta} < 1$.

1.5.2 Elastic Misfit Stress

The second contribution to the Gibbs energy that needs attention in precipitation systems is the effect of volumetric lattice mismatch between matrix and precipitate. Consider a case where the molar volumes of matrix and precipitate are different. When transforming one mole of atoms from the matrix to the precipitate phase, the volume change is given as

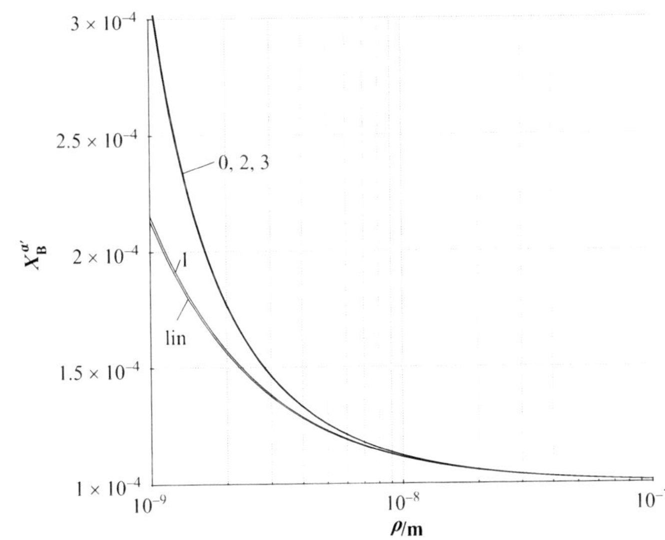

Figure 1.17. Equilibrium mole fraction of B in matrix predicted by Gibbs–Thomson equation for $X_B^\alpha = 0.001$, $X_B^\beta = 0.99$, $\gamma = 0.5$ J/m^{-2}, and $v^\beta = 9.5 \times 10^{-6}$.

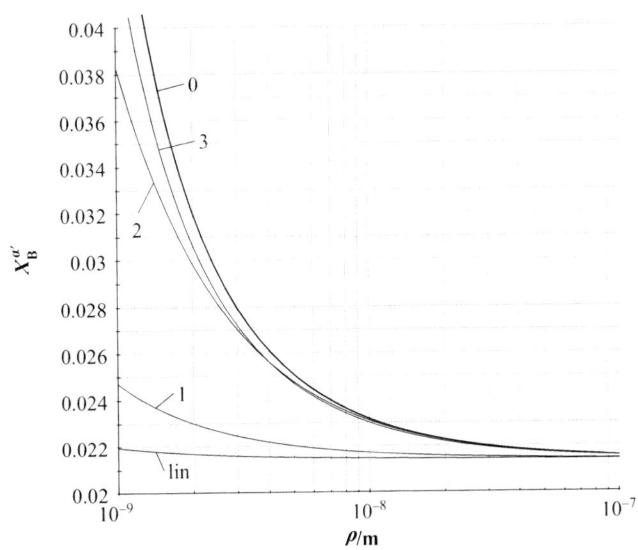

Figure 1.18. Equilibrium mole fraction of B in matrix predicted by Gibbs–Thomson equation for $X_B^\alpha = 0.0214$, $X_B^\beta = 0.2$, $\gamma = 0.063$ J/m^{-2}, and $v^\beta = 9.5 \times 10^{-6}$. These values correspond to the parameters used in Figure 1 of ref. [12].

$$v^* = \frac{v^\beta - v^\alpha}{v^\alpha}, \tag{1.72}$$

with v^α and v^β being the molar volumes of matrix and precipitate, respectively. In a system where all phases were free to expand and contract, volumetric misfit would not cause an effect on the free energy of the system. However, a small precipitate is constrained by the surrounding matrix and elastic stress is generated. Applying the Eshelby concept [13], the elastic energy, $\Delta G_{\text{vol}}^{\text{el}}$, associated with the elastic stress field around an ellipsoidal inclusion is given as

$$\Delta G_{\text{vol}}^{\text{el}} = \frac{E^{\alpha}}{1-\nu^{\alpha}}\left(\varepsilon^*\right)^2 = \frac{E^{\alpha}}{9\left(1-\nu^{\alpha}\right)}\left(v^*\right)^2, \tag{1.73}$$

with E^{α} and ν^{α} being the elastic modulus and Poisson's ratio of the matrix, respectively, and ε^* representing the linear elastic misfit given as

$$\varepsilon^* \approx \frac{v^*}{3}. \tag{1.74}$$

Elastic misfit stress behaves thermodynamically very similar to curvature-induced pressure and its effect on the phase diagram is of the same quality. Since ε^* occurs in the energy expression (1.73) in quadratic form, this energy contribution is always positive. Consequently, the Gibbs energy curve of b is also shifted upward in the G-X diagram, and the equilibrium composition of the two phases changes accordingly.

We shall finally note that capillarity and elastic misfit stress can impose large effects onto the kinetic behavior of precipitation systems. Both mechanisms are very pronounced in the early stages of precipitate life, where the radius of the precipitate is small and interfaces are often coherent. Misfit stress can, for instance, suppress homogeneous nucleation of precipitates and make nucleation possible only at heterogeneous sites where the misfit stress is partially compensated by mechanical stresses with opposite sign. This situation occurs at dislocations, where local compressive and tensile stresses around the dislocation core compensate part of the tensile or compressive hydrostatic stress from volumetric misfit.

In the later stages of precipitate life, the generation of misfit dislocations can accommodate part of the elastic stress around larger precipitates. Additionally, at semicoherent or incoherent phase boundaries, as well as large-angle grain boundaries, generation or annihilation of lattice vacancies can relax the volumetric misfit stress almost entirely. Lattice vacancies are sufficiently effective to make volumetric misfit stress a negligible effect in the coarsening stage of precipitation and, in all stages, for precipitates located at large-angle grain boundaries.

1.6 Equilibrium Structural Vacancies

Structural vacancies are inherent to a crystal lattice at temperatures above absolute zero. They are an *equilibrium* feature of crystalline materials stabilized by the configurational entropy gain in the process of introducing vacant lattice sites in an otherwise perfect crystal. Vacancies are created or annihilated at sources and sinks for vacancies, such as free surfaces, grain boundaries, dislocation jogs, or incoherent phase boundaries. One possible mechanism of vacancy creation is presented schematically in Figure 1.19. The sketch shows a regular lattice atom that jumps onto the surface of the crystal. The atom underneath moves to the position of the former atom and, thus, creates a bulk vacancy. This vacancy can now travel freely through the volume by repeated position exchange with any of the neighboring atoms.

When an atom from within the crystal moves to a position on the surface, the energy of this atom changes drastically due to a reduction of the number of interatomic bonds. As the surface position represents an energetically unfavorable location compared to a position inside the crystal,

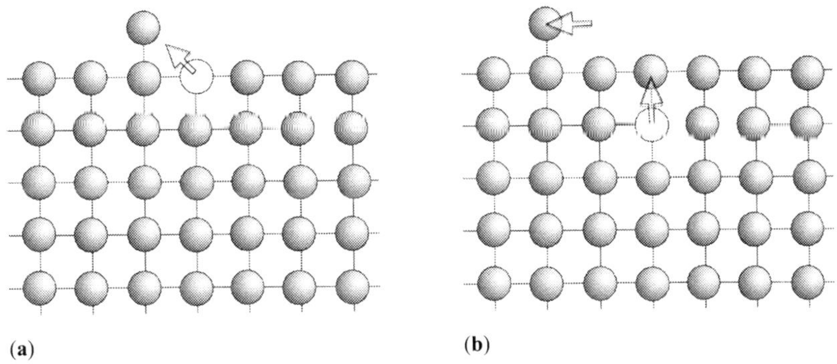

Figure 1.19. Structural vacancy generation on a free surface. A lattice atom escapes from a regular lattice position to a position on the surface (a) leaving an empty site in the top layer. The lattice atom below this site jumps into the vacant position, thus creating a bulk lattice vacancy inside the crystal (b).

the process of removing an atom from the bulk volume is connected to an expense of energy. The change, due to the formation of a single point defect, is called the *Gibbs energy of vacancy formation*, $\Delta G_{\mathrm{Va}}^{\mathrm{f}}$.

The Gibbs energy of vacancy formation can be expressed in terms of enthalpy and entropy according to the thermodynamic relation as

$$\Delta G_{\mathrm{Va}}^{\mathrm{f}} = \Delta H_{\mathrm{Va}}^{\mathrm{f}} - T\Delta S_{\mathrm{Va}}^{\mathrm{f}}, \quad (1.75)$$

where T is the absolute temperature and $\Delta H_{\mathrm{Va}}^{\mathrm{f}}$ and $\Delta S_{\mathrm{Va}}^{\mathrm{f}}$ are the vacancy formation enthalpy and entropy, respectively. It is important to note that the vacancy formation entropy is different from the configurational entropy of the system.

The formation enthalpy term can be expressed as

$$\Delta H_{\mathrm{Va}}^{\mathrm{f}} = \Delta U_{\mathrm{Va}}^{\mathrm{f}} + PV_{\mathrm{Va}}^{\mathrm{f}}, \quad (1.76)$$

where $\Delta U_{\mathrm{Va}}^{\mathrm{f}}$ and $V_{\mathrm{Va}}^{\mathrm{f}}$ are the internal vacancy formation energy and the vacancy formation volume and P is pressure. The term $PV_{\mathrm{Va}}^{\mathrm{f}}$ is small compared to the internal energy of formation and is often neglected at atmospheric pressure. In literature, the enthalpy of vacancy formation is, therefore, often written in terms of the internal energy of vacancy formation, $\Delta U_{\mathrm{Va}}^{\mathrm{f}}$, or simply the energy, $\Delta E_{\mathrm{Va}}^{\mathrm{f}}$.

From the Gibbs energy of vacancy formation, $\Delta G_{\mathrm{Va}}^{\mathrm{f}}$, the equilibrium site fraction of monovacancies in the crystal can finally be calculated with

$$X_{\mathrm{Va}}^{\mathrm{eq}} = \exp\left(-\frac{\Delta G_{\mathrm{Va}}^{\mathrm{f}}}{k_{\mathrm{B}}T}\right) = \exp\left(-\frac{\Delta H_{\mathrm{Va}}^{\mathrm{f}}}{k_{\mathrm{B}}T}\right)\exp\left(\frac{\Delta S_{\mathrm{Va}}^{\mathrm{f}}}{k_{\mathrm{B}}}\right), \quad (1.77)$$

or

$$X_{\mathrm{Va}}^{\mathrm{eq}} = \exp\left(-\frac{\Delta g_{\mathrm{Va}}^{\mathrm{f}}}{RT}\right) = \exp\left(-\frac{\Delta h_{\mathrm{Va}}^{\mathrm{f}}}{RT}\right)\exp\left(\frac{\Delta s_{\mathrm{Va}}^{\mathrm{f}}}{R}\right). \quad (1.78)$$

In equation (1.78), we have used molar quantities, which are indicated by lowercase symbols.

The entropy of vacancy formation, $\Delta S_{\mathrm{Va}}^{\mathrm{f}}$, or the molar entropy of vacancy formation, $\Delta s_{\mathrm{Va}}^{\mathrm{f}}$, refer to the disorder introduced into the crystal by a single monovacancy, or one mole of monovacancies. The missing atom changes the vibrational properties of the neighboring atoms and, thus, the vibrational properties of the entire crystal. It can be calculated in the high-temperature (Einstein) harmonic approximation as

$$S_{\mathrm{Va}}^{\mathrm{eq}} = k_{\mathrm{B}} \sum_{n} \ln\left(\frac{\omega_{0,n}}{\omega_n}\right), \tag{1.79}$$

where $\omega_{0,n}$ and ω_n are the eigen-frequencies of the crystal without and with vacancies, respectively. Atoms having a vacancy as neighbor tend to vibrate at lower frequencies because some bonds (springs) are missing. These atoms are, therefore, less well localized than atoms being entirely surrounded by other atoms. These atoms can be considered as being in a less ordered state.

The formation entropy is a measure for the spatial extension of a vacancy or, more general, of a zero-dimensional defect. The larger the value of $\Delta S_{\mathrm{Va}}^{\mathrm{f}}$ is, the larger is the extension of the defect and the more atoms are affected by the defect. Atoms in the defect region change their vibration frequency. An entropy value in the order of one k_{B} is representative for a point defect. Higher values are typical for defect clusters or line defects such as dislocations.

The following statements are generally accepted in the context of vacancy formation entropy:

- Compression of the solid increases the vacancy formation entropy.
- Vacancy formation entropies can vary widely within the identical crystal structure.
- Vacancy formation entropies are somewhat higher in bcc compared to fcc solids.

The first assertion is directly related to the relaxation of atoms around a vacancy. The second one indicates that other material properties, such as the exact shape of the atomic potentials, also influence the vacancy formation entropy. The reason for higher formation entropies in bcc crystals roots in the packing density, which is 0.68 for bcc and 0.74 for fcc. Consequently, it is easier for atoms in a bcc structure to relax compared to the close-packed fcc structure.

Some typical values of $\Delta h_{\mathrm{Va}}^{\mathrm{f}}$ and $\Delta s_{\mathrm{Va}}^{\mathrm{f}}$ are summarized in Table 1.1 and graphically depicted in Figure 1.20. For further analysis of the interplay between excess vacancies and diffusion, see Section 7.4 where a quantitative model for describing the influence of excess vacancies on diffusion, as well as a model for describing the excess vacancy evolution in a polycrystalline material, is presented and discussed.

Figure 1.20 reveals a few interesting features of the equilibrium vacancy density. Firstly, the vacancy site fraction in most metals adopts values between 10^{-5} and 10^{-4}, close to the liquidus temperature. Secondly, the concentration of structural vacancies *at room temperature* varies widely with the type of atomic species. For instance, $X_{\mathrm{Va}}^{\mathrm{eq}}$ is around 10^{-12} in Al, whereas $X_{\mathrm{Va}}^{\mathrm{eq}}$ in W is 56 (!!!) orders of magnitude smaller (see Table 1.1). Having in mind that diffusion in crystalline materials occurs by the exchange of vacancies and lattice atoms, one would thus expect that there are virtually no diffusive reactions occurring in W at room temperature.[25]

[25] As unit volume of a crystal contains typically 10^{28}–10^{29} substitutional lattice sites, a value $X_{\mathrm{Va}}^{\mathrm{eq}} \approx 10^{-68}$ means that, in 1 m^3 under equilibrium conditions, there are virtually all lattice sites occupied by atoms. We must be aware, however, that this situation will only occur after infinitely long time, when all excess vacancies have become annihilated at appropriate sinks.

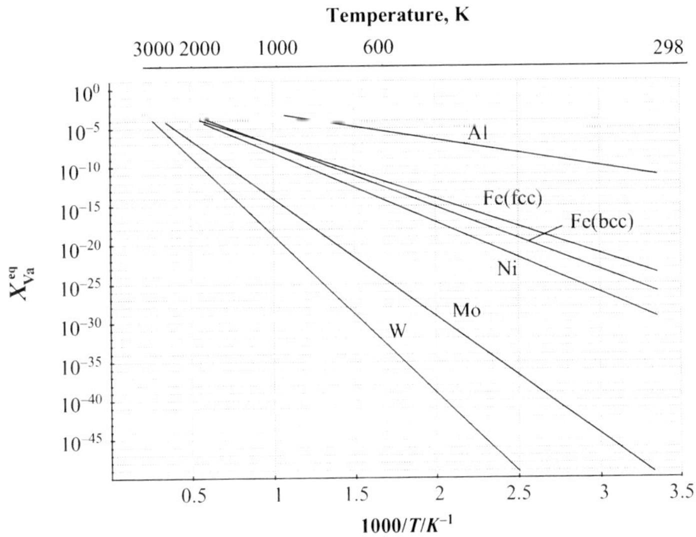

Figure 1.20. Equilibrium site fraction of structural vacancies in some metals.

Table 1.1. Typical values for the molar vacancy formation enthalpy, Δh_{Va}^{f}, and entropy, Δs_{Va}^{f}, as well as equilibrium vacancy site fraction, X_{Va}^{eq}, at 298 K for some metallic systems (Δh_{Va}^{f} from ref. [14], Δs_{Va} from various sources).

Element	Structure	Δh_{Va}^{f} / kJ/mol	$\Delta s_{Va}^{f}/k_B$	X_{Va}^{eq} at 298 K
Al	fcc	65	0.7	8.0×10^{-12}
Fe	bcc	154	2.17	8.8×10^{-27}
Fe	fcc	135	—	2.2×10^{-24}
Ni	fcc	172	1.96	5.0×10^{-30}
Mo	bcc	289	2.2	2.0×10^{-50}
W	bcc	386	3.2	5.4×10^{-67}

In Al-alloys, the equilibrium vacancy density of 10^{-12} suggests that at least some diffusive transport and diffusion-controlled reactions are possible in this material even at room temperature.[26] This conclusion is in accordance with experimental evidence.

[26] It is well known that Al at room temperature is a rather "busy" substance. Diffusion-controlled reactions, such as precipitation of metastable phases, can occur within minutes or hours after quenching, due to the accelerating effect of "frozen in" excess vacancies on diffusion.

CHAPTER 2

Precipitate Nucleation

The life cycle of a precipitate begins with the nucleation process. Nucleation is the initial, the "creation" step in precipitate life, comprising the time span starting with either a random or equilibrium solid solution to the time when the precipitate has developed to a stable or "critical" size, from where on its further evolution is governed by the (deterministic) laws of precipitate growth.

The process of precipitate nucleation resembles, in several aspects, the biological process of "creation" of a living being.[1] A major similarity is from the observation that nucleation is a process that is inherently related to the term *probability*. One can never say whether a single nucleation event will occur exactly at a given location and at a given time. Nucleation will only occur with a certain probability. Therefore, nucleation is considered as a *stochastic* process that, in any quantitative consideration, requires treatment in a statistical framework. Reasonably accurate prognosis of the nucleation rate, the number of "newly born" precipitates per unit volume and unit time, can only be made for a sufficiently large number of potential nucleation events. In social science, we can give rather accurate prognoses on birth rates prevailing in certain societies, but we cannot tell when or if a selected couple of human beings will give birth to a new human being.

Another similarity is the observation that nucleation is a process that is linked to an energy barrier that must be overcome to make nucleation happen. Nucleation is a process that is often difficult, in thermodynamic systems as well as in social. It is a well-known fact that, the higher the nucleation barrier is, the higher must be the effort (motivation) of the individuals to overcome it. The relation between the height of the barrier and the necessary effort is not linear, however. It is often observed that a rather small increase in the energy barrier leads to an almost complete stop of nucleation, making this measure a central quantity.

In social systems, at least a minimum attraction between two individuals is necessary for the occurrence of a "nucleation" event to be feasible. The higher the attractive force is, the higher is the probability that a biological nucleation event will occur. In solid-state systems, an attractive force between atoms forming the new phase is necessary to make nucleation possible.

[1] The picture developed here should of course not be taken too literally.

The higher this so-called *driving force* is, the higher is the probability for the single nucleation event and the higher is the nucleation rate. The interested reader will certainly find other similarities between thermodynamic and social systems.

Looking back in history, we find the first theoretical treatments of nucleation as early as in the mid-1920s. These original treatments deal with the formation of liquid droplets from supersaturated vapor, making use of fundamental ideas of Josiah Willard Gibbs, Sir William Thomson, Ludwig Boltzmann, Marian von Smoluchovski, and Albert Einstein. As we will see, the concepts developed therein are of general validity also in solid-state systems.

The contours of *Classical Nucleation Theory* (CNT) are treated in the first part of this chapter. After introducing the original concepts for addressing the problem of droplet formation from supersaturated vapor in the thermodynamic framework, a continuums-mechanical approach is developed. Then, the problem of nucleation in multicomponent systems is addressed, which is found to be significantly more complex than liquid droplet formation or solid-state nucleation of precipitates in two-component alloys, when treated on a rigorous basis. Nevertheless, useful approximate solutions to the multicomponent, solid-state nucleation problem have been developed and they will be introduced in this section.

2.1 Paving the Way for Nucleation Theory

In 1870, Sir William Thomson (Lord Kelvin), a British physicist born in Ireland (1824–1907), published his treatment of the equilibrium between free surfaces of liquid and its vapor in a closed vessel [15]. Thomson assumed that surfaces having different radii of curvature in different regions of the vessel, for instance in a small capillary, are characterized by different pressure of the vapor covering the surface, and he related the difference to the principle curvatures of these surfaces.

Approximately at the same time, Josiah Willard Gibbs, an American physicist (1839–1903), introduced his seminal work "On the Equilibrium of Heterogeneous Substances," [16] where he developed the fundamentals of modern equilibrium thermodynamics in a mathematical framework, which is used today in almost unchanged form. The most relevant part of this work in the context of nucleation is his treatment of a small droplet being in equilibrium with its supersaturated vapor. Accordingly, each size of droplet is related to a characteristic supersaturation, such that the energy, which is gained by condensing the droplet, balances exactly the work that must be expended in formation of the new interface between the droplet and the vapor. For this work, W, Gibbs gives the expression

$$W = \frac{1}{3}\sigma A, \tag{2.1}$$

where σ is the surface tension and A is the surface area of the droplet.

Between 1866 and 1878, Ludwig Boltzmann, an Austrian physicist and philosopher (1844–1906), manifested the irreversibility of elementary physical processes on a macroscopic scale, initially in a treatment of the evolution of colliding gas particles in the framework of kinetic gas theory. Best known is certainly his famous formulation of the relation between entropy and the number of possibilities of how a system can be constituted under the constraint of constant energy, commonly written as

$$S = k_B \ln \Omega. \tag{2.2}$$

S is the configurational entropy, k_B the Boltzmann constant, and Ω the number of permutations of possible system configurations. It is important to realize that Boltzmann introduced a concept claiming that even in equilibrated systems a certain amount of fluctuation is always present. For the present treatment, it is not even necessary to specify the type of fluctuation a priori, it is sufficient to know that the system can produce fluctuations, either in temperature, in composition, in density, or in structure. The probability of finding the system in a particular state is related to its entropy, according to equation (2.2).

This idea was picked up by Marian von Smoluchowski [17], an Austrian–Polish physicist (1872–1917), and Albert Einstein [18], (1879–1955),[2] as the conceptual basis for a treatment of the opalescence[3] of liquids near the critical point. The formation mechanism of colloidal particles causing the opalescence phenomenon has been explained by microscopic *fluctuations* in energy and constitution, providing the means to form colloidal *nuclei* within the otherwise macroscopically homogeneous liquid. Einstein started his treatment with equation (2.2), and by claiming that this relation does not have any physical meaning without specifying the quantity Ω by some elementary theory. However, he continued, the Boltzmann principle as manifested in equation (2.2) becomes relevant by itself when considering the irreversibility of physical processes as an *apparent* irreversibility. Even if a more likely state will commonly follow on a less likely state, when investigating individual system changes, the opposite process is also possible with a given probability. Consequently, in infinite time, an equilibrated system will run through all possible states of identical energy and, consequently, even adopt highly unlikely states with some probability at some instance in time.

Einstein next points out that the relation between S and Ω in equation (2.2) is not an exact one, as it only delivers an order of magnitude relation between the two quantities—at least, unless Ω is exactly specified by some theory. However, relation (2.2) can be used to derive some statistical properties of a system if the sample volume is made infinitely small. In other words, although the total energy of the system remains constant on a macroscopic scale and it appears to be static to the macroscopic observer, the system continuously and reversibly changes its constitution on a microscopic scale. And this is exactly what is needed in nucleation theory, where these microscopic variations in temperature, density, chemical composition, crystal structure, and others have been clearly identified as the source of energy for overcoming the nucleation barrier. From equation (2.2), the probability for the occurrence of some state (fluctuation) is given with

$$\Omega = \exp\left(\frac{S}{k_B}\right). \tag{2.3}$$

[2] It is difficult to settle the exact citizenship of Albert Einstein because he had many: He was born in the Kingdom of Württemberg and, therefore, was citizen of the German empire. He was stateless between 1896 and 1901 and became citizen of Switzerland afterward. For a short period of time, he also had the citizenship of the Austrian–Hungarian empire and the Free State of Prussia. He laid back the German citizenship with the rise of Adolf Hitler and, in 1940, adopted the US citizenship in addition to the Swiss one.

[3] Opalescence describes the phenomenon by which otherwise transparent substances appear colorful under certain circumstances due to scattering of light at extremely fine objects.

On the other hand, the variation in total energy of a closed system at constant pressure and constant temperature is given in differential form with

$$dU = dW + TdS. \tag{2.4}$$

U is the internal energy, W is the work done on the system, T is temperature, and S is entropy. In thermodynamic equilibrium, the internal energy is in its minimum and dU is zero. Thus, we have

$$dW = -TdS, \tag{2.5}$$

and equation (2.3) can be rewritten as

$$\Omega = \exp\left(-\frac{W}{k_B T}\right). \tag{2.6}$$

This relation manifests the observation that the probability of finding the system in a given state has a negative exponential dependence from the corresponding work of formation W of this state divided by $k_B T$. Consequently, any fluctuation in a system occurs with a probability in the order of magnitude given by equation (2.6).

Sixteen years later, Volmer and Weber [19] would use this result when deriving their fundamental relations of nucleation theory. This issue is the topic of the following sections where the present results are taken as a basis for developing the classical nucleation theory framework.

2.2 Nucleation of Liquid Droplets from Supersaturated Vapor

The theory of nucleation is concerned with the question of how many nuclei of a new phase are created in unit volume and unit time as a function of the state of the system under consideration. The quantity describing the nucleation process is the nucleation rate J, with its SI-unit being [$m^{-3}s^{-1}$].

On developing a quantitative expression for J, we are confronted with three major questions:

1. What is the energy change related to the formation of an embryo of the new phase?
2. Which process provides the energy for overcoming the nucleation barrier?
3. What is the rate of formation of precipitates of a new phase in the nucleation process?

In answering these questions, first, the thermodynamics of nucleus formation is revisited in more detail. Then, the treatments of Volmer and Weber [19] and Becker and Döring [20] are introduced which are commonly considered as the initiating works of classical nucleation theory. The presentation of results in this section is based on their original treatments within the *thermodynamic* framework, utilizing the example of nucleation of liquid droplets from supersaturated vapor or cavitation, where voids form within a liquid under negative pressure. The problem of nucleation in the solid state is worked out in the treatment of nucleation kinetics given later in the *continuums-mechanical* framework.

2.2.1 Thermodynamics of the Critical Nucleus

Consider a spherical liquid droplet with radius, r, which is in equilibrium with its supersaturated vapor at a pressure, P_r. If P_∞ denotes the pressure of the saturated vapor, according to the considerations of Thomson [15] and Gibbs [16], this equilibrium condition can be formulated as

$$W_1 = -\frac{4\pi r^3}{3}\frac{d}{M}RT + \frac{4\pi r^3}{3}\frac{d}{M}RT\ln\frac{P_r}{P_\infty}, \tag{2.7}$$

where M is the mass of the droplet, d is its density, R is the universal gas constant, T is temperature, and σ is the surface tension. Equation (2.7) manifests that each level of supersaturation is related to a specific size of droplet such that the supersaturation given by the excess pressure $(P_r - P_\infty)$ is exactly balanced by the work, W^*, to form this liquid droplet of size r. The droplet, which is in equilibrium with its surrounding supersaturated vapor, is called the *critical droplet* or *critical nucleus*.

On evaluating the work, W^*, to form the critical nucleus, we start with specifying the work that is necessary to form a liquid phase droplet at pressure P_r from the molecules of the vapor at pressure P_∞. Following the treatment of Volmer and Weber [19], in a first step, the gas molecules are assumed to be evaporated at pressure P_∞ and compressed to P_r. The required work, W_1, for this process is

$$W_1 = -\frac{4\pi r^3}{3}\frac{d}{M}RT + \frac{4\pi r^3}{3}\frac{d}{M}RT\ln\frac{P_r}{P_\infty}. \tag{2.8}$$

The second step, condensation at P_r and formation of the liquid phase, is related to the work W_2, which is given as

$$W_2 = W^* + \frac{4\pi r^3}{3}\frac{d}{M}RT\ln\frac{P_r}{P_\infty}. \tag{2.9}$$

On the other hand, the difference in free energy between a droplet and the same amount of homogeneous phase is $4\pi r^2\sigma$, which must balance the works W_1 and W_2 with

$$4\pi r^2\sigma = W_1 + W_2 = W^* + \frac{4\pi r^3}{3}\frac{d}{M}RT\ln\frac{P_r}{P_\infty}. \tag{2.10}$$

Inserting equation (2.7) yields

$$4\pi r^2\sigma = W^* + \frac{4\pi r^3}{3}\frac{d}{M}\cdot\frac{2\sigma M}{rd}. \tag{2.11}$$

The work required to form a critical nucleus is obtained with

$$W^* = \frac{1}{3}\sigma\cdot A, \tag{2.12}$$

with A being the surface area of the droplet. On inserting equation (2.7), Volmer and Weber arrived at the critical nucleation energy with

$$W^* = \frac{16\pi}{3} \frac{\sigma^3}{\left(RT \ln \frac{P_r}{P_\infty}\right)^2} \left(\frac{M}{d}\right)^2. \tag{2.13}$$

The authors emphasize that equation (2.13) has been derived from the balance between the work necessary to create the surface area of the droplet and the energy change of the bulk volume. We will return to this later when discussing the continuums-mechanical approach to precipitate nucleation.

2.2.2 Overcoming the Nucleation Barrier

According to equation (2.7), any supersaturated vapor of pressure $P_r > P_\infty$ is in equilibrium with droplets of size r. The energy of formation of the critical droplet is given by equation (2.13). Smaller droplets are inherently unstable and will always decay when treated on a purely thermodynamic basis. The question now arises, how a critical nucleus can be formed at all having in mind that the process of nucleus formation is apparently connected to the repeated expense of energy on attaching new molecules at least until the critical size is reached?

The answer to this question is founded in the physical principles relating microscopic and macroscopic systems as outlined in Section 2.1. Accordingly, any macroscopic system in equilibrium is subject to microscopic fluctuations in pressure, temperature, chemistry and/or crystal structure, and so forth. The probability for any such fluctuation is given by the exponential relation manifested in equation (2.6). Consider, first, the case of an unsaturated vapor containing N_0 molecules. Following the arguments of Volmer and Weber [19], there exists a discrete, stationary distribution of clusters with sizes $n < n^*$, where n is the number of molecules in the cluster and n^* the number of molecules in the critical cluster with

$$N_n = N_0 \cdot \exp\left(-\frac{W_n}{kT}\right). \tag{2.14}$$

W_n is the work of formation of a cluster containing n molecules. Volmer and Weber assumed that the elementary process of condensation or evaporation of single molecules causes the transition of clusters from one size class to another. For an unsaturated vapor, where no formation of supercritical, stable nuclei can occur, the individual embryos will permanently change from one size class to another by attachment or detachment of monomers, while the entire distribution of embryos maintains a stationary state. This situation is reflected in Figure 2.1 for temperatures T_1 and T_2.

Consider now a system that is transferred from temperature T_2 into a supersaturated state at temperature T_3. After reaching T_3, the distribution of embryos adapts to the modified conditions and tends toward a stationary equilibrium cluster distribution (ECD) again. However, in contrast to the unsaturated state, the process of attachment and detachment of molecules for subcritical and supercritical embryos will now be governed by different laws: Whereas addition of a molecule to a subcritical embryo ($n < n^*$) is still an energetically unfavorable process, it is

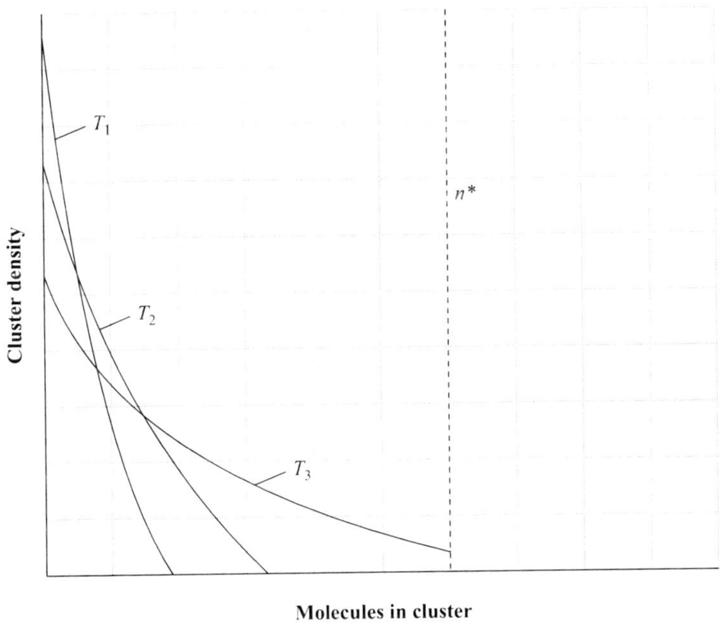

Figure 2.1. Equilibrium cluster distribution at temperatures $T_1 > T_2 > T_3$ for an unsaturated (T_1, T_2) and supersaturated vapor (T_3) according to Volmer and Weber [19]. n^* is the number of molecules in the critical cluster.

an energetically favorable process for supercritical embryos, now. Consequently, supercritical embryos will continue to grow spontaneously by attachment of colliding molecules with a rapidly decreasing probability for decaying again. This situation is sketched by the discontinuous line for temperature T_3 in Figure 2.1 at the critical nucleus size n^*.

As the number of clusters having critical size n^* is determined by the exponential expression (2.14), Volmer and Weber concluded that the nucleation rate, J_s, at which new liquid droplets are formed must be proportional to the cluster density at the critical size given with

$$J_s = K \cdot \exp\left(-\frac{W^*}{k_B T} \right). \tag{2.15}$$

K is a proportionality constant that will be investigated further in the next section. The subscript to the nucleation rate J_s indicates *steady-state* conditions, that is, time-independent nucleation. Time-dependence in the *transient* nucleation rate will be dealt with later.

Before proceeding, it must be emphasized that equation (2.15) exhibits a most important feature: In nucleation processes governed by microscopic fluctuations, nucleation will only occur at a conceivable rate if the value of W^* is of the order of the kinetic energy $k_B T$ of the atoms or molecules involved in the nucleation process. For increasing values of W^*, the nucleation rate goes rapidly toward extremely small values because the probability that a fluctuation of sufficient amplitude will occur becomes almost zero. Practically, this means that in many systems of practical interest, nucleation is greatly promoted or made possible at all only if the *homogeneous* nucleation barrier, W^*, is lowered by heterogeneous nucleation.

Think, for instance, of the comfortable finger warming pads in winter. When your fingers become cold, you take the pad with the liquid solution inside, click the piece of metal immersed in the pad and the liquid starts to crystallize. In this process, the pad releases comfortable warmth for several minutes by which you can warm up your fingers again. After using the pad, you put it into boiling water and the crystals will melt again; the pad is ready for another use. The basic principle of the warming pad is a salt solution, which has a high energy barrier for crystal nucleation. This is why the pad can be stored at temperatures far below the crystallization point for a long time. Crystallization is only stimulated mechanically by clicking the piece of metal and releasing a pressure wave into the crystal.

In solid-state matter, heterogeneous nucleation is occurring on, for example, dust particles or surfaces (for droplets); or lattice defects, such as dislocations, inclusions, or grain boundaries; for precipitate nucleation in solid-state matter. This will be discussed later.

2.2.3 The Kinetics of Droplet Formation

After setting the theoretical foundation for precipitate nucleation in terms of the elementary relation between the steady-state nucleation rate, J_s, and the nucleation barrier, W^*, in this section, the kinetic constant, K, is addressed and quantitative expressions for this quantity are developed.

In the original work of Volmer and Weber [19], further numerical analysis of the value for WK has been avoided with the argument that the exponential term in equation (2.15) is widely dominant over K and that it might, therefore, be sufficient, for many practical applications, to know whether nucleation occurs at all or whether it is suppressed. In fact, these authors succeeded to explain several phenomena related to nucleation by simply inspecting the exponential term and, thus, emphasizing its central significance. In many practical applications, knowledge of K is still important, and it will be elucidated next.

The first attempts to derive the kinetic coefficient, K, for droplet nucleation on a rigorous basis are due to Farkas [21] and Becker[4] and Döring[5] [20]. Both arrive at a similar result; however, we will follow the methodology of Becker and Döring, here, because it appears to be clearer and more consistent. A later comparison of their results with experiment, carried out by Volmer and Flood [22], has shown good agreement with the kinetic coefficient derived by these authors. These results are considered as good evidence for the validity of CNT.

Consider the condensation and evaporation of single molecules as the elementary mechanism of cluster transition. Let O_n be an embryo of size n. The processes of growth and shrinkage can be formally written as

$$O_n + O_1 \rightarrow O_{n+1}$$
$$O_n - O_1 \rightarrow O_{n-1}. \tag{2.16}$$

Let a_0 be the attachment frequency and $a_0 \cdot dt$ be the probability of a molecule condensing on unit area of cluster surface, $q_n \cdot dt$ be the probability that a molecule is evaporated from unit area of cluster surface, and A_n be the cluster surface area of an embryo with size n.

[4] Richard Becker, German physicist (1887–1955).

[5] Werner Döring, German physicist (1911–2006).

With Z_n representing the number of clusters containing n molecules, the total surface area of clusters of size n is given as $Z'_n = Z_n A_n$. The flux J of clusters from size n to $n - 1$ is then obtained with

$$J = a_0 Z'_{n-1} - q_n Z'_n. \tag{2.17}$$

For convenience, we further introduce the quantity b_n with

$$b_n = \frac{a_0}{q_n}. \tag{2.18}$$

For clusters below the critical size n^*, the numbers of b_n are monotonically increasing with increasing n. At the critical cluster size n^*, we just have $b_n = 1$. When substituting b_n into equation (2.17), we obtain the governing equation for cluster transition with

$$Z'_n = Z'_{n-1} b_n - \frac{J}{a_0} b_n. \tag{2.19}$$

The quantity b_n can be further detailed having in mind that the probabilities of attachment and detachment of molecules are proportional to the pressure on the droplet surface and, with the help of equation (2.7), we can write

$$\beta_n = \frac{P_{n^*}}{P_n} = \exp\left(\frac{2\sigma}{RT}\frac{M}{d}\left(\frac{1}{r_{n^*}} - \frac{1}{r_n}\right)\right). \tag{2.20}$$

The governing equation (2.19) represents a system of equations for the unknown quantities $Z'_n, Z'_{n+1}, \ldots Z'_s$, which reads

$$\begin{aligned} Z'_{n+1} &= Z'_n b_{n+1} - \frac{J}{a_0} b_{n+1} \\ Z'_{n+2} &= Z'_{n+1} b_{n+2} - \frac{J}{a_0} b_{n+2} \\ &\ldots \\ Z'_s &= Z'_{s-1} b_s - \frac{J}{a_0} b_s. \end{aligned} \tag{2.21}$$

The index s is a number for which we only require $s > n^*$. It is not necessary to specify the value of s in more detail. For the present analysis, it is sufficient that it be larger than the number of molecules in the critical cluster.

The quantities $Z'_{n+1}, Z'_{n+2}, \ldots Z'_{s-1}$ can now be eliminated by dividing the first equation by b_{n+1}, the second equation by $b_{n+1}b_{n+2}$, and so forth, and the last one by $b_{n+1}b_{n+2}\ldots b_s$. After summation of all corresponding equations, all values Z' between Z'_n and Z'_s vanish and we obtain

$$\frac{Z'_s}{b_{n+1}b_{n+2}\cdots b_s} = Z'_n - \frac{J}{a_0}\left(1 + \frac{1}{b_{n+1}} + \frac{1}{b_{n+1}b_{n+2}} + \cdots + \frac{1}{b_{n+1}b_{n+2}\cdots b_{s-1}}\right). \tag{2.22}$$

According to equation (2.22), the nucleation rate of droplets is known as soon as one Z'_n is determined.

In their treatment, Becker and Döring continue evaluating the droplet formation kinetics by utilizing an analogy to the flux of electric charges through a series of resistors. Accordingly, equations (2.21) adopt a formal structure of

$$\Phi_{i+1} = \Phi_i - JR_i, \tag{2.23}$$

with

$$\Phi_i = \frac{Z'_i}{b_2 b_3 \cdots b_i} \quad \text{and} \quad R_i = \frac{1}{a_0 b_2 b_3 \cdots b_i}. \tag{2.24}$$

In the analogy with electric current, equation (2.23) can be interpreted as the definition of the flux, J, between points i and i+1 caused by the potential difference $\Phi_i - \Phi_{i+1}$ and constrained by the resistance, R_i. The flux of stable clusters created by the nucleation process can thus be interpreted as a general flux, J, going through a series of resistors as

$$J(R_n + R_{n+1} + \cdots + R_{s+1}) = \Phi_n - \Phi_s. \tag{2.25}$$

In our case, Φ_1 is equal to Z'_1 and the potential at s is zero. The task of evaluating the nucleation rate, J, thus reduces to the task of summing up all partial resistances, R_i.

From equation (2.22), we can identify the expression for the partial resistance, R_i, as

$$R_i = \frac{1}{a_0} \exp\left(\alpha \left(\frac{1}{r_2} + \frac{1}{r_3} + \cdots + \frac{1}{r_i} - \frac{i-1}{r_{n*}} \right) \right), \tag{2.26}$$

with

$$\alpha = \frac{2\sigma}{RT} \frac{M}{d}. \tag{2.27}$$

For evaluation of this expression, the discrete sum of inverse radii can be substituted by an integral over the value x, with

$$x_n = \frac{r_n}{r_{n*}} = \left(\frac{n}{n*} \right)^{\frac{1}{3}}. \tag{2.28}$$

This new variable represents the size of the cluster in relation to the critical cluster. Furthermore, we have

$$n = n^* x_n^3 \quad \text{and} \quad dn = 3n^* x^2 dx. \tag{2.29}$$

This expression can now be substituted into the exponent of the partial resistance expression and we obtain

$$\left(\frac{1}{r_2}+\frac{1}{r_3}+\cdots+\frac{1}{r_i}\right)\cong\frac{1}{r_{n^*}}\int_{n=1}^{i}\frac{dn}{x_n}=\frac{3n^*}{r_{n^*}}\int_{x_1}^{x_i}x\,dx=\frac{3}{2}\frac{n^*}{r_{n^*}}\left(x_i^2-x_1^2\right). \tag{2.30}$$

For the last term in the exponent of the partial resistance, we have

$$i-1=n^*\left(x_i^3-x_1^3\right), \tag{2.31}$$

which leads to

$$R_i=\frac{1}{a_0}\exp\left(A'\left\{\left(3x_i^2-2x_i^3\right)-\left(3x_1^2-2x_1^3\right)\right\}\right). \tag{2.32}$$

The quantity A' has been introduced with

$$A'=\frac{\alpha n^*}{2r_{n^*}}=\frac{\sigma}{3}\frac{A_{n^*}}{k_BT}. \tag{2.33}$$

In a next step, the summation of all partial resistances is also replaced by an integral and we get

$$\int_1^s R_n dn=\int_{x_1}^{x_s}R(x)x^2dx=\frac{3n}{a_0}\exp\left(-A'\left(3x_1^2-2x_1^3\right)\right)\int_{x_1}^{x_s}\exp\left(A'\left(3x^2-2x^3\right)\right)x^2dx. \tag{2.34}$$

For further evaluation of the total resistance, R, we inspect the quantity x in more detail. From equation (2.28), we recognize that x is monotonically increasing with increasing cluster size until it reaches unity for the critical nucleus size. We have $x_1 << 1$ and, thus, can neglect the exponential expression in the prefactor to the integral. The total resistance can, therefore, be simplified with this approximation as

$$R\cong\frac{3n}{a_0}\int_{x_1}^{x_s}\exp\left(A'\left(3x^2-2x^3\right)\right)x^2dx. \tag{2.35}$$

The exponent in the resistance expression equation (2.35) has some interesting characteristics, which is shown in Figure 2.2. Firstly, since x is monotonically increasing, the exponent expression in the integral is also monotonically increasing. Secondly, the functions in x show a pronounced maximum around a value of $x = 1$. Based on this observation, Becker and Döring introduced the coordinate transformation $x = 1 + \xi$ and we obtain

$$\begin{aligned}R&\cong\frac{3n}{a_0}\int_{x_1-1}^{x_s-1}\exp\left(A'\left(1-3\xi^2-2\xi^3\right)\right)(1+\xi)^2\,d\xi\\&=\frac{3ne^{A'}}{a_0}\int_{x_1-1}^{x_s-1}\exp\left(-A'\left(3\xi^2+2\xi^3\right)\right)(1+\xi)^2\,d\xi.\end{aligned} \tag{2.36}$$

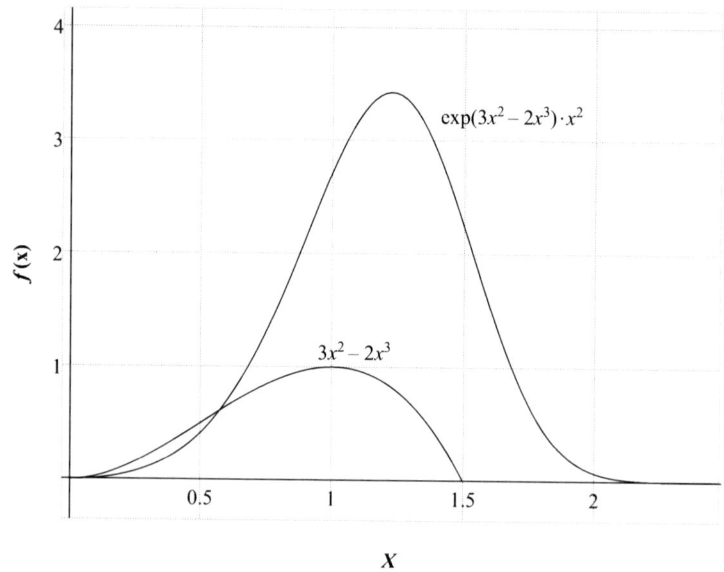

Figure 2.2. Shape of the functions in the total resistance expression, equation (2.35).

The value of A' has often values of the order of 20–50. For larger values of A', the exponent in the integral has an increasingly sharp maximum around $x = 0$. In this case, the integral can be approximated by the simple expression

$$R \cong \frac{3ne^{A'}}{a_0} \int_{x_1-1}^{x_s-1} \exp\left(-3A'\xi^2\right) d\xi, \tag{2.37}$$

without significantly changing the value of the integral around the maximum value. The corresponding functions are shown in Figure 2.3 for a low value of $A' = 3$.[6] The functions coincide increasingly better for higher values of A' as stated above. After the transformation and substitution of the function in the integral, we recognize that the exponent is symmetric and vanishes away from the maximum at $x = 0$. The integration limits can thus be extended and we obtain

$$R \cong \frac{3ne^{A'}}{a_0} \int_{-\infty}^{+\infty} \exp\left(-3A'\xi^2\right) d\xi, \tag{2.38}$$

with the solution

$$R \cong \frac{3n}{a_0} \sqrt{\frac{\pi}{3A'}} \exp(A'). \tag{2.39}$$

[6]The value of $A' = 3$ is unfavorably low for Becker and Dörings approximation of the function in the integral. It is simply chosen to demonstrate the principal shape of the curves. For more realistic values, the shape of the exponential functions is significantly narrower and the agreement between original and approximated functions is also significantly better.

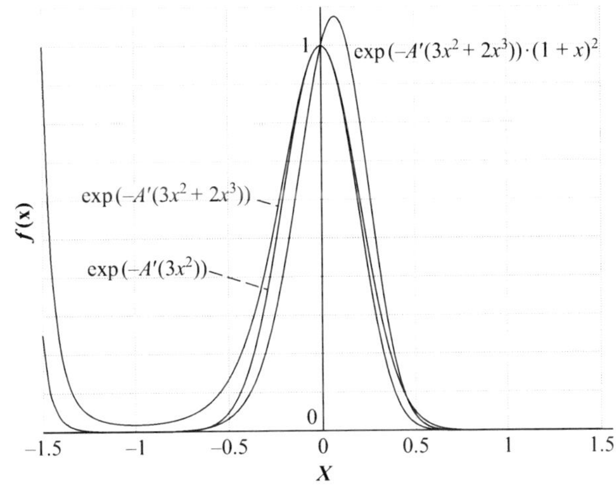

Figure 2.3. Shape of the functions in the total resistance expression after the coordinate transformation.

With this result, the nucleation rate of droplets from the supersaturated vapor finally becomes

$$J = J_{\mathrm{s}} = \frac{a_0 Z_1'}{n^*}\sqrt{\frac{A'}{3\pi}}\exp(-A') \quad \text{and} \quad A' = \frac{\sigma A_{n^*}}{3k_{\mathrm{B}}T} = \frac{W^*}{k_{\mathrm{B}}T}. \tag{2.40}$$

Again, the subscript "s" emphasizes steady-state nucleation conditions.

It is interesting that the treatment of Becker and Döring elegantly avoids the conceptual problem that the surface energy of very small clusters lacks physical meaning as soon as the number of atoms becomes small. Recalling the shape of the functions in the integration of the total resistance (Figure 2.2), it is apparent that the main contribution to the integral comes from the clusters close to the critical size. The result is thus only marginally affected by eventual errors in the evaluation of the partial resistance of the smallest clusters.

The treatment of Becker and Döring [20] followed some eight years after the pioneering work of Farkas [21], delivering essentially the same result. The major problem that was solved in Becker and Döring's work was an undetermined integration constant persistent in Farkas work. In the following chapter, another important treatment is presented, which is due to Zeldovich and which is carried out in a rather elegant mathematical framework.

2.2.4 The Zeldovich Factor

In 1943, Yakov Borisovich Zeldovich, a Belarusian physicist (1914–1987), introduced an elegant treatment [23] of the nucleation kinetics of voids,[7] which is briefly reviewed in this chapter. The major contribution to nucleation theory made in this work is the so-called Zeldovich factor, Z, which is an essential part of classical nucleation theory taking into account

[7] Although Zeldovich's work addresses the nucleation of voids, it is equally applicable to the nucleation of droplets.

the thermally activated states of critical clusters in the derivation of the fundamental equations of nucleation. The Zeldovich factor, Z, is later also utilized in the continuums treatment of nucleation introduced in Section 2.3.

Consider a large ensemble of n-mers,[8] of different size in a closed environment at constant supersaturation. In analogy to the treatment of Becker and Döring, the evolution of the cluster size distribution is controlled by the absorption and evaporation of monomers from the cluster surfaces. On considering a single class of size n, the change in concentration $c_{n,t}$ of clusters at any time t can be expressed as

$$\frac{\partial c_{n,t}}{\partial t} = -c_{n,t}\left[a_0 A_n + q_n A_n\right] + c_{n-1,t} a_0 A_{n-1} + c_{n+1,t} q_n A_{n+1}. \tag{2.41}$$

In this representation, the same symbols are used as previously in Becker and Döring's derivation in Section 2.2.3. Accordingly, a_0 is the frequency by which monomers attach to a flat surface of unit area, q_n is the detachment frequency of monomers from clusters of size n, and A_n is the surface area of a cluster of size n. The physical meaning of the right hand side of equation (2.41) is easily recognized as the net flux of clusters in cluster size space.

For further evaluation, assume that a supersaturated system be in equilibrium. As supersaturation leads to phase decomposition, an equilibrium steady-state distribution of clusters can only be maintained if clusters exceeding the critical size are continuously removed from the system and replaced again by the same amount of monomers. Once the steady-state is reached, the principle of detailed balance must hold, which means that the forward transition rates of particles growing by addition of monomers balances exactly the backward transition rates of particles shrinking by detachment of monomers. This principle can be written as

$$c_{\mathrm{eq},n} a_0 A_n = c_{\mathrm{eq},n+1} q_{n+1} A_{n+1}, \tag{2.42}$$

where the concentrations, $c_{\mathrm{eq},n}$, are time-independent. This fact is emphasized by the subscript "eq." When inserting this expression into equation (2.41), the following differential equation is obtained:

$$\frac{\partial c}{\partial t} = a_0 A_n c_{\mathrm{eq},n}\left(\frac{c_{n+1}}{c_{\mathrm{eq},n+1}} - \frac{c_n}{c_{\mathrm{eq},n}}\right) - a_0 A_{n-1} c_{\mathrm{eq},n-1}\left[\frac{c_n}{c_{\mathrm{eq},n}} - \frac{c_{n-1}}{c_{\mathrm{eq},n-1}}\right]. \tag{2.43}$$

Zeldovich argues that this differential equation can also be represented in terms of a differential equation, after Taylor expansion and ignoring all higher-order terms except for the first nonvanishing second-order term, as

$$\frac{\partial c}{\partial t} = \frac{\partial}{\partial n}\left(\beta c_{\mathrm{eq}} \frac{\partial c / c_{\mathrm{eq}}}{\partial n}\right). \tag{2.44}$$

Utilizing the principle of physical correspondence, one easily observes that this result resembles Fick's second law, except that the spatial derivatives are here performed in cluster size space. The quantity β is to be determined in order to solve the problem, with β describing the rate at which molecules attach to the surface of a cluster of given size.

[8] An n-mer denotes a polymer with n monomers.

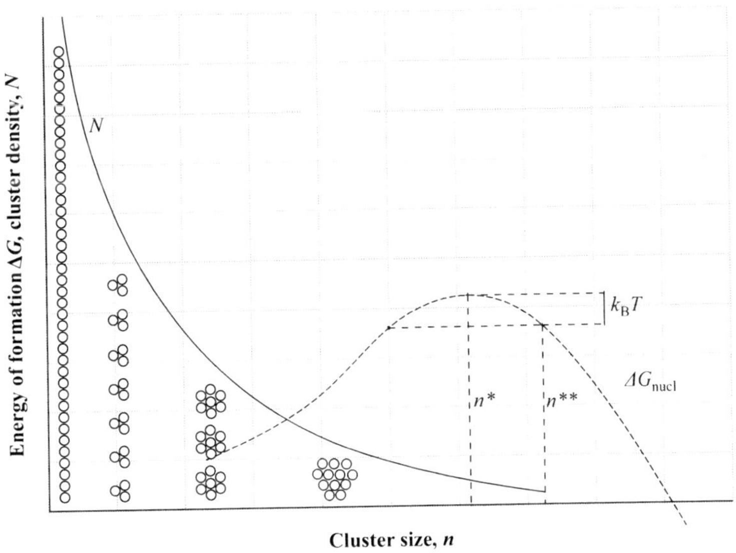

Figure 2.4. Schematic representation of the equilibrium cluster distribution (solid line) and free energy of nucleus formation (dashed line) around the region of critical size, n^*.

Under steady-state conditions, $\partial c/\partial t = 0$ and the steady-state flux, J_s, is

$$J_s = -\beta c_{eq} \frac{\partial c/c_{eq}}{\partial n} = \text{constant}. \tag{2.45}$$

The equilibrium concentration of clusters, c_{eq}, can now be substituted by the thermally activated concentration as

$$c_{eq} \approx c_1 \exp\left(\frac{\Delta G_{nucl}}{k_B T}\right). \tag{2.46}$$

The corresponding distribution is schematically shown in Figure 2.4 as solid line. The number density for higher values of the nucleation rate is drastically reduced according to the exponent expression for the size distribution. Moreover, the distribution is cut off at the super-critical size n^{**}, since we maintain a supersaturated steady-state distribution by artificially removing particles larger than n^{**} and replacing them by the same amount of monomers. The curve for the free energy of cluster formation is plotted as a dashed line.[9]

Equation (2.45) can be rewritten in the form

$$J_s \int_1^{\infty} \frac{dn}{\beta c_{eq}} = -\int_0^1 d\left(c/c_{eq}\right). \tag{2.47}$$

[9] Note that it is not necessary a priori to specify the exact shape of this function in Zeldovich's treatment.

In analogy to Becker and Döring's treatment, Zeldovich argues that the integrand has significant contributions only in the vicinity of the critical size. Therefore, the equilibrium concentration of clusters around the critical size can be expanded into a Taylor series as

$$\Delta G_{\mathrm{nucl}}(n) = \Delta G_{\mathrm{nucl}}(n^*) + \frac{(n-n^*)^2}{2}\frac{\partial^2 \Delta G_{\mathrm{nucl}}(n)}{\partial n^2}. \tag{2.48}$$

With this substitution and extension of the integration limits, the integration can be carried out and we obtain the final result for the steady-state nucleation rate of the form

$$J_{\mathrm{s}} = Z\beta^* c_{\mathrm{eq}}(n^*), \tag{2.49}$$

with the equilibrium density of critical clusters, c_{eq}, given by equation (2.46).

Equation (2.49) contains three main components describing the kinetics of nucleation, which are essentially

i. a term, Z, denoted as the *Zeldovich factor* reading

$$Z = \left[-\frac{1}{2\pi k_{\mathrm{B}} T} \left(\frac{\partial^2 \Delta G_{\mathrm{nucl}}}{\partial n^2} \right)_{n^*} \right]^{\frac{1}{2}}; \tag{2.50}$$

ii. a term β^* designating the attachment rate of monomers to the critical nucleus; and
iii. the density of clusters of critical size, n^*.

The Zeldovich factor, Z, might be interpreted as a nonequilibrium factor that accounts for the thermal activation of critical nuclei with energy, $k_{\mathrm{B}}T$, where the probability of an exactly critical cluster for continued growth is only 0.5, as the probability of decaying again and dissolving is just as high. This is elaborated in the following section on time lag in nucleation.

2.2.5 The Time Lag

In the previous section, the nucleation rate has been evaluated under the assumption of steady-state conditions. In a real system, this premise can only be satisfied after infinitely long time, when all transient processes have decayed and nucleation proceeds with a constant flux of supercritical clusters. Before reaching steady-state conditions, as given in equation (2.49), within a certain period of time, the flux of supercritical clusters increases from zero to the point of stationary conditions. This time lag is commonly denoted as the *induction period* or the *incubation time*, τ.

In order to derive a quantitative expression for the incubation time, we first reproduce the estimate of Wakeshima [24] in the form as reviewed and summarized by Feder et al. [25]. Consider a generalization of equation (2.47) first, which is valid for all times, given by

$$\int_0^1 d\left(c/c_{\mathrm{eq}}\right) = -\int_1^\infty \frac{J(n)}{\beta(n) c_{\mathrm{eq}}(n)} dn. \tag{2.51}$$

According to Wakeshima, the integrals can be solved with the assumptions that (*a*) the ratio $c/c_{eq} \cong 1$ for small *n*, (*b*) $c/c_{eq} \cong 0$ for values of $n >> n^*$, and (*c*) for times when the flux at the saddle point, n^*, has reached an appreciable fraction of the steady-state value, the integrand has a sharp maximum at n^*.[10] If $\beta(n)$ is assumed to be constant in this region, $J(n)$ can be expanded around the saddle point, n^*, to second order and one obtains

$$J(n^*) - J_s = -\frac{1}{4\pi^2 Z^2}\left(\frac{\partial^2 J}{\partial n^2}\right)_{n^*}. \tag{2.52}$$

With the continuity equation, the second derivative at the critical point, n^*, can be approximated as

$$\left(\frac{\partial^2 J}{\partial n^2}\right)^* = -\left\{\frac{\partial}{\partial n}\left(\frac{\partial c}{\partial t}\right)\right\}^* = -\left\{\frac{\partial}{\partial t}\left(\frac{\partial c}{\partial n}\right)\right\}^* \cong -\frac{d}{dt}\left(\frac{\partial c}{\partial n}\right)^*. \tag{2.53}$$

At the saddle point, we thus obtain

$$J(n^*) = -\beta^*\left(\frac{\partial c}{\partial n}\right)_{n^*}, \tag{2.54}$$

after inserting for the nucleation rate and taking advantage of the fact that the higher-order derivatives vanish at the critical size. One finally arrives at the following differential equation

$$\frac{\partial J(n^*)}{\partial t} + \frac{1}{\tau} J(n^*) = \frac{1}{\tau} J_s, \tag{2.55}$$

with the solution for the incubation time, τ, given as

$$\tau = \frac{1}{4\pi\beta^* Z^2}. \tag{2.56}$$

Integration of equation (2.55) delivers the equation for the *transient nucleation rate*, *J*, given as

$$J = J_s\left(1 - \exp\left(\frac{t}{\tau}\right)\right). \tag{2.57}$$

The shape of the exponential function describing the approach from the random solution to steady-state conditions is shown in Figure 2.5.

In the review of Feder et al. [25], an alternative derivation of the incubation time, τ, is given, which is based on a random walk treatment of the cluster evolution in cluster size space

[10] For justification of this assumption, compare the treatment of Becker and Döring given in Section 2.2.3.

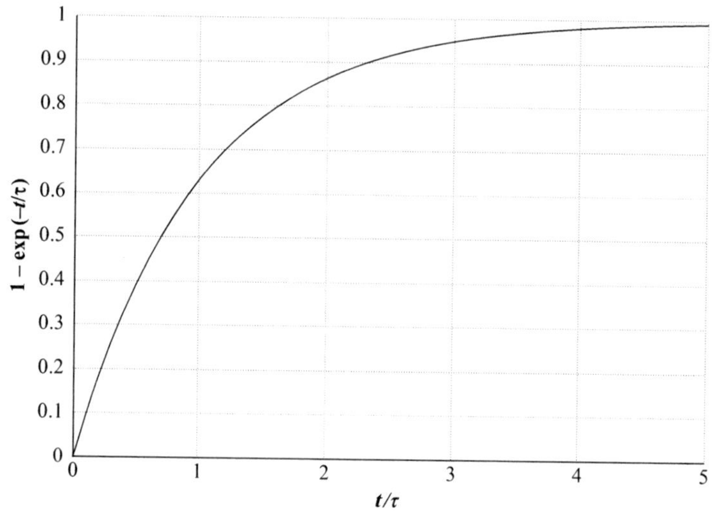

Figure 2.5. Transient nucleation rate as a function of the incubation time, τ, and time, t, after Wakeshima [24], equation (2.57).

together with the principle of microscopic time reversibility. The treatment can be started with rewriting the flux of clusters from size n to $n+1$ as

$$J = -\beta \frac{\partial c}{\partial n} + \dot{n}c, \tag{2.58}$$

where equation (2.45) is taken as a starting point. The quantity $\dot{n}$ defines the "drift velocity" or growth rate, given as

$$\dot{n} = \beta \frac{\partial \ln c_{\mathrm{eq}}}{\partial n} = -\frac{\beta}{k_B T} \frac{\partial \Delta G_{\mathrm{nucl}}}{\partial n}. \tag{2.59}$$

In this picture, the flux of clusters is seen as consisting of two terms. The first is proportional to the gradient of cluster density due to the "random walk" or diffusion of clusters. The second term represents the systematic growth of clusters due to the transient nature of the nonsteady-state part of the nucleation procedure.

With ΔG_{nucl} representing the nucleation barrier, one can define an interval Δ around the critical size, n^*, as

$$\Delta G_{\mathrm{nucl}}(n^*) - \Delta G_{\mathrm{nucl}}(n^* + 1/2\,\Delta) \cong \Delta G_{\mathrm{nucl}}(n^*) - \Delta G_{\mathrm{nucl}}(n^* - 1/2\,\Delta) = k_B T. \tag{2.60}$$

In this expression, the interval $n^* \pm 1/2\,\Delta$ may be viewed as the region where the random attachment and detachment of monomers govern the growth and decay processes of clusters. In other words, the evolution of clusters can be viewed as being given by the random walk of clusters in cluster size space.

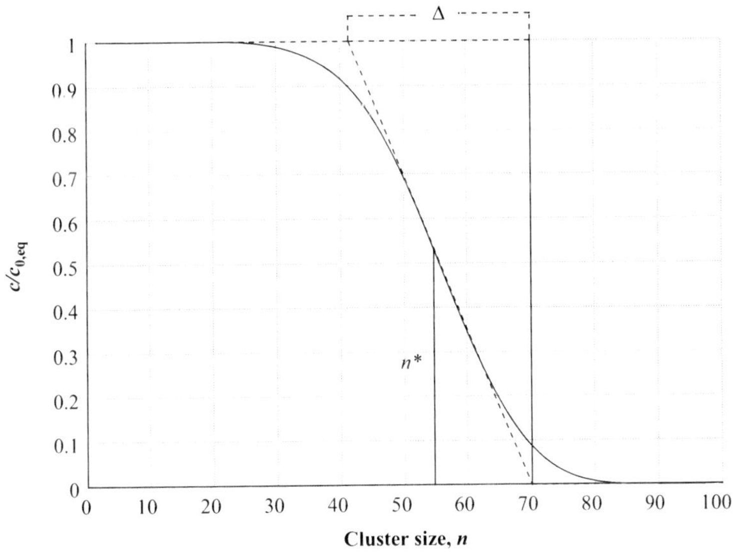

Figure 2.6. Equilibrium distribution of clusters in a distance, Δ, around the critical size, n^* (schematic).

Assume that clusters smaller than $n^* - 1/2\,\Delta$ will decay immediately to the equilibrium concentration of the monomers and the majority of clusters with sizes larger than $n^* + 1/2\,\Delta$ continue to grow and will be removed from the system according to the boundary conditions stated in the Zeldovich treatment. We thus observe a situation where approximately

$$\left(c/c_{0,\text{eq}}\right)_{n^*-1/2\Delta} \cong 1 \quad \text{and} \quad \left(c/c_{0,\text{eq}}\right)_{n^*+1/2\Delta} \cong 0. \tag{2.61}$$

When plotting this result schematically, the shape of the function $c/c_{0,\text{eq}}$ as a function of cluster size adopts a shape as depicted in Figure 2.6, where, at n^*, we have

$$\left(\frac{\partial\, c/c_{0,\text{eq}}}{\partial n}\right)_{n^*} \cong -1/\Delta. \tag{2.62}$$

With this approximation, the steady-state nucleation rate can be written as

$$J_s = \beta^* c_{0,\text{eq}}/\Delta. \tag{2.63}$$

Upon expanding this equation to second order around n^*, one obtains

$$1/\Delta = \sqrt{-\frac{1}{8}\left(\frac{\partial^2 \ln c_{0,\text{eq}}}{\partial n^2}\right)_{n^*}}. \tag{2.64}$$

This expression is very close to Z derived by Farkas [21] and Zeldovich [23] and we can thus identify the critical region as

$$\Delta = \frac{1}{Z}. \tag{2.65}$$

The Zeldovich factor, Z, is thus directly related to the region of the nucleation barrier that is bounded by the thermal agitation of the clusters with the energy, $k_B T$, which is the common interpretation of Z given in many treatments of nucleation theory.

From this result, we proceed with evaluating the expected time lag until steady-state nucleation conditions are established. Formally, the time needed for the creation of a critical cluster can be divided into the time, $\tau_<$, needed to grow a cluster from size $n = 1$ to size $n^* - 1/2\,\Delta$ and the time, τ_Δ, needed to grow the cluster from $n^* - 1/2\,\Delta$ to $n^* + 1/2\,\Delta$ as

$$\tau = \tau_< + \tau_\Delta. \tag{2.66}$$

After evaluating the two contributions to the total incubation time, τ, Feder et al. conclude that, at least for the case of water droplets, $\tau_< \ll \tau_\Delta$ and one can approximate the incubation period by only the time, τ_Δ, needed to grow through the region, Δ. Assuming that diffusion in this region is governed by random walk, that is the random attachment and detachment of monomers, we have

$$\Delta^2 = 2\beta^* \tau_\Delta. \tag{2.67}$$

From this result, one finally obtains for the incubation time

$$\tau = \frac{1}{2\beta^* Z^2}, \tag{2.68}$$

which differs from Wakeshima's estimate, equation (2.56), only in the factor 1/2 compared to $1/8\pi$.

In the case of droplet formation in supersaturated vapor, as for instance observed in expansion chamber experiments, the predicted incubation times are often found to be in the order of 10^{-4} to 10^{-6} s. These short incubation times are not easily verified experimentally. However, a good indication of the validity of the results of Wakeshima, equation (2.56), and Feder et al., equation (2.68), is shown later where excellent agreement of the incubation times derived in this section with lattice Monte Carlo simulations of solid-state precipitation is observed (see Section 2.3.4).

2.2.6 Note on Thermodynamic Properties of Small Clusters

In developing the concepts of CNT, it has been recognized early that there exist conceptual difficulties when applying continuums-mechanical quantities, such as surface tension or volume free energy, to clusters of only a few monomers. By definition, a thermodynamic system is sufficiently large to contain so many individual constituents that one can define "average" properties of the system that can be considered as homogeneous and constant over the entire spatial extension of the system. Typical examples of such properties are temperature, pressure, or chemical potentials. Even though we observe local microscopic fluctuations of these quantities, they can be considered as being homogeneous over the sample extension when viewed on a macroscopic level.

In the treatment of nucleation, the objects involved in the physical processes are often very small, sometimes extending down to the range of only a few atoms or monomers. In these situations, application of thermodynamic concepts must be considered as being, at least, critical. Since the properties of condensed matter are mainly determined by the atomic bonds between the monomers, and having in mind that these forces can extend over relatively wide distances, the assumption that a cluster of a few atoms has the same properties as a cluster of macroscopic dimensions must be checked carefully. This statement applies to bulk properties of the cluster as well as surface properties such as the interfacial energy.[11]

The scientists making the seminal contributions to CNT have been well aware of these problems. They have tried to avoid this difficulty in their theoretical treatments on different grounds, with the key element in most treatments being the observation that the flux, J, of clusters from critical to supercritical size[12] has a sharp maximum in the vicinity of the critical size, n^*. On this basis, for instance, Becker and Döring [20] have argued that the errors in treating the thermodynamic properties of the smallest clusters are less relevant for a correct integration of the cluster size fluxes/resistances compared to errors introduced when defining the properties of clusters at the critical size—an argument, which is certainly justified if the critical cluster size is sufficiently large.

Still, the conceptual difficulties of ascribing macroscopic thermodynamic properties to microscopic clusters of only a few monomers persists in those cases, where nucleation occurs under conditions of high supersaturation/high chemical driving forces. The critical cluster size itself is then in the range of only a few atoms, and the above arguments miss justification.

In the subsequent continuums-mechanical approach to solid-state nucleation, these considerations are not accounted for explicitly due to the absence of general theoretical concepts[13] addressing this problem. The interested reader should be aware of them and, hopefully, use the models presented in the following sections with sufficient care.

2.3 Solid-State Nucleation

After introducing the fundamental concepts of CNT as developed originally for the cases of liquid droplet formation from supersaturated vapor and void formation (cavitation) in liquids under pressure, in this section, it is shown that a quantitative theory of solid-state nucleation can be widely based on the classical relations derived for droplet formation. Still, nucleation in crystalline solids has a few characteristic features which must be properly accounted for. These are based on the observation that in contrast to molecules of a vapor or liquid which have the ability to move continuously in space, monomers in a crystalline material move in discrete

[11] This issue is further discussed, for instance, in Section 4.4 on the interfacial energy size correction.

[12] This flux is viewed as the particular property of a precipitation system, which determines the nucleation rate (compare Sections 2.2.3 and 2.2.4 on the kinetics of droplet formation).

[13] It must be emphasized, however, that atomistic simulation methods, such as Molecular Dynamics or lattice Monte Carlo, are becoming increasingly popular. These methods have the potential for delivering qualitative and quantitative answers to these fundamental questions.

spatial portions, determined by the spacing of sites on the crystal lattice. Some important features of solid-state nucleation[14] are:

- A well-defined interface exists between the nucleus and the parent phase, which can be coherent, semicoherent, or incoherent. This interface can be sharp as well as diffuse.
- If the product and parent phase have different molar volume, nucleation can involve long-range elastic stress, which generally makes nucleation more difficult.
- Nucleation can occur homogeneously in the undisturbed crystal lattice, or heterogeneously at lattice defects, such as dislocations, grain boundaries, inclusions, and so forth.
- Structural vacancies, which are annihilated or generated at appropriate sinks and sources, can relax elastic stress.
- The kinetics of solid-state nucleation is governed by the ability of monomers to exchange position with an empty neighboring lattice site (atom–vacancy exchange mechanism).
- Diffusion by atom–vacancy exchange is proportional to the density of vacancies. In crystals, several processes and mechanisms can introduce excess vacancies, which must be taken into account properly in the kinetic simulation.
- Crystal structures of phases with similar chemical composition can exhibit different degrees of crystal structure complexity. These structures compete with each other in the nucleation process and lead to crystal structure selection. This phenomenon is governed by the configurational entropy of the crystal structures.
- Existence of metastable precursor phases that can lead to complex precipitation sequences involving temporary formation and dissolution of some of the precipitates of the sequence.

There exist several additional aspects that are particular to solid-state nucleation. The above-mentioned mechanisms are, however, the most relevant ones and will be dealt with in this book.

Subsequently, the necessary steps in developing a quantitative theory of nucleation in the solid state are outlined and the expressions for multicomponent solid-state nucleation are derived on the basis of CNT.

2.3.1 The Precipitate–Matrix Interface

In crystalline matter, atoms are located on discrete lattice positions. The migration of atoms through the crystal occurs predominantly by the atom–vacancy exchange mechanism (compare Section 7.1) where atoms repeatedly exchange position with an empty, nearest-neighbor[15] lattice site. For the nucleation of new phases, this constraint implies that small clusters of a new phase will often show some crystallographic relationship to the lattice of the parent phase.

[14] In the context of this book, *solid-state* refers to the *crystalline* state. Amorphous structures resemble in many aspects that of undercooled liquids, which are not in the focus of the present book.

[15] In systems with chemical short-range ordering, nearest-neighbor positions are less favorable for an atom–vacancy exchange than second-nearest sites. In typical short-range ordered structures, atoms of one type prefer to have atoms of the other type in the nearest-neighbor positions. Therefore, diffusion occurs predominantly by second-nearest neighbor jumps because these jumps conserve the short-range chemical order.

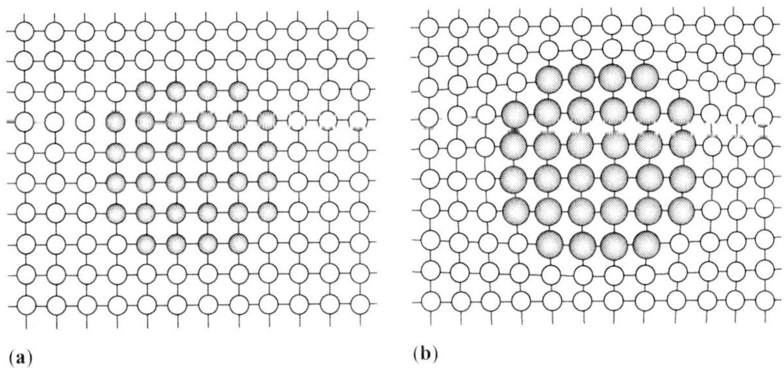

Figure 2.7. Schematic representation of a coherent precipitate without (a) and with (b) volumetric misfit.

Figure 2.7 shows schematically two precipitates with *coherent* interfaces and identical crystal structure. In Figure 2.7(a), the molar volumes of the parent and product phases are assumed to be identical. In this case, coherency of the interface is easily maintained even to large precipitate sizes [26].[16] The observation that coherent particles conserve the continuity and periodicity of the lattice planes across the precipitate has important implications for the nature of interaction between precipitates and dislocation under mechanical loading. This issue is a very important ingredient in modeling precipitation strengthening.

Figure 2.7(b) displays a situation where the molar volumes of precipitate and matrix do not match exactly. Upon nucleation of the new phase, the apparent *volume misfit* must be accommodated by local atomic displacements which introduces a long-range elastic stress field around the precipitate. For these situations, coherency of the precipitate, matrix interface, can be maintained only up to some given critical size. Above this limit, missing or additional atomic bonds interrupt the perfect matching of the lattice planes of precipitate and matrix. As the lattice matching of the coherent interface is periodically interrupted, this type of interface is denoted as *semicoherent*.

A typical example of semicoherent interface is shown schematically in Figure 2.8(a). The mismatch on the top and bottom faces of the (presumably rigid) precipitate requires that one of the bonds across the interface is broken (see the missing bond approximately in the middle of the bottom interface). This configuration introduces an elastic stress concentration around the mismatching lattice plane. On the top interface, the lattice singularity is located two atomic layers away from the interface.[17] Assuming that the matrix is the elastically softer material, this configuration allows for some reduction of the total elastic stress energy in the system.

Figure 2.8(b) shows a precipitate with entirely incoherent phase boundaries. The difference between the lattice orientations and/or lattice spacing of precipitate and matrix impairs any kind

[16] A typical example for precipitates with minimal lattice mismatch is found, for instance, in the Ni-base superalloy UDIMET 720Li. The mismatch between the fcc Ni-matrix and the $L1_2$-ordered γ' precipitates can be as small as 0.01% [26], which can be measured using synchrotron radiation. γ' precipitates can then maintain coherency with the matrix up to sizes >700 nm.

[17] This configuration can be achieved by, for example, dislocation climb, which is enabled by the continued condensation or evaporation of vacancies at the jogs of an edge dislocation.

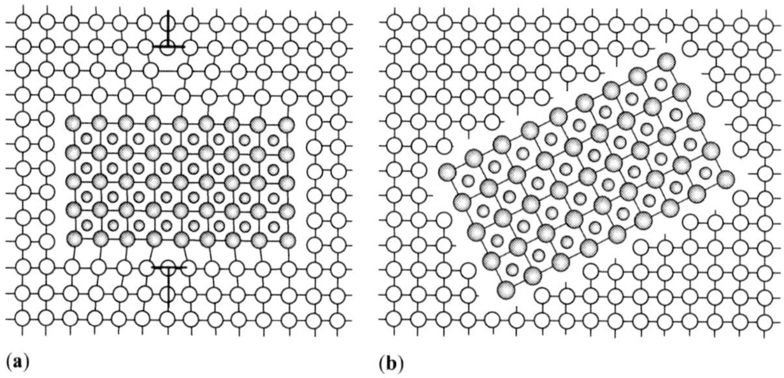

Figure 2.8. Precipitates with semicoherent ([a] top and bottom interface) and incoherent interfaces ([a] right and left sides; [b] all sides). See text for discussion.

of periodic lattice matching. As this type of phase boundary represents almost ideal sources and sinks for vacancies, incoherent interfaces are prone to rapid elastic stress relaxation. In contrast to coherent clusters, the elastic stress field surrounding incoherent precipitates is often only of weak amplitude despite the fact that the precipitate would exhibit considerable volumetric misfit when only comparing molar volumes.

2.3.2 Free Energy of Nucleus Formation

In the development of CNT, the concept of a droplet being in thermodynamic equilibrium with its surrounding vapor has been utilized for describing the energy of formation of a spherical nucleus. As this concept is based on thermodynamic equilibrium considerations, this treatment might be denoted as the *thermodynamic* approach to CNT. In the present section, an approach is introduced which starts with a continuums description of the total free energy change associated with the occurrence of a single nucleus. It arrives at the same expression for the nucleation rate and it is further denoted as *continuums-mechanical* CNT.

Consider a spherical cluster of a phase β forming in a parent matrix phase α. Figure 2.9 schematically shows a critical nucleus of β surrounded by other clusters of subcritical size. For the following treatment it is important to emphasize that every single nucleation event

i. occurs randomly by *heterophase fluctuations*;
ii. involves transfer of sufficiently small amount of matter such that the thermodynamic properties of the surrounding matrix are not changed.

On inspecting the implications of the first statement (i), we recall that, in CNT, nucleation is reasoned on the basis of microscopic fluctuations within an otherwise macroscopically homogeneous and equilibrated system of constant energy. This issue is elaborated further in Section 2.3.3, where the steady-state nucleation rate expression is developed for solid-state matter.

Statement (i) also implies that there exist no systematic and necessary composition gradients around the nucleus, because the fluctuations are assumed to be entirely random. The mechanism of nucleus formation is different in this aspect from the mechanism governing the

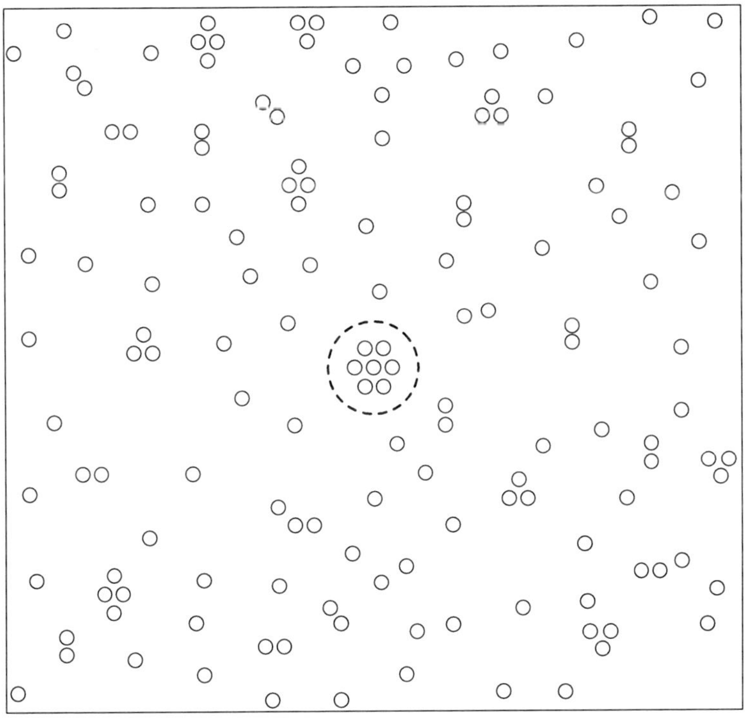

Figure 2.9. Schematic distribution of monomers, dimers, and so forth, during nucleation, with a critical cluster marked by a dashed circle.

diffusion-controlled growth of precipitates, which is dealt with in Chapter 3. Statement (ii) emphasizes that the reservoir of molecules for the nucleation event is sufficiently large such that the total energy of the system remains constant. The energy change accompanying the nucleation event can then be described in terms of constant properties of the parent matrix phase.

When forming the new phase, there are two contributions to the total free energy change that must be accounted for:

i. The volume free energy change which is related to the formation of a cluster with n molecules built from the same amount of parent phase; this energy contribution is proportional to the volume of the cluster given by $n\Omega$.
ii. The energy that is required to create the interfacial area between cluster and matrix; this contribution scales with the surface area of the cluster which is proportional to the volume to the power of 2/3 with $\sim(n\Omega)^{\frac{2}{3}}$.

Ω represents the partial volume of a single molecule or atom. The total energy change on nucleus formation, ΔG_{nucl}, can be written as[18]

$$\Delta G_{\text{nucl}} = n\Omega \cdot \Delta G_{\text{vol}} + 4\pi\left(\frac{3}{4\pi}n\Omega\right)^{\frac{2}{3}} \cdot \Delta G_{\text{surf}}, \tag{2.69}$$

[18] In this expression, the nucleus is assumed to be of spherical shape.

where ΔG_{vol} is the specific volume free energy change on transforming unit volume of the parent phase to the precipitate phase, and ΔG_{surf} is the energy required to create unit area of precipitate–matrix interface.

The latter quantity, ΔG_{surf}, is readily identified as the interfacial energy, γ, which is treated in detail in Chapter 4. For the following analysis, it is sufficient to assume that the interfacial energy, γ, between cluster and matrix is a function of the thermodynamic properties of the two phases: of temperature and pressure and of the chemical composition difference and gradient across the interface.

The specific volume free energy change, ΔG_{vol}, can be further detailed by splitting it up into chemical and mechanical contributions, both stemming from the transformation of unit volume of matrix to the precipitate phase. The specific volume free energy change then reads

$$\Delta G_{vol} = \Delta G_{vol}^{chem} + \Delta G_{vol}^{el}. \tag{2.70}$$

The former contribution is easily identified as the chemical driving force, $d_{chem}^{\beta}/v^{\alpha}$, which has been defined earlier in equation (1.45) and where v^{α} is the molar volume. For supersaturated systems, the chemical driving force is a positive quantity and its contribution to the volume free energy must therefore be taken as negative.

The other contribution to the volume free energy change is associated with the structural change from α to β. It has been shown in the previous sections that cluster formation is generally accompanied by a long-range stress field, the energy of which, ΔG_{vol}^{el}, has been given in equation (1.73).

Finally, for the specific free energy change on formation of a cluster of phase β, we obtain

$$\Delta G_{vol} = -\frac{d_{chem}^{\beta}}{v^{\alpha}} + \Delta G_{vol}^{el}. \tag{2.71}$$

It is important, now, to recognize that the volumetric and surface part of the free energy contributions have particular properties with respect to the sign of the functions and the dependence on cluster size. On one hand, nucleation can occur only if the volume free energy change is negative and bulk free energy is gained in the nucleation process. For positive volume free energy change, the nucleation process would not occur spontaneously because it is energetically unfavorable. On the other hand, the process of creating surface area in supersaturated systems is always associated with an expense of energy. The specific interfacial energy is always a positive quantity, as otherwise the cluster would try to achieve maximum surface area by entirely dissolving into isolated monomers. The resulting function describing the free energy change on formation of a nucleus of size n is shown schematically in Figure 2.10.

If the two contributions to the free energy of cluster formation are plotted as a function of the number of molecules in the cluster, the volume contribution scales linearly with n and, consequently, plots as a straight line in the figure. As outlined previously, the volume free energy change is always negative. The contribution from surface area scales with $n^{\frac{2}{3}}$ and is always positive, which appears in the figure as a curved line with positive slope. Finally, the sum of both curves which represents the total cluster formation energy, ΔG_{nucl}, exhibits a distinct maximum at the critical cluster size, n^*, above which the addition of any single molecule decreases the cluster energy, and below which, the addition of a molecule increases the cluster energy.

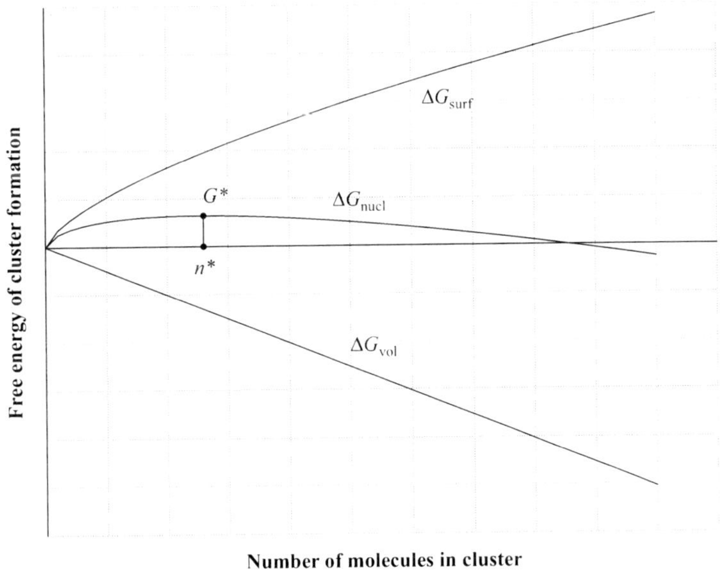

Figure 2.10. Free energy change accompanying cluster formation as a function of cluster size.

Before further investigating the basic features of the cluster formation energy, let us transform ΔG_{nucl} from number of molecules coordinates, n, into cluster radius space. On replacing the corresponding terms in equation (2.69), we obtain

$$\Delta G_{nucl} = \frac{4}{3}\pi\rho^3 \cdot \Delta G_{vol} + 4\pi\rho^2 \cdot \Delta G_{surf}, \tag{2.72}$$

where ρ denotes the radius of the cluster. The corresponding functions are shown schematically in Figure 2.11.

In radius space, the volume and surface contributions to the cluster formation energy of a spherical particle are now both represented by nonlinear functions, with the volume contribution scaling with ρ^3 and the interfacial energy contribution with ρ^2. Again, the total cluster formation energy, ΔG_{nucl}, exhibits a distinct maximum at the *critical radius*, r^*, and the *critical nucleation energy*, G^*.

The exact values of r^* and G^* can straightforwardly be evaluated after setting the derivative of ΔG_{nucl} equal to zero. Accordingly, we have

$$\frac{\partial}{\partial\rho}\Delta G_{nucl} = \frac{\partial}{\partial\rho}\left(\frac{4}{3}\pi\rho^3 \cdot \Delta G_{vol} + 4\pi\rho^2 \cdot \Delta G_{surf}\right) = 4\pi\rho^2 \cdot \Delta G_{vol} + 8\pi\rho \cdot \Delta G_{surf} = 0, \tag{2.73}$$

with the extremal values given as

$$r^* = -\frac{2\Delta G_{surf}}{\Delta G_{vol}} = \frac{2\gamma}{\dfrac{d^{\beta}_{chem}}{\nu^{\alpha}} - \Delta G^{el}_{vol}} \tag{2.74}$$

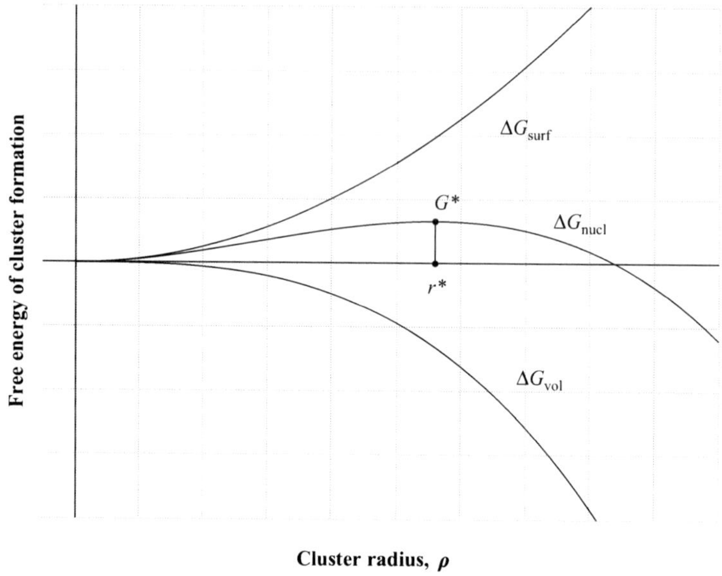

Figure 2.11. Free energy of cluster formation as a function of cluster radius.

and

$$G^* = \frac{16\pi}{3}\frac{(\Delta G_{surf})^3}{(\Delta G_{vol})^2} = \frac{16\pi}{3}\frac{\gamma^3}{\left(-\frac{d^{\beta}_{chem}}{v^{\alpha}} + \Delta G^{el}_{vol}\right)^2}. \tag{2.75}$$

The value of the critical nucleation energy, G^*, represents the key quantity that is needed to evaluate the steady-state nucleation rate, J_s, in solid-state systems. This is elaborated subsequently.

2.3.3 Steady-State Nucleation Rate in Crystalline Solids

In the treatment of condensation of liquid droplets from supersaturated vapor, the existence of microscopic fluctuations in an otherwise equilibrated macroscopic system has been identified as the physical source providing the nucleation energy for the formation of supercritical nuclei of a new phase. Although we find that the formation of a critical cluster by repeated addition of monomers is continuously energetically unfavorable and, thus, a very unlikely process by itself, the huge number of possible nucleation attempts facilitates the macroscopically observable nucleation rates when viewed on a sufficiently large scale. Even if we cannot give any quantitative prediction for the time and location of an individual nucleation event, it is possible to quantitatively predict the nucleation rate of a new phase in a large enough system on a statistical basis.

In principle, the physical relations obtained for droplet condensation can straightforwardly be applied to solid-state nucleation. Turnbull and Fisher [27], in their analysis of solid-state nucleation, arrive at a similar expression for the nucleation rate as previously in liquid droplet condensation. However, as already mentioned in the introduction to this section, some particularities of crystalline materials must be accounted for properly.

In a first step, the fundamental nucleation rate expression as developed in Section 2.2 for vapor condensation is formulated. In subsequent steps, the quantities entering this expression will be assigned their physical correspondence in crystalline materials.

The steady-state nucleation rate, J_s, for the formation kinetics of new phases in thermodynamic systems can be described by the relation [28]

$$J_s = NZ\beta^* \exp\left(-\frac{G^*}{k_B T}\right), \tag{2.76}$$

where N is the number of potential nucleation sites, Z is the Zeldovich factor, β^* is the atomic attachment rate, G^* is the critical nucleation energy, k_B is the Boltzmann constant, and T is the absolute temperature.

When inspecting equation (2.76) in detail, we observe that this relation is of rather general nature and, in this sense, has the capability of qualitatively and quantitatively describing nucleation processes in various different practical applications. In order to adapt the theory for each special case, the proportionality coefficients N, Z, and β^* must be given their correct physical representation. This is elucidated in the following section, after first revisiting shortly the physical meaning of the nucleation barrier, G^*.

The one ingredient to nucleation theory, which has the vastly dominating influence on the nucleation process, is the exponential term $\exp(-G^*/k_B T)$. Conceptually, it roots in the Boltzmann configurational entropy describing the probability of microscopic fluctuations as a function of the energy amplitude. In 1939, the Russian–Soviet physicist Yakov Il'ich Frenkel,[19] (1894–1952), ref. [29], presented a general theory which suggests that the exponential expression in G^* is valid for any kind of *heterophase fluctuation*.[20] In Frenkel's analysis, the precipitation system is viewed as a system containing a distribution of clusters of the new phase with the total free energy given by

$$G = \sum_n \Delta G_{\text{nucl}} f(n) - TS_{\text{cl,mix}}. \tag{2.77}$$

The function $f(n)$ denotes the number of clusters of size n per unit volume. ΔG_{nucl} is the cluster formation energy as introduced in the previous section and $S_{\text{cl,mix}}$ is the configurational entropy of mixing of clusters on the available lattice sites. The total energy, G, of the system can be minimized under the constraint of constant temperature, pressure, and mass. The equilibrium distribution of heterophase fluctuations in a condensed system is then given with

$$c(n) = N_0 \exp\left(\Delta G_{\text{nucl}}(n)/k_B T\right). \tag{2.78}$$

The factor N_0 in this equation stems from the partial entropy of mixing of clusters on the available lattice positions, which is $k_B \ln(f(n)/N_0)$, irrespective of the size of the cluster, n, and the details of the operative nucleation mechanism. For homogeneous nucleation, each single lattice

[19] In English-language literature, Yakov Frenkel put his name often down as J. Frenkel, the "J" standing for Jakob.

[20] The term "heterophase" emphasizes that, in addition to the microscopic fluctuations in local chemical composition, which are usually required for the creation of nuclei of a new phase, the nucleation process must also be accompanied by simultaneous fluctuations in crystal structure from the lattice type of the parent phase to that of the product phase.

position represents a potential nucleation site and N_0, thus, corresponds to the total number of lattice sites in the crystal.

Without any large effort [28], the treatment can be generalized for heterogeneous nucleation. For nucleation on dislocations, N_0 must be substituted by the number of potential nucleation sites at the dislocations which is usually approximated by the number of atoms located along the dislocation core. For grain boundary nucleation, the same procedure applies, and N_0 then corresponds to the number of atoms located in a monoatomic layer along the grain boundary. A convenient mathematical framework for quantifying the number of potential nucleation sites in a polycrystalline microstructure as a function of the dislocation density and the mean grain size is described in Section 2.4.2.

With N_0 being a well-defined function of the type of potential nucleation site, the value of the Zeldovich factor, Z, is investigated next. According to the treatment of Section 2.2.4, equation (2.50), this quantity is defined as

$$Z = \left[-\frac{1}{2\pi k_B T} \left(\frac{\partial^2 \Delta G_{\text{nucl}}}{\partial n^2} \right)_{n^*} \right]^{1/2}. \tag{2.79}$$

For the further treatment, the derivative of ΔG_{nucl} with respect to the number of molecules, n, in the cluster shall be evaluated using the expression for ΔG_{nucl} for spherical nuclei in terms of radius coordinates, equation (2.72). It is then easily possible to derive a closed expression for the Zeldovich factor based on the interfacial energy and total driving force. After replacing the volume of the sphere $4/3\pi\rho^3$ with the sum of molecular volumes $n\Omega$, we obtain

$$\begin{aligned} \Delta G_{\text{nucl}} &= \frac{4}{3}\pi\rho^3 \cdot \Delta G_{\text{vol}} + 4\pi\rho^2 \cdot \Delta G_{\text{surf}} \\ &= n\Omega \cdot \Delta G_{\text{vol}} + 4\pi \left(\frac{3}{4\pi} n\Omega \right)^{\frac{2}{3}} \cdot \Delta G_{\text{surf}} = na^3 \cdot \Delta G_{\text{vol}} + 4\pi \left(\frac{3}{4\pi} na^3 \right)^{\frac{2}{3}} \cdot \Delta G_{\text{surf}}, \end{aligned} \tag{2.80}$$

with a being the interatomic distance.

For the first derivative of ΔG_{nucl} with respect to the number of molecules in the cluster, we get

$$\frac{\partial \Delta G_{\text{nucl}}}{\partial n} = a^3 \cdot \Delta G_{\text{vol}} + 2a^2 \left(\frac{3}{4\pi} n \right)^{-\frac{1}{3}} \cdot \Delta G_{\text{surf}}, \tag{2.81}$$

and from the extremum condition

$$\frac{\partial \Delta G_{\text{nucl}}}{\partial n} = 0, \tag{2.82}$$

the number of molecules, n^*, in a spherical critical cluster is found with

$$\begin{aligned} n^* &= \frac{4\pi}{3} \frac{1}{a^3} \left(-\frac{2\Delta G_{\text{surf}}}{\Delta G_{\text{vol}}} \right)^3 \\ &= -\frac{32\pi}{3\Omega} \left(\frac{\Delta G_{\text{surf}}}{\Delta G_{\text{vol}}} \right)^3. \end{aligned} \tag{2.83}$$

For the second derivative, we obtain

$$\frac{\partial^2 \Delta G_{\text{nucl}}}{\partial n^2} = \frac{\partial}{\partial n}\left(a^3 \cdot \Delta G_{\text{vol}} + 2a^2 \left(\frac{3}{4\pi} n \right)^{-\frac{1}{3}} \cdot \Delta G_{\text{surf}} \right)$$

$$= 2a^2 \left(\frac{4\pi}{3} \right)^{\frac{1}{3}} \cdot \Delta G_{\text{surf}} \cdot \left(-\frac{1}{3} \right) n^{-\frac{4}{3}} \tag{2.84}$$

In the Zeldovich factor, the second derivative is evaluated at the critical size, n^*, and we can thus insert n^*, equation (2.83), for n. The second derivative then reads

$$\frac{\partial^2 \Delta G_{\text{nucl}}}{\partial n^2} = 2a^2 \left(\frac{4\pi}{3} \right)^{\frac{1}{3}} \cdot \Delta G_{\text{surf}} \cdot \left(-\frac{1}{3} \right) \left[\frac{4\pi}{3} \frac{1}{a^3} \left(-\frac{2\Delta G_{\text{surf}}}{\Delta G_{\text{vol}}} \right)^3 \right]^{-\frac{4}{3}}$$

$$= -\frac{a^6}{32\pi} \cdot \frac{\left(\Delta G_{\text{vol}}\right)^4}{\left(\Delta G_{\text{surf}}\right)^3}, \tag{2.85}$$

For the Zeldovich factor, expressed in terms of the total energy change on nucleation ΔG_{nucl} as introduced in equation (2.72), we finally obtain

$$Z = \left[\frac{a^6}{64\pi^2 k_B T} \cdot \frac{\left(\Delta G_{\text{vol}}\right)^4}{\left(\Delta G_{\text{surf}}\right)^3} \right]^{1/2}. \tag{2.86}$$

With the Zeldovich factor determined for nucleation of a spherical particle, the last coefficient in the nucleation rate expression, equation (2.76), that must be adapted to crystalline solids is the atomic attachment rate, β^*. This quantity represents the rate at which molecules forming the precipitate attach to the critical cluster. In literature, various different expressions for β^* are reported. Russell [28], for instance, in a two-flux analysis of nucleation, gives a value of

$$\beta^* = (4\pi)^{\frac{1}{3}} \left(3v^\alpha n^*\right)^{\frac{2}{3}} D_B \frac{X_B^\alpha}{a^4}. \tag{2.87}$$

where D_B and X_B^α are the effective diffusion coefficient and mole fraction of solute atoms B, respectively. An alternative and generalized expression for this quantity is given later in Section 2.5 on multicomponent nucleation.

From the analysis presented in this section, we find that the nucleation rate is vastly determined by the exponential factor containing the critical nucleation energy, G^*. Substantial nucleation is observed only if G^* is in the order of the kinetic energy of the cluster, $k_B T$. The proportionality between the exponential factor and the steady-state nucleation rate, J_S, is conveniently described by the three quantities N_0, β^*, and Z, with N_0 being related to the number of potential nucleation sites in the microstructure of the solid, Z being a thermodynamic quantity, and β^* incorporating the transport properties of the material in the form of the diffusional mobility of the atoms involved in the precipitation process.

2.3.4 Time-Dependent Nucleation

In the treatment of classical nucleation theory, the steady-state nucleation rate, J_s, has been evaluated as the flux of clusters in cluster size space measured at the critical size n^* and p^*. Though the expression for J_s as given in equation (2.76) is of general nature, it is still based on the assumption that an equilibrium distribution of clusters is present at time, $t = 0$. In many situations of practical relevance, this presumption is hardly fulfilled, as the equilibrium distribution of clusters sizes is a function of temperature, and practical heat treatment of phase separating alloys relies on variations in temperature throughout the treatment process. This situation is depicted in Figure 2.12 where a schematic representation is shown for typical size distributions of clusters as a function of temperature, T.

In Figure 2.12, the cluster size distribution at T_1 represents an equilibrium distribution at the highest assumed temperature. As the influence of entropy is most pronounced at high T, the existence of large unstable clusters is rather unlikely compared to the existence of monomers and dimers, and the distribution shows a very narrow extension with a pronounced peak at very small cluster size. At T_2, the distribution is wider because the entropic effect of destabilization of clusters becomes weaker relative to the phase separating effect of chemical attractive and repulsive forces between the atoms.

The dashed line in Figure 2.12 indicates the critical cluster size, n^*, above which the clusters are thermodynamically stable and continue to grow in a deterministic manner. For temperatures T_1 and T_2, the largest nuclei are still smaller than the critical size and all nuclei reaching the largest sizes decay again on average. At T_3, the situation is different, because the largest nuclei of the equilibrium cluster size distribution exceed the critical size and any cluster becoming overcritical will continue to grow. The observation that overcritical clusters do not decay again and continuously grow is responsible for the shape of the cluster size distribution with a

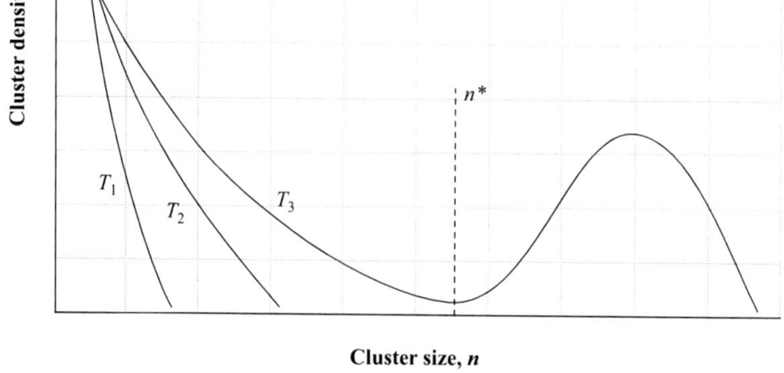

Figure 2.12. Equilibrium cluster size distributions for a hypothetical thermodynamic system at temperatures T_1 and T_2. Size distribution at temperature T_3, with a transient distribution of stable growing clusters.

maximum located above the critical size. The flux of clusters into the direction of larger sizes has been interpreted previously as the steady-state nucleation rate.

If we assume that, in some heat treatment process, the material is annealed at a higher temperature, T_1, and then rapidly quenched to a lower temperature, T_3, in order to establish equilibrium and reach steady-state nucleation conditions, the cluster size distribution established at T_1 and $t = 0$ must evolve toward the equilibrium size distribution at temperature, T_3. This cannot be achieved instantaneously as the redistribution of clusters is driven by atomic attachment and detachment, as well as diffusion through the crystal. The time that is necessary to establish the new equilibrium size distribution given at a specific temperature is denoted as the *induction period* or *incubation time*, τ.

In order to derive expressions for τ, equation (2.44) must be solved analytically, which is a task that has proven to be rather difficult. In literature, several solutions have been presented, all of them involving serious approximations and, according to Russell [28], "none of them being convincing."

In a treatment of this problem, Wakeshima [24] derived a relation between the *transient nucleation rate*, J, and the *steady-state nucleation rate*, J_S, given as

$$J = J_S\left(1 - \exp\left(-\frac{t}{\tau}\right)\right). \tag{2.88}$$

See Section 2.2.5 on the time lag in nucleation for details. Based on work of Turnbull [30], the following expression relating steady-state and transient nucleation is nowadays most often used

$$J = J_S \exp\left(-\frac{\tau}{t}\right). \tag{2.89}$$

The two functional relationships are shown in Figure 2.13 for an incubation time of $\tau = 1$s. Accordingly, the expression proposed by Wakeshima predicts a significantly faster transition

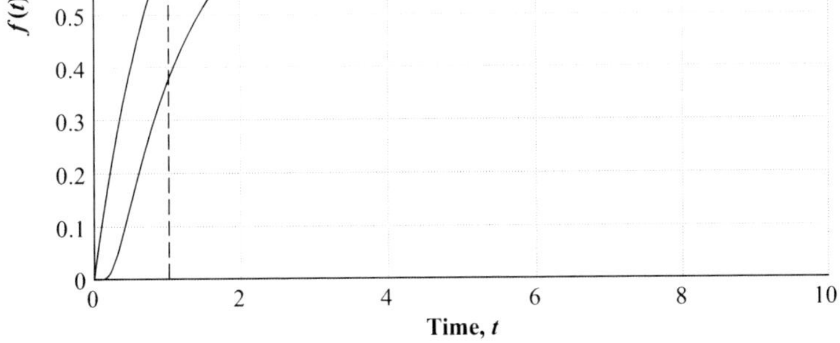

Figure 2.13. Functional relationships between steady-state nucleation and transient nucleation rate.

toward steady state compared to the expression based on Turnbull. Still, both functions have the same basic behavior, starting at a value of zero and approaching the final value of one.

When accepting the expression going back to Turnbull, and commonly used in most CNT treatments, the *transient nucleation rate*, J, is finally given as

$$J = J_s \exp\left(-\frac{\tau}{t}\right) = NZ\beta^* \exp\left(-\frac{G^*}{k_B T}\right)\exp\left(-\frac{\tau}{t}\right). \tag{2.90}$$

In recent times, Soisson and Martin [31] have observed excellent confirmation of the estimates of Wakeshima [24] and Feder et al. [25] for the value of the induction period, τ, in carrying out kinetic Monte Carlo simulations of solid-state nucleation. These authors compare the evolution of the number of precipitates, N_P, from simulations starting either from (*a*) a fully random solid solution or (*b*) after equilibration of their system close to the precipitate solution temperature. In the first case, Soisson and Martin find that the nucleation rate of clusters follows a kinetics that is bounded by the time lag estimates of Wakeshima and Feder et al. In the second case, nucleation follows a straight line in the number density of supercritical clusters versus time plot. The results of this analysis are shown in Figure 2.14.

In the interpretation of these results, Soisson and Martin find that nucleation occurs almost instantaneously if the distribution of clusters is close to the equilibrium distribution (open circles in Figure 2.14). This condition was achieved by a simulated annealing of the system close to the

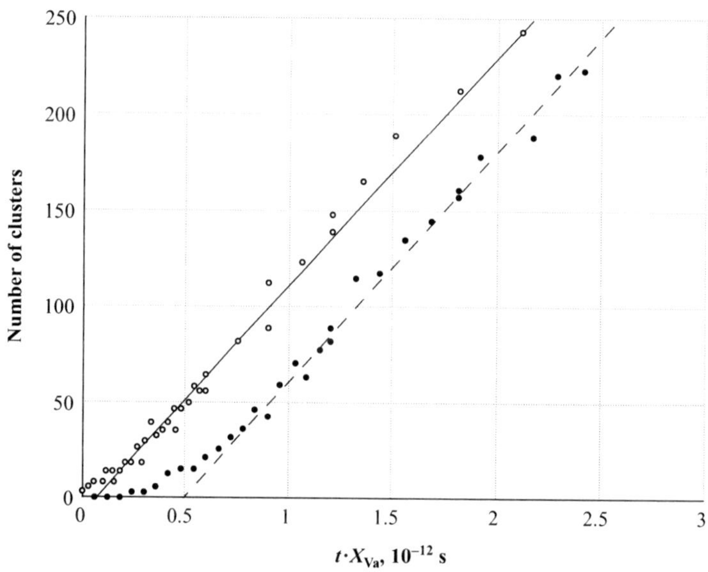

Figure 2.14. Analysis of solid-state nucleation kinetics: Kinetic Monte Carlo study of the evolution of supercritical clusters versus time on a $(256)^3$ lattice with one vacancy. t is the real time coupled to Monte Carlo steps and X_{Va} is the vacancy site fraction.[21]

[21] The X-axis scaling with $t \cdot X_{Va}$ is used by Soisson and Martin in order to be able to compare Monte Carlo simulations with different box sizes. The simulations are carried out with one vacancy per simulation box.

phase boundary prior to quenching to test temperature at ΔT below the phase boundary. After equilibration, the size distribution contained a sufficiently high number of larger clusters already close to the critical size, such that nucleation could set in almost immediately after reaching the test temperature. In this case, the extrapolated line crosses the X-axis close to the origin.

In the case where the simulation was started from a random solution (filled circles in Figure 2.14), a clear shift of the extrapolated line is visible representing the effect of the incubation time, τ, which is required to establish the equilibrium cluster distribution. In their further analysis, Soisson and Martin also find good *quantitative* agreement of incubation times predicted from Wakeshima and Feder et al., with the numerical results from the Monte Carlo simulation, their results commonly being bounded by the CNT results.

Another interesting result of this analysis is the observation that the incubation time, τ, is a function of supersaturation, both in the analytical as well as the numerical simulation cases. The larger the incubation time is, the lower is the supersaturation. At high supersaturation, nucleation starts very rapidly and the incubation times become rather short and undetectable from the numerical simulations.

2.3.5 The Volume Misfit Stress

If the molar volumes of the new phase and the matrix are different, and if the precipitate is coherent, a stress field evolves around the cluster which deforms the lattices of both phases. The energy associated with the formation of a mismatching, coherent cluster can be evaluated in an elegant way on the basis of Eshelby's concept [13] for determination of the elastic stress field around an ellipsoidal inclusion.

Eshelby's treatment is based on the assumption of an isotropic elastic medium of infinite size. In the center of the volume, an unstrained particle is located which is assumed to transform into a structure with a different specific volume, that is, a certain volumetric misfit. The procedure for evaluating the elastic energy associated with this transformation starts in the unconstrained state and comprises of the following steps:

- Remove the untransformed particle from the unstrained matrix. This step leaves a hole in the matrix region.
- Let the particle transform outside the region without any constraints. The transformation alters the volume of the particle but leaves its shape unchanged.
- Apply surface forces on the particle to restore its original volume and shape. Put the particle back into the hole and rejoin it to the matrix.
- Apply body forces on the particle that just balance the surface forces applied in the previous step. This step will restore mechanical equilibrium. It will also lead to straining of both the particle as well as the matrix.

If all elastic constants are isotropic and identical in the precipitate and the matrix, a simple relation[22] is obtained for the elastic energy, $\Delta G_{\mathrm{vol}}^{\mathrm{el}}$, of a strained inclusion with[23]

[22] Eshelby [13] presents solutions also for situations where the elastic properties of particle and matrix are different. These formulae are rather lengthy, though, and in reasonable approximation, this simple version can be used in practical evaluation of the elastic energy associated with coherent precipitation.

[23] Compare equation (1.73), Section 1.5.

$$\Delta G_{vol}^{el} = \frac{E^{\alpha}}{1-\nu_{P}^{\alpha}}(\varepsilon^*)^2. \tag{2.91}$$

E^{α} and ν_{P}^{α} are the Young's modulus and Poisson's ratio of the matrix. ε^* is the linear elastic misfit which is related to the volume misfit, v^*, by $\varepsilon^* \approx v^*/3$. We have referred the elastic properties to the properties of the matrix, as these are most often known with good accuracy.

It is important to recognize that ε^* occurs in this equation in quadratic form. As a consequence, the energy contribution from elastic stress is always positive. Elastic misfit stress will thus always make nucleation less favorable, irrespective of whether the misfit produces a compressive or tensile stress field.[24] In several cases of technical relevance, the elastic energy accompanying the nucleation process is so high that it compensates the chemical driving force given by the supersaturation of the precipitation system. The net driving force for nucleation then becomes zero or negative and formation of the new phase by coherent homogeneous nucleation is entirely suppressed even though substantial supersaturation might be present.

2.3.6 Excess Structural Vacancies

The evolution of the excess vacancy concentration and its influence on diffusion and precipitation kinetics are extremely important issues when dealing with diffusion-controlled reactions. The major impact of the vacancy concentration on the materials behavior is twofold: Excess vacancies impact severely on the diffusion kinetics of atoms. This part of vacancy interaction represents the *kinetic* aspect, which is dealt with in detail in Section 7.4. In the present chapter, we will be concerned with the *thermodynamic* effect.

In technological applications, a supersaturation of structural vacancies can be realized through various processes: Firstly, if a material is annealed at elevated temperature and subsequently quenched, an excess amount of vacancies can be "frozen-in" if the mean diffusional distance for vacancies to reach the next vacancy source/sink is too large for a given temperature and time. The quenched-in excess vacancies temporarily speed up all diffusional processes until they finally annihilate at some vacancy sink and the equilibrium vacancy site fraction value given by equation (1.78) is reached. Secondly, excess vacancies can be introduced into the material by deformation processes, such as rolling or forging. The forced movement of dislocations through the material in the course of the deformation process can induce substantial vacancy supersaturation, which is comparable in amplitude with annealing and rapid quenching processes. Some types of phase transformations involve the generation of large stresses and local plasticity, for example, the martensitic transformation. This effect can also introduce considerable amount of excess vacancies.

Excess vacancies interact with precipitation processes in two ways. If the precipitate incorporates vacancies on its crystal lattice, the vacancies can be considered as a chemical component which is structurally required to build up the precipitate. From this viewpoint, an excess of vacancies increases the chemical driving force and, thus, facilitates nucleation.

On the other hand, excess vacancies can be annihilated or generated at incoherent or semicoherent phase boundaries. Excess vacancies can thus relax volumetric misfit strain and

---

[24] This conclusion is in contrast to the effect of excess vacancies on the nucleation process where the sign of the size mismatch also determines the sign of the energy contribution. See Section 2.3.6 for details.

compensate the elastic stress that is present around misfitting incoherent precipitates. In cases where precipitates have large volume misfit such that the elastic energy overcompensates the available chemical driving force for coherent nucleation, vacancy annihilation or generation at incoherent phase boundaries makes nucleation possible at all.

Russell [32] has analyzed the effect of excess vacancies on nucleation quantitatively. In the introduction of his treatment, he states that "excess vacancies always increase the pre-exponent in the nucleation rate equation,[25] but in addition may also lower the nucleation barrier so as to make homogeneous incoherent nucleation possible." The second part of the statement refers to the *thermodynamic* effect of excess vacancies.

Consider a cluster of n atoms forming in a parent phase with a volumetric misfit given by v^*. The cluster formation process can be described by the following sequence:

1. Create a cluster with n atoms from the matrix.
2. Annihilate or generate just as many vacancies at the precipitate–matrix interface, which are necessary to relax all misfit strain.
3. Account for a number x of vacancies that are annihilated or generated in excess to the vacancies used in step 2 due to vacancy supersaturation.

Mathematically, the reaction to form an embryo,[26] O_n, containing n atoms can be written as

$$O_n = nO_1 + \left(nv^* + x\right)O_{\text{Va}}, \tag{2.92}$$

where O_1 represents a single atom (monomer) and O_{Va} is a vacancy. The individual terms in this equation correspond to the three steps listed above. The first term refers to the formation of the n-mer in the absence of vacancies. The second term corresponds to the relaxation step where vacancies are annihilated/created at the precipitate–matrix interface in order to compensate the elastic strain. The third term accounts for the excess or shortage of vacancies compared to the fully relaxed state. A sketch of this situation is shown in Figure 2.15.

For the total energy change on nucleation, ΔG_{nucl}, we can thus write [32]

$$\begin{aligned}\Delta G_{\text{nucl}} &= n\Omega \cdot \Delta G_{\text{vol}}^{\text{chem}} + \left(n\Omega v^* + x\Omega\right) \cdot \Delta G_{\text{vol}}^{\text{chem,va}} \\ &\quad + n\Omega \cdot \Delta G_{\text{vol}}^{\text{el}} + 4\pi\left(\frac{3}{4\pi}n\Omega\right)^{\frac{2}{3}} \cdot \Delta G_{\text{surf}},\end{aligned} \tag{2.93}$$

where Ω represents the atomic volume. The elastic strain energy, equation (2.91), can be written in terms of the excess/shortage of vacancies, x, with

$$\Delta G_{\text{vol}}^{\text{el}} = \frac{E^\alpha}{9\left(1 - \nu_{\text{P}}^\alpha\right)}\left(\frac{x}{n}\right)^2, \tag{2.94}$$

[25] Which is related to the kinetic aspect of the impact of excess vacancies on phase transformations: excess vacancies accelerate substitutional diffusion.

[26] Compare also to the reaction scheme without vacancies as used earlier in equation (2.16).

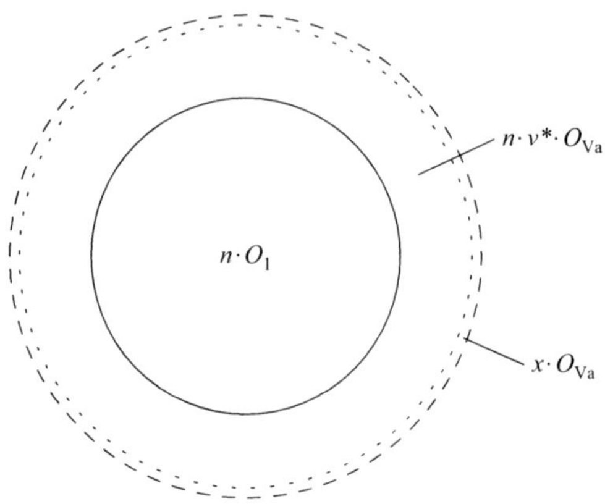

Figure 2.15. Nucleation in the presence of excess vacancies. Precipitate containing n atoms (solid line), region of the matrix which compensates the volumetric misfit, ν^* (dotted line), and excess/shortage of vacancies, x, around the nucleus (dashed line).

where we have used the relation

$$\varepsilon^* = \frac{\nu^*}{3} = \frac{1}{3}\frac{x}{n}. \tag{2.95}$$

E^α and ν_{P}^α are the Young's modulus and Poisson's ratio of the matrix. ε^* is the linear elastic misfit and ν^* is the volumetric misfit. For the total energy change on nucleation we finally obtain

$$\Delta G_{\mathrm{nucl}} = -\frac{n\Omega}{\nu^\alpha} d_{\mathrm{chem}}^{\beta} - \frac{n\Omega\nu^* + x\Omega}{\nu^\alpha} d_{\mathrm{chem}}^{\mathrm{Va}} + \frac{E^\alpha x^2 \Omega}{9n\left(1-\nu_{\mathrm{P}}^\alpha\right)} + 4\pi\left(\frac{3}{4\pi}n\Omega\right)^{\frac{2}{3}} \cdot \Delta G_{\mathrm{surf}}, \tag{2.96}$$

with the chemical contribution of vacancies to the chemical driving force given as

$$d_{\mathrm{chem}}^{\mathrm{Va}} = -\Delta G_{\mathrm{vol}}^{\mathrm{chem,va}} = RT \ln\left(\frac{X_{\mathrm{Va}}}{X_{\mathrm{Va}}^{\mathrm{eq}}}\right). \tag{2.97}$$

Equation (2.96) has two independent variables, the number of monomers in the nucleus, n, and the excess or shortage of vacancies, x, compared to the unstrained state. The critical nucleation energy is obtained at the extremum of this function after setting the first derivatives to zero and insertion back into equation (2.96). Accordingly, we have

$$\frac{\partial}{\partial x}\Delta G_{\mathrm{nucl}} = -\frac{\Omega}{\nu^\alpha} d_{\mathrm{chem}}^{\mathrm{Va}} + \frac{2xE^\alpha\Omega}{9n\left(1-\nu_{\mathrm{P}}^\alpha\right)} = 0, \tag{2.98}$$

with the solution

$$x^* = \frac{d_{\mathrm{chem}}^{\mathrm{Va}}}{\nu^\alpha} \cdot \frac{9n\left(1-\nu_{\mathrm{P}}^\alpha\right)}{2E^\alpha}. \tag{2.99}$$

Using this expression in the derivative with respect to n yields

$$\frac{\partial}{\partial n}\Delta G_{\mathrm{nucl}} = \frac{\partial}{\partial n}\left[\begin{array}{l} -\dfrac{n\Omega}{v^{\alpha}} d_{\mathrm{chem}}^{\beta} - \dfrac{n\Omega v^{*}}{v^{\alpha}} d_{\mathrm{chem}}^{\mathrm{Va}} + \dfrac{9n\Omega\left(1-v_{\mathrm{P}}^{\alpha}\right)}{2E^{\alpha}\left(v^{\alpha}\right)^{2}}\left(d_{\mathrm{chem}}^{\mathrm{Va}}\right)^{2} \\ + \dfrac{E^{\alpha}\Omega}{9n\left(1-v_{\mathrm{P}}^{\alpha}\right)} \cdot \left(\dfrac{9n\left(1-v_{\mathrm{P}}^{\alpha}\right)}{2E^{\alpha}\left(v^{\alpha}\right)^{2}}\right)^{2}\left(d_{\mathrm{chem}}^{\mathrm{Va}}\right)^{2} + 4\pi\left(\dfrac{3}{4\pi}n\Omega\right)^{\frac{2}{3}} \cdot \Delta G_{\mathrm{surf}} \end{array}\right]$$

$$= -\frac{\Omega}{v^{\alpha}} d_{\mathrm{chem}}^{\beta} - \frac{\Omega v^{*}}{v^{\alpha}} d_{\mathrm{chem}}^{\mathrm{Va}} + \frac{9\Omega\left(1-v_{\mathrm{P}}^{\alpha}\right)}{4E^{\alpha}\left(v^{\alpha}\right)^{2}}\left(d_{\mathrm{chem}}^{\mathrm{Va}}\right)^{2} + \frac{8}{3}\pi\left(\frac{3}{4\pi}\Omega\right)^{\frac{2}{3}} \cdot n^{-\frac{1}{3}} \cdot \Delta G_{\mathrm{surf}} = 0, \tag{2.100}$$

with the solution[27]

$$n^{*} = -\frac{32\pi}{3\Omega}\left(\frac{\Delta G_{\mathrm{surf}}}{\Delta G'_{\mathrm{vol}}}\right)^{3}, \tag{2.101}$$

and

$$\Delta G'_{\mathrm{vol}} = -\frac{d_{\mathrm{chem}}^{\beta}}{v^{\alpha}} - v^{*}\frac{d_{\mathrm{chem}}^{\mathrm{Va}}}{v^{\alpha}} + \frac{9\left(1-v_{\mathrm{P}}^{\alpha}\right)}{4E^{\alpha}\left(v^{\alpha}\right)^{2}}\left(d_{\mathrm{chem}}^{\mathrm{Va}}\right)^{2}. \tag{2.102}$$

The critical energy, G^{*}, for nucleation of a spherical particle of a new phase is finally given by

$$G^{*} = \frac{16\pi \cdot \gamma^{3}}{3 \cdot \left[-\dfrac{d_{\mathrm{chem}}^{\beta}}{v^{\alpha}} - v^{*} \cdot \dfrac{d_{\mathrm{chem}}^{\mathrm{Va}}}{v^{\alpha}} + \dfrac{9}{4}\dfrac{(1-v)}{E^{\alpha}} \cdot \left(\dfrac{d_{\mathrm{chem}}^{\mathrm{Va}}}{v^{\alpha}}\right)^{2}\right]^{2}}. \tag{2.103}$$

In interpreting this result, we identify several possibilities for the role of excess vacancies in the nucleation process. In the trivial case without vacancy supersaturation, $d_{\mathrm{chem}}^{\mathrm{Va}} = 0$, and the critical nucleation barrier expression reduces to the usual form for strain-free transformation given earlier in equation (2.75).[28]

If the precipitation system contains an excess of vacancies, we have $d_{\mathrm{chem}}^{\mathrm{Va}} > 0$. The second term in the denominator is then either negative or positive, depending on the sign of the volumetric misfit value, v^{*}. If the specific volume of the precipitate is larger than the matrix

[27] Compare also equation (2.83), which reads $n^{*} = -\dfrac{32\pi}{3\Omega}\left(\dfrac{\Delta G_{\mathrm{surf}}}{\Delta G_{\mathrm{vol}}}\right)^{3}$.

[28] Please notice that equation (2.75) also contains the elastic strain energy for coherent nucleation. For incoherent precipitates, this quantity is commonly taken as zero due to rapid stress relaxation by structural vacancy generation or annihilation at the incoherent interface.

value, we have $v^* > 0$. Excess vacancies then generate a negative energy contribution to the volume free energy change and the effective chemical driving force is larger than in the case without vacancy excess. For precipitates with larger molar volume, excess vacancies assist the nucleation process as they are annihilated simultaneously to the nucleation process and reduce the vacancy supersaturation, thus decreasing the total energy of the precipitation system. If the specific volume of the precipitate is smaller than the matrix, excess vacancies effectively reduce the chemical driving force and, thus, act against precipitate nucleation.

The third term in the denominator, describing the energy contribution from excess or shortage of vacancies, x, compared to the unstrained state, is always positive and, thus, hinders the nucleation process. With this term, the strain energy is taken into account that stems from the annihilation process of more vacancies[29] than needed for a stress-free transformation.

Finally, it is necessary to emphasize that the effect of excess vacancies on nucleation as described in this section occurs only in the course of *incoherent homogeneous* nucleation. Since this type of nucleation usually also requires the existence of appropriate heterophase fluctuations, which represent fluctuations in crystal structure, incoherent nucleation often takes place at heterogeneous nucleation sites, such as grain boundaries. In the vicinity of grain boundaries, the vacancy site fraction is commonly close to the equilibrium value and we, thus, hardly observe vacancy supersaturation. Application of equation (2.103) in nucleation kinetics simulations must, therefore, be reasonably well justified by experimental evidence.

2.4 Heterogeneous Nucleation

The nucleation of a new phase within the undisturbed crystal lattice of a parent phase can be substantially impeded by several factors. The two most relevant ones are the elastic stress field around a coherent precipitate with volumetric misfit and the requirement of heterophase fluctuations for nucleation of phases with a crystal structure different than the matrix. For some types of precipitates, these factors represent severe enough restrictions such that *homogeneous nucleation* is entirely suppressed.

Luckily, real materials are far from being perfect. Unless we are concerned with small single crystals that can exist as objects with virtually perfect periodicity, real solid-state matter offers several kinds of defects which act as *heterogeneous* nucleation sites. If clusters of a new phase are formed along these sites, this process is denoted as *heterogeneous nucleation.*[30]

---

[29] The annihilation of too many vacancies than needed for the stress-free transformation is driven by the vacancy supersaturation. The incoherent phase boundary represents a free surface, where the supersaturation of the matrix can be reduced, however, at the cost of creating misfit strain.

[30] Heterogeneous nucleation is not a phenomenon that is restricted to the solid state. It is well known from practical observation that heterogeneous objects, such as dust particles, can substantially accelerate the nucleation of liquid droplets, or even snow, from water-supersaturated air. The latter phenomenon often occurs in the vicinity of industrial production sites where an increased amount of air pollution is produced, which provides the heterogeneous nucleation sites for so-called *industrial snow*. Although the region around the production site does not experience real snowfall, the immediate area around the pollution site is often covered with a thin layer of snow nucleated heterogeneously on dust particles.

2.4.1 Heterogeneous Nucleation Sites

Assume that a cluster of a new phase is being formed in a crystalline matrix. If nucleation is made difficult due to elastic misfit stress or crystal structure complexities, heterogeneous lattice sites can assist the nucleation process with a partial compensation of the misfit stress by an opposing inherent stress field, by rapid stress relaxation made possible by the presence of ideal sources and sinks for vacancies, or by the reduction of system energy by replacement of interfacial area of the defect by the new phase.

Figure 2.16 shows a schematic representation of typical microstructural features for heterogeneous nucleation. Figure 2.16(a) represents a cubic lattice with a single vacancy and a divacancy.[31] Both defects lead to some distortion of the lattice and, therefore, an elastic stress field. In addition, two types of solute atoms are shown, which are either larger in size, or smaller, compared to the regular lattice atoms. This size misfit also leads to a distortion of the lattice and an elastic stress field, similar to that of vacancies. In real microstructures, vacancies often occur as nearest neighbor to a relatively larger atom, thus forming a vacancy–solute pair, where the stress field of the larger component compensates, in part, the stress field of the smaller one. Single solute atoms as well as monovacancies are often denoted as *zero-dimensional defects*, as they only occupy one single lattice site.

The microstructure displayed in Figure 2.16(b) contains a single edge dislocation. The compressive and tensile stress fields around the dislocation are efficient heterogeneous nucleation sites for precipitates with both larger and smaller molar volume. This is discussed in more detail subsequently. As dislocations extend linearly in one spatial dimension, they are denoted as *one-dimensional lattice defects*.

Figure 2.16(c) shows a typical high-angle grain boundary where the lattice orientations of the two neighboring grains impair any coherency of lattice planes. The grain boundary itself is clearly visible as the region without periodic atomic bonding between the two crystals.[32] It is widely accepted that the disturbance of crystal order in the grain boundary has a thickness of a few atomic layers. The structure of the grain boundary itself is often viewed as being amorphous instead of crystalline, therefore providing ideal conditions for nucleation of clusters with high volumetric misfit (stress relaxation by rapid vacancy condensation or evaporation) as well as clusters with largely different crystal structure. In the amorphous grain boundary region, no severe restrictions in terms of coherency exist and nucleation is thus strongly facilitated. Grain boundaries are planar objects, which is why they are commonly denoted as *two-dimensional lattice defects*.

---

[31] It is important to recognize that vacancies are equilibrium features of a real microstructure. Although the missing atomic bonds increase the total energy of the system, the entropy of mixing of vacant lattice sites compensates this effect. This issue is treated in more detail in Chapter 7 on diffusion.

[32] This picture might be misleading to some extent, as it suggests that there exist no, or only weak, bonds across the interface. In Chapter 4, dealing with the interfacial energy, however, it will be shown that atomic interactions are not restricted to first and second nearest neighbors. Instead, the bonds extend rather far into the region across the interfaces. Otherwise, polycrystalline materials would never be able to withstand severe mechanical loading, which they obviously do in practical experience. Fine-grained metals usually show an excellent combination of high strength and high ductility, although they contain plenty of interfacial area.

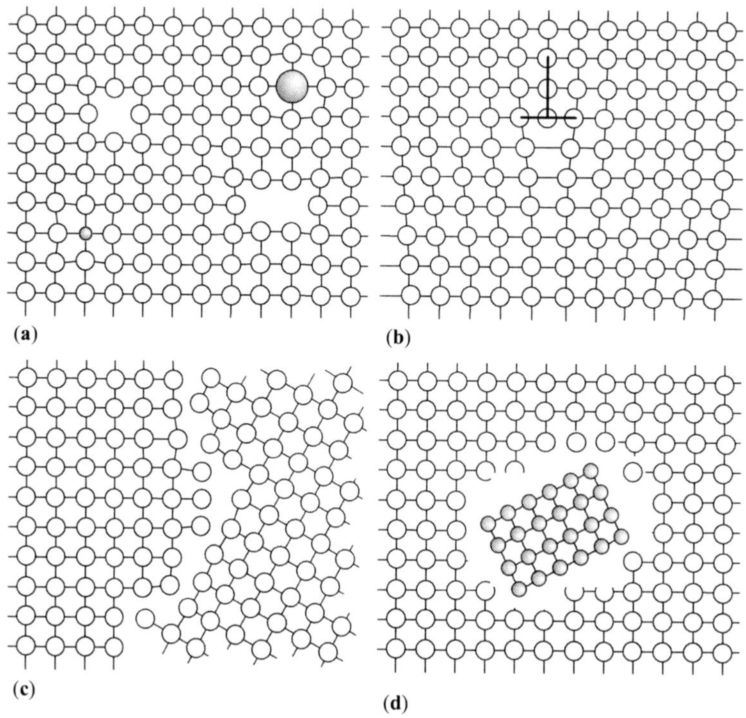

Figure 2.16. Typical lattice defects representing heterogeneous nucleation sites in a crystalline microstructure. Mono and divacancy, larger and smaller atom (a), edge dislocation (b), high-angle grain boundary (c), and incoherent inclusion (d).

Finally, Figure 2.16(d) sketches a typical *three-dimensional lattice defect* represented by an incoherent inclusion. In metal processing, these objects are often unwanted remnants of the production process. They are typically oxides, silicates, aluminates, and so forth, which are introduced during the casting process or during welding operations. Inclusions often act as heterogeneous nucleation sites for other types of precipitates or phases. They can be introduced into the material on purpose, for instance, to stimulate phase transformations, or accidentally, as a consequence of imperfect production conditions. Unwanted, hard inclusions are typically problematic in applications where dynamic loading is a major design criterion, since they cause stress concentrations and, therefore, represent preferential nucleation sites for cracks.

The major mechanisms of heterogeneous nucleation on high-angle grain boundaries and incoherent phase boundaries are the amorphous nature of the interface facilitating heterophase fluctuations as well as rapid stress relaxation by structural vacancies. The dominating feature of (edge) dislocations is the presence of compressive and tensile stress along the dislocation core. Precipitates with positive volume mismatch (see Figure 2.17) preferentially form at the tensile part of the stress field, whereas precipitates with negative volume mismatch preferentially nucleate in the compressive region.

Quantification of the energetic benefit of heterogeneous nucleation in a general approach is difficult, as, for instance, grain boundaries exist with a wide spectrum of properties and energies depending on the misorientation of the two neighboring grains. The difficulty with dislocations is the fact that the stress field has only limited spatial extension and strongly interacts with the nucleus of the new phase. In practical simulations, it is often convenient to assume zero

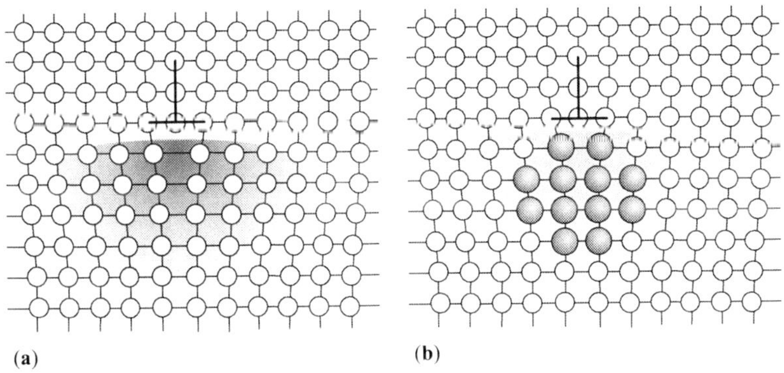

Figure 2.17. Schematic representation of a tensile stress field (dark background color) at the tip of an edge dislocation (a) and partial compensation of the tensile stress field by a cluster of atoms with larger atomic radius (b).

volumetric misfit during grain boundary nucleation, and a significant reduction in misfit strain for nucleation at dislocations. Models for the incorporation of heterogeneous nucleation site energies into an *effective* interfacial energy are outlined in Section 2.4.4.

2.4.2 Potential Nucleation Sites in a Heterogeneous Microstructure

The general nucleation theory developed by Frenkel [29], which is adopted here in the treatment of solid-state nucleation, clearly emphasizes the direct proportionality between the steady-state nucleation rate, J_S, with the exponential term containing the nucleation barrier, G^*. One important proportionality factor in this relation, equation (2.76), is the number of lattice sites, N_0, where nucleation events can potentially occur. This parameter is investigated in this section.

When considering homogeneous nucleation in a crystalline material, it is straightforward to assume that each single atom in the structure represents a potential nucleation site. From knowledge of the molar volume of the matrix phase α, the number of potential homogeneous nucleation sites, N_{hom}, in unit volume is obtained as

$$N_{\mathrm{hom}} = \frac{N_{\mathrm{A}}}{v^{\alpha}}, \qquad (2.104)$$

with N_{A} being the Avogadro number and v^{α} the molar volume. Typical values for N_{hom} in some metallic materials are summarized in Table 2.1.

As briefly outlined already in the previous section, the homogeneous nucleation of precipitates with nonzero volume misfit in the matrix is often suppressed in favor of heterogeneous nucleation sites. Apart from grain boundaries and incoherent phase boundaries, which will be discussed subsequently, the compressive and tensile stress fields around dislocations represent ideal nucleation sites for precipitates with small to medium volume mismatch. The stress field of the precipitate is partially compensated by the stress field of the dislocation and nucleation can thus be substantially facilitated.[33]

[33] See also Section 2.3.5 where the elastic energy around mismatching coherent spherical inclusions is treated.

Table 2.1. Number of atoms in unit volume for estimation of the number of potential homogeneous nucleation sites in various crystal lattices at room temperature.

Element	Structure	Molar volume/m^{-3}	Atoms/m^3
Al	fcc	9.99×10^{-6}	8.61×10^{28}
Fe	bcc	7.09×10^{-6}	8.49×10^{28}
Fe	fcc	6.58×10^{-6}	9.15×10^{28}
Ni	fcc	6.59×10^{-6}	9.17×10^{28}
Mo	bcc	10.60×10^{-6}	5.68×10^{28}
Ti	bcc	9.33×10^{-6}	6.45×10^{28}

Table 2.2. Potential nucleation sites, N_{disl}, in Fe-base alloys as a function of dislocation density, ρ, and conditions for which these values typically apply.

ρ/m^{-2}	Typical for ...	N_{disl}/m^3
10^{11}	well-annealed fcc	4.4×10^{20}
10^{12}	well-annealed bcc	4.4×10^{21}
10^{13}	hot deformation (low $\dot{\varepsilon}$)	4.4×10^{22}
10^{14}	hot deformation (high $\dot{\varepsilon}$)	4.4×10^{23}
10^{15}	medium-C martensite	4.4×10^{24}
10^{16}	fresh martensite (high-C)	4.4×10^{25}

When estimating the number density of heterogeneous nucleation sites at dislocations, in a first approximation, we assume that each single atom along the dislocation core represents a potential nucleation site. With the dislocation density, ρ, the nucleation site density, N_{disl}, is given as

$$N_{disl} = \frac{\rho}{a}, \tag{2.105}$$

where a is the mean atomic distance. Some representative values of N_{disl} are summarized in Table 2.2.

The potential number of nucleation sites at grain boundaries can be estimated on the basis of a suitable and convenient geometrical model representing the polycrystalline grain morphology. In Figure 2.18, scanning electron microscopy (SEM) images of grain morphologies of fractured polycrystalline microstructures are shown, as observed typically after tensile testing or component failure.

There exist several classes of geometrical objects that can resemble the appearance of polycrystalline microstructures with more or less accuracy. As outlined by Lord Kelvin (Sir William Thomson) [33] already in 1887, however, any such object must fulfill several criteria to be physically plausible and useful in practical application. For instance, any such

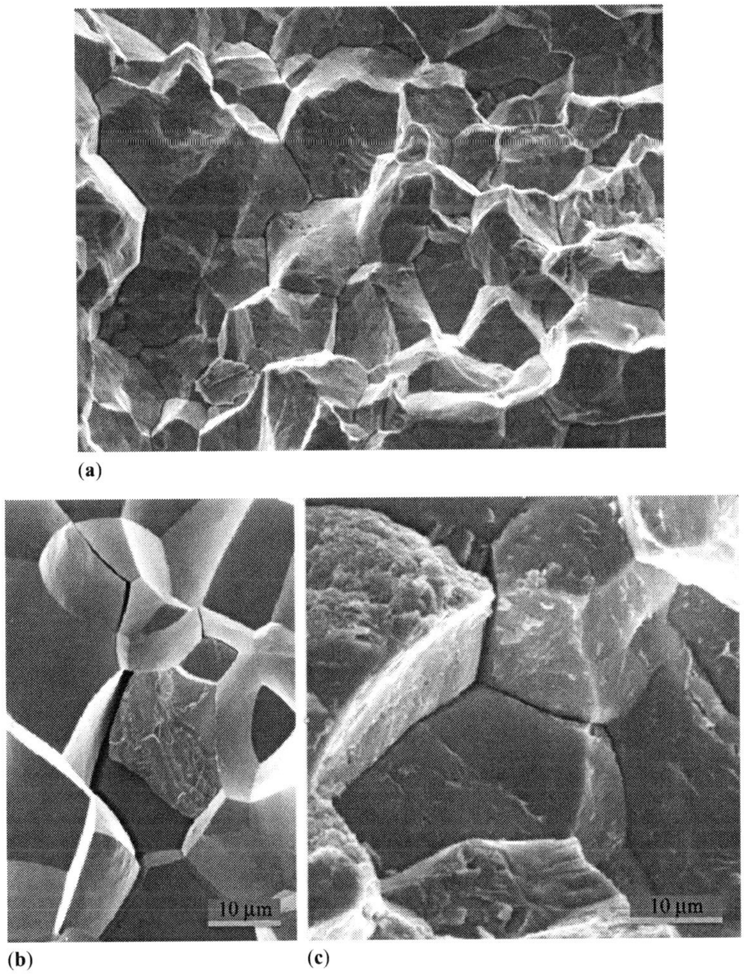

Figure 2.18. Scanning electron microscopy images of (fractured) polycrystalline microstructure of Cu (a) and low-alloy steel (b,c). Courtesy of Ch. Zaruba, Vienna University of Technology (2011) and B. Sonderegger, Graz University of Technology (2006).

geometrical object must be space filling, which means that the periodic arrangement in a three-dimensional array must exactly fill the total specimen volume without overlapping or leaving uncovered regions. Many simple objects, such as cubes, tetrahedra, or more complex geometrical bodies, fulfill this first requirement. However, for a realistic picture of practically observed microstructures, some additional constraints apply. One of these is the requirement of minimal total energy, which can be satisfied by configurations with minimum separating area between the individual grains.[34] The number of potential geometrical figures is significantly reduced by this condition. Still, several potential candidates exist that fulfill these first two conditions. Sir William Thomson, in his discussion, emphasizes that another constraint must hold, which is the equilibrium between three surfaces being in contact in one single

[34] This condition applies strictly only if the interfacial energy is isotropic.

line. In each point, where the faces of the body meet, the angles between the planes must be 120°, otherwise there would exist some net pressure on the interface making the configuration unstable.

An object that fulfills all three requirements is the tetrakaidecahedron, also known as the *Kelvin cell.* This object is

i. space filling,
ii. has minimum separating area, and
iii. has 120° angles between the faces.

The tetrakaidecahedron is a body with eight hexagons and six quadrilaterals. Interestingly, the faces of the Kelvin cell are not exactly planar but slightly curved, in order to meet condition (iii). In the present treatment, for the benefit of simplicity, we nevertheless consider a version of the tetrakaidecahedron with planar faces. The 14-sided space-filling structure is shown in Figure 2.19, both, as a single cell and in a multicell arrangement.

The geometrical parameters describing the extension of the tetrakaidecahedron are shown in Figure 2.20. Figure 2.20(a) shows the classical equiaxed tetrakaidecahedron with the single parameter, *d*, describing uniquely its spatial extension. The dash-dotted line in this figure indicates the symmetry axis.

Interestingly, the equiaxed tetrakaidecahedron can easily be geometrically elongated in one spatial dimension by insertion of an octagonal prism. This operation has been proposed, for instance, by J. Rajek [34]. His treatment is reproduced here. In the modified geometry, the tetrakaidecahedron is capable of simultaneously representing equiaxed and elongated structures within one geometrical figure. This is a convenient property which can be used to describe the features of stretched grains, for instance, after wire drawing or other forming operations producing elongated grains. The modified tetrakaidecahedron has also proved to be particularly useful when dealing with martensitic substructures as observed, for instance, in quenched steel [34]. A corresponding sketch is depicted in Figure 2.21. The complex microstructure of these materials can be conveniently represented by a combination of large equiaxed tetrakaidecahedra for the prior austenite grain structure and elongated tetrakaidecahedra representing the subgrains of the martensite lath packet structure.

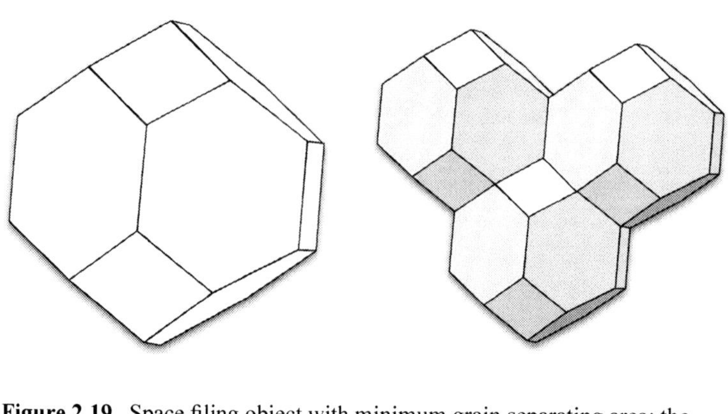

Figure 2.19. Space filing object with minimum grain separating area: the tetrakaidecahedron.

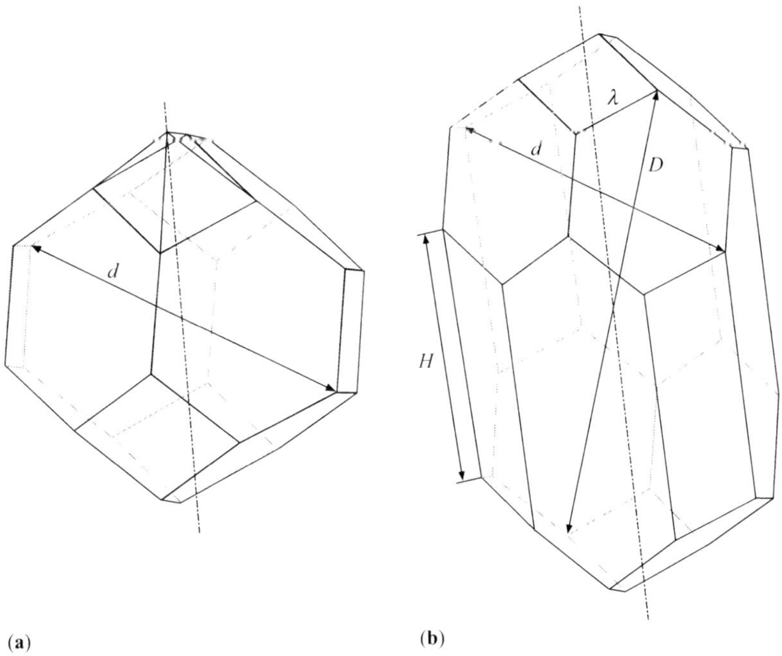

Figure 2.20. Geometrical parameters describing the equiaxed (a) and the elongated (b) tetrakaidecahedron.

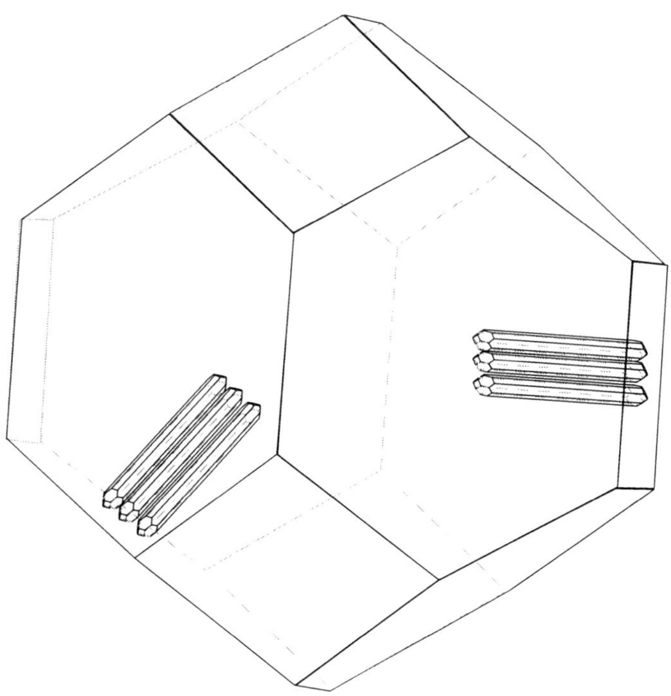

Figure 2.21. Geometrical representation of martensitic substructure in steel using elongated tetrakaidecahedra.

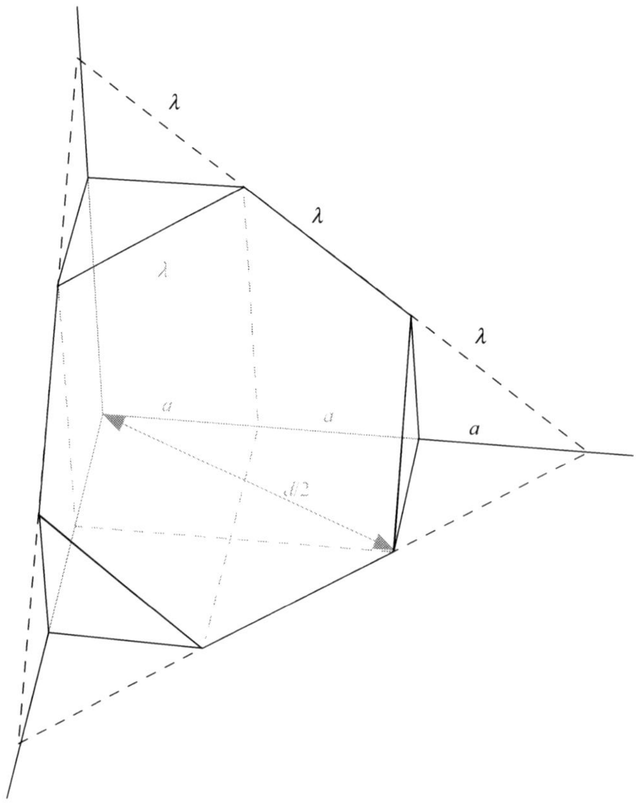

Figure 2.22. On the geometrical properties of a tetrakaidecahedron.

For further analysis of the geometrical properties of the tetrakaidecahedron, it is advantageous to extract a symmetric portion of the tetrakaidecahedron, first. This is shown in Figure 2.22 where one-eighth of the entire figure is plotted together with the geometrical variables d, a, and λ. From the figure, we can identify the following basic relations:

$$\lambda = \frac{d}{\sqrt{10}}, \quad \lambda = \sqrt{2} \cdot a \quad \text{and} \quad a = \frac{d}{\sqrt{20}}. \tag{2.106}$$

For representation of elongated grains, as shown in Figure 2.20(b), it is convenient to define the ratio between d and D in the form of an elongation factor, s, with

$$s = \frac{D}{d}. \tag{2.107}$$

On stretching the equiaxed tetrakaidecahedron, two half-cells are created with a prism located in between them. The height of the prism can be computed from the relation

$$H = 2\left[\sqrt{\left(\frac{D}{2}\right)^2 - \frac{d^2}{20}} - \frac{d}{\sqrt{5}}\right]. \tag{2.108}$$

With these geometric relations, we can proceed to evaluate the basic parameters describing a polycrystalline microstructure being composed of equally sized tetrakaidecahedra. The number of tetrakaidecahedral grains, n_G, filling unit volume of matter is obtained from the relation

$$n_G = \frac{1}{\bar{V}}, \tag{2.109}$$

with $\bar{V}$ being the volume of a single grain.

From the basic geometrical relations identified, the volume of a single tetrakaidecahedral grain, $\bar{V}$, is found to be[35]

$$\bar{V} = V = 8\sqrt{2} \cdot \lambda^3 + 6\lambda^2 \cdot H. \tag{2.110}$$

In the present nomenclature, the quantities carrying a bar correspond to isolated grains, whereas the quantities without bar correspond to the same quantities in the multigrain ensemble.

The total grain boundary area in a polycrystalline microstructure can be calculated from the surface area of an isolated grain, $\bar{A}$, multiplied with the total number of grains, n_G, and divided by a factor 2 taking into account that each face is shared by two grains. Accordingly, we have

$$\bar{A} = 6\left(1 + 2\sqrt{3}\right)\lambda^2 + 2\lambda(2H + \lambda), \tag{2.111}$$

and

$$A = \frac{n_G}{2}\bar{A}. \tag{2.112}$$

Analogously, the total length of grain boundary edges, $\bar{L}$, corresponding to an isolated grain is

$$\bar{L} = 40\lambda + 8H \tag{2.113}$$

and the total grain boundary edge length in the polycrystalline microstructure is

$$L = \frac{n_G}{3}\bar{L}, \tag{2.114}$$

as each edge is shared by three grains.

Finally, the number of grain corners of an isolated grain is

$$\bar{C} = 24, \tag{2.115}$$

and the total number of grain boundary corners in a polycrystalline structure is

$$C = \frac{n_G}{4}\bar{C}, \tag{2.116}$$

as each corner is shared by four grains.

---

[35] The value of H is zero for equiaxed grains and nonzero for elongated grains.

Table 2.3. Typical number of potential homogeneous (bulk) and heterogeneous nucleation sites in a polycrystalline microstructure.

Typical values for ...	$N_{\mathrm{nucl}}/\mathrm{m}^3$
bulk nucleation fcc	9.0×10^{28}
dislocations (fresh martensite)	4.4×10^{25}
dislocations (hot-deformed fcc)	4.4×10^{23}
dislocations (well-annealed fcc)	4.4×10^{20}
grain boundaries (10 μm)	7.8×10^{24}
grain boundaries (100 μm)	7.8×10^{23}
subgrain boundaries (1 μm, $s = 100$)	2.1×10^{25}
subgrain boundaries (10 μm, $s = 100$)	2.1×10^{24}
grain boundary edges (10 μm)	5.2×10^{20}
grain boundary edges (100 μm)	5.2×10^{18}
grain boundary corners (10 μm)	1.7×10^{16}
grain boundary corners (100 μm)	1.7×10^{13}

With the relations discussed, it is straightforward to estimate the density of potential heterogeneous nucleation sites in a polycrystalline microstructure. After normalizing these expressions to the atomic distance, a, we finally obtain

$$N_{\mathrm{GB}} = \frac{A}{a^2}, \quad N_{\mathrm{GBE}} = \frac{L}{a}, \quad N_{\mathrm{GBC}} = C. \tag{2.117}$$

Typical values of the number of potential heterogeneous nucleation sites along grain boundaries are summarized in Table 2.3 together with values obtained at dislocations and for homogeneous nucleation.

2.4.3 Nucleation Site Saturation

Nucleation is governed by random heterophase fluctuations. Individual nucleation events occur on a more or less random basis, both in position as well as time. Once a supercritical nucleus has formed, the deterministic laws of particle growth govern its further evolution. Attachment and detachment probabilities of monomers are then widely different and the vast majority of atoms arriving at the precipitate will remain attached to it. As a consequence, a diffusive flux occurs toward the precipitate and the vicinity of the precipitate becomes depleted of solute atoms forming the precipitate. This issue is dealt with in detail in Chapter 3.

In nucleation theory, the surrounding of a critical nucleus is treated as being unaffected by the nucleation process. No concentration profile is assumed to evolve around it. However, this assumption is valid only for a single, isolated nucleation event. For nucleation in sufficiently large systems, many individual nuclei are formed at different times and the nucleation stage of one particle can overlap significantly with the growth stage of another one. Consequently, nucleation and growth of precipitates must be treated as two processes that intersect with each

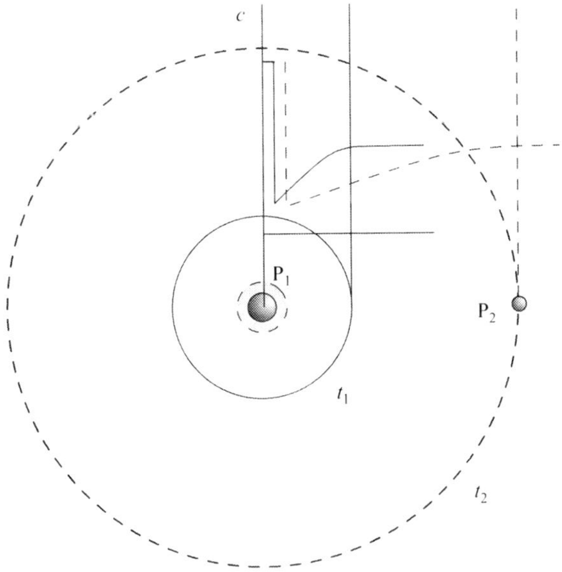

Figure 2.23. Overlapping diffusion field and concentration profile of precipitate, P_1, with nucleation of particle, P_2. See text.

other and occur at the same time. In most precipitation systems, the overlap between the two mechanisms is substantial.

Consider a precipitate, P_1, that has nucleated and grown until time, t_1. This precipitate is shown in the center of Figure 2.23 together with the concentration profile that has developed around this particle. Precipitate and diffusion field are shown with solid lines. A second particle, P_2, is about to nucleate in the vicinity of P_1. If this second nucleation event occurs at t_1, it will be entirely unaffected by the growth of P_1. However, if the event has not taken place until time t_2, the diffusive flux toward P_1 gradually decreases the amount of solute available to form P_2 and the probability that the nucleation event takes place after t_2 decreases rapidly. In fact, the entire spherical diffusion field of P_1 might be considered as a region where nucleation of additional precipitates becomes rather unlikely. This fact must be taken into account properly in the evaluation of the total number of potential nucleation sites in the nucleation rate expression (2.76).

A pragmatic approach to this effect is realized in the software *MatCalc*,[36] where the precipitate radius is multiplied with a constant factor (typically in the order of 5) and the number of potential nucleation sites covered by this "extended" precipitate is evaluated. The total number of available nucleation sites is then reduced by this number and thus the effect of site saturation is accounted for.

2.4.4 Effective Interfacial Energies in Heterogeneous Nucleation

In the treatment of nucleation thus far, we have considered the volume free energy change due to the transformation of bulk volume of the parent phase to the structure and composition of the transformation phase, and the energy which is necessary to create the new interphase area.

[36] See Section 9.3.

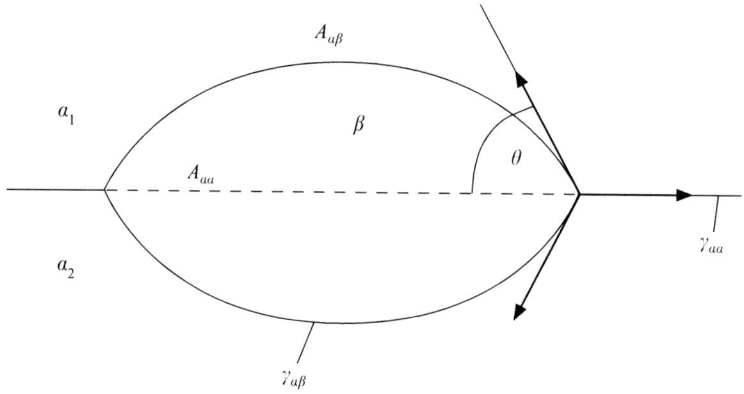

Figure 2.24. Equilibrium shape of precipitates at a planar grain boundary determined by the force balance of interfacial energies.

In heterogeneous nucleation, however, simultaneously, some part of the heterogeneous nucleation region is eliminated by the new phase, thus, releasing some energy and aiding to the formation of the new phase.

For instance, if nucleation occurs at dislocations, the volume of the new phase replaces part of the dislocation core volume, releasing the corresponding amount of dislocation line energy, Γ, with Γ being roughly given as

$$\Gamma = \frac{1}{2} G b^2 . \tag{2.118}$$

G is the shear modulus of the matrix phase and b is the Burger's vector. On formation of a precipitate with radius, r, the dislocation core region is eliminated[37] along a length of $2r$ and the total gain in dislocation energy is

$$\Delta G_{\text{disl}} = G b^2 r . \tag{2.119}$$

For nucleation of precipitates at grain boundaries, the energy balance considerations are similar to that of nucleation at dislocations, except that, now, the new phase eliminates some portion of the grain boundary area. Clemm and Fisher [35] have analyzed this issue quantitatively and the essence of their treatment is recapitulated here.

If nucleation of precipitates occurs at grain boundaries, the equilibrium shape of the particles usually deviates from that of spherical precipitates. At grain boundaries, the shape of precipitates is often lens-shaped due to a tensile force exerted by the grain boundary on the equator of the precipitate (see Figure 2.24).

[37] The term "eliminated" applies only to situations where the precipitate nucleates incoherently and the dislocation is entirely replaced by the new phase. In many practical cases, nucleation occurs coherently even at dislocations and the effective energy gain is less than the "ideal" value.

If all forces are in equilibrium, the tensile force from the grain boundary with (isotropic) interfacial energy, $\gamma_{\alpha\alpha}$, exactly balances twice the projection of the tensile force from the phase boundary, $\gamma_{\alpha\beta}$. This condition is expressed mathematically by the equation

$$\cos\theta = \frac{\gamma_{\alpha\alpha}}{2\gamma_{\alpha\beta}}. \tag{2.120}$$

Let the grain boundary area, $A_{\alpha\alpha}$, which is replaced by the nucleating precipitate, the cap surface area, $A_{\alpha\beta}$, and the total volume of the lens, V, be

$$A_{\alpha\alpha} = ar^2,\ A_{\alpha\beta} = br^2, \text{ and } V = cr^3, \tag{2.121}$$

with the unknown coefficients a, b, and c describing the grain boundary junctions between two, three, or four contacting grains.

The work done in forming a nucleus of volume, V, can then be generalized by taking into account that the precipitate eliminates some part of the grain boundary area with

$$\Delta G_{\text{nucl}} = A_{\alpha\beta}\gamma_{\alpha\beta} - A_{\alpha\alpha}\gamma_{\alpha\alpha} + V\Delta G_{\text{vol}}, \tag{2.122}$$

where the identity holds $\gamma_{\alpha\beta} = \Delta G_{\text{surf}}$. With the relations (2.121), we further have

$$\Delta G_{\text{nucl}} = br^2\gamma_{\alpha\beta} - ar^2\gamma_{\alpha\alpha} + cr^3\Delta G_{\text{vol}}. \tag{2.123}$$

The critical nucleation energy, G^*, is obtained from the calculus of the extremal value of ΔG_{nucl} as

$$\frac{\partial \Delta G_{\text{nucl}}}{\partial r} = 2br\cdot\gamma_{\alpha\beta} - 2ar\cdot\gamma_{\alpha\alpha} + 3cr^2\Delta G_{\text{vol}} = 0. \tag{2.124}$$

The critical energy, G^*, is finally found as

$$G^* = \frac{4}{27}\frac{\left(b\cdot\gamma_{\alpha\beta} - a\cdot\gamma_{\alpha\alpha}\right)^3}{\left(c\cdot\Delta G_{\text{vol}}\right)^2}. \tag{2.125}$$

The task that is left, now, is to evaluate the coefficients a, b, and c for the desired precipitation geometry, where we notice that a, b, and c are functions of the grain boundary energy, $\gamma_{\alpha\alpha}$, and the interfacial energy, $\gamma_{\alpha\beta}$, according to equation (2.120).

In first instance, consider a precipitate located on the contact plane between two grains α_1 and α_2. This configuration corresponds to a planar *two-grain junction*. The equilibrium shape of this precipitate is a lens with a total volume being described by the intersecting volume of two spheres, as shown in Figure 2.25.

For the eliminated grain boundary area, we obtain

$$A_{\alpha\alpha} = ar^2 = \pi\sin^2\theta r^2, \tag{2.126}$$

and the coefficient, a, becomes

$$a = \pi\sin^2\theta. \tag{2.127}$$

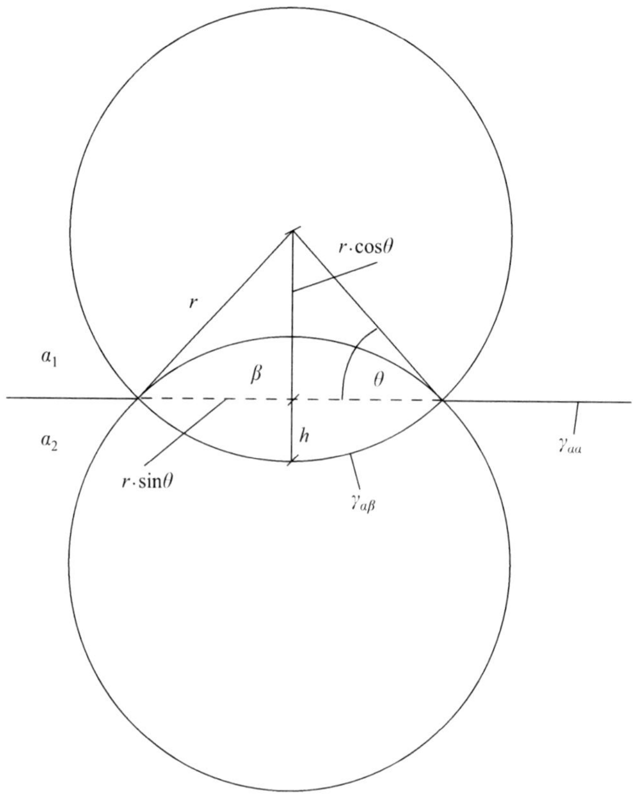

Figure 2.25. Lens-shaped precipitate at a grain boundary (two-grain junction).

Analogously, the surface area of the spherical cap is

$$A_{\alpha\beta} = br^2 = 2(2\pi r)h = 2\pi(1-\cos\theta)r^2, \tag{2.128}$$

with the coefficient, b, given as

$$b = 4\pi(1-\cos\theta). \tag{2.129}$$

For the volume of the two-grain junction precipitate, one obtains

$$V = cr^3 = \pi h^2\left(r - \frac{h}{3}\right) = \frac{2\pi}{3}\left(2 - 3\cos\theta + \cos^2\theta\right)r^3. \tag{2.130}$$

The coefficient, c, is then given with

$$c = \frac{2\pi}{3}\left(2 - 3\cos\theta + \cos^2\theta\right). \tag{2.131}$$

Clemm and Fisher have presented this type of analysis for three- and four-grain junctions also, however, obtaining considerably more complex formulae. These are summarized in Table 2.4 where the parameter k has been introduced with

Table 2.4. Coefficients *a*, *b*, and *c* for evaluation of the effective interfacial energy of grain boundary precipitates [35].

Nucleus location	**Coefficients**
Two-grain junction 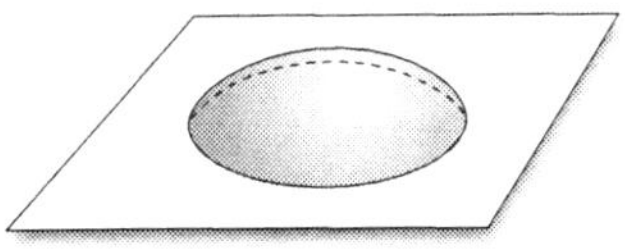	$a=\pi\left(1-k^2\right)$ $b=4\pi\left(1-k\right)$ $c=\frac{2\pi}{3}\left(2-3k+k^2\right)$
Three-grain junction 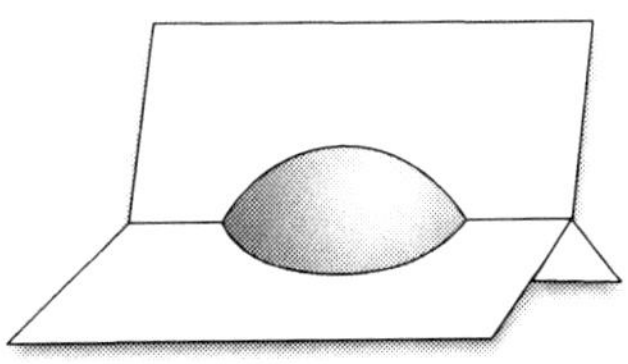	$a=3\beta\left(1-k^2\right)-k\left(3-4k^2\right)^{\frac{1}{2}}$ $b=12\left(\frac{\pi}{2}-\alpha-k\beta\right)$ $c=2\left(\pi-2\alpha+k^2\left(3-4k^2\right)^{\frac{1}{2}}-\beta k\left(3-k^2\right)\right)$ with $\alpha=\arcsin\frac{1}{2\left(1-k^2\right)^{\frac{1}{2}}}$ $\quad\beta=\arccos\frac{k}{\left[3\left(1-k^2\right)\right]^{\frac{1}{2}}}$
Four-grain junction 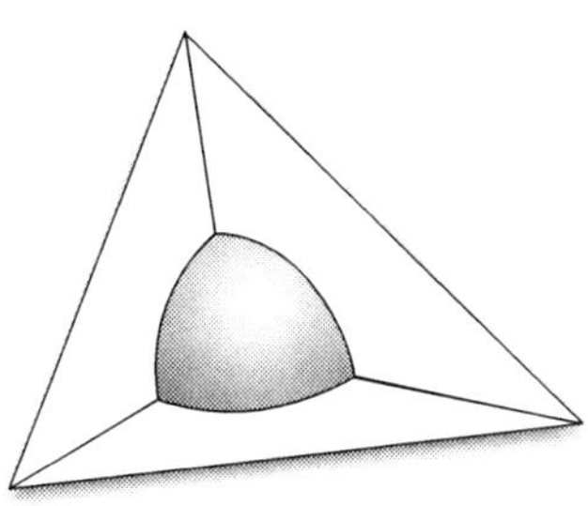	$a=3\left\{2\phi\left(1-k^2\right)-K\left[\left(1-k^2-\frac{K^2}{4}\right)^{\frac{1}{2}}-\frac{K^2}{\sqrt{8}}\right]\right\}$ $b=24\left(\frac{\pi}{3}-k\phi-\delta\right)$ $c=2\left\{4\left(\frac{\pi}{3}-\delta\right)+kK\left[\left(1-k^2-\frac{K^2}{4}\right)^{\frac{1}{2}}-\frac{K^2}{\sqrt{8}}\right]-2k\phi\left(3-k^2\right)\right\}$ with $K=\frac{4}{3}\left(\frac{3}{2}-2k^2\right)^{\frac{1}{2}}-\frac{3}{2}k$ $\phi=\arcsin\frac{K}{2\left(1-k^2\right)^{\frac{1}{2}}}$ $\quad\delta=\arccos\frac{\sqrt{2}-k\left(3-K^2\right)^{\frac{1}{2}}}{K\left(1-k^2\right)^{\frac{1}{2}}}$

$$k = \cos\theta = \frac{\gamma_{\alpha\alpha}}{2\gamma_{\alpha\beta}}. \tag{2.132}$$

For convenience, an effective interfacial energy, γ', can be introduced by comparing the expression for the nucleation barrier, G^*, given by Clemm and Fisher, equation (2.125), with the usual form as given by

$$G^* = \frac{16\pi}{3}\frac{(\gamma')^3}{(\Delta G_{\text{vol}})^2}. \tag{2.133}$$

For γ', one obtains

$$\gamma' = \left(\frac{1}{36\pi}\right)^{\frac{1}{3}} \cdot \frac{b\gamma_{\alpha\beta} - a\gamma_{\alpha\alpha}}{c^{\frac{2}{3}}}, \tag{2.134}$$

which can straightforwardly be utilized in the CNT framework for evaluation of nucleation barriers.

One critical issue in a quantitative treatment of nucleation on grain boundaries is the knowledge of the interfacial energy, $\gamma_{\alpha\beta}$, and the grain boundary energy, $\gamma_{\alpha\alpha}$. A model for estimation of the former quantity, $\gamma_{\alpha\beta}$, is presented in Chapter 4, where the precipitate–matrix interfacial energy is dealt with in detail. For the latter quantity, the grain boundary energy, $\gamma_{\alpha\alpha}$, the situation is less clear and one has to mainly rely on the available experimental data or data from atomistic simulations. In the next section, this issue is briefly reviewed.

Based on the treatment of Clemm and Fisher, in Figure 2.26, the ratio between the coherent, homogeneous interfacial energy and the effective interfacial energies, γ'/γ, is evaluated for different values of the grain boundary energy. For zero grain boundary energy, $\gamma_{\alpha\alpha} = 0$, the effective interfacial energy recovers the value of the coherent, homogeneous case. For finite grain boundary energy values, the effective interfacial energy, γ', decreases continuously for all three cases of two-grain, three-grain, and four-grain junctions. Consequently, the higher the effect of grain boundaries on nucleation is, the higher the grain boundary energy is, and the lower the precipitate–matrix interfacial energy.

In many metallic systems, typical values of the grain boundary energy range from 0.3 to 1.5 J/m^2 (see Section 2.4.5). In phase transformation systems, where the precipitate–matrix interfacial energy is clearly below these values, we find a very strong preferential nucleation at grain boundaries compared to dislocations or homogeneous nucleation. In this context, a typical example of wide technological relevance is the Fe–C system (unalloyed steel), where austenite (fcc-Fe) transforms into ferrite (bcc-Fe) below approximately 800–900°C. At low C-contents, the interfacial energy between fcc- and bcc-Fe is rather small, $\gamma_{\alpha\beta} << 0.2$, and experiment shows that nucleation occurs almost instantaneously after passing the austenite/ferrite phase boundary. Closer investigation of this phenomenon [36] has demonstrated that this effect becomes pronounced at very low C-content, where the growth rate of the ferrite can adopt very high values. Cyclic transformation from austenite to ferrite leads to a steady-state ferrite grain size, indicating that nucleation of the ferrite grains occurs predominantly at the

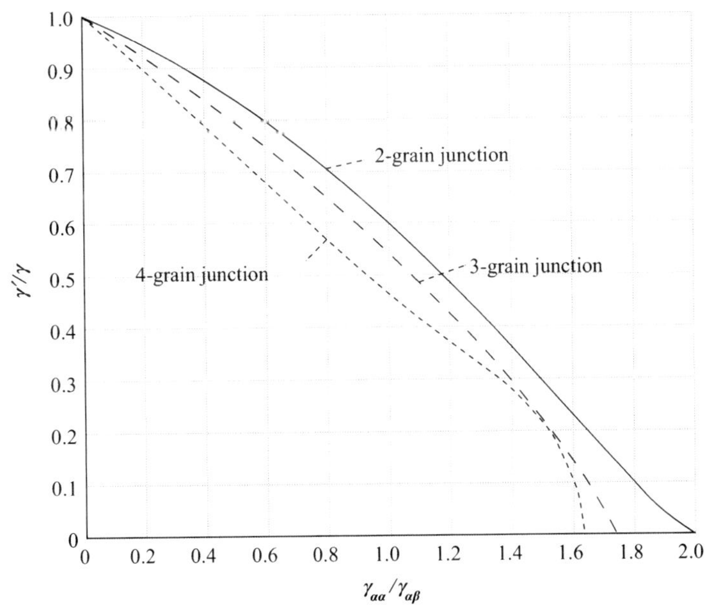

Figure 2.26. Dependence of the effective interfacial energy, γ', from the ratio between grain boundary energy, $\gamma_{\alpha\alpha}$, and precipitate–matrix interfacial energy, $\gamma_{\alpha\beta}$, at two-, three-, and four-grain junctions. Values according to Table 2.4 and equation (2.134).

austenite four-grain junctions and the total number of grains is approximately conserved during the repeated phase transformations. This effect becomes less pronounced with increasing C-content, as the interface velocity becomes lower in this case, and nucleation on three-grain and two-grain junctions becomes competitive at higher supersaturation.

The three curves in Figure 2.26 reach a value of zero at critical values of $\gamma'/\gamma = 2$ for two-grain junctions, $\gamma'/\gamma = 1.732$ for three-grain junctions, and $\gamma'/\gamma = 1.634$ for four-grain junctions. Below these values, a nucleus of the new phase can form instantaneously without any nucleation barrier to be overcome. The nucleation and growth rates in the early stages of such transformation are then mainly controlled by the availability of grain boundary area and the intrinsic mobility of the phase boundary.

2.4.5 Grain Boundary Energy

Grain boundaries are commonly distinguished into low-angle grain boundaries (LAGB) and high-angle grain boundaries (HAGB), each of them showing very particular features. Low-angle boundaries can be viewed as being composed of a periodic array of dislocations. Up to a tilt angle of approximately 15° the individual dislocations of a low-angle boundary can be clearly distinguished. The individual dislocations are widely enough spaced such that their stress fields do not severely overlap.

The energy of a LAGB is described by the Read–Shockley [37] model where the energy of the grain boundary, $\gamma_{\alpha\alpha}^{\mathrm{LA}}$, is evaluated as a function of the tilt angle, Θ, with

$$\gamma_{\alpha\alpha}^{\mathrm{LA}} = \gamma_{\alpha\alpha}^{0}\Theta\left(A - \ln\Theta\right), \tag{2.135}$$

where A is a constant in the order of 0.23, and $\gamma_{\alpha\alpha}^{0}$ is a function depending on the mechanical properties of the materials as

$$\gamma_{\alpha\alpha}^{0} = \frac{G}{4\pi\left(1-\nu_{\mathrm{P}}\right)}. \tag{2.136}$$

G is the shear modulus and ν_{P} is Poisson's ratio.

For HAGBs, the situation is much less clear. The interface of a HAGB is often treated as an amorphous region without evidence of an orientation relationship between the two contacting grains. Special orientation relationships are possible, though, where periodic coincidence between atoms exists. For these special boundaries, the grain boundary energy can be drastically reduced due to particular symmetry relationships. Figure 2.27 sketches the typical shape of the grain boundary energy as a function of the tilt angle. In the case of pure metals (dashed line), the special coincident lattice site orientations are clearly visible as pronounced energy minima.

In alloys, the grain boundary energy values are generally lower due to segregation of impurities into the grain boundary (solid line). These elements effectively reduce the energy of the grain boundary. Interestingly, the coincident lattice site orientations are much less affected by segregation because the tendency for enrichment of solute atoms in LAGB is also significantly reduced.

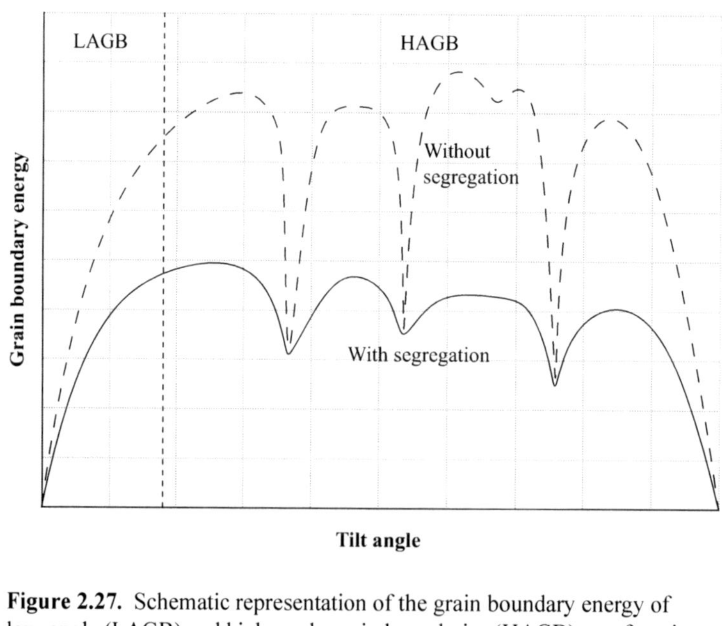

Figure 2.27. Schematic representation of the grain boundary energy of low-angle (LAGB) and high-angle grain boundaries (HAGB) as a function of the tilt angle between the two grains.

Table 2.5. Typical values for HAGB energies.

Element	Structure	$\gamma_{\alpha\alpha}$ /mJ/m^2	T/K	Ref.
Ag	fcc	375		[38] exp
Al	fcc	324		[38] exp
Al	fcc	300–550	0	[39] comp*
Au	fcc	378		[38] exp
Au	fcc	250–500	0	[39] comp*
Co	fcc	1010	1728	[40] exp
Cu	fcc	625		[38] exp
Cu	fcc	500–1000	0	[39] comp*
Cu–30at%Zn	fcc	595		[38] exp
Fe	fcc	756		[38] exp
Fe	fcc	985	1753	[40] exp
Fe–Cr	bcc	800–1600, Figure 2.28	1298	[41] comp*
Fe–3at%Si	bcc	617		[38] exp
Stainless steel (304)	fcc	835		[38] exp
Mo	bcc	1274–2124	0	[42] comp
Nb	bcc	980–1585	0	[42] comp
Ni	fcc	866		[38] exp
Ni	fcc	930	1643	[40] exp
Ni	fcc	700–1400	0	[39] comp*
Sn	fcc	164		[38] exp
Zn	fcc	340		[38] exp

* Values for *general* boundaries, no coincident site lattice orientations considered.
exp refers to experimental and comp refers to computed.

Table 2.5 summarizes typical values of grain boundary energies for various metals and alloys as observed experimentally and from computation. These values are valid for pure substances, without impurity segregation, which must be taken into account when using these numbers for the grain boundary energy, $\gamma_{\alpha\alpha}$.

Figure 2.28 presents the grain boundary energy calculated with Molecular Dynamics simulations [41] for the bcc Fe–Cr system. The plot shows the dependence of the grain boundary energy from the tilt angle, as well as the alloy composition.

The evaluation of a grain boundary energy value for a specific alloy becomes even more involved when taking into account that $\gamma_{\alpha\alpha}$ is also a function of temperature, with the grain boundary energy generally decreasing with temperature. In several metals, the energy evaluated at 0 K reduces to approximately one-third of its value close to the melting point, due to entropic contributions [43]. The values given in Table 2.5 for HAGBs should therefore be considered as upper limits.

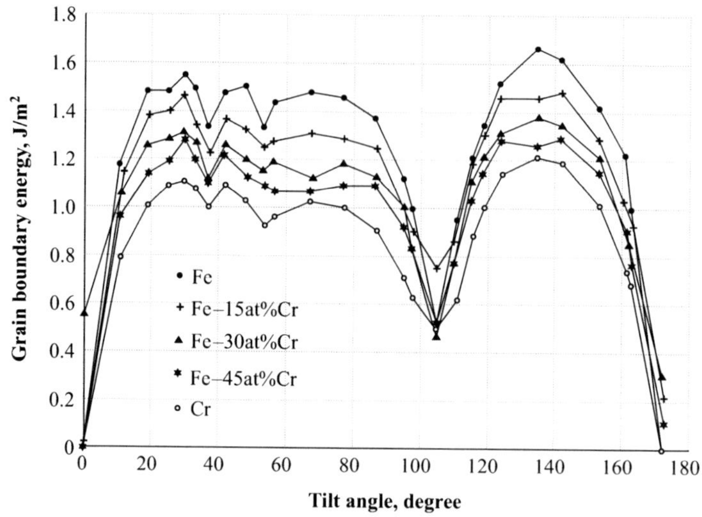

Figure 2.28. Grain boundary energy of HAGBs in Fe–Cr as a function of the tilt angle. The pronounced minimum is found at the {112} orientation. Experimental data from Shibuta, Takamoto, and Suzuki [41].

2.5 Nucleation in Multicomponent Environment

In most solid-state phase transformations, CNT has proven to be a sound and useful theory, both qualitatively as well as quantitatively. The nucleation of new phases can reasonably well be described when treating individual nucleation events on a statistical basis, that is, in terms of the nucleation rate, J.

So far, we have accounted for the *kinetics* of the nucleation events in terms of the mobility of a single, representative diffusing species and a single, effective, diffusion coefficient, D_{eff}, for the relevant precipitate-forming element.[38] In multicomponent systems, however, precipitates are commonly composed of several atomic species with variable diffusivity and the kinetics of the nucleation event is then governed by the diffusion kinetics of all participating components. This issue is dealt with in the first part of this section, where the CNT expressions for binary solid-state systems are generalized for an arbitrary number of components.

Another premise that we have made so far is that the chemical composition of the critical nucleus is known, either experimentally or in some other way prescribed. Only for this case, the general transient nucleation rate expression, equation (2.90), can be evaluated unambiguously. Unfortunately, in most practical cases, this important information is lacking either due to experimental difficulties when determining the composition of nanometer and subnanometer-sized particles or due to the lack of theoretical or simulation concepts.

[38] Compare equation (2.87).

2.5.1 CNT in Multicomponent Environment

When dealing with nucleation in solid-state systems, the *transient nucleation rate*, J, describing the number of nucleation events per unit volume and unit time, has been expressed as[39]

$$J = NZ\beta^* \exp\left(-\frac{G^*}{k_B T}\right)\exp\left(-\frac{\tau}{t}\right). \tag{2.137}$$

For an evaluation of the quantities occurring in this expression *in a general multicomponent environment*, we make the presumption that all thermodynamic quantities are known as a function of temperature, pressure, and chemical composition. Furthermore, we assume that the chemical composition of the critical nucleus is somehow given.[40]

Under these conditions, we find that

- the total number of potential nucleation sites is independent of the number of chemical components and it is, thus, identical to the value in two-component systems;
- the Zeldovich factor, Z, only contains thermodynamic quantities,[41] that is, the second derivative of the total energy of nucleus formation, ΔG_{nucl}, and Z can thus be evaluated for any given chemical composition of the nucleus and parent matrix phase and given size of the critical nucleus;
- the critical nucleation energy, G^*, as expressed by equation (2.75), is given as a function of the interfacial energy, γ, and the volume free energy change, ΔG_{vol}, which can both be evaluated from thermodynamic consideration (see Chapter 1 for the free energy and Chapter 4 for the interfacial energy) and the premises made above;
- the incubation time, τ, is given by equation (2.56) or (2.68) in terms of the Zeldovich factor, Z, and the atomic attachment rate, β^*.

The main quantity which is not fully defined in a general multicomponent environment remains the atomic attachment rate, β^*, the reason being the use of a single representative diffusion coefficient, D_B, of component B in equation (2.87) instead of taking into account the diffusion kinetics of all participating components.

The issue of determining a generalized expression for the atomic attachment rate, β^*, has been addressed by Svoboda et al. [44], who suggest that the product of mole fraction, X_B, and diffusivity, D_B, of the representative component B be replaced by a weighted effective diffusion coefficient in the form

$$D_{\mathrm{eff}} = \left[v^{\alpha} \sum_{i=1}^{n} \frac{(c_{ki} - c_{0i})^2}{c_{0i} D_{0i}} \right]^{-1}. \tag{2.138}$$

39 Compare equation (2.90), Section 2.3.4.

40 This is generally not the case. This issue will be discussed in detail in the next section.

41 Compare equation (2.47).

Table 2.6. Expressions for evaluation of the nucleation rate, J, in multicomponent environment.

Quantity/ dimension	Description	Expression	Comment
N (m^{-3})	Potential number of nucleation sites		Section 2.4.2
Z (dimensionless)	Zeldovich factor	$Z = \left[-\frac{1}{2\pi k_B T} \left(\frac{\partial^2 \Delta G_{nucl}}{\partial n^2} \right)_{n^*} \right]^{\frac{1}{2}}$ (general)	Equation (2.50)
		$\left[\frac{a^6}{64\pi^2 k_B T} \cdot \frac{(\Delta G_{vol})^4}{(\Delta G_{surf})^3} \right]^{\frac{1}{2}}$ (spherical)	Equation (2.86)
β^* (s^{-1})	Atomic attachment rate	$\frac{4\pi (\rho^*)^2}{a^4 v^\alpha} \left[\sum_{i=1}^{n} \frac{(c_{ki} - c_{0i})^2}{c_{0i} D_{0i}} \right]^{-1}$ (spherical)	Equation (2.139)
G^* (J)	Critical nucleation energy	$\frac{16\pi}{3} \frac{(\Delta G_{surf})^3}{(\Delta G_{vol})^2}$ (spherical)	Equation (2.75)
ρ^* (m)	Critical radius	$-\frac{2\Delta G_{surf}}{\Delta G_{vol}}$ (spherical)	Equation (2.74)
n^* (dimensionless)	Number of monomers in critical nucleus	$-\frac{32\pi}{3\Omega} \left(\frac{\Delta G_{surf}}{\Delta G_{vol}} \right)^3$ (spherical)	Equation (2.83)
ΔG_{vol} (J/m^3)	Volume free energy change	$-\frac{d^{\beta}_{chem}}{v^\alpha} + \Delta G^{el}_{vol}$	Equation (2.71)
ΔG_{surf} (J/m^2)	Surface free energy/interfacial energy	$\frac{n_S \cdot z_{S,eff}}{N_A \cdot z_{L,eff}} \cdot \Delta h_{sol}$	Equation (4.21)
τ (s)	Incubation time	$\frac{1}{4\pi \beta^* Z^2}$ (Wakeshima)	Equation (2.56)
		$\frac{1}{2\beta^* Z^2}$ (Feder)	Equation (2.68)

With this substitution, a generalized expression for the atomic attachment rate is obtained, which reads

$$\beta^* = \frac{4\pi (\rho^*)^2}{a^4 v^\alpha} \left[\sum_{i=1}^{n} \frac{(c_{ki} - c_{0i})^2}{c_{0i} D_{0i}} \right]^{-1}. \tag{2.139}$$

c_{ki} are the concentrations of element i in the precipitate, c_{0i} are the concentrations in the matrix, and D_{0i} are the tracer diffusivities of element i in the matrix. Table 2.6 summarizes the expressions that are required for evaluation of the multicomponent CNT nucleation rate expression for J.

2.5.2 The Composition of the Critical Nucleus

If the composition of the critical nucleus is known a priori, the nucleation rate expression, J, can be evaluated straightforwardly. Unfortunately, these cases are rare. Typical examples for phases where the chemical composition is given unambiguously are stoichiometric phases, such as Mg_2Si in the Al–Mg–Si system or Fe_3C in the Fe–C system.

In most practically relevant cases, the composition of the precipitate is not known a priori and this important information must be obtained from either experimental evidence or by some theoretical concept or approximation. Typical examples where these considerations apply are all kinds of solution phases, or phases with at least some solubility for other elements on any one sublattice, such as, for instance, bcc-Cu precipitates in Fe–Cu ferrite, where the Cu-nucleus can incorporate substantial amount of Fe, or $L1_2$-ordered $Ni_3(Al,Ti)$ γ' precipitates in Ni–Al–Ti systems, where some of the Al is substituted by Ti on the Al-sublattice.

In order to fix the additional degrees of freedom, several concepts exist and can be applied. Any of these is characterized by the specific computational effort that is required to solve, or derive, the governing equations.

The most common assumption in this context and, at the same time, the model requiring least extra computational effort is the assumption of maximum chemical driving force, or *ortho-equilibrium nucleus composition*. In this approach, the composition of the nucleus is assumed to be such that the tangent to the Gibbs energy curve of the precipitate phase in the G-X diagram is parallel to the tangent of the parent matrix phase at the given matrix composition.

To illustrate the maximum driving force concept for the nucleus composition, the metastable bcc Fe–Cu system is taken as an example. In this system, a miscibility gap exists between Cu and Fe giving the possibility of precipitating nanometer-sized bcc Cu-particles from supersaturated Fe-rich ferrite. Figure 2.29 shows the molar Gibbs energy, g, of the bcc Fe–Cu system as a function of the total Cu-mole fraction, X_{Cu}, at a temperature of 773 K. The figure shows the common tangent to the free energy curve to indicate equilibrium conditions, as well as two tangents for a supersaturated state with $X_{Cu} = 0.02$, one for the parent matrix phase and the other for the precipitate. The latter two tangents can be used to determine the chemical driving force for precipitation of bcc-Cu from the supersaturated ferrite.[42] The inset in the top left corner of the figure shows a magnification of the Fe-rich side of this diagram with the corresponding tangents and an additional vertical line for the equilibrium Cu-composition of the Fe-matrix.

[42] For further details on the tangent construction and the evaluation of driving forces, see Section 1.4.

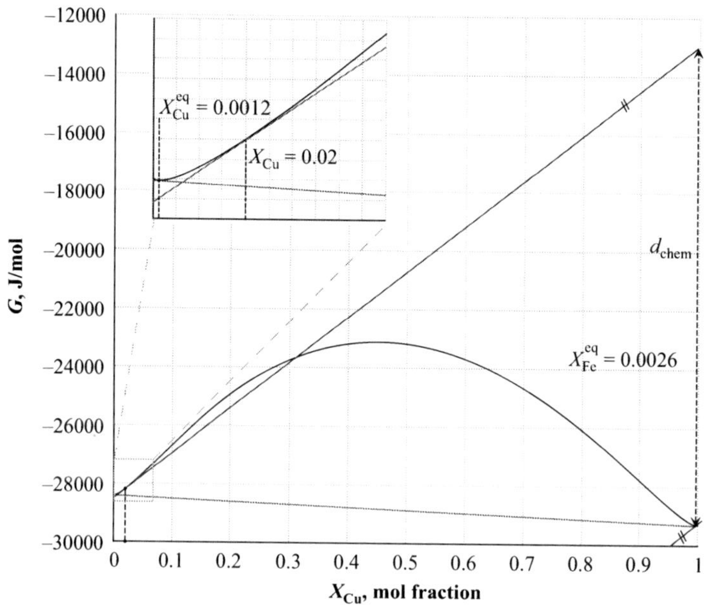

Figure 2.29. Gibbs energy versus composition diagram for bcc Fe–Cu at 773 K. For discussion, see text.

The concept of maximum driving force is often justified on the basis that the critical nucleation energy, G^*, becomes minimal in this case, which is correct if the interfacial energy, γ, is assumed to be constant and independent of chemical composition.[43] When accepting a constant composition-independent value for the interfacial energy, the minimum of G^* leads to a maximum in the nucleation rate, J. One could thus argue that precipitates with *ortho-equilibrium nucleus composition* will nucleate most frequently and they will thus achieve the highest number density compared to particles with lower chemical driving force.

Unfortunately, the interfacial energy is not composition-independent, but a quadratic function of the composition difference between precipitate and matrix. This statement is investigated in Figure 2.30, showing the chemical driving force and the interfacial energy as functions of the Cu content of the nucleus. The supersaturation of the matrix is assumed to be $X_{Cu} = 0.02$.

The chemical driving force shown in Figure 2.30 is evaluated as the distance of each point of the G-X curve from the extrapolated tangent at $X_{Cu} = 0.02$ as shown in Figure 2.29. Accordingly, we find that the chemical driving force adopts positive values for all compositions with approximately $X_{Cu} = 0.31$ and more. Theoretically, all of these compositions are thermodynamically stable and must be accounted for in any selection process for the optimum nucleus composition.

Figure 2.30 also shows the dependency of the interfacial energy from the nucleus composition. As the interfacial energy is approximately proportional to the square of differences in chemical composition between precipitate and matrix, we find that this quantity increases steadily for all values $X_{Cu,nucl} > 0.02$. Note that the discontinuity in the curve stems from the transition of the nucleus from the ferromagnetic to the paramagnetic state at around $X_{Cu} = 0.27$.

[43] This is generally not the case, however, which will be shown later.

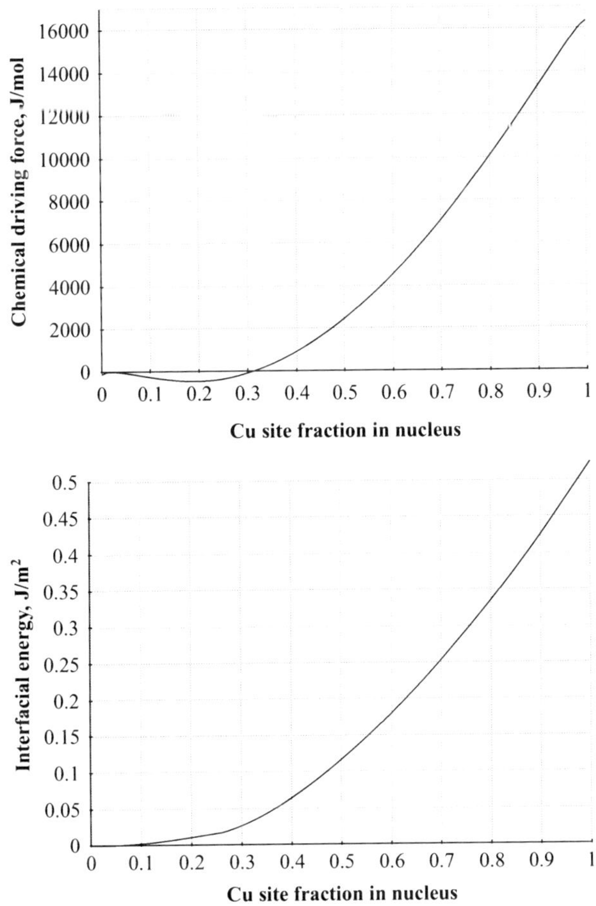

Figure 2.30. Chemical driving force and interfacial energy for bcc-Cu precipitation in Fe–2at%Cu at 773 K.

It is interesting, now, that the two plots in Figure 2.30 suggest different composition regions where the ideal nucleus composition with minimum G^* can be expected. According to the driving force plot, the ideal composition should be given at high Cu-content of the nucleus, where the driving force reaches a maximum. In contrast, the interfacial energy plot suggests that the minimum interfacial energy is achieved at exactly the composition of the matrix, which is $X_{Cu} = 0.02$ in the present case. A solution to this problem is quickly identified when plotting the critical nucleation energy, G^*, according to equation (2.75).

Figure 2.31 presents the critical nucleation energy, G^*, as a function of the Cu-content of the nucleus. Apparently, this quantity exhibits a pronounced minimum at a nucleus composition of approximately $X_{Cu} = 0.5$. Although at high Cu-content the chemical driving force for precipitation has a maximum, simultaneously, the interfacial energy has also a maximum, as this quantity scales quadratically with the difference in chemical composition of precipitate and matrix. The required total nucleation energy can be minimized if a composition is chosen which is less rich in Cu, thus decreasing the interfacial energy contribution to the nucleation barrier. At some point, this value reaches an optimum, which is a minimum in G^*, where the loss in driving

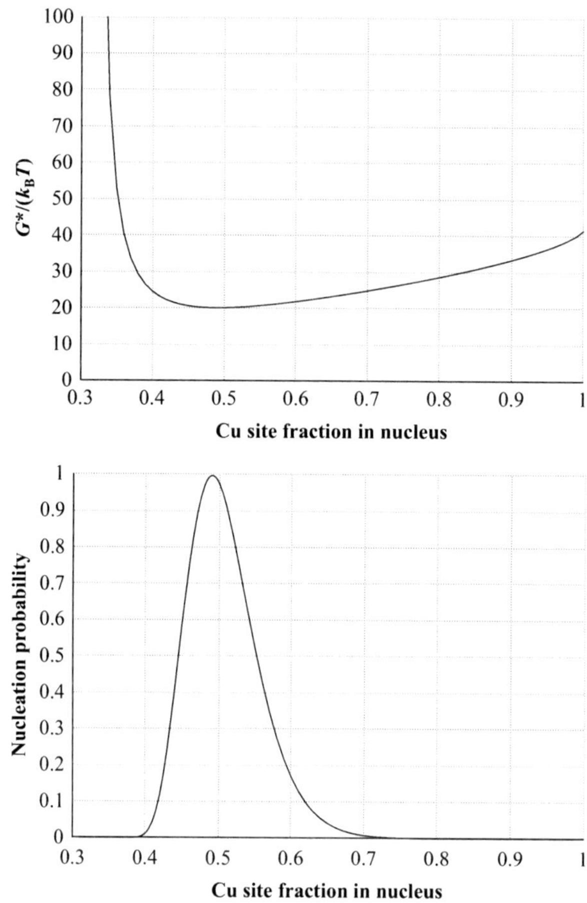

Figure 2.31. Critical nucleation energy, G^*/k_BT, and normalized nucleation probability for bcc Fe–Cu precipitation at 773 K.

force is best compensated for by the advantage in interfacial energy. At even lower Cu-content, the interfacial energy continues to decrease, however, the also steadily decreasing driving force makes nucleation less and less energetically favorable and the nucleation barrier increases again.

The concept of a "most likely" nucleus composition, which is characterized by a minimum nucleation barrier, G^*, was introduced first by J. N. Hobstetter [45] in 1949. It was recently put on quantitative ground and applied to bcc-Cu nucleation in bcc Fe–Cu by Kozeschnik [46], providing good evidence for the applicability of this concept.

The concept of minimum G^* provides a convenient approach for fixing the degree of freedoms with respect to the nucleus composition because it is a purely thermodynamic concept and does not require evaluation of kinetic quantities. In a more rigorous approach, one should also take into account the diffusive mobilities of all precipitate-forming elements and maximize the nucleation rate instead of only minimizing G^*. This issue has been addressed in part by Kozeschnik, Svoboda, and Fischer [47] in a discussion of the nucleation kinetics of Cr-carbides in steel as a function of the alloy content of the precipitate. The authors carried out numerous precipitation simulations with precipitate populations of different initial chemical composition being in direct nucleation and growth competition during the simulation. After analysis of the results and attempts to identify a single representative chemical composition to derive the precipitation process, the authors conclude that the common choice of ortho-equilibrium composition does not provide a suitable representative nucleus composition for a reproduction of the system behavior. They emphasize finally that useful concepts addressing this issue are still to be developed and not yet available.

2.6 Summary

In this chapter, the basic concepts shaping the framework of Classical Nucleation Theory (CNT) are introduced. Based on the observation that the work of formation of a critical cluster is provided on a natural basis by microscopic fluctuations [17,18], in 1926, Volmer and Weber [19] show that the nucleation rate of supercritical clusters is proportional to the density of clusters at the critical size, n^*. This quantity, in turn, is given by an exponential expression containing the nucleation energy as the essential physical property.

The kinetic constant, K, linking the steady-state nucleation rate, J_s, to the critical cluster density has been investigated by, for example, Farkas [21], Becker and Döring [20], and Zeldovich [23]. In essence, K is found to be dependent on the density of monomers, the Zeldovich factor, Z, and the attachment rate of monomers, β^*, to the critical cluster. The predictions of CNT, as presented in this section, have received strong experimental support by experiments of Volmer and Flood [22], who compared the predictions of CNT to numerous experiments on water, ethanol, and other fluids. The observed agreement between experiment and prediction can be considered as a confirmation of the validity of CNT.

The early approaches of nucleation theory are mainly focused on the formation of liquid droplets from supersaturated vapor, or the nucleation of voids within liquids. Apart from the analyses reproduced here, there exist several further developments and treatments attempting to improve the predictive capability of CNT. However, none of these have succeeded in promoting the basic concepts substantially, nor have they brought significant quantitative improvement. For a detailed analysis of the earlier work on this subject, which is clearly going beyond the scope of this book, the reader is referred to the extensive review of Feder et al. [25].

From the results obtained for liquid droplet nucleation, it is shown that the fundamental relations of CNT apply in almost unchanged form to solid-state nucleation processes. The works of Frenkel [29] and Turnbull and Fisher [27] indicated that any kind of heterophase equilibrium fluctuation has the potential to provide the necessary energy to overcome the nucleation barrier. The nucleation rate expression adapts a very similar form compared to the one for droplet nucleation, and it can straightforwardly be modified to apply to solid-state nucleation conditions.

This task is carried out in the chapters describing nucleation in the solid state. The relations developed for liquid droplet formation are generalized to multicomponent nucleation and a consistent framework for simulation of precipitate nucleation is presented. The influence of heterogeneous nucleation sites on the energetics of the nucleation process is outlined and strategies are presented on how to determine the chemical composition of the nucleus in multicomponent systems.

Although promising concepts are available for even quantitative simulation of solid-state nucleation and no serious gaps in the qualitative and quantitative understanding of this process persist, still some effort is needed in developing strategies for an efficient treatment of multi-component nucleation. Particularly, question relating to the nucleus composition has not yet been answered satisfactorily.

Other open issues in the field of precipitate nucleation that have not been treated at all in this book are questions related to the selection of crystal structures in cases where different types of precipitates have similar nucleation barriers, contain approximately the same alloying elements but have different crystal structure and structure complexity. Experiments show that, in this case, the precipitate with higher crystal complexity is often suppressed and appears in precipitation sequences only after very long heat treatment times.

A typical example of such a situation is given in the competition between CrV-nitrides of the MX-type and the Z-phase in 9–12% Cr–steel, where the Z-phase is suppressed for times exceeding thousands and ten thousands of hours at annealing temperatures around 600°C. Although Z-phase has a similar chemical composition compared to the MX-nitride, and even higher thermodynamic stability, experiment shows that the vast majority of CrVN-containing precipitates are of the simple fcc-structured MX-nitride type compared to the double-layer structure of Z-phase.

Finally, sound metallurgical concepts to address issues such as precipitation in martensite, precipitation in bainitic microstructure or the issue of para-equilibrium nucleation of cementite in steel are presently lacking. The hope remains, however, that the rapid progress in this research field continues and that the near future will deliver some answers to these open questions.

CHAPTER 3

Diffusion-Controlled Precipitate Growth and Coarsening

The first stage of a phase transformation is controlled by the rate of generation of supercritical nuclei of the new phase. The nucleation of particles is a *stochastic* process which can be quantitatively described on the basis of statistical thermodynamics and mechanics. The concepts of Classical Nucleation Theory (CNT) have been outlined in the previous chapter with special emphasis on solid-state nucleation. It is important to emphasize that the formulation of nucleation theory is founded on the average behavior of large thermodynamic systems. CNT does not deliver answers to the spatial and temporal occurrence of *individual* nucleation events, nor can it do so. However, CNT does deliver *quantitative* information on the number of newly formed supercritical nuclei in a sufficiently large volume and given time interval. This information can conveniently be used in any consecutive treatment of the evolution of precipitation systems.

In contrast to the stochastic nature of the first stage of a phase transformation, the successive stage of precipitate growth is governed by the migration of phase boundaries between the transforming phases. The laws controlling this process are *deterministic* in nature, as the growth and shrinking processes of the new phase can (unambiguously) be described by appropriate evolution equations for each individual precipitate. This aspect of precipitation is the focus of this chapter.

3.1 Problem Formulation

Consider a supercritical volume of a transforming phase in a supersaturated[1] chemical environment. With the continuous arrival of new precipitate-forming atoms and attachment to the precipitate surface, the precipitate undergoes a steady growth process, the kinetics of which is controlled by the diffusional fluxes, J_i, of components i in the matrix phase and in the precipitate.

[1] In a *supersaturated* thermodynamic system, the probability for the forward process of monomer attachment to the supercritical precipitate is higher than the backward process of monomer detachment. This condition leads to precipitate growth.

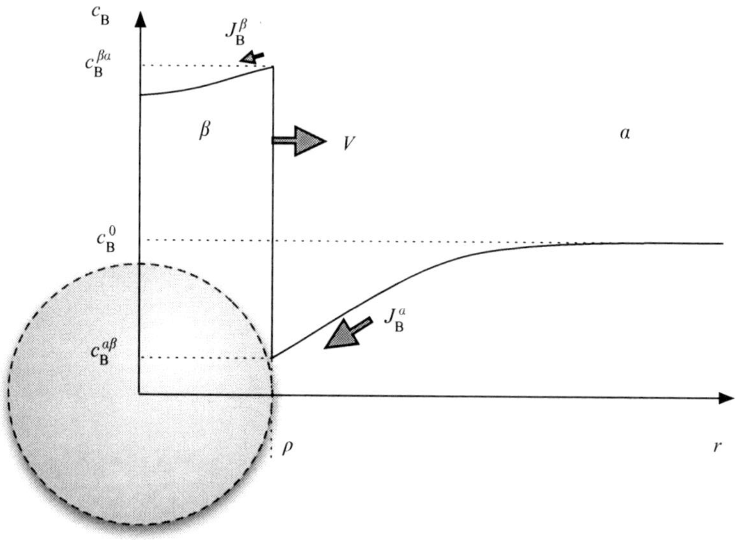

Figure 3.1. Schematic representation of a B-rich spherical precipitate growing in a supersaturated matrix.

Figure 3.1 sketches the composition profiles for the precipitate-forming alloying element B during *diffusion-controlled growth* of a B-rich spherical precipitate in an A-rich matrix. The radius of the precipitate is ρ and the concentration of precipitate-forming atoms is higher inside the precipitate compared to the concentration of elements in the surrounding matrix. These are indicated by the concentration variables c_B^0 for the nominal B-content infinitely far away from the precipitate, $c_B^{\beta\alpha}$ for the B-content in the precipitate at the precipitate–matrix interface, and $c_B^{\alpha\beta}$ for the concentration of B at the matrix side of the precipitate–matrix interface.

The gradients in concentration, c, at the left and right sides of the precipitate–matrix interface induce diffusional fluxes, J, which are given by Fick's first law[2] as

$$J_B^\alpha = -D_B^\alpha \frac{\partial c_B^\alpha}{\partial r} = -\frac{D_B^\alpha}{\nu^\alpha}\frac{\partial X_B^\alpha}{\partial r}, \tag{3.1}$$

$$J_B^\beta = -D_B^\beta \frac{\partial c_B^\beta}{\partial r} = -\frac{D_B^\beta}{\nu^\beta}\frac{\partial X_B^\beta}{\partial r}, \tag{3.2}$$

with D_B^α being the diffusion coefficient of element B in phase α, r being the spherical radius coordinate, c_B^α being concentration, and X_B^α the mole fraction of B in phase α.

Using the gradient operator (Nabla operator) with $\nabla = \partial / \partial r$, the steady-state diffusion equations for the case of binary diffusion can also be written as

$$J_B^\alpha = -D_B^\alpha \nabla c_B^\alpha = -\frac{D_B^\alpha}{\nu^\alpha}\nabla X_B^\alpha, \tag{3.3}$$

[2]Note that the exact relation between diffusional flux and diffusion force is given by the chemical potential gradient instead of the concentration gradient. Incorporation of multicomponent thermodynamic effects into diffusion coefficients are outlined in Chapter 7 on diffusion.

$$J_{\mathrm{B}}^{\beta} = -D_{\mathrm{B}}^{\beta}\nabla c_{\mathrm{B}}^{\beta} = -\frac{D_{\mathrm{B}}^{\beta}}{v^{\beta}}\nabla X_{\mathrm{B}}^{\beta}. \tag{3.4}$$

In multicomponent systems, the scalar diffusion coefficients, D, are substituted by diffusivity matrices relating the flux of component, i, to all elements, j, in the system. The fluxes on both sides of the interface then read

$$J_i^{\alpha} = -\sum_{j=1}^{n-1} D_{ij}^{\alpha}\nabla c_{ij}^{\alpha} = -\frac{1}{v^{\alpha}}\sum_{j=1}^{n-1} D_{ij}^{\alpha}\nabla X_j^{\alpha}, \tag{3.5}$$

$$J_i^{\beta} = -\sum_{j=1}^{n-1} D_{ij}^{\beta}\nabla c_{ij}^{\beta} = -\frac{1}{v^{\beta}}\sum_{j=1}^{n-1} D_{ij}^{\beta}\nabla X_j^{\beta}. \tag{3.6}$$

The situation depicted in Figure 3.1 already indicates that the fluxes, J_{B}^{α} and J_{B}^{β}, on the two sides of the interface do not necessarily balance. This is symbolized by the different size of arrows for the diffusional fluxes. If the fluxes into the phase boundary, $J_{\mathrm{B}}^{\alpha\beta}$, and out of the phase boundary, $J_{\mathrm{B}}^{\beta\alpha}$, are unbalanced, due to variations in the diffusion coefficients in the two phases and/or differences in the composition gradients, mass conservation requires that this imbalance is compensated by a movement of the phase boundary with a velocity, v, and a concentration jump at the interface given as

$$[[c_i^{\alpha\beta}]] = c_i^{\beta\alpha} - c_i^{\alpha\beta}. \tag{3.7}$$

The mass balance equation for the diffusion-controlled migration of a phase boundary finally reads

$$J_i^{\alpha\beta} + J_i^{\beta\alpha} = [[c_i^{\alpha\beta}]]\cdot v. \tag{3.8}$$

With this fundamental relation, the diffusional growth of a new phase is generally described. Equation (3.8) represents a system of equations of dimension n–1, where n is the number of components in the system.[3] Equation (3.8) must be solved for the interface velocity, v, and the unknown, but not independent, fluxes, J, and interface compositions, c. This type of problem is denoted as the *moving boundary problem* or *Stefan problem*.[4] Strategies for a solution of this problem are discussed in the following sections.

3.2 Diffusion-Controlled Growth with Local Thermodynamic Equilibrium

For n components, the moving boundary problem of a diffusion-controlled phase transformation has 2(n–1) unknown variables for the chemical composition at the phase interface, as well

[3]The dimension is n–1 because the flux of one (reference) component can always be expressed in terms of the fluxes of all other components.

[4]after the Austrian–Slovene physicist and mathematician Jošef Stefan (1835–1893). He investigated the simultaneous liquid–solid interface movement and heat transfer during solidification of water (ice formation).

as one variable for the interface velocity, v. With n–1 equations of type (3.8), the system is underdetermined and, for a unique solution, the 2(n–1)+1 degrees of freedom must be reduced to n–1 by appropriate assumptions. As a most convenient and often used approximation in this context, the *local equilibrium hypothesis* is introduced in this section.

3.2.1 Local Equilibrium and Composition Profiles

Consider a thermodynamic system where two phases are in contact with each other across a phase boundary. Consider further that the movement of the phase boundary is determined by the inward and outward diffusional fluxes at the interface. As phase boundaries are often of incoherent type, comparable in nature to a high-angle grain boundary, the diffusion kinetics in the interface is substantially faster than the diffusion kinetics in the solid phases being in contact. One can therefore assume that instantaneously after being brought into contact, the two phases establish a state of local equilibrium at the interface which is characterized by negligibly small chemical potential gradients across the interface due to the assumed rapid interfacial diffusion.[5]

In the local equilibrium framework, the *contact conditions* for the chemical potentials on the two sides of the interface are given as

$$\mu_i^{\beta\alpha} = \mu_i^{\alpha\beta}, \tag{3.9}$$

for all i, and with $\mu_i^{\beta\alpha}$ and $\mu_i^{\alpha\beta}$ representing the chemical potentials of components, i, at the two sides of the phase boundary.

In the special case of a binary system A-B, the local equilibrium assumption delivers a straightforward relation between the interfacial concentrations, $c_{\mathrm{B}}^{\beta\alpha}$ and $c_{\mathrm{B}}^{\alpha\beta}$ with the corresponding A-B phase diagram. Figures 3.2 and 3.3 sketch the composition profiles around a moving phase boundary with the composition jump, $[[c_i^{\alpha\beta}]]$, indicated by a bold line and open circles in the left image. The right image displays the corresponding section in the binary phase diagram. Figure 3.2 shows the composition profiles for the case of a B-rich precipitate in an A-rich matrix, whereas Figure 3.3 displays the situation for a B-poor precipitate in a B-rich matrix. For both cases, at a given temperature, the interfacial compositions can directly be read from the conode shown in the phase diagram sections with open circles.

The composition profile shown in Figure 3.2 indicates that the solute atoms B migrate toward the B-rich precipitate in order to stimulate growth. This type of phase transformation is often denoted as *positive precipitation.* In contrast, the B-atoms diffuse away from the interface in the case shown in Figure 3.3, and leave a B-depleted region behind the interface. As the B-content of the product phase is lower than the B-content of the parent phase, this type of transformation is also denoted as *negative precipitation.*

The local equilibrium model delivers unique values for the interfacial compositions, $c_{\mathrm{B}}^{\beta\alpha}$ and $c_{\mathrm{B}}^{\alpha\beta}$, in the binary case. Consequently, the interface velocity, v, is given explicitly from equation (3.8) with

$$v = \frac{J_{\mathrm{B}}^{\alpha\beta} + J_{\mathrm{B}}^{\beta\alpha}}{[[c_{\mathrm{B}}^{\alpha\beta}]]}. \tag{3.10}$$

[5]One consequence of the local equilibrium hypothesis is that no energy is dissipated inside the interface during local equilibrium diffusion-controlled phase transformations.

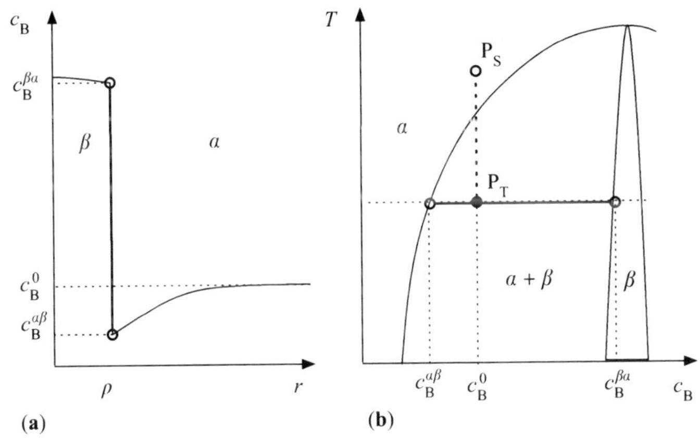

Figure 3.2. Relation between phase diagram (b) and composition profiles (a) at a moving phase boundary for *B-rich* precipitates in the local equilibrium approach.

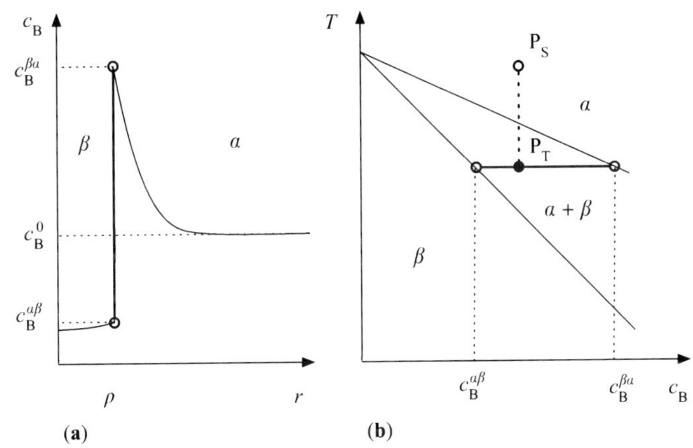

Figure 3.3. Relation between phase diagram (b) and composition profiles (a) at a moving phase boundary for *B-poor* precipitates in the local equilibrium approach.

Equation (3.10) can be solved with knowledge of the fluxes, $J_B^{\alpha\beta}$ and $J_B^{\beta\alpha}$. In the following sections, analytical solutions to the problem of binary diffusion-controlled phase transformations utilizing the local equilibrium approach are presented for different geometrical boundary conditions. General multicomponent solutions with simultaneous diffusion of multiple elements are substantially more difficult and must be performed on a numerical basis. Corresponding computer code is realized, for instance, in the software DICTRA (DIffusion Controlled phase TRAnsformation) or MatCalc (MATerials CALCulator), which are introduced later in Chapter 9.

3.2.2 Binary Diffusion-Controlled Growth—the Zener Model

Consider a stoichiometric phase β growing into a parent phase α, where the transformation rate, that is, the interface velocity, v, is controlled by the long-range diffusive transport of element B.

In 1949, Clarence Melvin Zener, an American physicist (1905–1993), introduced a simple and elegant solution to the binary moving boundary problem under the assumption of local equilibrium at the interface [48]. This theory represents a generalization of the one-dimensional model developed earlier by Arthur Dube (see ref. [2] in [48]) for the diffusion-controlled growth of plates.

According to equation (3.10), the velocity, v, of the phase boundary can be evaluated with knowledge of the diffusive fluxes, $J_{\mathrm{B}}^{\alpha\beta}$ and $J_{\mathrm{B}}^{\beta\alpha}$, of element B into the phase boundary and the interfacial compositions, $c_{\mathrm{B}}^{\alpha\beta}$ and $c_{\mathrm{B}}^{\beta\alpha}$. For a stoichiometric target phase β, the complexity of the problem is slightly reduced as the flux of component B in the target phase becomes $J_{\mathrm{B}}^{\beta\alpha} = 0$. The governing equation then reads

$$[[c_{\mathrm{B}}^{\alpha\beta}]]\cdot v = J_{\mathrm{B}}^{\alpha\beta} = D_{\mathrm{B}}^{\alpha}\frac{\partial c_{\mathrm{B}}}{\partial x}, \tag{3.11}$$

with D_{B}^{α} representing the diffusion coefficient of B in the parent phase α.

For solution of equation (3.11), Zener made the following assumptions:

- The concentrations at the moving interface, $c_{\mathrm{B}}^{\alpha\beta}$ and $c_{\mathrm{B}}^{\beta\alpha}$, are determined by the condition of *local thermodynamic equilibrium*.
- No dissipation of energy occurs by interfacial friction, that is, the interface is sufficiently mobile such that the migration of the interface is solely determined by the diffusive processes in the matrix phase α.
- The composition gradient in the parent phase can be approximated by a linear profile.

In his treatment, Zener introduces a spatial coordinate, $s(t)$, describing the position of the interface between the parent and product phases in n-dimensional space. For $n = 1$, the situation corresponds to the growth of a plate with the half-thickness, ξ, $n = 2$ refers to the growth of a cylinder with radius r, and $n = 3$ corresponds to the growth of a spherical precipitate with radius r. The general growth relation is then given as

$$s(t) = \alpha_n \left(D_{\mathrm{B}}^{\alpha}\cdot t\right)^{\frac{1}{2}}, \tag{3.12}$$

where t is time and a_n is a dimensionless constant depending on both, the dimensionality, n, of the problem and the concentrations, $c_{\mathrm{B}}^{\alpha\beta}$, c_{B}^{β}, and c_{B}^{0}. As a_n is found to be a complex function of these variables, Zener only provides solutions for two limiting cases introduced here, for $n = 1$ and $n = 3$. These are

$$\alpha_1 = K_1 \frac{\left(c_{\mathrm{B}}^{0} - c_{\mathrm{B}}^{\alpha\beta}\right)}{\left(c_{\mathrm{B}}^{\beta} - c_{\mathrm{B}}^{0}\right)^{\frac{1}{2}}\left(c_{\mathrm{B}}^{\beta} - c_{\mathrm{B}}^{\alpha\beta}\right)^{\frac{1}{2}}}, \tag{3.13}$$

for the one-dimensional case of a plate with K_1 varying between 1.13 and 1.41 as c_{B}^{0} varies from $c_{\mathrm{B}}^{\alpha\beta}$ to c_{B}^{β}, and

$$\alpha_3 = K_3 \frac{\left(c_{\mathrm{B}}^{0} - c_{\mathrm{B}}^{\alpha\beta}\right)^{\frac{1}{2}}}{\left(c_{\mathrm{B}}^{\beta} - c_{\mathrm{B}}^{0}\right)^{\frac{1}{2}}}, \tag{3.14}$$

for the three-dimensional case of a spherical precipitate with K_3 varying between 1.4 and 2.4 as c_B^0 varies from $c_B^{\alpha\beta}$ to c_B^{β}. The variations in the constants, K_n, describe the gradual transition from the asymptotic cases of c_B^0 being located close to the interfacial composition, $c_B^{\alpha\beta}$, on the matrix side of the interface, and c_B^0 being located close to the composition, c_B^{β}, of the precipitate. For further details on Zener's analysis and his solutions for other situations, the interested reader is referred to the original publication [48], as well as the general treatment provided in the following section.

It is instructive, however, to reproduce the approximate treatment of Zener for evaluation of the composition gradients, $\partial c_B/\partial s$, for the one- and three-dimensional cases which delivers simple closed expressions for the growth rates of plates and spheres. A sketch of Zener's approximation is depicted in Figure 3.4 for the planar case with $n = 1$.

With the spatial coordinate, x, and the variable, ξ, denoting the half-width of a plate-shaped precipitate, the figure suggests that the composition gradient in the matrix phase α at the interface can be approximated by

$$\left.\frac{\partial c_B}{\partial x}\right|_{x=\xi} \approx \frac{\Delta c_B}{\Delta x}. \tag{3.15}$$

The origin of this relation is founded in the requirement of mass conservation for the amount of atom B transferred from the parent to the product phase. In the figure, this condition is expressed in the identity of the grayed areas, A_1 and A_2. Upon substitution of

$$\Delta x = \frac{2\left(c_B^{\beta} - c_B^{0}\right)\xi}{\Delta c}, \tag{3.16}$$

and

$$\Delta c_B = \left(c_B^{0} - c_B^{\alpha\beta}\right), \tag{3.17}$$

into equation (3.11), the interface velocity, v_1, in the one-dimensional planar case is obtained as

$$v_1 = \frac{\partial \xi}{\partial t} = \frac{D_B^{\alpha}}{2\xi} \frac{\left(c_B^{0} - c_B^{\alpha\beta}\right)^2}{\left(c_B^{\beta} - c_B^{\alpha\beta}\right)\left(c_B^{\beta} - c_B^{0}\right)} = \frac{D_B^{\alpha}}{2\xi}\left(\alpha_1'\right)^2. \tag{3.18}$$

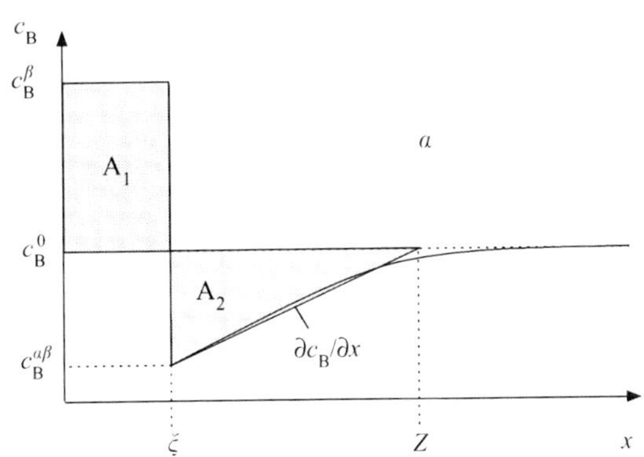

Figure 3.4. Zener's approximation for the diffusion-controlled growth of a spherical precipitate. See text for discussion.

After separation of variables and integration, the interface position, $\xi(t)$, is found with

$$\xi(t) = \alpha_1' \left(D_{\mathrm{B}}^{\alpha} t \right)^{\frac{1}{2}}, \tag{3.19}$$

and the growth constant, α_1', given as

$$\alpha_1' = \frac{\left(c_{\mathrm{B}}^{0} - c_{\mathrm{B}}^{\alpha\beta} \right)}{\left(c_{\mathrm{B}}^{\beta} - c_{\mathrm{B}}^{\alpha\beta} \right)^{\frac{1}{2}} \left(c_{\mathrm{B}}^{\beta} - c_{\mathrm{B}}^{0} \right)^{\frac{1}{2}}}. \tag{3.20}$$

For the corresponding analysis of the three-dimensional case of spherical precipitates, Zener distinguishes two limiting cases depending on the location of the nominal B-content, c_{B}^{0}, with respect to the concentrations, $c_{\mathrm{B}}^{\alpha\beta}$ and c_{B}^{β}. These are shown schematically in Figure 3.5.

The first case, corresponding to the inequality

$$\left(c_{\mathrm{B}}^{\beta} - c_{\mathrm{B}}^{0} \right) / \left(c_{\mathrm{B}}^{\beta} - c_{\mathrm{B}}^{\alpha\beta} \right) << 1, \tag{3.21}$$

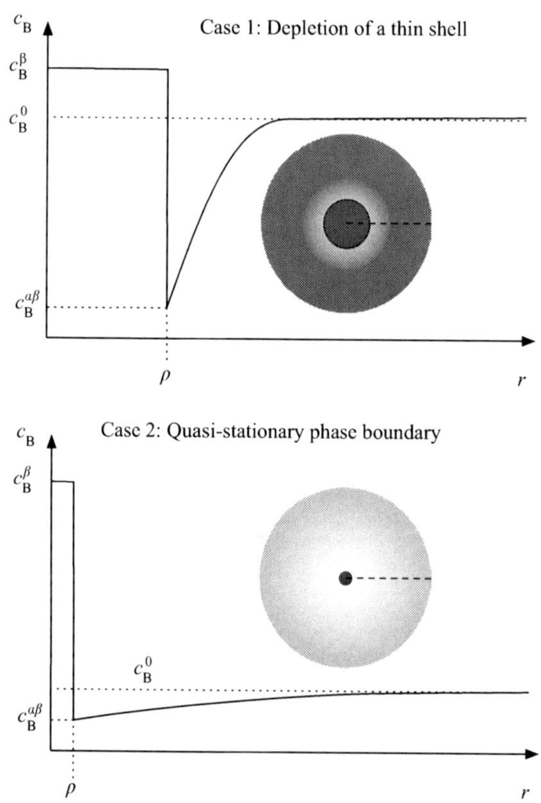

Figure 3.5. Schematic composition profiles for the two limiting cases in Zener's analysis of the growth rates of positive precipitation of spherical particles.

describes a situation where the B-content of the precipitate is close to the nominal B-content of the matrix. In the diffusion-controlled growth of the precipitate, a narrow solute-depleted area develops which is located in front of the migrating phase boundary. The growth constant for this case is found to be

$$\alpha'_{3,1} = 3^{\frac{1}{2}}\alpha'_1 = 3^{\frac{1}{2}} \frac{\left(c_B^0 - c_B^{\alpha\beta}\right)}{\left(c_B^{\beta} - c_B^{\alpha\beta}\right)^{\frac{1}{2}}\left(c_B^{\beta} - c_B^{0}\right)^{\frac{1}{2}}}, \tag{3.22}$$

with the radius coordinate, r, and

$$v_3 = \frac{\partial r}{\partial t} = \frac{D_B^{\alpha}}{2r}\left(\alpha'_3\right)^2. \tag{3.23}$$

Under these conditions, the composition profile resembles that of the planar case aside from the factor $3^{1/2}$ in the growth constant, α'. The rate of interface migration, $v_{3,1}$, for this case is

$$v_{3,1} = \frac{\partial r}{\partial t} = \frac{3D_B^{\alpha}}{2r} \frac{\left(c_B^0 - c_B^{\alpha\beta}\right)^2}{\left(c_B^{\beta} - c_B^{\alpha\beta}\right)\left(c_B^{\beta} - c_B^{0}\right)}. \tag{3.24}$$

For the other limiting case, the inequality

$$\left(c_B^0 - c_B^{\beta}\right) / \left(c_B^{\beta} - c_B^{\alpha\beta}\right) << 1 \tag{3.25}$$

holds, representing situations where the nominal B-content of the matrix is close to the equilibrium value of $c_B^{\alpha\beta}$. The depleted region then extends far into the matrix phase and it is large compared to the radius of the precipitate. The growth constant is obtained with

$$\alpha'_{3,2} = 2^{\frac{1}{2}} \cdot \left(\frac{c_B^0 - c_B^{\alpha\beta}}{c_B^{\beta} - c_B^{\alpha\beta}}\right)^{\frac{1}{2}}. \tag{3.26}$$

For the condition (3.25), that is, $c_B^0 \approx c_B^{\alpha\beta}$, we find that $\alpha'_{3,2}$ given in equation (3.26) differs from the value of $\alpha'_{3,1}$ given in equation (3.22) by the factor $(3/2)\frac{1}{2}$, as well as the weighted contributions from the composition differences.

For the growth rate of a spherical precipitate, under conditions where the composition difference between matrix and precipitate is large and the diffusion field extends far into the matrix compared to the precipitate radius, Zener obtained

$$v_{3,2} = \frac{\partial r}{\partial t} = \frac{D_B^{\alpha}}{r} \cdot \frac{c_B^0 - c_B^{\alpha\beta}}{c_B^{\beta} - c_B^{\alpha\beta}}. \tag{3.27}$$

Zener's solutions can straightforwardly be analyzed further on the basis of a supersaturation parameter, S, given as

$$S = \frac{c_B^0 - c_B^{\alpha\beta}}{c_B^{\beta} - c_B^{\alpha\beta}}. \tag{3.28}$$

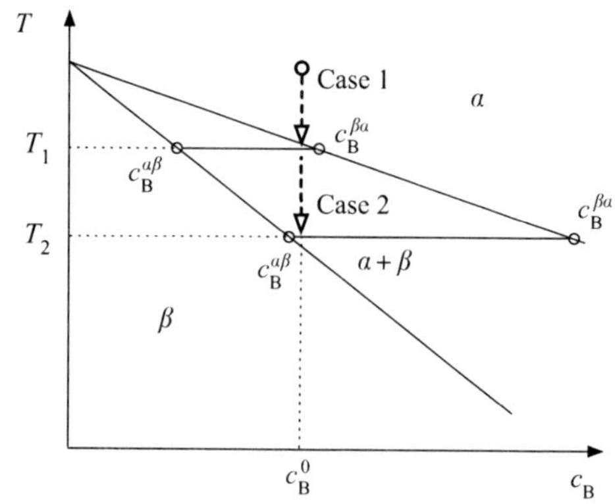

Figure 3.6. Relation between Zener's two extreme cases and a typical phase diagram for phases α and β with negative precipitation of a B-poor product phase.

Large values of S, for which the nominal composition of the matrix is located close to the B-content of the precipitate, correspond to situations where the diffusion field is small and extends only over a narrow region in front of the migrating interface. Small values of S represent situations where the transformation is characterized by a wide diffusion field around a comparably small precipitate.

Both situations are depicted in Figure 3.6 as cases 1 and 2, representing the two extreme cases of the Zener treatment with S<< and S>>. In case 1, the B-content of the product phase, $c_B^{\beta\alpha}$, is located close to the nominal content of the alloy, c_B^0. Figure 3.7(a) displays the corresponding composition profile around the moving phase boundary. During the transformation, a precipitate is formed with substantially less B-atoms than the parent matrix. The transformation is driven by the diffusion of B-atoms in the matrix phase ahead of the interface. It ceases as soon as the enrichment of the matrix phase is homogeneous and reaches the equilibrium value given by the phase diagram. This situation is indicated with dashed lines. The diagram emphasizes again that mass conservation in the transformation process requires identity of the two areas A_1 and A_2.

Figure 3.7(b) displays the composition profile for case 2, where the B-content of the precipitate, $c_B^{\beta\alpha}$, is located close to the nominal content, c_B^0, of the matrix. This type of transformation is characterized by a steep composition gradient in front of the interface. The product phase β grows far into the parent matrix, again with identity of the areas A_1 and A_2 due to the mass conservation constraint.

Figure 3.8 summarizes the growth rates for Zener's solutions for the planar interface as well as the spherical solutions as a function of the parameter S. The spherical solutions are drawn as solid lines within the regions of validity and as dashed lines outside these limits. The solutions show that spherical growth is generally faster than growth in planar geometry. For low values of the supersaturation parameter, S, the differences are significant. For values of S approaching unity, the differences in the growth constant, a, become continuously smaller. The growth constants finally diverge at $S = 1$, representing the particular situation where the composition of the matrix phase is identical to the composition of the precipitate.

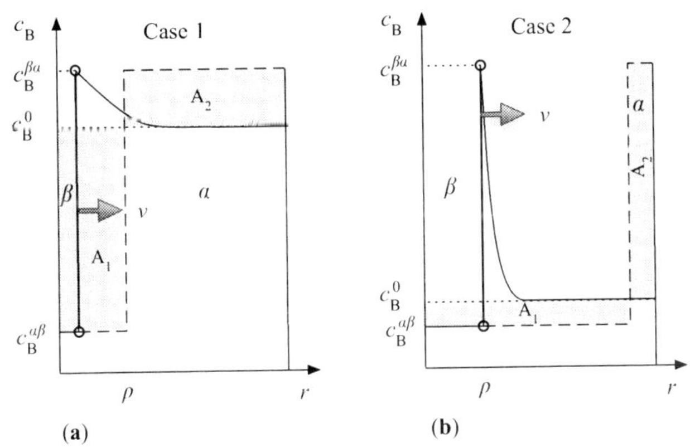

Figure 3.7. Composition profiles around a moving phase boundary between phases α and β for negative precipitation of a B-depleted product phase with (a) Zener's limiting case 1 and (b) Zener's case 2.

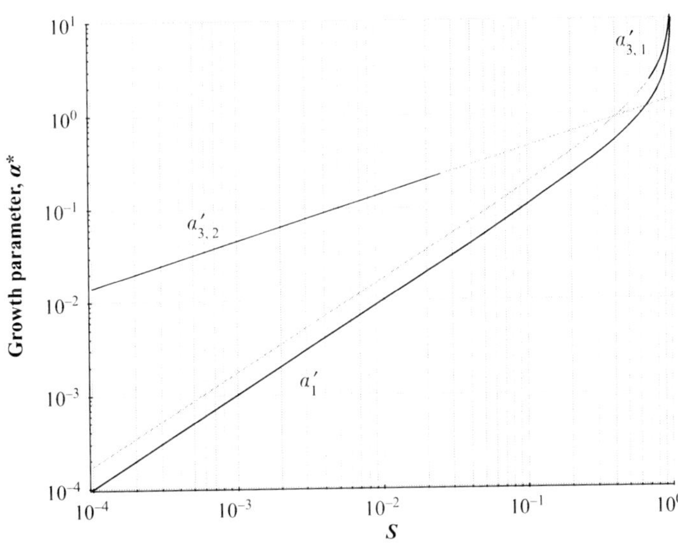

Figure 3.8. Rates of diffusion-controlled growth in dependence of the composition parameter, S.

3.2.3 The Quasi-Stationary Solution for Spherical Precipitates

If the B-content of the precipitate phase is high compared to the B-content of the matrix, the diffusion field surrounding the precipitate extends far into the matrix phase. In this special case, an alternative derivation of the growth relation can be given starting from Fick's second law written in spherical coordinates. As the spatial extension of the precipitate is significantly smaller than the diffusion field supporting the diffusion-controlled growth, the position of the precipitate–matrix interface can be approximated as being stationary with time. This assumption is commonly denoted as the *quasi-static* or *stationary* approximation of the moving boundary problem. In this special case, corresponding to Zener's treatment with a growth constant

$\alpha'_{3,2}$ given by equation (3.26), solution of the moving boundary problem can alternatively be obtained from the transient diffusion equation in spherical coordinates[6] with the boundary condition

$$\frac{\partial c_{\mathrm{B}}}{\partial t} = D_{\mathrm{B}}^{\alpha} \frac{\partial}{\partial r}\left[r^2 \frac{\partial c_{\mathrm{B}}}{\partial r} \right] = 0. \tag{3.29}$$

This equation has the general solution

$$c_{\mathrm{B}} = A + \frac{B}{r}. \tag{3.30}$$

With the local equilibrium hypothesis, the constants A and B can be specified and we obtain a solution for the concentration of B-atoms as a function of r as

$$c_{\mathrm{B}} = c_{\mathrm{B}}^{0} - \left(c_{\mathrm{B}}^{0} - c_{\mathrm{B}}^{\alpha\beta} \right) \frac{\rho}{r}, \tag{3.31}$$

where ρ is the radius of the precipitate. Derivation of this equation with respect to the spatial coordinate, r, delivers

$$\frac{\partial c_{\mathrm{B}}}{\partial r} = \left(c_{\mathrm{B}}^{0} - c_{\mathrm{B}}^{\alpha\beta} \right) \frac{\rho}{r^2}, \tag{3.32}$$

and the flux, $J_{\mathrm{B}}^{\alpha\beta}$, into the interface at $r = \rho$ becomes

$$J_{\mathrm{B}}^{\alpha\beta}(\rho) = D_{\mathrm{B}}^{\alpha} \frac{\partial c_{\mathrm{B}}}{\partial r}(\rho) = \left(c_{\mathrm{B}}^{0} - c_{\mathrm{B}}^{\alpha\beta} \right) \frac{D_{\mathrm{B}}^{\alpha}}{\rho}. \tag{3.33}$$

On inserting the mass conservation law, equation (3.11), with

$$\left(c_{\mathrm{B}}^{\beta\alpha} - c_{\mathrm{B}}^{\alpha\beta} \right) \cdot v = J_{\mathrm{B}}^{\alpha\beta}(\rho) = D_{\mathrm{B}}^{\alpha} \frac{\partial c_{\mathrm{B}}}{\partial r}, \tag{3.34}$$

we obtain the expression

$$\left(c_{\mathrm{B}}^{\beta\alpha} - c_{\mathrm{B}}^{\alpha\beta} \right) \cdot v = \left(c_{\mathrm{B}}^{0} - c_{\mathrm{B}}^{\alpha\beta} \right) \frac{D_{\mathrm{B}}^{\alpha}}{\rho}. \tag{3.35}$$

As $v = \partial\rho / \partial t$, the growth rate of a spherical precipitate in the quasi-stationary approach finally reads

$$\frac{\partial \rho}{\partial t} = \frac{\left(c_{\mathrm{B}}^{0} - c_{\mathrm{B}}^{\alpha\beta} \right)}{\left(c_{\mathrm{B}}^{\beta\alpha} - c_{\mathrm{B}}^{\alpha\beta} \right)} \frac{D_{\mathrm{B}}^{\alpha}}{\rho}. \tag{3.36}$$

Using the dimensionless supersaturation parameter, S, defined in the previous section, we arrive at

---

[6] See also Section 7.2.1.

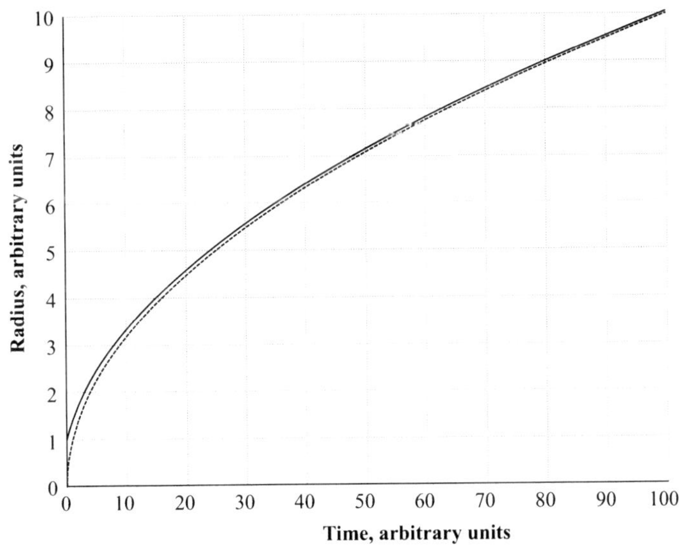

Figure 3.9. Parabolic law, radius versus time, for diffusion-controlled growth of spherical precipitates. Solid line is drawn for $\rho_0 = 1$, dashed line for $\rho_0 = 0$.

$$\frac{\partial \rho}{\partial t} = S\frac{D_B^\alpha}{\rho}, \tag{3.37}$$

$$\rho\dot{\rho} = D_B^\alpha S, \tag{3.38}$$

$$\rho(t) = \sqrt{\rho_0^2 + 2SD_B^\alpha t}, \tag{3.39}$$

with ρ_0 being the radius of the precipitate at $t = 0$.

From this treatment it is apparent that the precipitate radius scales with the square root of time. This behavior is generally known as the *parabolic law* for diffusion-controlled growth. Figure 3.9 displays the generic evolution of the precipitate radius versus time for $S = 1$ and $D = 1$. Figure 3.10 displays the radius of the precipitate versus the square root of time, yielding a straight line for the initial radius $\rho_0 = 0$.

3.2.4 Analytical Solution for High and Low Dimensionless Supersaturation

In the previous section, closed analytical solutions of the moving boundary problem were introduced considering the limiting cases of either the nominal B-content of the matrix being close to the precipitate B-content, $c_B^{\beta\alpha}$, or close to the matrix equilibrium B-content, $c_B^{\alpha\beta}$. In this section, a general solution for the moving boundary problem involving one diffusing species will be introduced. For a complete derivation, see, for instance, the treatment of M. Glicksman in his textbook [14] on solid-state diffusion.

Consider the growth of a spherical precipitate with homogeneous chemical composition, $c_B^{\beta\alpha}$. As there will be no diffusive fluxes occurring inside the precipitate, Fick's second law must be defined only outside the precipitate. For the spherical case, one has

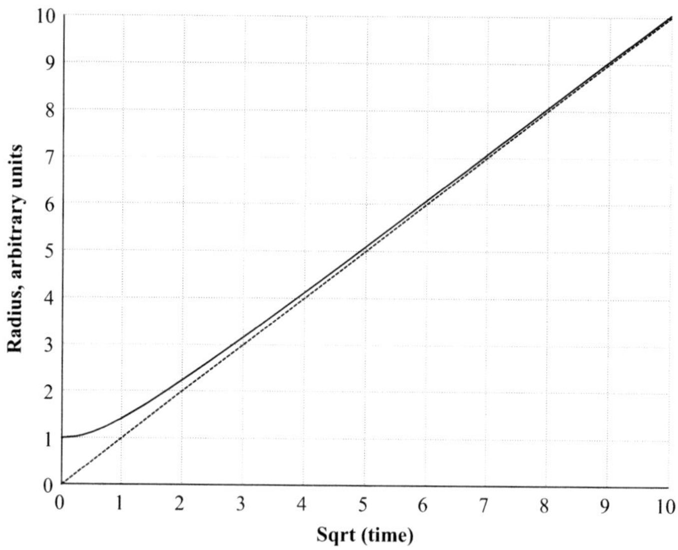

Figure 3.10. Parabolic law, radius versus square root of time, for diffusion-controlled growth of spherical precipitates. Solid line is drawn for $\rho_0 = 1$, dashed line for $\rho_0 = 0$.

$$\frac{\partial c_{\mathrm{B}}}{\partial t} = D_{\mathrm{B}}^{\alpha} \frac{\partial}{\partial r}\left[r^2 \frac{\partial c_{\mathrm{B}}}{\partial r}\right], r > \rho. \tag{3.40}$$

At the moving interface, mass conservation holds with

$$\left(c_{\mathrm{B}}^{\beta\alpha} - c_{\mathrm{B}}^{\alpha\beta}\right) \cdot \frac{\partial \rho}{\partial t} = D_{\mathrm{B}}^{\alpha} \frac{\partial c_{\mathrm{B}}}{\partial r}. \tag{3.41}$$

For a general solution of the moving boundary problem, equations (3.40) and (3.41) must be solved simultaneously. As the position of the boundary is not stationary in the general case, a coordinate transformation into a moving frame of reference is introduced with

$$\eta = \frac{r - \rho}{\sqrt{4 D_{\mathrm{B}}^{\alpha} t}}. \tag{3.42}$$

After performing this substitution and some arithmetic, which is elaborated in ref. [14], the governing equation (3.40) adopts the relatively simple form

$$-2\left[\eta + \frac{1}{\eta}\right] \frac{\partial c_{\mathrm{B}}}{\partial \eta} = \frac{\partial^2 c_{\mathrm{B}}}{\partial \eta^2}. \tag{3.43}$$

On solution of this equation by partial integration and back substitution of the radius coordinates, the position of the interface can be expressed in terms of a coefficient, K, with

$$\rho(t) = \sqrt{\rho_0^2 + 4K^2 D_{\mathrm{B}}^{\alpha} t}. \tag{3.44}$$

On comparison of this equation with the quasi-stationary solution, equation (3.39), we recognize that the supersaturation parameter, S, has been replaced by the quantity $2K^2$.

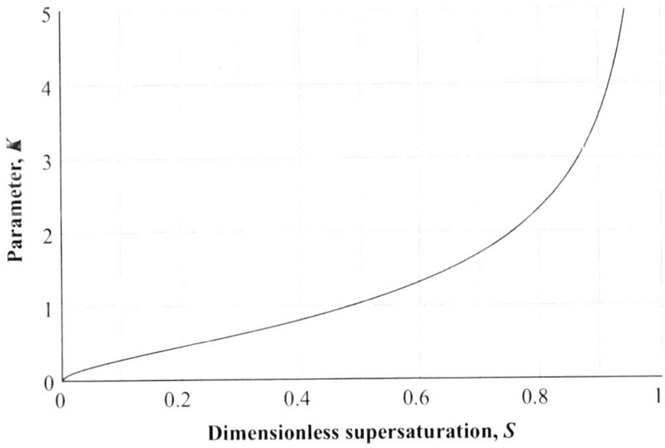

Figure 3.11. Relation between parameter, K, and dimensionless supersaturation, S, in exact solution of the moving boundary problem.

Unfortunately, K is not an explicit expression in the composition variables, $c_B^{\beta\alpha}$, $c_B^{\alpha\beta}$, and c_B^0, but it is implicitly given as

$$2K^2\left[1-\sqrt{\pi}Ke^{K^2}\,erfc(K)\right]=S. \tag{3.45}$$

In this form, the value of K can only be obtained numerically for given values of S. In Figure 3.11, the relation between the dimensionless supersaturation and the parameter, K, is displayed for $K \le 5$.

For a comparison with the growth rate expressions given in the previous section, an additional substitution can be introduced with

$$\Phi = 2K^2. \tag{3.46}$$

The parabolic growth law then becomes

$$\rho(t)=\sqrt{\rho_0^2+2\Phi D_B^\alpha t}, \tag{3.47}$$

and

$$\Phi\left[1-\sqrt{\pi\frac{\Phi}{2}}\cdot e^{\frac{\Phi}{2}}\cdot erfc(\sqrt{\frac{\Phi}{2}})\right]=S. \tag{3.48}$$

Figure 3.12 displays the growth parameter, α'_{exact}, of the exact solution

$$\alpha'_{\text{exact}}=\sqrt{2\Phi}, \tag{3.49}$$

together with Zener's solutions derived earlier in Section 3.2.2 as a function of the dimensionless supersaturation, S.

It is important to recognize that the present solution to the moving boundary problem is valid for all supersaturations. In the limiting case of $S \to 0$, which corresponds to the case of low supersaturation, one obtains

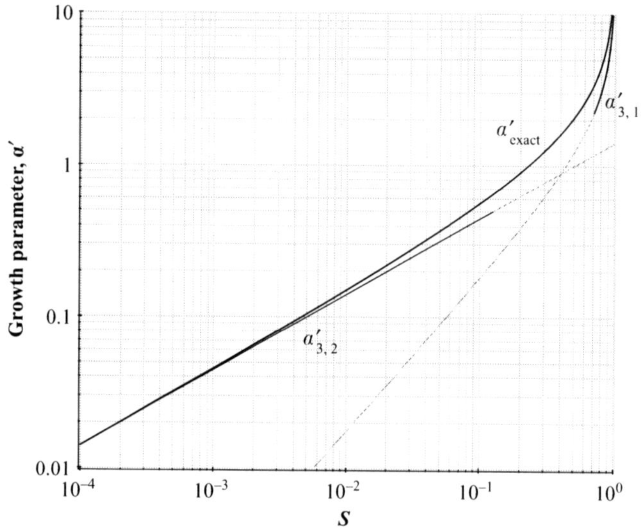

Figure 3.12. General solution of diffusion-controlled growth of spherical particle compared to Zener's solution in dependence of the composition parameter, S.

$$K\big|_{S\to 0} = \sqrt{\frac{S}{2}}. \tag{3.50}$$

In this case, the well-known quasi-stationary solution of the moving boundary solution given in Section 3.2.3, equation (3.36), is recovered.

For $S \to 1$, K approaches a value of

$$K\big|_{S\to 1} = \sqrt{\frac{3}{2(1-S)}}. \tag{3.51}$$

With K diverging as $S \to 1$, we find that the growth rate becomes infinitely fast in the limiting case of a dimensionless supersaturation of 1.

3.2.5 Influence of Capillarity on Precipitate Growth

The previous treatment of precipitate growth has been developed on the basis of known interfacial compositions, $c_{\mathrm{B}}^{\beta\alpha}$ and $c_{\mathrm{B}}^{\alpha\beta}$, in both the precipitate and the matrix phase. With the local equilibrium hypothesis and for the case of binary alloys, these values are read directly from the phase diagram. For small precipitates, however, the composition values at the interface can deviate substantially from the thermodynamic equilibrium values due to interfacial pressure. This observation is known as the Gibbs–Thomson effect and it is discussed in Section 1.5 on the effect of interfacial curvature and pressure on the Gibbs energy.

On basis of the Gibbs energy versus composition (G-X) diagram, equation (1.66) has been derived, describing the composition, $X_{\mathrm{B}}^{\alpha\beta'}$, of the interface on the matrix side as a function of the radius, ρ, of the precipitate with

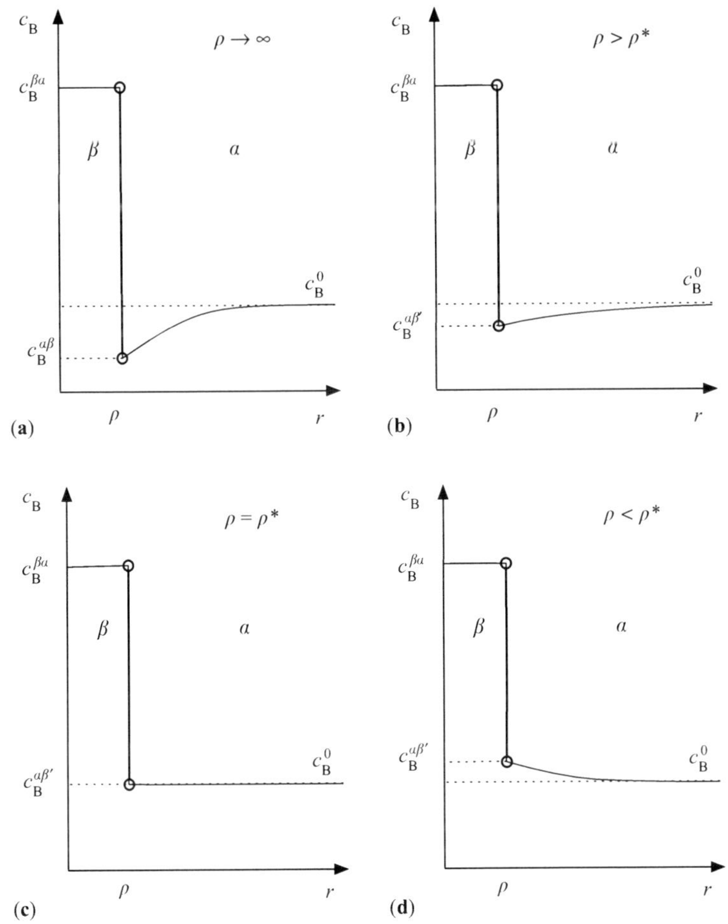

Figure 3.13. Influence of precipitate radius, ρ, on the composition of the interface, $X_B^{\alpha\beta'}$, on the matrix side. ρ^* is the critical radius where the precipitate is in equilibrium with the supersaturated matrix.

$$X_B^{\alpha\beta'} \approx X_B^{\alpha\beta} \cdot \exp\left(\frac{2\gamma v^\beta}{\left(X_B^{\beta\alpha} - X_B^{\alpha\beta}\right)RT} \cdot \frac{1}{\rho} \right). \tag{3.52}$$

According to this equation, the modified interfacial composition, $X_B^{\alpha\beta'}$, is identical to the local equilibrium value, $X_B^{\alpha\beta}$, in the limit of $\rho \to \infty$. In that case, the exponential term becomes unity. For smaller values of ρ, the exponential term adopts values larger than unity and the interfacial composition, $X_B^{\alpha\beta'}$, increases compared to the equilibrium value.[7] It is important to recognize the dependency of the interfacial composition, $X_B^{\alpha\beta'}$, with $1/\rho$, which emphasizes that the Gibbs–Thomson effect will be progressively more important with decreasing precipitate size, while it will be negligibly small for comparably large particles.

Figure 3.13 sketches the influence of interfacial curvature on the composition profiles around a stoichiometric precipitate. The four graphs display situations where the radius, ρ, of

---

[7] We will encounter the Gibbs–Thomson effect later again in the discussion of the Ostwald ripening process in precipitate coarsening.

the precipitate has typical values in comparison to the critical radius, ρ^*. The latter quantity denotes the particular radius where the precipitate is in equilibrium with the supersaturated matrix.[8] Figure 3.13(a) depicts the situation corresponding to the equilibrium case of an infinitely large precipitate in the absence of interfacial curvature. The composition variables, $c_B^{\beta\alpha}$ and $c_B^{\alpha\beta}$, are read directly from the equilibrium phase diagram in this case.

In Figure 3.13(b), the composition profile is shown for a precipitate with a radius that is larger than the critical radius, ρ^*, however, already small enough such that the Gibbs–Thomson effect becomes significant. The matrix composition, $X_B^{\alpha\beta'}$, has increased and has come closer to the nominal composition, X_B^0. As a consequence, the composition gradient at the interface is smaller now and the flux of atoms into the interface is reduced. According to the mass balance equation (3.41), the interface velocity $\partial\rho/\partial t$ reduces proportionally and growth of the precipitate is retarded.

Figure 3.13(c) displays the situation where the Gibbs–Thomson effect just compensates the composition difference between equilibrium content, $c_B^{\alpha\beta}$, and nominal B-content, X_B^0. In this case, the composition gradient in the matrix vanishes and the precipitate neither grows nor shrinks. It is in equilibrium with the supersaturated matrix. Figure 3.13(d), finally, shows a precipitate with radius smaller than the critical radius, ρ^*. In this case, a negative composition gradient develops in the matrix which causes B-atoms to diffuse away from the precipitate–matrix interface. The preferential removal of atoms leads to a dissolution of the precipitate which will be even accelerated with decreasing size.

P. Rivera-Diaz-del-Castillo and H. K. D. H. Bhadeshia [49] have analyzed the impact of capillarity on the growth of spherical precipitates. They follow the treatment of Zener, which was outlined before, and define a growth parameter, α_ρ, with

$$\rho = \alpha_\rho \sqrt{D_B^\alpha t}. \tag{3.53}$$

For the size-dependent growth parameter, α_ρ, they obtain the asymptotic solution for S<<, with S given by equation (3.28), as

$$\alpha_\rho = \left(\frac{2S\left[1 - \frac{\rho^*}{\rho}\right]}{1 - S\frac{\rho^*}{\rho}} \right)^{\frac{1}{2}}. \tag{3.54}$$

Figure 3.14 shows the normalized growth parameter, α_ρ, calculated for a supersaturation parameter, $S = 10^{-4}$. The critical radius is assumed to be 10^{-9} m. The influence of curvature becomes significant as soon as the radius of the precipitate is lower than approximately 10 times the critical radius, ρ^*. Close to ρ^*, the growth constant goes quickly toward zero, emphasizing the importance of capillarity for smaller particles.

---

[8] Compare also Section 2.2.1 on the thermodynamics of the critical nucleus.

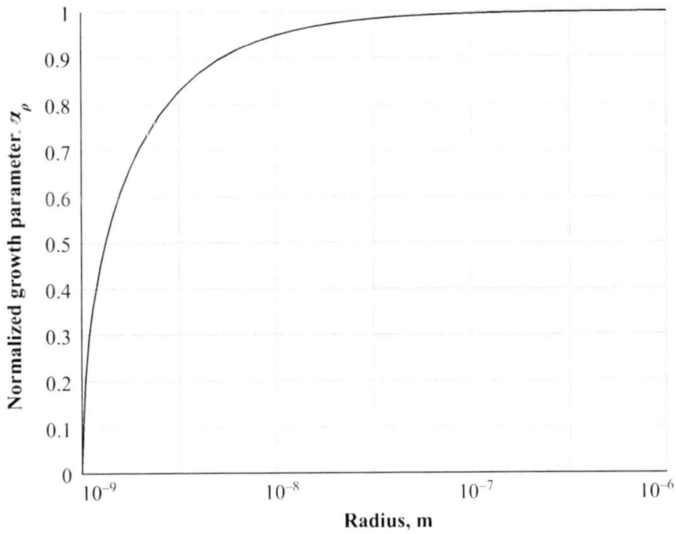

Figure 3.14. Normalized growth parameter in the limit of S<< and taking into account the effect of capillarity. The critical radius is assumed to be 10^{-9} m.

3.3 Multicomponent Diffusion-Controlled Growth

In the previous sections, solutions to the moving boundary problem were sought for cases where the interfacial compositions could be read directly from the underlying binary phase diagram. In this special case, the moving boundary problem for diffusion-controlled growth can be straightforwardly solved. For higher-order alloys, the situation at the moving phase boundary is considerably more complex and requires sophisticated numerical solution in the general case. In this section, the multicomponent problem will be formulated, first, and correspondence with the local equilibrium hypothesis and the phase diagram is investigated. Finally, approximate solutions for special cases are introduced and discussed which can be utilized in practical phase transformation simulations.

3.3.1 The Multicomponent Local Equilibrium Tie-Lines

The general formulation of the multicomponent moving boundary problem has been elaborated already in equation (3.10), which represents a system of coupled, non-linear equations with $2(n-1)+1$ unknown parameters. These are the velocity of the phase boundary, v, as well as two interfacial composition variables for each diffusing species. The number of compositional variables is $2(n-1)$ because one element can always be expressed in terms of the other components.

For the case of a ternary system, A-B-C, the two governing equations read

$$\left(c_{\mathrm{B}}^{\beta\alpha} - c_{\mathrm{B}}^{\alpha\beta}\right) \cdot v = J_{\mathrm{B}}^{\alpha\beta} + J_{\mathrm{B}}^{\beta\alpha}, \tag{3.55}$$

$$\left(c_{\mathrm{C}}^{\beta\alpha} - c_{\mathrm{C}}^{\alpha\beta}\right) \cdot v = J_{\mathrm{C}}^{\alpha\beta} + J_{\mathrm{C}}^{\beta\alpha}. \tag{3.56}$$

Component A is assumed to be the reference species. With Fick's first law, these equations can be rewritten as

$$\left(c_{\mathrm{B}}^{\beta\alpha} - c_{\mathrm{B}}^{\alpha\beta}\right) \cdot v = -D_{\mathrm{B}}^{\alpha} \frac{\partial c_{\mathrm{B}}^{\alpha\beta}}{\partial r} - D_{\mathrm{B}}^{\beta} \frac{\partial c_{\mathrm{B}}^{\beta\alpha}}{\partial r}, \tag{3.57}$$

$$\left(c_{\mathrm{C}}^{\beta\alpha} - c_{\mathrm{C}}^{\alpha\beta}\right) \cdot v = -D_{\mathrm{C}}^{\alpha} \frac{\partial c_{\mathrm{C}}^{\alpha\beta}}{\partial r} - D_{\mathrm{C}}^{\beta} \frac{\partial c_{\mathrm{C}}^{\beta\alpha}}{\partial r}. \tag{3.58}$$

This nonlinear system of coupled equations has five unknowns, which are the interfacial compositions, $c_{\mathrm{B}}^{\beta\alpha}$ and $c_{\mathrm{C}}^{\beta\alpha}$, on the precipitate side of the interface, the interfacial compositions, $c_{\mathrm{B}}^{\alpha\beta}$ and $c_{\mathrm{C}}^{\alpha\beta}$, on the matrix side of the interface, and the interface velocity, v. For higher-order systems, equations for each additional element are simply appended, keeping the basic solution procedure otherwise identical.

A solution of this system of equations must be sought, for the boundary condition that v obtained from each equation delivers the same value. The interfacial compositions can be determined having in mind that based on the local equilibrium hypothesis these values are not independent, but coupled through the thermodynamic properties of the system. The fundamental relation between the interfacial compositions and the interfacial diffusive fluxes, or in general, the diffusion properties of the components, is outlined on the examples of precipitation of a B-rich and C-poor phase a from a phase γ and another example for precipitation of a B- and C-rich phase, Θ, in an A-B-C system.

For the first example, we follow the treatment of Coates [50,51], who considered the precipitation of body-centered cubic ferrite from face-centered cubic austenite in the Fe-rich corner of the Fe-Mn-C system. Since the treatment of Coates is of a rather general nature, we continue using the generic symbols A, B, and C for the three components, where A corresponds to the matrix element Fe, and B and C represent the substitutional and interstitial minority components Mn and C. It must be mentioned, though, that there exist numerous other treatments of multicomponent local equilibrium phase transformations, some of them being the works of M. Hillert [52,53], G. Purdy and coworkers [54], and H. Aaronson and coworkers [55].

Let an alloy be annealed in the γ one-phase region and subsequently be quenched into the γ/a two-phase field. The corresponding phase diagram is depicted in Figure 3.15 for a temperature where the alloy with nominal composition, c_{B}^{0} andc_{C}^{0}, is located in the two-phase region. A circle filled in gray marks this point. Between the phase boundaries, straight *tie-lines* connect points of chemical composition where the two phases are in equilibrium. In the binary phase diagram, these lines have appeared as horizontal lines at a single temperature from which the interfacial compositions could be identified in a unique and direct manner. In ternary and higher-order alloys, it is no longer possible to identify a single interfacial composition. Instead, an infinite number of tie-lines can be found that potentially satisfy the local equilibrium condition.

In Figure 3.15, the bold curved line sketches a possible choice of interfacial composition (IC) contour between the points 1 and 2 on the phase boundary. Naturally, this curve passes through the point of nominal composition. It can be drawn, however, for any choice, which is located inside the isoactivity lines of the two components indicated by dashed horizontal and vertical lines.[9]

[9]The isoactivity lines are drawn as horizontal and vertical lines for the benefit of simplicity. The slope can deviate from this simple configuration, however, in systems with strong thermodynamic interaction between the components.

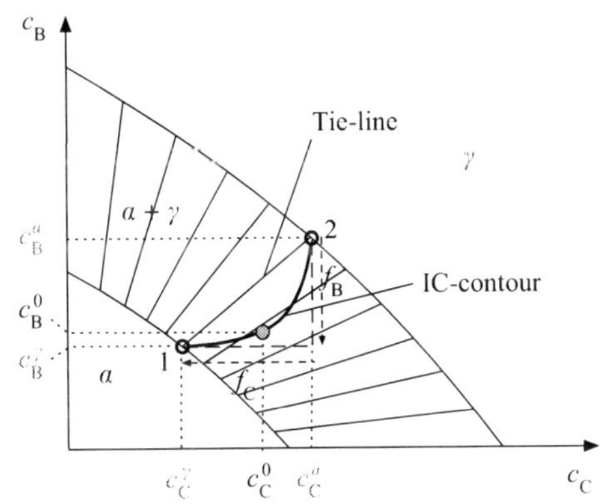

Figure 3.15. Ternary phase diagram for the transformation of a phase γ into a phase α in the ternary system A-B-C.

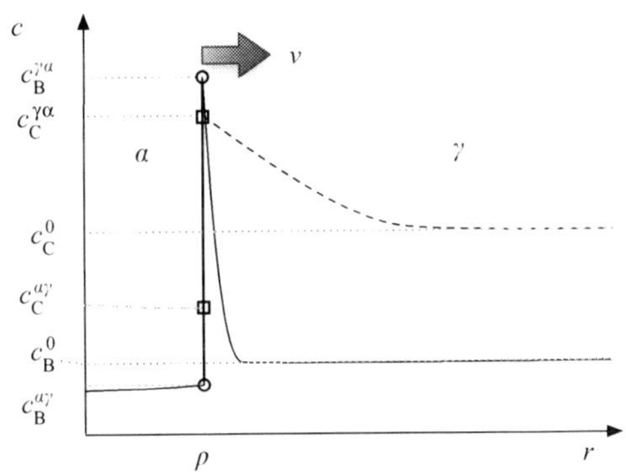

Figure 3.16. Composition profile around a moving phase boundary corresponding to the case shown in Figure 3.15.

Figure 3.16 displays composition profiles corresponding to this particular phase transformation. The image shows that long-range diffusion of both, components B and C, is necessary to progress the transformation. An arrow indicates the direction of movement of the phase boundary. As the diffusion profiles in the γ-phase exhibit significantly different slopes, it is evident that the diffusivity of component C is assumed to be considerably higher than the diffusivity of component B in this example. The interface composition of the former component is marked by open rectangles, and that of the latter component by open circles.

The choice of IC-contour is determined by the diffusional characteristic of the two components and the local composition gradients at the phase boundary. With the previously claimed requirement of identical interface velocity, v, for each of the governing equations (3.10), for solution of the multicomponent moving boundary problem, an IC-contour must be sought which satisfies the mass balance conditions and simultaneously delivers compatible interface velocities, v. For identical diffusivities of components B and C, the operating IC-contour, for instance, corresponds exactly to the tie-line passing through the point of nominal composition.

Since any choice of IC-contour inside the isoactivity limits of the two elements satisfies the local equilibrium condition, it is important to investigate the consequences of choice of IC-contour on the kinetics of the transformation process. When starting at the point of nominal composition located on the IC-contour in Figure 3.15 and looking at the composition profiles in Figure 3.16, we find that diffusion of both, components B and C, is necessary in order to progress the phase transformation front. The interfacial composition jumps have nonzero values for both components. This situation is typical for two diffusing species with similar diffusion coefficients.

If the diffusivities of the two diffusing species are widely different, as it is the case, for instance, if the transformation involves substitutional and interstitial components,[10] there exist two possibilities for how mass conservation can be maintained in equations (3.10) with compatible interface velocities and local thermodynamic equilibrium:

- If the faster diffusing component controls the transformation, the slower components can compensate for sluggish diffusion kinetics (low value of the diffusion coefficient, D) by minimizing the effective composition jump across the interface. Even if the transformation proceeds at a fast rate, only minimum *long-range* diffusion ahead of the moving interface occurs in this case, as parent and product phase have close to identical chemical composition for the slow diffusing species. Under these conditions, a compositional *spike* develops for the slow diffusing components. Diffusion occurs mainly *inside* the interface and in a very small region in front of it. This case is typical for the initial transformation stage where interface migration occurs under full driving force.
- If the slower diffusing component controls the transformation, the faster component compensates for its fast diffusion kinetics with a minimization of the chemical potential gradient across the interface. This situation typically occurs after equilibration of the chemical potentials of the faster constituents, that is, in the later stages of the reaction, or under conditions of low transformation driving force.

Typical composition profiles for both cases are sketched in Figure 3.17. In Figure 3.17(a), the transformation is controlled by the faster component C and the interface migration rate is assumed to be high. This is indicated by a large arrow in the direction of interface movement. The composition gradient of the sluggish component B adapts a very steep slope and has only minimum extension ahead of the interface. Diffusion occurs mostly within the interface and a very narrow region ahead of it. The most important ingredient of this type of transformation is the (almost) identical content of B in the parent and the product phases; this is indicated by bold letters in the figure. Under these conditions, no long-range diffusion of the sluggish component B occurs during the transformation.

In Figure 3.17(b), the transformation is controlled by diffusion of the sluggish component B. In order to maintain mass conservation during transformation, only a small amount of mass of the fast diffusing component C is transferred across the interface. This requirement is

[10]The difference between diffusion coefficients of interstitial and substitutional elements is often as large as six to eight orders of magnitude. In many treatments of diffusion-controlled phase transformations, the slowly diffusing substitutional components are therefore considered as the rate-controlling species in the transformation process, whereas the diffusion of interstitial components is approximated as being infinitely fast, so that it need not be considered at all.

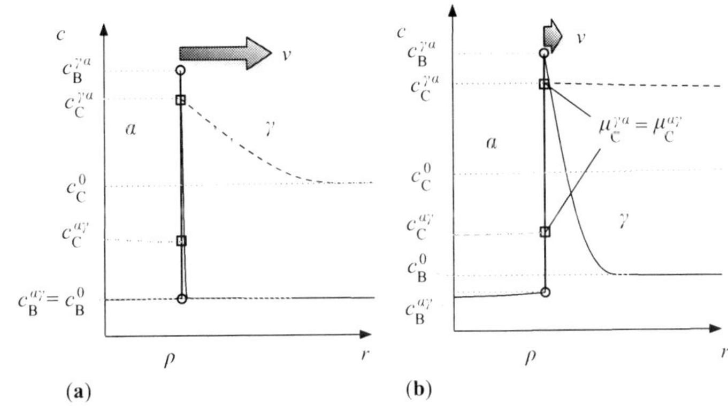

Figure 3.17. Composition profiles for the transformation of a phase γ into a phase α in the A-B-C ternary system with different limiting cases for the diffusion kinetics of the two species B and C with (a) Zener's limiting case 1 and (b) Zener's case 2.

attained by minimizing the chemical potential gradient between the two sides of the interface. Although the diffusion kinetics of C is rapid, the diffusional flux, J, is kept at a minimum because the driving force for diffusion, the chemical potential gradient, is minimized.[11] The flux, J, is just large enough to compensate the mass transfer of C due to interface migration with a composition jump across the interface.

Coates [50,51] has further investigated the relation between interface composition and diffusivities of the participating atomic species using the *fractional composition* variables, f_B and f_C, as defined in Figure 3.15. The fractional compositions of both components, f_B and f_C, are zero at the α side of the interface and unity at the γ side. If one plots the IC-contours in dependence of the ratio of diffusivities, D_B/D_C, over the fractional compositions, f, the graph shown in Figure 3.18 is obtained. For identical diffusivities of components B and C, the IC-contour is a straight line connecting the limiting compositions. The line is gradually more distorted until the line running along the axes is obtained for a value of infinity.

3.3.2 Fast and Slow Local Equilibrium Transformation Regions

With the concept of IC-contours as developed in the previous section, it is possible to further investigate the relation between composition and interface velocity. Consider the situation shown in the schematic phase diagram in Figure 3.15. For each composition given along the IC-contour, a characteristic interface velocity can be calculated, where v gradually goes from $v = 0$, for $f_B = f_C = 0$ to $v = \infty$, for $f_B = f_C = 1$. Consequently, for all possible compositions within the two-phase region of the phase diagram, lines of identical interface velocity can be drawn. This is shown in the schematic graph in Figure 3.19.

If the diffusivities of the two components, B and C, are identical, the IC-contours coincide with the tie-lines and the interface velocity IV- contours are distributed equidistantly within the

[11] The diffusive flux, J, is the product of diffusional mobility, M, and the chemical potential gradient, $\partial\mu_C/\partial x$.

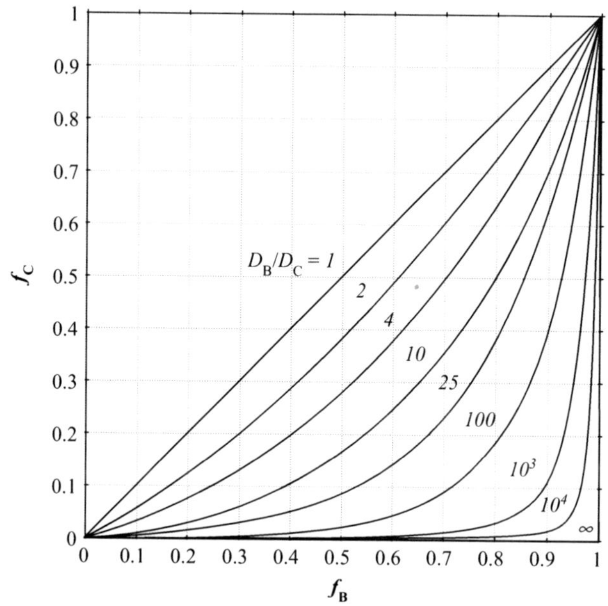

Figure 3.18. IC-contours as a function of fractional compositions, *f*, calculated for different ratio of diffusivity of components B and C.

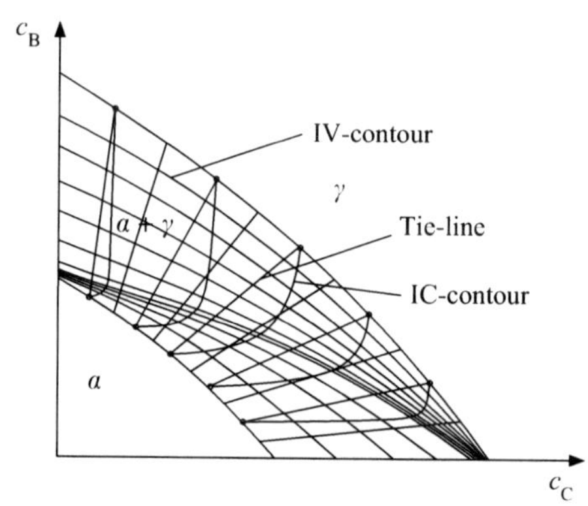

Figure 3.19. Ternary phase diagram for the transformation of a phase γ into a phase α in the A-B-C ternary system.

two-phase field. The IV-contours shown in Figure 3.19 correspond to a situation where the diffusivity of component C is higher than the diffusivity of B.

The relation between IV-contours and the diffusivity ratios between B and C has been determined in a work by Coates [50,51], the results of which are summarized in Figure 3.20. In Figure 3.20(a), the diffusivity of both components is identical, with the result that the IV-contours are straight parallel lines. Already, for rather small differences in diffusivity, for example, a factor of 4 as shown in Figure 3.20(b), the lines change slope and are no longer parallel.

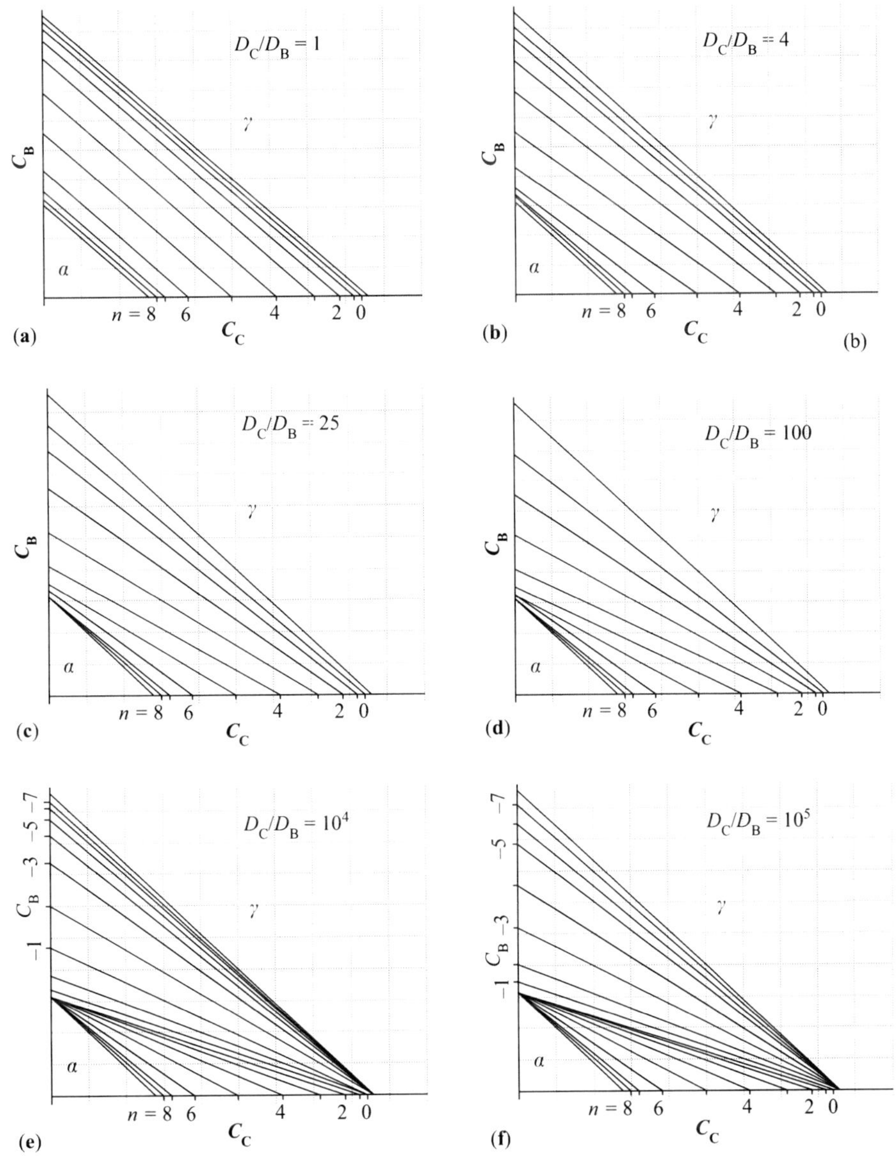

Figure 3.20. IV-contours for ternary phase transformation with different ratio of diffusion coefficients for components B and C.

The numbers n in the figure denote IV-contours that are different in value by a factor of 2, where $n = 0$ denotes a velocity contour of some given (small) value, v_0. The interface velocity of all other contours can be calculated relative to this reference velocity by the relation $v_n = v_0 \cdot 2^n$. For $D_C / D_B \geq 10^4$, the IV-contours for values smaller than the reference velocity, v_0, are also drawn, with negative values of n in this case.

Interestingly, the IV-contours for larger diffusivity differences clearly separate two regions of the phase diagram, where the transformation proceeds either in a fast or a slow mode. In the fast transformation region, interface migration is controlled by the diffusion of the fast component C. IC-contours are selected, in this case, which provide identical B-content of the parent and the product phase. In the slow transformation region, the transformation kinetics is controlled by the slow component B. The fast component C is characterized by a minimum chemical potential gradient across the interface, which provides compatible diffusional flux values to maintain mass conservation.

3.3.3 Local Equilibrium Controlled Precipitation in Multicomponent Systems

The evolution of precipitation processes in multicomponent systems is usually a complex process where the precipitation reaction often changes its appearance throughout the transformation. The complex character of multicomponent precipitation reactions is demonstrated on the example of $(Fe,Cr)_{23}C_6$ precipitation in the ternary Fe–Cr–C system. This case has been investigated in detail by Schneider and Inden [56], and it is reproduced in part in this section.

Consider the ternary system Fe–12wt%Cr–0.1wt%C with the phase α representing the ferritic steel matrix, and the phase θ representing the $M_{23}C_6$-type Fe–Cr carbide. Assume, for this example, that there exist no metastable precursor phases to the thermodynamically stable $(Fe,Cr)_{23}C_6$ precipitate. The phase diagram for this reaction is shown in Figure 3.21.

Let us assume that the reaction starts with a supercritical precipitate having the chemical composition given by thermodynamic equilibrium and yielding the maximum chemical driving force.[12] Figure 3.22 displays the evolution of the interface Cr-composition in the θ precipitate as well as in the α matrix as a function of the precipitate radius. The Cr-content of the nucleus yielding maximum driving force is around 66 wt% in the very early stage of the reaction. Interestingly, very short time after growth has started, the interface Cr-content on the precipitate side of the interface decreases to a value around 32 wt% and reaches a temporary plateau at around 10 s. Growth of the $(Fe,Cr)_{23}C_6$ precipitate then proceeds with this Cr-content until the radius of the precipitate increases above 0.5 μm.

The behavior of the precipitation system, as evaluated by Schneider and Inden, is interesting because using the operating tie-line with a reduced Cr-content in the precipitate, compared to the full equilibrium value with maximum chemical driving force, allows the system to grow the precipitate at a much faster rate. Less Cr-diffusion is necessary to grow the precipitate to the same radius because the Fe/Cr ratio in this case is closer to the Fe/Cr ratio of the parent α matrix. Only after supersaturation of the matrix has substantially decreased at later stages (10,000 s and more), growth of the precipitate occurs with an operating tie-line that corresponds to the full equilibrium composition of the precipitate.

If only the formation of $(Fe,Cr)_{23}C_6$ is considered in the simulation, the reaction would come to a stop shortly after 10,000 s. In the work of Schneider and Inden, however, simultaneous precipitation of $(Fe,Cr)_{23}C_6$, $(Fe,Cr)_7C_3$, and $(Fe,Cr)_3C$ has been simulated. The last stage of the reaction shown in Figure 3.22 involves the dissolution of metastable phases and must, thus, be considered as an additional reaction. For further details, the interested reader is referred to the original article [56].

[12] As the precipitate phase fraction is not the equilibrium value in this stage, the maximum driving force is evaluated by the tangent construction. See Section 1.4 on the chemical driving force for details.

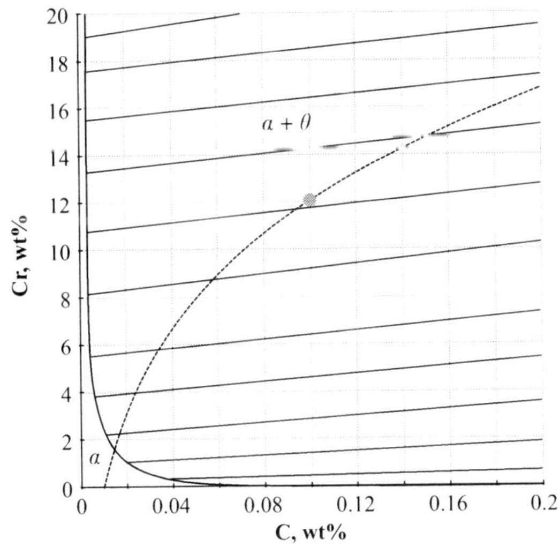

Figure 3.21. Phase diagram for precipitation of θ, $(Fe,Cr)_{23}C_6$, in a Fe–12wt%Cr–0.1wt%C. Nominal composition indicated by a filled gray circle. The dashed line displays the isoactivity line for C.

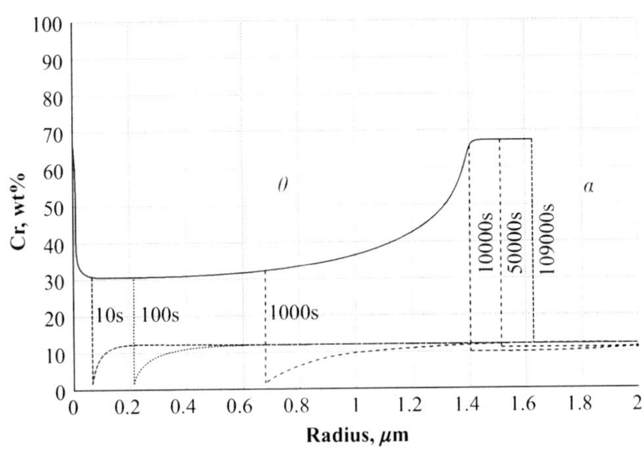

Figure 3.22. Simulated evolution of the Cr-content during precipitation of θ, $(Fe,Cr)_{23}C_6$, in the Fe–12wt%Cr–0.1wt%C system [56]. The numbers on the Cr-profiles indicate different times throughout the reaction.

For the time being, it is important to recognize that multicomponent precipitation reactions and the identification of operative tie-lines can be rather involved. It is essential to understand that the operative tie-line usually evolves throughout the reaction and does not remain constant in the general precipitation case. For this reason, there exist no general closed analytical solutions for the multicomponent growth of precipitates. In the next section, an approximate solution developed at the KTH Stockholm by J. Ågren and coworkers is introduced, which allows for a treatment of diffusion-controlled reactions without explicitly considering the diffusion fields around the interface. An entirely different model is presented later in Section 3.5, where a mean-field treatment of precipitate growth is discussed, which is developed on basis of the thermodynamic extremal principle.

3.3.4 Approximate Treatment of Multinary Diffusional Transformations

In the previous section, an example for the multicomponent moving boundary problem, equation (3.10), had been presented and discussed. Solution of this problem has been obtained numerically using the software DICTRA, which solves the coupled problem of simultaneous multicomponent long-range diffusion and interface movement. Unfortunately, this procedure is computationally rather expensive and treatment of precipitate growth in this approach is practically restricted to a comparably small number of interacting precipitates. In ref. [56], DICTRA cell simulations have been presented for the interaction of up to three precipitates in multicomponent environment. For many applications, where a higher number of precipitates should be treated, simplified models are required which can be solved more rapidly by numerical methods.

One possibility for simplification is to replace the time-consuming explicit treatment of concentration profiles around the precipitate and inside it by suitable analytical profiles. This task has been performed in the group of J. Ågren [57], whose treatment is reviewed here. The solution process starts with the binary case of a spherical precipitate. This result is then reformulated by introducing an effective diffusion distance factor. In the last step, a set of multicomponent growth rate expressions is derived, which takes into account all individual effective diffusion distances and local thermodynamic equilibrium at the interface. Main features of this solution are that the final growth rate expression is valid for small and large supersaturation equally and that it contains all multicomponent diffusion cross terms.

Consider a spherical precipitate with homogeneous composition in an infinite matrix with elements A and B and a nominal B-concentration, c_{B}^{0}. The compositions at the precipitate–matrix interface shall be given with $c_{\mathrm{B}}^{\beta\alpha}$ and $c_{\mathrm{B}}^{\alpha\beta}$. The growth rate of a precipitate in a binary system has been derived earlier in Section 3.2.4. Accordingly, the exact analytical solution, as given in equation (3.44), reads

$$\rho(t) = \sqrt{\rho_0^2 + 4K^2 D_{\mathrm{B}}^{\alpha} t}, \tag{3.59}$$

with the parameter K given implicitly by equation (3.45). Upon derivation, the growth rate is obtained as

$$v = \frac{2K^2 D_{\mathrm{B}}^{\alpha}}{\rho}. \tag{3.60}$$

Simultaneously, the growth rate can be obtained from the general form of the mass conservation law, equation (3.8), as

$$[[c_{\mathrm{B}}^{\alpha\beta}]] \cdot v = J_{\mathrm{B}}^{\alpha\beta} = D_{\mathrm{B}}^{\alpha} \nabla c_{\mathrm{B}}^{\alpha\beta}, \tag{3.61}$$

where use has been made of Fick's first law and the assumption of zero diffusional flux inside the precipitate. This condition implies a homogeneous precipitate composition. On replacing the gradient at the interface by a difference expression, analogously to Zener's treatment discussed in Section 3.2.2, one obtains

$$\left(c_{\mathrm{B}}^{\beta\alpha} - c_{\mathrm{B}}^{\alpha\beta}\right) \cdot v = D_{\mathrm{B}}^{\alpha} \frac{\left(c_{\mathrm{B}}^{0} - c_{\mathrm{B}}^{\alpha\beta}\right)}{d_{\mathrm{B}}}, \tag{3.62}$$

with d_{B} representing the characteristic diffusion length of element B. With the dimensionless supersaturation, S, as defined in equation (3.28), this expression becomes

$$v = D_{\mathrm{B}}^{\alpha} \frac{S}{d_{\mathrm{B}}}. \tag{3.63}$$

When comparing this result with equation (3.60), we observe the direct correspondence

$$d_{\mathrm{B}} = \frac{S}{2K_{\mathrm{B}}^{2}} \rho = \xi_{\mathrm{B}} \rho. \tag{3.64}$$

The parameter $\xi_{\mathrm{B}} = S / \left(2K_{\mathrm{B}}^{2}\right)$ represents a quantity that specifies an effective diffusion distance in relation to the precipitate radius as a function of the dimensionless supersaturation, S.

In multicomponent systems, we have (n–1) independent diffusing elements and, thus, the same number of equations for the flux balances with

$$\left(c_{i}^{\beta\alpha} - c_{i}^{\alpha\beta}\right) \cdot v = \sum_{j=1}^{n-1} D_{ij}^{\alpha} \nabla c_{j}^{\alpha}. \tag{3.65}$$

In this expression, full account is taken of the multicomponent nature of the diffusion coefficient which is expressed as a sum over all element gradients in the system.[13] These equations determine (n–1) conditions for the (2n–1) unknown variables and n represents the number of components in the system. With the local equilibrium hypothesis, the other n conditions can be fixed with

$$\mu_{i}^{\beta\alpha} = \mu_{i}^{\alpha\beta}. \tag{3.66}$$

In analogy to the approximation for one diffusing component shown earlier, the concentration gradients, ∇c_{j}^{α}, can be approximated by the effective diffusion distances, d_i, thus linking the interfacial compositions to the effective diffusion distances, $d_j = \xi_j \rho$, in the form of equation (3.62) with

$$\left(c_{i}^{\beta\alpha} - c_{i}^{\alpha\beta}\right) \cdot v = \sum_{j=1}^{n-1} \frac{D_{ij}^{\alpha} \left(c_{j}^{0} - c_{j}^{\alpha\beta}\right)}{\xi_{j} \rho}. \tag{3.67}$$

With equations (3.67) and (3.66), the precipitation problem is fully determined and can be solved numerically based on appropriate computational algorithms.

[13] See Section 7.2.5 on diffusion for details.

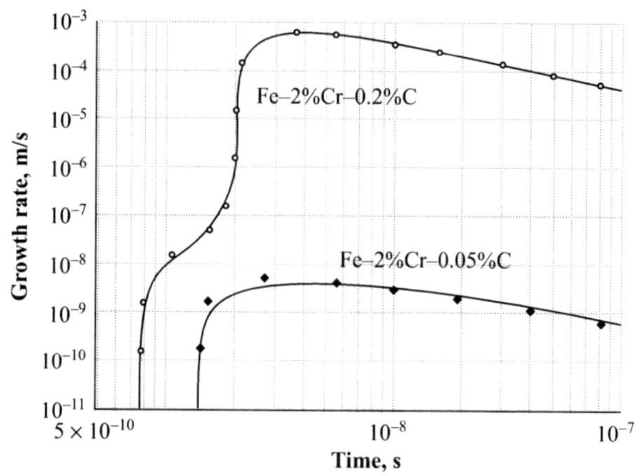

Figure 3.23. Growth rates for $M_{23}C_6$ precipitates obtained from the approximate solution (solid lines) and solution of the moving boundary problem with DICTRA (symbols) for two different alloys (composition in wt%).

Chen, Stepsson, and Ågren. [57] apply the model to the precipitation of $M_{23}C_6$ carbide particles in alloys with Fe–2wt%Cr and 0.2 and 0.05wt%C, respectively. In one case, they observe slow growth of the metastable precipitates in the slow growth mode controlled by the diffusion of substitutional Cr, in the other case they find fast growth controlled mainly by the fast diffusion of C. Their simulations are reproduced in Figure 3.23.

The solid lines in Figure 3.23 depict the results of the analytical approximation introduced earlier. The symbols represent solutions of the full diffusion problem performed with the software DICTRA, which was utilized for multicomponent precipitate growth simulations already in the previous Section 3.3.3, and which is introduced later in more detail in the software Section 9.1. In the simulations, an interfacial energy contribution of 0.4 J/m^2 has been used, which is responsible for the slow growth rates in the early stages of the reactions of the Fe–2wt%Cr–0.2wt%C alloy. The critical radius has been evaluated with 0.78 nm in the case of high supersaturation and around 1.3 nm in the case of low supersaturation.

If one considers the DICTRA calculation as the rigorous solution to the problem, these results demonstrate the capabilities of the approximate solution. On the basis of the example reproduced here, the authors of the computational study [57] show that the approximate solution is capable of reproducing the basic features of the considered reaction in good accuracy, being significantly less computationally expensive, though. Figure 3.24 displays the evolution of the interfacial Cr-composition throughout the precipitation reaction. The graph shows the composition variables in terms of u-fractions, which makes the results independent of the C-content and allows for a direct comparison of the substitutional element content of the individual phases.

For high supersaturation, indicated by the dashed lines, the Cr-content of the precipitate starts at a relatively high value. Before reaching 2 nm-radius, the Cr-content at the interface drops to the value given by the nominal composition of the matrix, however. From this point on, the growth rate of the precipitate is controlled by fast diffusion of the interstitial component C. The reaction does not involve the long-range transport of Cr, since both phases share the same Fe to Cr ratio.

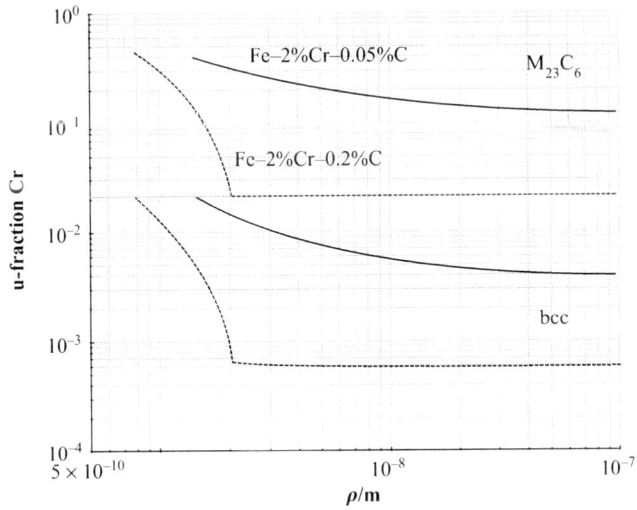

Figure 3.24. Evolution of the interfacial composition in the approximate multicomponent solution of $M_{23}C_6$ precipitation in Fe–2wt%Cr–x wt%C. The solid lines are calculated for low supersaturation (0.05wt%C) and the dashed lines for high supersaturation (0.2wt%C).

At low supersaturation, the critical nucleus is almost twice as large. Again, the reaction starts with the highest Cr-content in the precipitate, however, in this case, the precipitate grows with significant uptake of slow diffusing substitutional Cr. The growth rate is, therefore, several orders of magnitude slower, as shown previously in Figure 3.23. The chemical driving force and the mass conservation requirements at the moving interface do not allow for a fast transformation controlled by C.

3.4 Energy Dissipation at a Moving Phase Boundary—the Mixed-Mode Model

A phase transformation that involves different chemical composition of the parent and product phase is severely affected by the diffusion kinetics of components from one phase across the interface into the other phase. The process, where parent and product phases become enriched or depleted of certain elements is known as *partitioning*. A rather popular and convenient assumption to describe diffusion-controlled phase transformations is the local equilibrium concept and it has been introduced in the previous sections. In the local equilibrium hypothesis, one assumes that the diffusivity inside the interface is so rapid compared to the surrounding crystal that the two phases getting into contact immediately establish a state of *local* thermodynamic equilibrium. In binary systems, the interfacial composition on both sides of the interface is then straightforwardly read directly from the phase diagram. This observation implies, however, that there occurs no dissipation of energy inside the interface by any other process such as, for instance, frictional forces.

The assumption of zero energy dissipation inside the interface is a reasonable and convenient simplification for cases where the majority of the free energy dissipation occurs outside the interface by long-range diffusion. In the general case, however, one must also take into account that

- the process of transferring atoms from one crystal structure to another generally occurs under expense of energy. To progress the interface, atoms must be detached from the parent phase crystal lattice, transferred across the interface, and attached again to the product lattice. This process is similar in its nature to macroscopic friction.
- atoms can be "trapped" inside the interface and "dragged" along with the migrating phase boundary under expense of energy. The trapped atoms prefer to remain inside the phase boundary instead of being incorporated into the crystal lattice of the product phase. This process is known as *solute drag.*
- if product and parent phase have different molar volumes, condensation or evaporation of excess vacancies, or vacancies in a shortage can play a major role in relaxing volumetric misfit stresses. The annihilation or generation of vacancies dissipates energy and alters the interfacial migration rate.
- the migrating phase boundary encounters obstacles, such as small precipitates, and is either partially pinned or drags the particles along during migration. This process impacts the phase transformation rate by providing a retarding pressure on the phase boundary that causes a deviation of interface conditions from local equilibrium.

Quantitative treatment of these processes is often complex because it requires knowledge of several (essentially unknown) interaction parameters that are difficult to calculate on a theoretical basis, and equally difficult to measure. Thornton, Ågren, and Vorhees [58] have reviewed some of these effects and their impact on moving phase boundaries. These authors reference several comprehensive reviews going into more detail in specific aspects of solid-state phase transformations and moving boundary problem modeling.

When inspecting the total driving force acting on the phase boundary, Hillert [6] demonstrated that an expression for the *integral driving force* per mole of growing phase, $d_{\mathrm{F}}^{(m)}$, can be derived with

$$d_{\mathrm{F}}^{(m)} = -\sum_{i=0}^{n} X_i \left(\mu_i^{\beta} - \mu_i^{\alpha} \right), \tag{3.68}$$

where n is the number of components, X_i is the mole fraction, and μ is the chemical potential in phases α and β. During a phase transformation, the integral driving force is dissipated by various processes, such that one can write

$$d_{\mathrm{F}}^{(m)} = d_{\mathrm{F}}^{I} + d_{\mathrm{F}}^{II} + \ldots, \tag{3.69}$$

with the superscripts *I*, *II*, and so forth, denoting the particular process occurring in the transformation. The nature of these partial dissipation processes is explored subsequently.

In phase transformation modeling with moving boundaries, it is often convenient to consider two distinct mechanisms of energy dissipation. One represents long-range diffusional processes in the bulk volume surrounding the interface which dissipate energy by entropy production. The other mechanism involves the dissipation of energy by migration of the phase boundary against a retarding pressure. The integral driving force can then be split into two parts reading

$$d_{\mathrm{F}}^{(m)} = d_{\mathrm{F}}^{\mathrm{D}} + d_{\mathrm{F}}^{\mathrm{IF}} . \tag{3.70}$$

The total molar driving force, $d_{\mathrm{F}}^{(m)}$, is thus represented by a long-range diffusion related part, $d_{\mathrm{F}}^{\mathrm{D}}$, and an interface movement related part, $d_{\mathrm{F}}^{\mathrm{IF}}$. In the limit of $d_{\mathrm{F}}^{\mathrm{IF}} = 0$, we recover the purely *diffusion-controlled* reaction treated in the previous sections.

For the other extreme case with $d_{\mathrm{F}}^{\mathrm{D}} = 0$, the entire driving force is dissipated by interfacial processes and the phase transformation is entirely *interface-controlled.* In this case, a linear relation between the driving force and the interface velocity, v, holds with (see, for instance the comprehensive textbook of J. W. Christian [59])

$$v = M^{\mathrm{IF}} \cdot d_{\mathrm{F}}^{\mathrm{IF}} , \tag{3.71}$$

where M^{IF} is the *effective interface mobility*, in units of m^4/Js.

Since neither the diffusion-related dissipation term nor the interface-related term is commonly zero and vanishes completely, in the general case, a phase transformation occurs under simultaneous control of both dissipative mechanisms. This type of transformation is denoted as *mixed-mode*. Its nature and phenomenological description have been investigated by J. Sietsma and S. van der Zwaag [60]. Their treatment is reviewed here.

Consider a partitioning transformation where a binary A-B alloy which is constituted by a phase α is quenched into the α/β two-phase region. β is assumed to be the transformation phase forming inside α. Immediately after reaching the transformation temperature, a sufficiently large supercritical nucleus of the β phase shall be formed. The interface between the new phase β and the parent phase α will experience the integral driving force, $d_{\mathrm{F}}^{(m)}$, for (*a*) migration into the direction of the transforming phase α and (*b*) for changing the composition of the transformation phase β toward its equilibrium value. As a consequence of this driving pressure, the parent phase α starts to decompose into the phase β and a modified parent phase a′.

If we assume that there exists no gradient in composition in the growing phase, either because it is a stoichiometric phase or because diffusion in β is significantly faster than in α, the parameter determining the character of the transformation is the interfacial composition, $c_{\mathrm{B}}^{\alpha\beta}$, ahead of the interface. If the mobility of the interface is high compared to the diffusion kinetics of the partitioning elements, the B-content on the α-side of the interface will be close to its equilibrium value and the transformation will be mostly diffusion-controlled. If, in contrast, the effective mobility of the interface is low and the diffusion kinetics is fast, the transformation will be mainly interface-controlled and the B-content at the interface is lower by some amount.

Figure 3.25 shows three typical conditions at the interface of a transforming phase β growing into a matrix phase α. In Figure 3.25(a), the interfacial composition is fully determined by local thermodynamic equilibrium. This is reflected in the fact that both, the B-content of the transformation phase β as well as the B-content of the transforming phase α, are identical to their local equilibrium values, $c_{\mathrm{B}}^{\beta\alpha}$ and $c_{\mathrm{B}}^{\alpha\beta}$. This reaction, thus, occurs under full diffusion control with the interface mobility sufficiently high to have no conceivable influence on the reaction.

As B-atoms are rejected from the β phase in a partitioning transformation of the type considered here, and as they accumulate in front of the interface, the phase transformation process is accompanied by a gradient in B-concentration in the α phase and a diffusional flux of B away

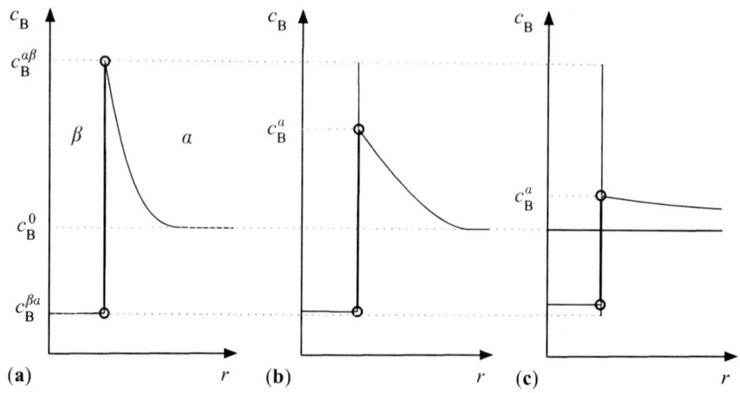

Figure 3.25. Evolution of the interfacial composition in the mixed-mode approach. In graph (a), full diffusion controll is assumed, the interfacial composition on the precipitate and matrix sides equal their local equilibrium values, $c_B^{\alpha\beta}$ and $c_B^{\beta\alpha}$. In graphs (b) and (c), the degree of interface-control is gradually increasing and the transformation is characterized by a decrease of the nonequilibrium matrix composition c_B^{α} toward c_B^0.

from the interface into α. In case (a), the interface mobility is high enough to immediately compensate the depletion of B-atoms due to this flux by movement into the direction of α and accumulation of additional B-atoms rejected from β. In cases (b) and (c), the interface cannot keep up with the diffusion process to compensate the depletion of B-atoms. Consequently, the B-content at the matrix side of the interface drops to a value, c_B^{α}, which is lower than the local equilibrium value, $c_B^{\alpha\beta}$. In the limit where the interface mobility goes toward very small values, the compositional gradient in the transforming phase α will vanish and the value of c_B^{α} is obtained from the mass-balance relation

$$c_B^{\alpha} = \frac{c_B^0 - f^{\beta} c_B^{\beta\alpha}}{1 - f^{\beta}}. \tag{3.72}$$

If only a small fraction of α forms, $f^{\beta} \approx 0$ and $c_B^{\alpha} \approx c_B^0$.

Similar to the dimensionless supersaturation parameter, S, defined in equation (3.28) of the local equilibrium treatment before as

$$S = \frac{c_B^0 - c_B^{\alpha\beta}}{c_B^{\beta} - c_B^{\alpha\beta}}, \tag{3.73}$$

the mixed-mode character of a partitioning phase transformation can conveniently be characterized by a supersaturation parameter, S', which reads

$$S' = \frac{c_B^{\alpha\beta} - c_B^{\alpha}}{c_B^{\alpha\beta} - c_B^0}. \tag{3.74}$$

If S' adopts a value of zero, the transformation is diffusion-controlled, whereas a value of unity denotes an interface-controlled reaction.

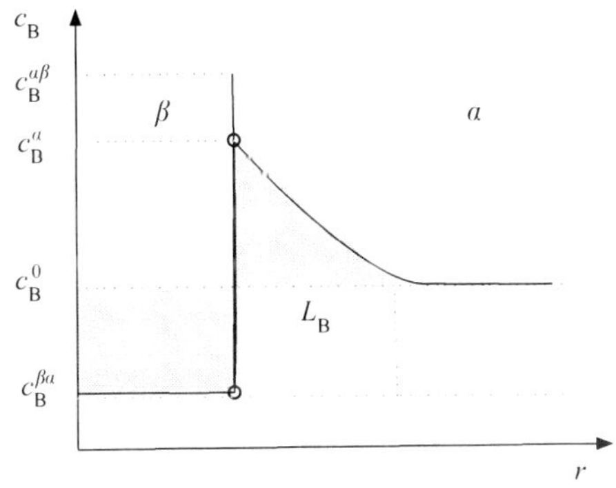

Figure 3.26. Approximation of the concentration profile ahead of the interface by the characteristic diffusion length, L_B.

In the next step, a relation is sought between the character of the transformation and the parameters diffusivity, D_B^α, of B, mobility, M^{IF}, of the interface, and interfacial composition, $c_B^{\beta\alpha}$ and $c_B^{\alpha\beta}$. From mass conservation, we obtain in the mixed-mode transformation model

$$\left(c_B^\alpha - c_B^{\beta\alpha}\right)\cdot v = -D_B^\alpha \frac{\partial c_B}{\partial x}. \tag{3.75}$$

In analogy to Zener's treatment in Section 3.2.2, the gradient $\partial c_B/\partial x$ at the interface is replaced by the approximation

$$\frac{\partial c_B}{\partial x} = \frac{\left(c_B^\alpha - c_B^0\right)}{L_B}, \tag{3.76}$$

and the mass conservation law reads

$$\left(c_B^\alpha - c_B^{\beta\alpha}\right)\cdot v = D_B^\alpha \frac{\left(c_B^\alpha - c_B^0\right)}{L_B}, \tag{3.77}$$

where L_B is the characteristic diffusion length for element B.

Consider a volume, V^β, which is formed during a mixed-mode phase transformation. The amount of solute transferred across the moving interface is given by the concentration jump, $\left(c_B^0 - c_B^{\beta\alpha}\right)$, at the interface and the transformation volume, V^β. The accumulation of solute B ahead of the interface is depicted schematically in Figure 3.26 where the identical areas of the two shaded regions emphasize the mass balance requirement.

The amount of solute on the α side is expressed in terms of the characteristic diffusion length, L_B. The global mass conservation condition then reads

$$V^\beta\left(c_B^0 - c_B^{\beta\alpha}\right) = \frac{1}{2} A^\beta L_B \left(c_B^\alpha - c_B^0\right), \tag{3.78}$$

where A^β denotes the interfacial area and the transformation occurs in planar geometry. Combination with equation (3.77) and elimination of L_B delivers

$$v\cdot\left(c_{\mathrm{B}}^{\alpha}-c_{\mathrm{B}}^{\beta\alpha}\right)=D_{\mathrm{B}}\frac{c_{\mathrm{B}}^{\alpha}-c_{\mathrm{B}}^{0}}{\dfrac{2V^{\beta}}{A^{\beta}}\left(\dfrac{c_{\mathrm{B}}^{0}-c_{\mathrm{B}}^{\beta\alpha}}{c_{\mathrm{B}}^{\alpha}-c_{\mathrm{B}}^{0}}\right)}. \tag{3.79}$$

For further investigation of the mixed-mode character of a phase transformation, the velocity, v, is replaced by the general relation between the interface velocity and the driving pressure given in equation (3.71). Before insertion, the driving pressure available for interface migration, $d_{\mathrm{F}}^{\mathrm{IF}}$, is replaced by a linearized approximation in the concentration differences between local equilibrium value, $c_{\mathrm{B}}^{\alpha\beta}$, and the mixed-mode value, c_{B}^{α}, with a proportionality constant, c′, as

$$v=M^{\mathrm{IF}}\cdot d_{\mathrm{F}}^{\mathrm{IF}}=\chi' M^{\mathrm{IF}}\left(c_{\mathrm{B}}^{\alpha\beta}-c_{\mathrm{B}}^{\alpha}\right). \tag{3.80}$$

With this approximation, equation (3.79) becomes

$$\chi' M^{\mathrm{IF}}\left(c_{\mathrm{B}}^{\alpha\beta}-c_{\mathrm{B}}^{\alpha}\right)\left(c_{\mathrm{B}}^{\alpha}-c_{\mathrm{B}}^{\beta\alpha}\right)=D_{\mathrm{B}}\frac{c_{\mathrm{B}}^{\alpha}-c_{\mathrm{B}}^{0}}{\dfrac{2V^{\beta}}{A^{\beta}}\left(\dfrac{c_{\mathrm{B}}^{0}-c_{\mathrm{B}}^{\beta\alpha}}{c_{\mathrm{B}}^{\alpha}-c_{\mathrm{B}}^{0}}\right)}, \tag{3.81}$$

or

$$\left(c_{\mathrm{B}}^{\alpha\beta}-c_{\mathrm{B}}^{\alpha}\right)\left(c_{\mathrm{B}}^{\alpha}-c_{\mathrm{B}}^{\beta\alpha}\right)\left(c_{\mathrm{B}}^{0}-c_{\mathrm{B}}^{\beta\alpha}\right)=\frac{D_{\mathrm{B}}}{\chi' M^{\mathrm{IF}}}\frac{A^{\beta}}{2V^{\beta}}\left(c_{\mathrm{B}}^{\alpha}-c_{\mathrm{B}}^{0}\right)^{2}. \tag{3.82}$$

If the molar volumes of the two phases are identical, the concentration variables, c_{B}, can be replaced by the mole fraction variables, X_{B}. With

$$v=\chi M^{\mathrm{IF}}\left(X_{\mathrm{B}}^{\alpha\beta}-X_{\mathrm{B}}^{\alpha}\right)\quad\text{and}\quad Z_{\mathrm{m}}=\frac{D_{\mathrm{B}}}{\chi M^{\mathrm{IF}}}\frac{A^{\beta}}{V^{\beta}}, \tag{3.83}$$

we arrive at

$$2\left(X_{\mathrm{B}}^{\alpha\beta}-X_{\mathrm{B}}^{\alpha}\right)\left(X_{\mathrm{B}}^{\alpha}-X_{\mathrm{B}}^{\beta\alpha}\right)\left(X_{\mathrm{B}}^{0}-X_{\mathrm{B}}^{\beta\alpha}\right)=Z_{\mathrm{m}}\left(X_{\mathrm{B}}^{\alpha}-X_{\mathrm{B}}^{0}\right)^{2}. \tag{3.84}$$

In this equation, all parameters except X_{B}^{α} are known. After substitution of $\Delta X_{\mathrm{B}}^{0}=X_{\mathrm{B}}^{0}-X_{\mathrm{B}}^{\beta\alpha}$, expansion, and rearrangement, a quadratic function for X_{B}^{α} is obtained with

$$\begin{aligned}&\frac{1}{2}\left(Z_{\mathrm{m}}+2\Delta X_{\mathrm{B}}^{0}\right)\left(X_{\mathrm{B}}^{\alpha}\right)^{2}-\left(\Delta X_{\mathrm{B}}^{0}\cdot X_{\mathrm{B}}^{\alpha\beta}+\Delta X_{\mathrm{B}}^{0}\cdot X_{\mathrm{B}}^{\beta\alpha}+Z_{\mathrm{m}}X_{\mathrm{B}}^{0}\right)\left(X_{\mathrm{B}}^{\alpha}\right)\\&+\frac{1}{2}\left(Z_{\mathrm{m}}\left(X_{\mathrm{B}}^{0}\right)^{2}+2\Delta X_{\mathrm{B}}^{0}\cdot X_{\mathrm{B}}^{\beta\alpha}X_{\mathrm{B}}^{\alpha\beta}\right)=0.\end{aligned} \tag{3.85}$$

Solution with standard methods finally delivers

$$\left(X_{\mathrm{B}}^{\alpha}\right)_{1,2}=\frac{1}{Z_{\mathrm{m}}+2\Delta X_{\mathrm{B}}^{0}}\cdot\left[\begin{aligned}&\left(\Delta X_{\mathrm{B}}^{0}\cdot\left(X_{\mathrm{B}}^{\alpha\beta}+X_{\mathrm{B}}^{\beta\alpha}\right)+Z_{\mathrm{m}}X_{\mathrm{B}}^{0}\right)\pm\\&\sqrt{\begin{aligned}&\left(\Delta X_{\mathrm{B}}^{0}\cdot\left(X_{\mathrm{B}}^{\alpha\beta}+X_{\mathrm{B}}^{\beta\alpha}\right)+Z_{\mathrm{m}}X_{\mathrm{B}}^{0}\right)^{2}-\\&\left(Z_{\mathrm{m}}\left(X_{\mathrm{B}}^{0}\right)^{2}+2\Delta X_{\mathrm{B}}^{0}\cdot X_{\mathrm{B}}^{\beta\alpha}X_{\mathrm{B}}^{\alpha\beta}\right)\left(Z_{\mathrm{m}}+2\Delta X_{\mathrm{B}}^{0}\right)\end{aligned}}\end{aligned}\right]. \tag{3.86}$$

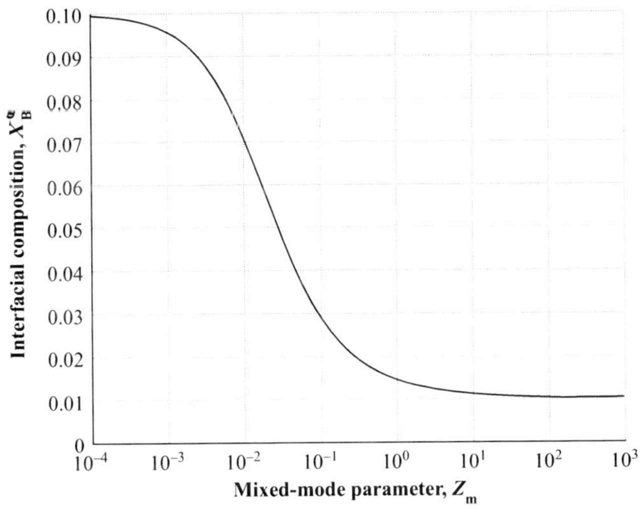

Figure 3.27. Interfacial composition in the mixed-mode model for $X_B^0 = 0.01$, $X_B^{\alpha\beta} = 0.001$, and $X_B^{\beta\alpha} = 0.10$.

In equation (3.86), the parameter Z_m determines the mixed-mode character of the transformation. In the limit of $Z_m \to 0$, which corresponds to situations of very high interfacial mobility, the interfacial composition on the side of the parent phase α approaches the equilibrium value with $X_B^\alpha \to X_B^{\alpha\beta}$. The transformation then occurs under diffusion control. In the opposite case, where the interface mobility is low compared to the diffusion kinetics of the partitioning elements, we have $Z_m \to \infty$ and $X_B^\alpha \to X_B^0$. The transformation is then interface-controlled.

The evolution of the interfacial composition as a function of the mixed-mode parameter, Z_m, is shown in Figure 3.27. For the values used in the evaluation, the B-content ahead of the moving boundary approaches the equilibrium value of $X_B^{\beta\alpha} = 0.10$ for the diffusion-controlled reaction. For interface control, the interfacial composition evolves toward the nominal B-content of $X_B^0 = 0.01$.

As the nature of a mixed-mode transformation is governed by the value of Z_m, it is necessary to inspect this quantity in more detail. The three major material-related quantities entering Z_m are the diffusivity, D_B, the interface mobility, M^{IF}, and the parameter, χ, which is related to the driving pressure. Whereas the former two parameters are of kinetic nature, the latter one is related to the alloy thermodynamics. The other part of Z_m contains the term A^β / V^β which is a geometry-related quantity that denotes the surface area to transformed volume ratio.

Among the materials-related parameters, the diffusivity, D_B, is often very different for interstitial and substitutional diffusion. With interstitial diffusion often being several orders of magnitude faster than substitutional diffusion, the Z_m values for interstitial diffusion are generally larger and the interface-controlled contribution to the mixed-mode transformation is generally more pronounced.

In their analysis, Sietsma and van der Zwaag [60] investigated the mixed-mode character of the austenite to ferrite reaction in steel with 0.1 to 1.0at%C. They conclude that the diffusion-controlled transformation mode is more pronounced for materials with higher content of alloying elements, in their case for the alloys with higher C-content. The lower the C-content, the faster is the reaction usually occurring, and the higher is the interface-controlled contribution.

Sietsma and van der Zwaag [60] also conclude that any kind of transformation will start off as an interface-controlled reaction due to the A^{β}/V^{β} term. Since the transformed volume is very small in the beginning of the reaction, Z_{m} is large at the early stages and the interface-controlled character of the mixed-mode transformation is most pronounced. In the later stages, the value of Z_{m} decreases and the transformation character evolves toward diffusion control.

3.5 Mean-Field Evolution Equations for Precipitate Growth

In the modeling of growth of precipitates, the fundamental relation used in all the previously described approaches has been the mass balance equation (3.8) relating the jump in interfacial composition variables with the interface velocity and the fluxes into and out of the interface given by the composition gradient at the interface. The corresponding theory is elaborated in Section 3.1.

The mass balance equation (3.8) at the interface constitutes a system of equations of dimension n–1, where n is the number of components in the system. It contains (2n–1) unknown parameters, however, it can only be solved with additional assumptions. In Sections 3.2–3.4, several models are discussed which use the *local equilibrium hypothesis* to fix the remaining degrees of freedom.

In the present section, an alternative approach will be utilized to derive evolution equations for the growth and dissolution of spherical precipitates in multicomponent systems. The derivation of these equations is based on the *thermodynamic extremal principle* (TEP), which suggests that a thermokinetic system evolves along the particular kinetic pathway with maximum entropy production. The characteristic parameters of the model are derived in the framework of a mean-field approach, which provides evolution equations that are comparably simple to evaluate and which makes the model particularly suitable for fast and robust simulations in higher-order multicomponent multiphase systems.

In the following subsection, the TEP will be introduced first. Afterward, the derivation of the evolution equations is elaborated for simple as well as complex lattices. Finally, the behavior of this model is compared to the classical local equilibrium-based approaches.

3.5.1 The Thermodynamic Extremal Principle

In 1991, J. Svoboda and I. Turek [61] presented a formulation of the *thermodynamic extremal principle* (TEP), which is utilized later in this section for the derivation of the evolution equations for precipitation systems. Accordingly, the TEP can conveniently be derived for the diffusion-controlled evolution of thermodynamic systems when starting with an analogy to mechanical dissipative systems.

In classical mechanics, the evolution of a closed mechanical system with dissipative terms can be described by the Lagrange equations as

$$\frac{\partial L}{\partial q_i} - \frac{\mathrm{d}}{\mathrm{d}t}\left(\frac{\partial L}{\partial \dot{q}_i}\right) = \frac{\partial F}{\partial \dot{q}_i}, \tag{3.87}$$

with the index i going from 1 to K, with K denoting the degrees of freedom of the system. The parameters, q_i, are designated as *independent generalized coordinates*, whereas the parameters,

$\dot{q}_i$, are *independent generalized velocities*. L is the Lagrangian of the mechanical system which is equal to the difference between potential and kinetic energy. F is denoted as the Rayleigh dissipative function, with

$$2F = Q, \tag{3.88}$$

and Q representing the total energy dissipation rate. For the validity of these equations, it is necessary that F is a positive definite quadratic form of $\dot{q}_i$. For details on the Lagrange equation (3.87), the interested reader is referred to the textbook by J. W. Leech [62] on classical mechanics.

In Section 1.1, the Gibbs energy, G, has been introduced as a convenient means for describing the properties of a thermodynamic system with

$$G = U + PV - TS. \tag{3.89}$$

At constant temperature T, constant pressure P, and constant volume V,[14] the term PV is a constant and the internal energy, U, and the entropy, S, of the system are functions of the configuration of the system only. As the configuration of the system is given by the mean positions of the atoms of the solid-state material, the Gibbs energy, G, of this system evolves in time only by the diffusive transport of atoms.[15]

Usually, the exact "atomic" configuration of macroscopic systems is not known in detail, which suggests the introduction of "macroscopic" parameters describing the constitution of the system. The choice of these parameters, q_i, is entirely dependent on the type of system and reaction that ought to be described. In diffusive transport simulations in a one-phase system, for instance, the independent parameters might be the concentration of elements in discrete regions of the sample. In the present context of precipitation modeling, the independent state parameters could be the size, number, and chemical composition of precipitates in the system.

It is important, now, to recognize that on the basis of the above assumptions, the Gibbs energy, G, of any possible configuration is a function of the independent state parameters, q_i, only with $G = G(q_i)$. In contrast, the *evolution* of the system away from the present configuration depends on the independent state parameters, q_i, as well as the generalized velocities, $\dot{q}_i$, with $Q = Q(q_i, \dot{q}_i)$.

In analogy to the Lagrange formalism for a mechanical system, one finds that the Lagrangian function, L, is related to the Gibbs energy, G, with

$$L = -G(q_i), \tag{3.90}$$

as the total kinetic energy of the thermodynamic system is assumed to be zero. When inserting the relations (3.88) and (3.90) into equation (3.87), one obtains the final result

[14] This assumption is reasonable for solid-state matter at constant T and P.

[15] In the treatment of Svoboda and Turek [61], the assumption is made that the kinetic energy of the atoms caused by the thermal vibrations around their equilibrium positions does not "perceptibly" contribute to the total kinetic energy of the system. Thus, the energy contribution of thermal vibrations is neglected in the formulation of the Gibbs energy.

$$\frac{\partial G}{\partial q_i} = -\frac{1}{2}\frac{\partial Q}{\partial \dot{q}_i}. \tag{3.91}$$

Equation (3.91) represents a unique relation between the Gibbs energy, G, of a solid-state thermodynamic system and the energy dissipation rate, Q, in terms of the independent state parameters, q_i, and the generalized velocities, $\dot{q}_i$.

In their work, Svoboda and Turek [61] further demonstrate that the fundamental relation (3.91) can also be derived from the assertion saying that a thermodynamic system evolves along the particular kinetic pathway corresponding to a maximum of the Gibbs energy dissipation rate, Q, where the relations

$$Q = -\frac{\partial G}{\partial t}, \tag{3.92}$$

or

$$\frac{\partial G}{\partial t} + Q = 0 \tag{3.93}$$

hold.

In the maximum of Q, the derivative of Q with respect to any generalized velocity, $\dot{q}_i$, corresponding to the "kinetic" parameters in the system must vanish according to

$$\frac{\partial Q}{\partial \dot{q}_i} = 0. \tag{3.94}$$

With standard mathematical methods, the above condition can be expressed as

$$\frac{\partial}{\partial \dot{q}_i}\left[Q + \lambda\left(\frac{\mathrm{d}G}{\mathrm{d}t} + Q\right)\right] = 0, \tag{3.95}$$

with λ being a Lagrange multiplier. Application of the chain rule for derivation yields

$$\frac{\mathrm{d}G}{\mathrm{d}t} = \sum_i \frac{\partial G}{\partial q_i}\cdot \dot{q}_i, \tag{3.96}$$

and substitution into equation (3.95) gives

$$\lambda\frac{\partial G}{\partial q_i} + (\lambda+1)\frac{\partial Q}{\partial \dot{q}_i} = 0. \tag{3.97}$$

Multiplication with $\dot{q}_i$ and summation of all equations finally delivers

$$\lambda\frac{\partial G}{\partial t} + 2(\lambda+1)Q = 0, \tag{3.98}$$

with utilizing equation (3.96) as well as the relation

$$\sum_i \frac{\partial G}{\partial \dot{q}_i}\cdot \dot{q}_i = 2Q, \tag{3.99}$$

which is valid if Q is a homogeneous quadratic form of the kinetic parameters $\dot{q}_i$ [62].

The Lagrange multiplier is obtained straightforwardly with $\lambda = -2$ from comparison of equation (3.98) with equation (3.93). Thus, it is obvious that equation (3.98) becomes identical to equation (3.91) and the above assertion saying that a closed solid-state thermodynamic system evolves with maximum Gibbs energy dissipation rate, Q, is verified.

Svoboda and Turek [61] further conclude that the assertion that a thermodynamic system evolves with maximum Gibbs energy dissipation rate subjected to the constraint (3.93) is more general than the original statement, equation (3.91), because it does not involve the explicit dependence of Q on the kinetic parameters, $\dot{q}_i$.

Finally, these authors [61] formulate the following principle:

If a closed solid-state thermodynamic system out of the thermodynamic equilibrium can be described by a set of macroscopic state parameters determining the total Gibbs energy, G, of the system, which does not explicitly depend on the kinetics of the system, and if the state of the system is being changed at constant temperature and pressure because of the linear diffusive processes causing the total Gibbs energy dissipation, then the time evolution of the set of macroscopic state parameters corresponds to the maximum of the total Gibbs energy dissipation rate, Q, under the constraint (3.93).

The above statement is also known as the *Onsager thermodynamic extremal principle*, as it was first stated in 1931 by Lars Onsager [63], a Norwegian chemical engineer and physicist, in his treatment of the well-known *reciprocal relations*.

In isothermal and isobaric systems, it can be shown that the maximum of the Gibbs energy dissipation rate, Q, is equivalent to the maximum of the entropy production, $\dot{S}^*$, with the asterisk emphasizing that the entropy is produced inside the system and does not involve entropy entering from the environment. At constant T and constant P, the relation

$$\dot{S}^* = -\frac{1}{T}\frac{\partial G}{\partial t} \tag{3.100}$$

can be derived, as elaborated by Svoboda, Turek, and Fischer [64]. With T being constant, comparison with equation (3.92) makes obvious that the constrained maximum of the Gibbs energy dissipation rate, Q, is equivalent to the constrained maximum of the entropy production, $\dot{S}^*$.

On this basis, it is possible to formulate an alternative version of the TEP, which is (*a*) shortened and (*b*) expressed in terms of entropy production instead of Gibbs energy dissipation as:

A closed solid-state thermodynamic system at constant T and P, which is out of equilibrium, evolves along the particular kinetic pathway with maximum entropy production, $\dot{S}^*$, *under the constraint* $\dot{G} + Q = 0$ *(3.93).*

In the past 20 years, the TEP has been successfully applied several times to derive well-known as well as original physical models of various types and application [65–70]. The large advantage of the TEP lies in the capability to straightforwardly deliver evolution equations describing the kinetic processes in a system, once the Gibbs energy and the dissipation rates are defined properly on the basis of a suitable set of characteristic parameters/state parameters. In the following section, the TEP will be used to develop evolution equations for the precipitation of spherical particles in a multicomponent matrix.

3.5.2 Mean-Field Evolution Equations for Substitutional/Interstital Phases

Svoboda et al. [44] have recently proposed a set of characteristic parameters for the evolution of precipitates in multicomponent multiphase multiparticle systems utilizing TEP. The model will be denoted as the *SFFK model* subsequently. The treatment of Svoboda et al. is closely followed here, before a generalized version of the original model is introduced, which allows taking into account compositional constraints from complex crystal structure phases, too.[16]

Consider a randomly distributed arrangement of homogeneous spherical precipitates which is embedded in a homogeneous multicomponent matrix. The term "homogeneous" emphasizes the model assumption that the composition of precipitates and matrix are described by mean-field composition variables instead of considering explicitly the concentration profiles inside and outside the precipitates. Assume further that there exist well-defined sharp interfaces between the precipitates and the embedding matrix. This situation is depicted schematically in Figure 3.28.

Each phase in the system is assumed to consist of s substitutional and p interstitial elements, with $n = s + p$. Let N_i be a fixed number of moles of component i, m be the number of precipitates in the system, ρ_k be the radius of precipitate k, u_{ki} be the site fraction of element i in the precipitate, and U_k be a structural factor defining the stoichiometry of the precipitate in terms of substitutional and interstitial components. For instance, in the case of an $M_{23}C_6$ precipitate, U_k will be 6/23, thus ensuring that the sum of all u_{ki} of substitutional components is unity.[17]

The mean concentrations in the precipitate, c_{ki}, are defined as

$$c_{ki} = \frac{u_{ki}}{v_S}, \tag{3.101}$$

with v_S being the partial molar volume of the substitutional components.

Once the concentration of elements in the precipitate is defined, the number of moles of component i in the matrix is given with

$$N_{0i} = N_i - \sum_{k=1}^{m} \frac{4\pi\rho_k^3 c_{ki}}{3}. \tag{3.102}$$

The mean concentration, c_{0i}, of element i in the matrix is given as

---

[16] The model discussed in this section has originally been derived for spherical precipitates. In Section 6.2, the treatment is extended to cigar-shaped (prolate) and disc-shaped (oblate) precipitates with the help of shape-factors.

[17] This quantity if often referred to as u-fraction variable. It has particular importance in the treatment of para-equilibrium conditions, since it defines the ratio of substitutional atoms independent of the concentration of interstitials.

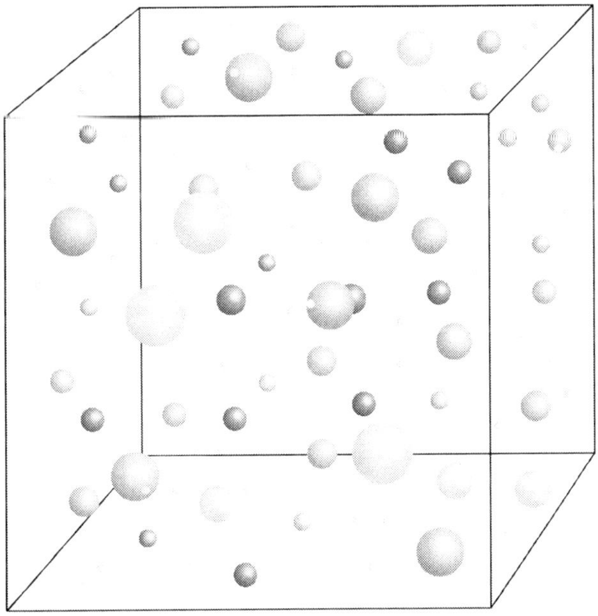

Figure 3.28. Spatial arrangement of homogeneous spherical precipitates with different size, composition, and phase type in a homogeneous multicomponent matrix as used in the SFFK model.

$$c_{0i} = \frac{N_{0i}}{v_S \sum_{j=1}^{s} N_{0j}}. \tag{3.103}$$

Throughout this section, the index "0" is used to designate properties of the matrix, while "k" is used for properties of the precipitates. With these definitions, the characteristic parameters for application of the TEP are determined as the radius, ρ_k, and the mean concentrations, c_{ki}, of the precipitates. The parameters of the matrix are fixed by mass conservation, equation (3.102). Let us further assume that an appropriate model, for example, the CALPHAD method (see Section 1.3.3), is at hand to deliver the chemical potentials of all elements, i,in all phases as functions of temperature, T, and composition, c_i.

The total Gibbs energy of the precipitation system can be written as

$$G = \sum_{i=1}^{n} N_{0i}\mu_{0i} + \sum_{k=1}^{m} \frac{4\pi\rho_k^3}{3}\left(\lambda_k + \sum_{i=1}^{n} c_{ki}\mu_{ki}\right) + \sum_{k=1}^{m} 4\pi\rho_k^2\gamma_k, \tag{3.104}$$

with the individual terms in this equation being directly associated with contributions of

1. the *Gibbs energy of the matrix*: The sum of the number of moles, N_{0i}, of components in the matrix multiplied with the corresponding chemical potentials, μ_{0i}, is performed over all elements, i, in the system.

2. the *Gibbs energy of the precipitates*: The outer sum is performed over all precipitates, weighted by the precipitate volume. The two terms in brackets account for (*a*) the specific mechanical energy, λ_k, associated with the elastic stress field caused by volume misfit and (*b*) the specific chemical energy given by the sum of the products of concentrations and chemical potentials in the precipitates.
3. the *total interfacial energy*: The sum is performed over all precipitates, weighted by the precipitate surface area. γ_k designates the mean specific interfacial energy separating the precipitate, *k*, from the matrix. This quantity is discussed in detail in Chapter 4.

Once the Gibbs energy of the system is defined in terms of the characteristic parameters, ρ_k and c_{ki}, the next step for application of the TEP is to define the possible mechanisms by which the excess Gibbs energy can be dissipated in the course of the system evolution. For the system at hand, Svoboda et al. [44] consider three dissipative processes where the Gibbs energy dissipation rates, Q_1, Q_2, and Q_3, are associated with

1. *interface migration*: If a precipitate grows or shrinks, the precipitate–matrix interface migrates with a velocity, *v*. The movement of the interface is accompanied by local rearrangement of atoms which proceeds against a retarding force. The nature of this force compares to mechanical friction and, in this sense, dissipates energy.
2. *diffusion inside the precipitate*: If the chemical composition of the precipitate evolves in the course of the precipitation reaction, this process is accompanied by diffusive fluxes inside the precipitate which dissipate energy by production of entropy.
3. *diffusion outside the precipitate*: The evolution of precipitate size or composition is generally accompanied by the long-range transport of atoms toward or away from the precipitate. The long-range diffusive transport of atoms occurs in the embedding matrix and dissipates energy by production of (configurational) entropy.

These processes are investigated in more detail below.

The total Gibbs energy dissipated by interface migration can directly be written as

$$Q_1 = \sum_{k=1}^{m} \frac{4\pi\rho_k^2}{M_k^{\mathrm{IF}}} \dot{\rho}_k^2 . \tag{3.105}$$

The expression contains the precipitate surface area, $4\pi\rho_k^2$, which is multiplied with $1/M_k^{\mathrm{IF}}$ and the square of the radial interface velocity, $\dot{\rho}_k^2$. In this relation, the inverse of the interface mobility, $1/M_k^{\mathrm{IF}}$, can be interpreted as a friction coefficient. Its product with the interface velocity square is the specific energy that is dissipated by interface movement over the precipitate surface.

The second process dissipating energy is determined by the diffusion of atoms inside the precipitate. Let the atoms entering or leaving the precipitate in the course of growth and shrinkage be deposited or removed uniformly over the precipitate volume. The radial flux, j_{ki}, is then given by

$$j_{ki} = -\frac{r\dot{c}_{ki}}{3}, \tag{3.106}$$

with $0 \le r \le \rho_k$, and the total dissipation rate due to diffusion in the precipitates follows as

$$Q_2 = \sum_{k=1}^{m} \sum_{i=1}^{n} \int_0^{\rho_k} \frac{RT}{c_{ki} D_{ki}} 4\pi r^2 j_{ki}^2 \mathrm{d}r. \tag{3.107}$$

The Gibbs energy dissipation rate expression again contains the precipitate surface area with $4\pi\rho_k^2$, and the square of the diffusive flux, j_{ki}, inside the precipitate. The inverse of the diffusion coefficient, $1/D_{ki}$, plays the role of a friction coefficient similar to the inverse of the interface mobility before.

The integral in equation (3.107) can be solved directly and the Gibbs energy dissipation rate due to diffusion inside the precipitate is obtained with

$$Q_2 = \sum_{k=1}^{m} \sum_{i=1}^{n} \frac{4\pi RT \rho_k^5}{45 c_{ki} D_{ki}} \dot{c}_{ki}^2. \tag{3.108}$$

Before analyzing the Gibbs energy dissipation rate caused by fluxes in the matrix, the change of composition of the precipitate due to interface migration must be investigated further. The rate of change of the number of moles of component, i, in the precipitate is given with

$$\dot{N}_{ki} = 4\pi\rho_k^2 \left(\dot{\rho}_k c_{0i} - J_{ki}^* \right). \tag{3.109}$$

The first term in the equation denotes the number of moles of atoms that are incorporated into the precipitate by interface migration, whereas the second term denotes the number of moles of atoms that enter or leave the precipitate due to the diffusive flux, J_{ki}^*, in the matrix at the position of the interface.

At the same time, the number of moles of component, i, in the precipitate is given by

$$N_{ki} = \frac{4\pi\rho_k^3}{3} c_{ki}. \tag{3.110}$$

Differentiation delivers

$$\dot{N}_{ki} = 4\pi\rho_k^2 c_{ki} \dot{\rho}_k + \frac{4\pi\rho_k^3 \dot{c}_{ki}}{3}, \tag{3.111}$$

and the transinterfacial diffusional flux is found by comparison of equations (3.109) and (3.111) with

$$J_{ki}^* = \dot{\rho}_k \left(c_{0i} - c_{ki} \right) - \frac{\rho_k \dot{c}_{ki}}{3}. \tag{3.112}$$

In analogy to the assumptions for diffusive transport inside the precipitate, the fluxes in the matrix are determined by uniform deposition or collection of atoms within a sphere of radius, Z. The diffusive flux, J_{ki}, in the matrix is then given by

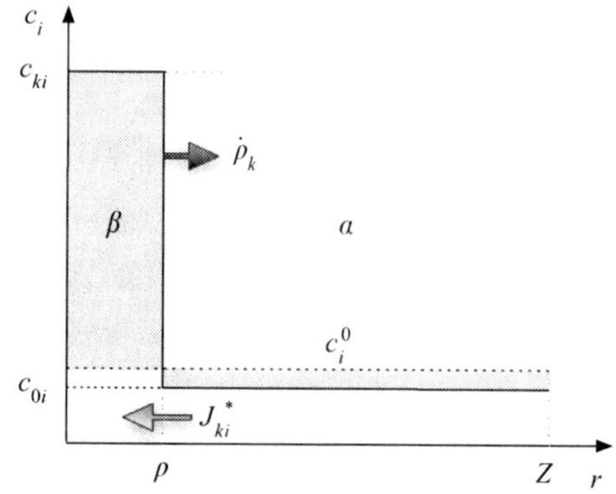

Figure 3.29. Mean-field description of the precipitation problem in the SFFK model. The precipitate β grows into the α matrix at a velocity, $\dot{\rho}_k$. At the same time, the diffusive flux, J_{ki}^*, passes through the interface. The shaded regions contain the same amount of solute atoms, thus assuring mass conservation (spherical geometry!).

$$J_{ki} = J_{ki}^* \frac{\rho_k^2}{r^2} \frac{Z^3 - r^3}{Z^3 - \rho_k^3}, \tag{3.113}$$

with $\rho_k < r \leq Z$.

The geometric arrangement used here is schematically depicted in Figure 3.29, showing the assumed mean concentrations and the moving phase boundary in the present mean-field *Ansatz*. It is important to emphasize that the mean-field model only uses the mean composition of matrix and precipitate to evaluate the evolution equations. There are no interfacial compositions entering the model, in contrast to all approaches founded in the local equilibrium assumption. The kinetics of the interface is only determined by the diffusive fluxes and the interface mobility under the assumption of homogeneous deposition of atoms.

With the approximation $Z >> \rho_k$, the total Gibbs energy dissipation rate due to diffusive fluxes in the matrix is obtained with

$$Q_3 = \sum_{k=1}^{m} \sum_{i=1}^{n} \int_{\rho_k}^{Z} \frac{RT}{c_{0i} D_{0i}} 4\pi r^2 J_{ki}^2 \mathrm{d}r \approx \sum_{k=1}^{m} \sum_{i=1}^{n} \frac{4\pi RT \rho_k^3 \left(\dot{\rho}_k \left(c_{ki} - c_{0i} \right) + \rho_k \dot{c}_{ki}/3 \right)^2}{c_{0i} D_{0i}}, \tag{3.114}$$

the interpretation of individual terms being analogous to the other two dissipation rate terms.

The total Gibbs energy dissipation rate, Q, is finally given as

$$Q = Q_1 + Q_2 + Q_3. \tag{3.115}$$

The next step in developing the evolution equations is to reformulate the equations in terms of independent variables. As the amounts of one substitutional component and one interstitial component (and taking into account empty interstitial sites as interstitial vacancies) can always be expressed in terms of the other components on the same sublattice, the following equations are obtained for the derivatives of G with respect to the characteristic parameters:

$$\frac{\partial G}{\partial \rho_k} = 8\pi\rho_k\gamma_k + 4\pi\rho_k^2 \left[\begin{array}{l} \lambda_k + \frac{1}{\Omega}\left\{\mu_{k1} - \mu_{01} + U_k\left(\mu_{ks+1} - \mu_{0s+1}\right)\right\} + \\ \sum_{i=2}^{s} c_{ki}\left(\mu_{ki} - \mu_{0i} - \mu_{k1} + \mu_{01}\right) + \sum_{i=s+2}^{n} c_{ki}\left(\mu_{ki} - \mu_{0i} - \mu_{ks+1} + \mu_{0s+1}\right) \end{array} \right],$$

$$k = 1, \ldots, m \tag{3.116}$$

$$\frac{\partial G}{\partial c_{ki}} = \frac{4\pi\rho_k^3}{3}\left(\mu_{ki} - \mu_{0i} - \mu_{k1} + \mu_{01}\right), \quad k = 1, \ldots, m, \; i = 2, \ldots, s, \tag{3.117}$$

$$\frac{\partial G}{\partial c_{ki}} = \frac{4\pi\rho_k^3}{3}\left(\mu_{ki} - \mu_{0i} - \mu_{ks+1} + \mu_{0s+1}\right), \quad k = 1, \ldots, m, \; i = s + 2, \ldots, n. \tag{3.118}$$

The reason for two equations existing for the derivative with respect to the composition variables is the assumption of one substitutional and one interstitial sublattice. This is expressed in the definition range of the index variables.

For the derivatives with respect to the kinetic parameters, the following equations are obtained:

$$\begin{aligned} \frac{\partial Q}{\partial \dot{\rho}_k} &= 8\pi\rho_k^2 \left[\frac{1}{M_k^{\mathrm{IF}}} + RT\rho_k \sum_{i=1}^{n} \frac{\left(c_{ki} - c_{0i}\right)^2}{c_{0i}D_{0i}} \right] \dot{\rho}_k \\ &+ \frac{8\pi RT\rho_k^4}{3} \left[\sum_{i=2}^{s} \left(\frac{c_{ki} - c_{0i}}{c_{0i}D_{0i}} - \frac{c_{k1} - c_{01}}{c_{01}D_{01}} \right) \dot{c}_{ki} + \sum_{i=s+2}^{n} \left(\frac{c_{ki} - c_{0i}}{c_{0i}D_{0i}} - \frac{c_{kss+1} - c_{0s+1}}{c_{0s+1}D_{0s+1}} \right) \dot{c}_{ki} \right], \end{aligned}$$

$$k = 1, \ldots, m \tag{3.119}$$

$$\begin{aligned} \frac{\partial Q}{\partial \dot{c}_{ki}} &= \frac{8\pi RT\rho_k^4}{3} \left(\frac{c_{ki} - c_{0i}}{c_{0i}D_{0i}} - \frac{c_{k1} - c_{01}}{c_{01}D_{01}} \right) \dot{\rho}_k \\ &+ \frac{8\pi RT\rho_k^5}{45} \left[\left(\frac{1}{c_{ki}D_{ki}} + \frac{5}{c_{0i}D_{0i}} \right) \dot{c}_{ki} + \sum_{j=2}^{s} \left(\frac{1}{c_{k1}D_{k1}} + \frac{5}{c_{01}D_{01}} \right) \dot{c}_{kj} \right], \end{aligned} \tag{3.120}$$

$$k = 1, \ldots, m, \; i = 2, \ldots, s$$

$$\begin{aligned} \frac{\partial Q}{\partial \dot{c}_{ki}} &= \frac{8\pi RT\rho_k^4}{3} \left(\frac{c_{ki} - c_{0i}}{c_{0i}D_{0i}} - \frac{c_{ks+1} - c_{0s+1}}{c_{0s+1}D_{0s+1}} \right) \dot{\rho}_k \\ &+ \frac{8\pi RT\rho_k^5}{45} \left[\left(\frac{1}{c_{ki}D_{ki}} + \frac{5}{c_{0i}D_{0i}} \right) \dot{c}_{ki} + \sum_{j=2}^{s} \left(\frac{1}{c_{ks+1}D_{ks+1}} + \frac{5}{c_{0s+s1}D_{0s+1}} \right) \dot{c}_{kj} \right]. \end{aligned} \tag{3.121}$$

$$k = 1, \ldots, m, \; i = s + 2, \ldots, n$$

With the derivatives of G and Q in terms of independent state variables at hand, the above equations can be inserted into the TEP, equation (3.91), which leads to a linear system of equations for the kinetic parameters, $\dot{\rho}_k$ and $\dot{c}_{ki}$. Interestingly, the matrix of coefficients, A_{ij}, is not dense and can be decomposed into m smaller sets of linear equations for m precipitates with of the form

$$\sum_{j=1}^{n-1} A_{ij}\dot{q}_j = B_i . \tag{3.122}$$

The rates of the characteristic parameters, $\dot{q}_j$, correspond to the evolution of radius, $\dot{\rho}_k$ and $\dot{c}_{ki}$, of the individual precipitates. For a fixed value of k, the coefficients of the matrix are given as

$$A_{11} = \frac{1}{M_k^{\mathrm{IF}}} + RT\rho_k \sum_{i=1}^{n} \frac{\left(c_{ki} - c_{0i}\right)^2}{c_{0i}D_{0i}}, \tag{3.123}$$

$$A_{1i} = A_{i1} = \frac{RT\rho_k^2}{3}\left(\frac{c_{ki} - c_{0i}}{c_{0i}D_{0i}} - \frac{c_{k1} - c_{01}}{c_{01}D_{01}}\right), \quad i = 2,\ldots,s, \tag{3.124}$$

$$A_{1i} = A_{i1} = \frac{RT\rho_k^2}{3}\left(\frac{c_{ki+1} - c_{0i+1}}{c_{0i+1}D_{0i+1}} - \frac{c_{ks+1} - c_{0s+1}}{c_{0s+1}D_{0s+1}}\right), \quad i = s+1,\ldots,n-1, \tag{3.125}$$

$$A_{ij} = \frac{RT\rho_k^3}{45}\left[\left(\frac{1}{c_{ki}D_{ki}} + \frac{5}{c_{0i}D_{0i}}\right)\delta_{ij} + \left(\frac{1}{c_{k1}D_{k1}} + \frac{5}{c_{01}D_{01}}\right)\right], \quad i = 2,\ldots,s, j = 2,\ldots,s, \tag{3.126}$$

$$A_{ij} = \frac{RT\rho_k^3}{45}\left[\left(\frac{1}{c_{ki+1}D_{ki+1}} + \frac{5}{c_{0i+1}D_{0i+1}}\right)\delta_{ij} + \left(\frac{1}{c_{ks+1}D_{ks+1}} + \frac{5}{c_{0s+1}D_{0s+1}}\right)\right], \tag{3.127}$$
$$i = s+1,\ldots,n-1, \; j = s+1,\ldots,n-1$$

$$A_{ij} = A_{ji} = 0, \; i = 2,\ldots,s, \; j = s+1,\ldots,n-1. \tag{3.128}$$

The symbol, δ_{ij}, stands for the Kronecker delta function. The right hand side of the equation system is given by

$$B_1 = -\frac{2\gamma_k}{\rho_k} - \lambda_k - \frac{1}{\Omega}\left\{\mu_{k1} - \mu_{01} + U_k\left(\mu_{ks+1} - \mu_{0s+1}\right)\right\}$$
$$-\sum_{i=2}^{s} c_{ki}\left(\mu_{ki} - \mu_{0i} - \mu_{k1} + \mu_{01}\right) - \sum_{i=s+2}^{n} c_{ki}\left(\mu_{ki} - \mu_{0i} - \mu_{ks+1} + \mu_{0s+1}\right), \tag{3.129}$$

$$B_i = -\frac{\rho_k}{3}\left(\mu_{ki} - \mu_{0i} - \mu_{k1} + \mu_{01}\right), \quad i = 2,\ldots,s, \tag{3.130}$$

$$B_i = -\frac{\rho_k}{3}\left(\mu_{ki+1} - \mu_{0i+1} - \mu_{ks+1} + \mu_{0s+1}\right), \quad i = s+1,\ldots,n-1. \tag{3.131}$$

The chemical potentials in equation (3.129) can be lumped together to yield the chemical driving force, $d_{k,\mathrm{chem}}^{\beta}$, for precipitation of β with

$$d^{\beta}_{k,\text{chem}} = -v^{\alpha}\sum_{i=1}^{n} c_{ki}\left(\mu_{ki} - \mu_{oi}\right), \tag{3.132}$$

where we assume that the molar volumes of precipitate, v^{β}, and matrix, v^{α}, are identical. The first coefficient of the right hand side of the equation system then adopts the simple form

$$B_1 = -\frac{2\gamma_k}{\rho_k} - \lambda_k + \frac{d^{\beta}_{k,\text{chem}}}{v^{\alpha}}. \tag{3.133}$$

As the Gibbs energy dissipation rate, Q, is a positive definite quadratic form of the independent kinetic parameters, the coefficient matrix, A_{ij}, is positive definite, too. In this case, the equation system can always be solved to deliver solutions for the rates of the characteristic parameters, $\dot{\rho}_k$ and $\dot{c}_{ki}$.

Time integration of the kinetic parameters can be performed with standard techniques. The evolution equations can also be used in integral models for nucleation growth and coarsening, such as the Wagner–Kampmann model, which is described later in Section 5.3.

3.5.3 Evolution Equations for General Sublattice Phases

The evolution equations derived in the previous section have been developed for phases with one substitutional and one interstitial sublattice. In many practical cases, precipitate phases can have rather complex crystal structures with three or four specific sublattices which constrain the stoichiometry of the phase. In order to make the evolution equations that are derived with the TEP applicable also for complex phases, an alternative formulation has been developed by Kozeschnik, Svoboda, and Fischer [71].

The basic equations for the general sublattice formulation are again the expressions for the Gibbs energy, G, and the Gibbs energy dissipation rate, Q, expressed in terms of the characteristic parameters. In contrast to the original treatment, however, the dependencies between element contents are not implicitly integrated into the equation system but they are added to the linear system of equations in the form of Lagrange multiplier constraints.

Consider a case, where the precipitate phase is characterized by p_k constraints that express the relations among elements on the individual sublattices. These constraints can be written as

$$\sum_{i=1}^{n} a_{kij}c_{ki} - U_{kj} = 0, \quad k = 1,\ldots,m;\ j = 1,\ldots,p_k. \tag{3.134}$$

Just like before, expressions are derived for the derivatives of Gibbs energy and energy dissipation rate. These read

$$\frac{\partial G}{\partial \rho_k} = 8\pi\rho_k\gamma_k + 4\pi\rho_k^2\left[\lambda_k + \sum_{i=1}^{n} c_{ki}\left(\mu_{ki} - \mu_{0i}\right)\right], \quad k = 1,\ldots,m, \tag{3.135}$$

$$\frac{\partial G}{\partial c_{ki}} = \frac{4\pi\rho_k^3}{3}\left(\mu_{ki} - \mu_{0i}\right), \quad k = 1,\ldots,m;\ i = 1,\ldots,n, \tag{3.136}$$

and $$\frac{\partial Q}{\partial \dot{\rho}_k} = 8\pi\rho_k^2\left[\frac{1}{M_k^{\mathrm{IF}}} + RT\rho_k\sum_{i=1}^{n}\frac{(c_{ki}-c_{0i})^2}{c_{0i}D_{0i}}\right]\dot{\rho}_k + \frac{8\pi RT\rho_k^4}{3}\sum_{i=1}^{n}\frac{c_{ki}-c_{0i}}{c_{0i}D_{0i}}\dot{c}_{ki},\quad k=1,\ldots,m, \tag{3.137}$$

$$\frac{\partial Q}{\partial \dot{c}_{ki}} = \frac{8\pi RT\rho_k^4}{3}\frac{c_{ki}-c_{0i}}{c_{0i}D_{0i}}\dot{\rho}_k + \frac{8\pi RT\rho_k^5}{45}\left(\frac{1}{c_{ki}D_{ki}} + \frac{5}{c_{0i}D_{0i}}\right)\dot{c}_{ki},\quad k=1,\ldots,m,\ i=1,\ldots,n; \tag{3.138}$$

In order to link the constraints, equation (3.134), to the evolution equations, the TEP is slightly reformulated. Again, the evolution equations are determined from the maximum Gibbs energy dissipation rate, Q, with the constraint, $\dot{G}+Q=0$. However, the extra constraints, equation (3.134), are now directly taken into account in the maximization. With the additional Lagrange multipliers, v_{kj}, the governing set of equations becomes

$$\frac{\partial}{\partial \dot{\rho}_k}\left[Q+\lambda\left(\dot{G}+Q\right)+\sum_{k=1}^{m}\sum_{j=1}^{p_k}v_{kj}\sum a_{kij}\dot{c}_{ki}\right]=0,\quad k=1,\ldots,m, \tag{3.139}$$

$$\frac{\partial}{\partial \dot{c}_{ki}}\left[Q+\lambda\left(\dot{G}+Q\right)+\sum_{k=1}^{m}\sum_{j=1}^{p_k}v_{kj}\sum a_{kij}\dot{c}_{ki}\right]=0,\quad k=1,\ldots,m. \tag{3.140}$$

Under these conditions, the system of linear equations becomes

$$\sum_{j=1}^{n+1+p_k} A_{ij}\dot{q}_j = B_i,\quad i=1,\ldots,n, \tag{3.141}$$

and the coefficients are given as

$$A_{ij} = \frac{RT\rho_k^3}{45}\left(\frac{1}{c_{ki}D_{ki}} + \frac{5}{c_{0i}D_{0i}}\right)\delta_{ij},\quad i=1,\ldots,n, j=1,\ldots,n; \tag{3.142}$$

$$A_{n+1i} = A_{in+1} = \frac{RT\rho_k^2}{3}\frac{c_{ki}-c_{0i}}{c_{0i}D_{0i}},\quad i=1,\ldots,n, \tag{3.143}$$

$$A_{n+1n+1} = \frac{1}{M_k^{\mathrm{IF}}} + RT\rho_k\sum_{i=1}^{n}\frac{(c_{ki}-c_{0i})^2}{c_{0i}D_{0i}}, \tag{3.144}$$

$$A_{in+1+j} = A_{n+1+ji} = a_{kij},\quad i=1,\ldots,n;\ j=1,\ldots,p_k, \tag{3.145}$$

$$A_{n+1i} = A_{in+1} = 0,\quad i=n+2,\ldots,n+1+p_k, \tag{3.146}$$

$$A_{ij} = 0,\quad i=n+2,\ldots,n+1+p_k;\ j=n+2,\ldots,n+1+p_k, \tag{3.147}$$

and the right hand side of the equation system is

$$B_i = -\frac{\rho_k}{3}\left(\mu_{ki}-\mu_{0i}\right),\quad i=1,\ldots,n, \tag{3.148}$$

$$B_{n+1} = -\frac{2\gamma_k}{\rho_k} - \lambda_k + \frac{d^{\beta}_{k,\text{chem}}}{v^{\alpha}}, \tag{3.149}$$

$$B_i = 0, \quad i = n+2, \ldots, n+1+p_k. \tag{3.150}$$

In this formulation, more equations are obtained compared to the original formulation of the SFFK model. However, the equations adopt a simpler form and can easily be applied to phases with arbitrarily complex crystal structure. The evolution equations derived in this section are implemented in the software *MatCalc*, which is described later in Section 9.2.

3.5.4 Comparison with Local Equilibrium Based Growth Models

For the diffusion-controlled growth of a stoichiometric precipitate with only one diffusing component, the mean-field evolution equations developed in the previous section can be compared to the parabolic growth rate equations derived in the framework of the local equilibrium approach (see Figure 3.12 in Section 3.2.4). For a single stoichiometric precipitate, the system of equations (3.141) reduces to a single equation reading

$$A\dot{\rho} = B. \tag{3.151}$$

The only independent parameter in this case is the precipitate radius, ρ, the evolution of which is determined by the above equation. From the treatment demonstrated in the previous sections and omitting the index, k, the two coefficients A and B are given as

$$A = \frac{1}{M^{\text{IF}}} + RT\rho\frac{\left(c_{k\text{B}} - c_{0\text{B}}\right)^2}{c_{0\text{B}}D_{0\text{B}}}, \tag{3.152}$$

and

$$B = -\frac{2\gamma}{\rho} - \lambda + \frac{d^{\beta}_{\text{chem}}}{v^{\alpha}}. \tag{3.153}$$

These coefficients can be further simplified, when neglecting interfacial friction ($M^{\text{IF}} \to \infty$), capillarity ($\gamma \to 0$), and mechanical energy from elastic strain ($\lambda \to 0$) in a first approximation. The governing equation for precipitate growth then reads

$$RT\rho\frac{\left(c_{k\text{B}} - c_{0\text{B}}\right)^2}{c_{0\text{B}}D_{0\text{B}}}v^{\alpha} \cdot \dot{\rho} = d^{\beta}_{\text{chem}}. \tag{3.154}$$

In a binary system, the chemical driving force, d^{β}_{chem}, is given as

$$d^{\beta}_{\text{chem}} = -\left[X^{\beta'}_{\text{A}}\left(\mu^{\beta'}_{\text{A}} - \mu^{\alpha}_{\text{A}}\right) + X^{\beta'}_{\text{B}}\left(\mu^{\beta'}_{\text{B}} - \mu^{\alpha}_{\text{B}}\right)\right]. \tag{3.155}$$

The chemical potentials evaluated in the β phase under the condition of thermodynamic equilibrium can equivalently be expressed in terms of the chemical potentials evaluated in the matrix phase α (see Figure 3.30). The chemical driving force then becomes

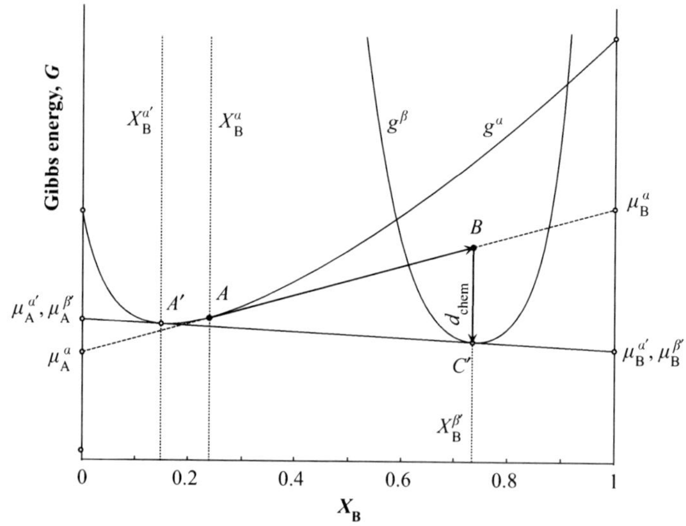

Figure 3.30. *G-X* diagram for a multiphase thermodynamic equilibrium sketching the procedure for evaluation of the chemical driving force, d_{chem}, based on the equilibrium compositions, $X_B^{\alpha'}$ and $X_B^{\beta'}$.

$$d_{chem}^{\beta} = -\left[X_A^{\beta'}\left(\mu_A^{\alpha'} - \mu_A^{\alpha}\right) + X_B^{\beta'}\left(\mu_B^{\alpha'} - \mu_B^{\alpha}\right)\right], \tag{3.156}$$

where the primed quantities refer to the thermodynamic equilibrium case.

If the content of B-atoms is much smaller than the content of A, corresponding to a dilute solution, for the chemical potential of component A in supersaturated state, μ_A^{α}, and in equilibrium, $\mu_A^{\alpha'}$,we have (Raoult's law)

$$\mu_A^{\alpha'} \approx \mu_A^{\alpha}. \tag{3.157}$$

The chemical potential of component B in the matrix α can be expressed in terms of the activity, a_B, equation (1.40), with

$$\mu_B^{\alpha} = g_B^{0,\alpha} + RT \ln a_B^{\alpha} = g_B^{0,\alpha} + RT \ln\left(f_B^{\alpha} X_B^{\alpha}\right). \tag{3.158}$$

With the assumption that the activity coefficients, f_B^{α} and $f_B^{\alpha'}$, are identical at the matrix compositions A′ and A, the chemical driving force becomes

$$d_{chem}^{\beta} = -\left[X_B^{\beta'}\left(\mu_B^{\alpha'} - \mu_B^{\alpha}\right)\right] = -RTX_B^{\beta'} \ln\left(\frac{X_B^{\alpha'}}{X_B^{\alpha}}\right). \tag{3.159}$$

For the coefficient *B*, we finally obtain

$$B = \frac{d_{chem}^{\beta}}{v^{\alpha}} = -RTc_B^{\beta'} \ln\left(\frac{c_B^{\alpha'}}{c_B^{\alpha}}\right), \tag{3.160}$$

where a conversion has been done into concentration variables, c, to allow for a straightforward comparison with the classical growth rate equations.

For the growth rate of the spherical precipitate in the mean-field approximation, we finally get

$$\dot{\rho} = \frac{B}{A} = \frac{-RTc_{\mathrm{B}}^{\beta'} \ln\left(\dfrac{c_{\mathrm{B}}^{\alpha'}}{c_{\mathrm{B}}^{\alpha}}\right)}{RT\rho \dfrac{\left(c_{\mathrm{B}}^{\beta'} - c_{\mathrm{B}}^{\alpha}\right)^2}{c_{\mathrm{B}}^{\alpha} D_{\mathrm{B}}^{\alpha}}}, \tag{3.161}$$

and

$$\rho\dot{\rho} = \frac{-c_{\mathrm{B}}^{\beta'} c_{\mathrm{B}}^{\alpha} D_{\mathrm{B}}^{\alpha} \ln\left(\dfrac{c_{\mathrm{B}}^{\alpha'}}{c_{\mathrm{B}}^{\alpha}}\right)}{\left(c_{\mathrm{B}}^{\beta'} - c_{\mathrm{B}}^{\alpha}\right)^2}. \tag{3.162}$$

After integration, the parabolic growth law is recovered with

$$\rho(t) = \sqrt{\rho_0^2 + 2D_{\mathrm{B}}^{\alpha} t \cdot \frac{-c_{\mathrm{B}}^{\beta'} c_{\mathrm{B}}^{\alpha} \ln\left(\dfrac{c_{\mathrm{B}}^{\alpha'}}{c_{\mathrm{B}}^{\alpha}}\right)}{\left(c_{\mathrm{B}}^{\beta'} - c_{\mathrm{B}}^{\alpha}\right)^2}}, \tag{3.163}$$

and the SFFK supersaturation parameter, S'', is given as

$$S'' = \frac{-c_{\mathrm{B}}^{\beta'} c_{\mathrm{B}}^{\alpha} \ln\left(\dfrac{c_{\mathrm{B}}^{\alpha'}}{c_{\mathrm{B}}^{\alpha}}\right)}{\left(c_{\mathrm{B}}^{\beta'} - c_{\mathrm{B}}^{\alpha}\right)^2}. \tag{3.164}$$

For comparison with the solutions from diffusion-controlled precipitate growth and local thermodynamic equilibrium, the growth parameter,

$$\alpha_{\mathrm{SFFK}} = \sqrt{2S''}, \tag{3.165}$$

is plotted together with the Zener approximation (Section 3.2.2) and the general solution for high and low supersaturation (Section 3.2.4) in Figure 3.31. The SFFK growth parameter is evaluated for a concentration of B in the precipitate, $c_{\mathrm{B}}^{\beta'} = 1/v^{\alpha}$, and an equilibrium B-content of the matrix, $c_{\mathrm{B}}^{\alpha'} = 10^{-5}/v^{\alpha}$.

The simplified evolution equations for stoichiometric precipitates from the mean-field approach and application of the TEP reproduce the general features of a diffusion-controlled transformation reasonably well over the entire range of supersaturation. The growth parameter, α_{SFFK}, is higher by a factor of approximately 2 in the region of low supersaturation and approximately 5 for high supersaturation. Keeping in mind the severe approximations used in the derivation of the SFFK model, the agreement must be considered as reasonably well, as the particular strength of the SFFK model lies in its efficiency in complex multicomponent systems.

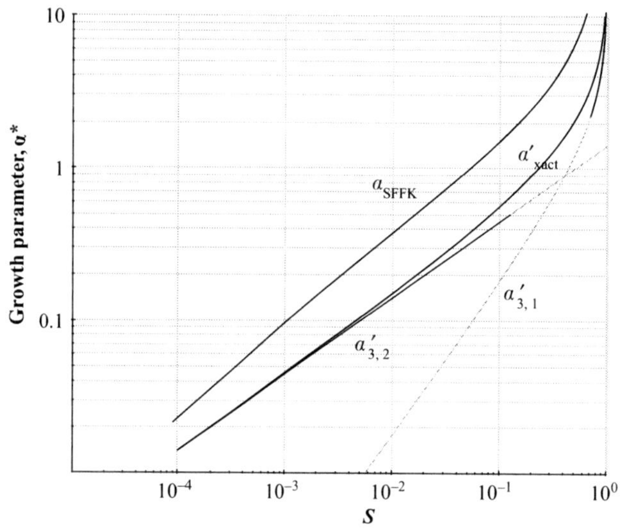

Figure 3.31. Comparison of mean-field growth rate parameter, α_{SFFK}, with solutions for diffusion-controlled growth of spherical particle obtained with the local equilibrium assumption in dependence of the supersaturation parameter, S.

3.6 Precipitate Coarsening

In 1896, the Baltic–German chemist Friedrich Wilhelm Ostwald (1853–1932) first described the phenomenon that dispersoid systems containing a size distribution of particles spontaneously evolve their constitution in time [72,73]. During *Ostwald ripening*, larger particles, which are thermodynamically more stable than smaller particles, continue to grow at the expense of the smaller particles. In this process, the mean radius of the ensemble of particles continuously increases while their number density decreases. The process of Ostwald ripening is also known as *coarsening*.

The driving force for Ostwald ripening, or coarsening, is the minimization of total interfacial area. Small particles have a higher surface area to volume ratio, and they are, therefore, energetically less favorable than large particles with a lower interfacial area to volume ratio. The total energy of an ensemble of many small particles is higher than the energy of an ensemble of fewer larger particles of the same total volume.

The process of coarsening can straightforwardly be understood on the basis of the Gibbs–Thomson equation (see Section 1.5). Consider two particles of the same phase and chemical composition, but different size, located inside a closed matrix of components A and B. Both particles are assumed to be in a state of local equilibrium with the adjacent matrix region. On each of the particles, a pressure, P_{curv}, is acting, which is given by

$$P_{\text{curv}} = \frac{2\gamma}{\rho}. \tag{3.166}$$

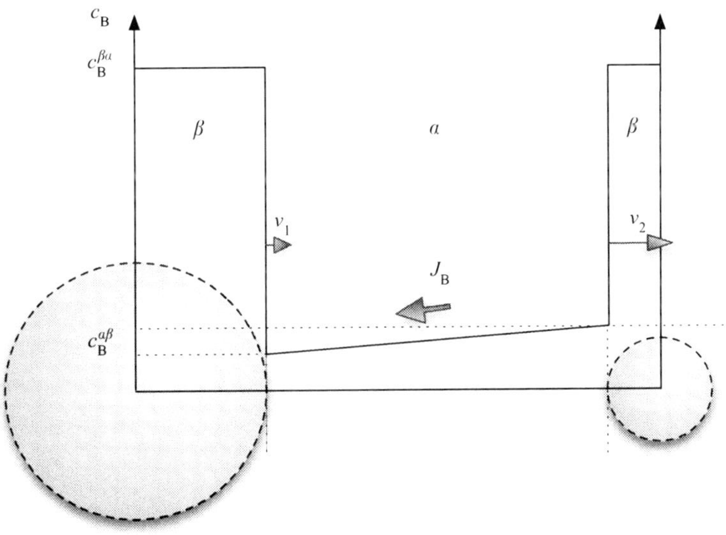

Figure 3.32. Chemical composition profile for two precipitates of different size during Ostwald ripening.

Obviously, this pressure scales inversely with the precipitate radius, ρ. The mole fraction of component B in the matrix, $X_B^{\alpha'}$, in contact with the particles is given by[18]

$$X_B^{\alpha'} \approx X_B^{\alpha}\left(1+\frac{2\gamma v^{\beta}}{\left(X_B^{\beta}-X_B^{\alpha}\right)RT}\cdot\frac{1}{\rho}\right), \tag{3.167}$$

where X_B^{α} and X_B^{β} are the equilibrium composition of B in matrix and precipitate, respectively, γ is the interfacial energy, and ρ is the radius of the precipitate. Equation (3.167) emphasizes that the matrix content of B-atoms achieves the lowest value for the second term in brackets being zero. In terms of concentration variables, we have

$$c_B^{\alpha'} \approx c_B^{\alpha}\left(1+\frac{2\gamma}{\left(c_B^{\beta}-c_B^{\alpha}\right)RT}\cdot\frac{1}{\rho}\right). \tag{3.168}$$

For constant thermophysical properties and chemical composition of the precipitate, this condition can be approached with increasing precipitate radius, ρ. With smaller precipitate radius, the equilibrium composition of the matrix continuously increases. The composition profile and the interface velocities corresponding to two particles of different size located in the vicinity of each other are schematically shown in Figure 3.32.

[18]This equation represents the linearized version of the Gibbs–Thomson equation which is valid for larger radii, ρ. See Section 1.5 for details.

The B-concentration in the matrix next to the smaller particle is higher, according to the Gibbs–Thomson equation. (3.167), compared to the larger particle. Consequently, a concentration gradient exists, inducing a net flux of B-atoms from the smaller to the larger particle. After some given time, the smaller particle will dissolve and the larger particle will incorporate the B-atoms of the smaller particle.

3.6.1 The LSW-Theory of Precipitate Coarsening

The problem described in the previous section has been investigated on a theoretical basis independently by Carl Wilhelm Wagner [74], a German physical chemist (1901–1977), Ilya Mikhailovich Lifshitz, a Soviet theoretical Physisist (1917–1982), and Vitaliy Valenitnovich Slyozov,[19] a Russian Scientist (1930–) at Kharkov University, Ukraine [75]. When treating the problem of precipitate coarsening on a theoretical basis, three issues are to be considered:

1. What is the distribution function of the dispersion of precipitates in the asymptotic limit of $t \to \infty$?
2. How does the mean radius, $\bar{r}$, of the precipitates evolve in time?
3. How does the number density, m, of precipitates evolve in time?

In the following analysis, the treatment of C. Wagner is reproduced in the derivation of the steady-state size distribution and evolution of precipitate parameters, $\bar{r}$ and m.

Consider a dispersion of precipitates given by their distribution function, $f(r,t)$, where r is the radius of the particles and the definition

$$f(r,t) = \lim_{\Delta r \to 0} \frac{N(r, r+\Delta r, t)}{\Delta r} \tag{3.169}$$

applies. The term $N(r, r+\Delta r, t)$ refers to the number of precipitates in the size interval, Δr. In a closed system, the number of precipitates reaching a given radius, or exceeding it, is equal to the product of the size distribution with the growth rate of the precipitate, $\partial \dot{r}$, corresponding to the differential equation

$$\frac{\partial f(r,t)}{\partial t} = \frac{\partial \left(f(r,t)\dot{r} \right)}{\partial r} \tag{3.170}$$

The total number of precipitates, m, in unit volume is given by the integral

$$m = \int_0^\infty f(r,t)\, dr. \tag{3.171}$$

The evolution of m is controlled by the number of precipitates, which approach a radius of $r = 0$ in unit time and unit volume, equal to

[19]His last name is also written as Slezov.

$$-\dot{m} = -\frac{\partial}{\partial t}\int_0^\infty f(r,t)\,dr = -\lim_{r\to 0}\left(f(r,t)\dot{r}\right). \tag{3.172}$$

The phase fraction of precipitates, f^β, is finally obtained by the integral

$$f^\beta = \int_0^\infty \frac{4\pi r^3}{3} f(r,t)\,dr. \tag{3.173}$$

In application to Ostwald ripening, f^β is commonly assumed to be constant and the change in phase fraction is assumed to be $\dot{f}^\beta = 0$. The above equations will be used in the subsequent analysis.

Equations (3.170) and (3.172) can be solved if an appropriate expression for the growth rate, $\dot{r}$, of the precipitates is found. For diffusion-controlled evolution of the precipitate radius, the rate of change of number of atoms, n, in the precipitate is proportional to the precipitate surface, $4\pi r^2$ the diffusion coefficient of the precipitate-forming element, D_B^α, and the concentration gradient at the precipitate surface. The latter quantity can be approximated in analogy to Zener's treatment (compare Section 3.2.2) with $\left(c_\mathrm{B}^{\alpha'} - \bar{c}_\mathrm{B}^\alpha\right)/r$, where $c_\mathrm{B}^{\alpha'}$ is the concentration of B-atoms at the precipitate surface and $\bar{c}_\mathrm{B}^\alpha$ is the mean concentration of B in the solution. For $\dot{n}$, we thus obtain

$$\dot{n} = -4\pi r D_\mathrm{B}^\alpha \left(c_\mathrm{B}^{\alpha'} - \bar{c}_\mathrm{B}^\alpha\right), \tag{3.174}$$

where the assumption has been made that the precipitate size is small compared to the interparticle spacing and the B-content of the precipitate is much higher than the mean B-content of the matrix. In the case of diffusion-controlled growth of the particle, local equilibrium can be assumed at the interface and the concentration of B at the precipitate surface is given by equation (3.168).

If the growth rate of the precipitate is controlled by the attachment of monomers to the surface, $\dot{n}$ can be described by the product of precipitate surface area and a kinetic constant, k, and the difference between the equilibrium concentration, c_B^α, and the actual concentration of B-atoms at the precipitate surface, $c_\mathrm{B}^{\alpha'}$, with

$$\dot{n} = -4\pi r^2 D_\mathrm{B}^\alpha \left(c_\mathrm{B}^\alpha - c_\mathrm{B}^{\alpha'}\right). \tag{3.175}$$

From equations (3.174) and (3.175), the concentration at the precipitate surface is obtained as

$$c_\mathrm{B}^{\alpha'} = \frac{krc_\mathrm{B}^\alpha + D_\mathrm{B}^\alpha \bar{c}_\mathrm{B}^\alpha}{kr + D_\mathrm{B}^\alpha}. \tag{3.176}$$

Insertion of this equation into either equation (3.174) or (3.175) delivers

$$\dot{n} = -\frac{4\pi r^2 k D_\mathrm{B}^\alpha}{kr + D_\mathrm{B}^\alpha}\left(c_\mathrm{B}^\alpha - \bar{c}_\mathrm{B}^\alpha\right). \tag{3.177}$$

With the time evolution of the volume of a single precipitate,

$$\frac{\mathrm{d}}{\mathrm{d}t}\left(\frac{4\pi r^3}{3}\right) = 4\pi r^2 \dot{r} = \dot{n} v^\beta, \tag{3.178}$$

where v^β is the molar volume of the particle. The evolution of the radius of a particle is obtained as

$$\dot{r} = \frac{\dot{n} v^\beta}{4\pi r^2} = -\frac{k D_\mathrm{B}^\alpha v^\beta}{kr + D_\mathrm{B}^\alpha}\left(c_\mathrm{B}^\alpha - \bar{c}_\mathrm{B}^\alpha\right). \tag{3.179}$$

For the rate of radius change of the precipitate, Wagner distinguishes two limiting cases for either diffusion-controlled or interface-controlled growth. In the first case, $kr >> D_\mathrm{B}^\alpha$ and equation (3.179) becomes

$$\dot{r} = -\frac{D_\mathrm{B}^\alpha v^\beta}{r}\left(c_\mathrm{B}^\alpha - \bar{c}_\mathrm{B}^\alpha\right). \tag{3.180}$$

If the reaction is controlled by the attachment rate of monomers to the precipitate surface, the condition $kr << D_\mathrm{B}^\alpha$ holds and equation (3.179) reads

$$\dot{r} = \frac{\dot{n} v^\beta}{4\pi r^2} = -k v^\beta \left(c_\mathrm{B}^\alpha - \bar{c}_\mathrm{B}^\alpha\right). \tag{3.181}$$

The distinction between the two cases will be elaborated later.

Consider the evolution of the B-concentration of the matrix phase, which is the negative sum of all changes of B-atoms in the precipitates per unit volume as

$$\frac{\partial \bar{c}_\mathrm{B}^\alpha}{\partial t} = -\int_0^\infty \dot{n}(r) f(r,t)\,\mathrm{d}r. \tag{3.182}$$

If the total amount of component B in the matrix is small compared to the amount of B in the precipitates, the assumption can be made that the amount of B released into the matrix by small particles is approximately the same as the amount of B picked up by the larger particles due to the net diffusional flux caused by the concentration gradient depicted in Figure 3.32. Consequently, the change of the amount of component B in the matrix is negligibly small compared to the change of the mean radius of the ensemble of particles. After insertion of equations (3.168) and (3.179) into equation (3.182), the governing equation for the evolution of the mean concentration of B in the matrix is obtained with

$$\frac{\bar{c}_\mathrm{B}^\alpha}{c_\mathrm{B}^\alpha} = 1 + \frac{2\gamma}{\left(c_\mathrm{B}^\beta - c_\mathrm{B}^\alpha\right)RT} \frac{\int_0^\infty f(r,t)\dfrac{r}{kr + D_\mathrm{B}^\alpha}\mathrm{d}r}{\int_0^\infty f(r,t)\dfrac{r^2}{kr + D_\mathrm{B}^\alpha}\mathrm{d}r}. \tag{3.183}$$

After inserting equations (3.183) and (3.168) into equation (3.179), the growth rate of a precipitate is obtained as

$$\dot{r} = -\frac{2\gamma c_B^\alpha v^\beta}{\left(c_B^\beta - c_B^\alpha\right)RT}\frac{kD_B^\alpha}{kr + D_B^\alpha}\left[1 - \frac{r\int_0^\infty f(r,t)\frac{r}{kr + D_B^\alpha}\mathrm{d}r}{\int_0^\infty f(r,t)\frac{r^2}{kr + D_B^\alpha}\mathrm{d}r}\right]. \tag{3.184}$$

From equation (3.184), the radius, r^*, of a particle in equilibrium with the matrix phase can be evaluated under the condition $\dot{r} = 0$, which yields

$$r^* = \frac{\int_0^\infty f(r,t)\frac{r^2}{kr + D_B^\alpha}\mathrm{d}r}{\int_0^\infty f(r,t)\frac{r}{kr + D_B^\alpha}\mathrm{d}r}. \tag{3.185}$$

Equation (3.184) can then be written in a simpler form as

$$\dot{r} = \frac{2\gamma c_B^\alpha v^\beta}{\left(c_B^\beta - c_B^\alpha\right)RT}\frac{kD_B^\alpha}{kr + D_B^\alpha}\left(\frac{r}{r^*} - 1\right). \tag{3.186}$$

Insertion of this equation into equation (3.170) delivers

$$\frac{\partial f(r,t)}{\partial t} = -\frac{2\gamma c_B^\alpha v^\beta}{\left(c_B^\beta - c_B^\alpha\right)r^* RT}\frac{\partial}{\partial r}\left[\frac{kD_B^\alpha}{kr + D_B^\alpha}\left(r - r^*\right)\frac{f(r,t)}{r}\right]. \tag{3.187}$$

For diffusion-controlled growth, $kr >> D_B^\alpha$. The critical radius, r^*, then becomes identical to the mean radius, $\bar{r}$, with

$$r^* = \frac{\int_0^\infty f(r,t)r\mathrm{d}r}{\int_0^\infty f(r,t)\mathrm{d}r} = \bar{r}, \tag{3.188}$$

and equation (3.187) simplifies to

$$\frac{\partial f(r,t)}{\partial t} = -\frac{2\gamma D_B^\alpha c_B^\alpha v^\beta}{\left(c_B^\beta - c_B^\alpha\right)r^* RT}\frac{\partial}{\partial r}\left[\left(r - r^*\right)\frac{f(r,t)}{r^2}\right]. \tag{3.189}$$

For an analysis of the evolution of the precipitate number density, the precipitate radius evolution as given in equation (3.186) is evaluated in the diffusion-controlled growth limit,

$kr >> D_B^\alpha$, first, and considering the condition $r << r^*$.[20] The radius change for the smallest particles is then approximately given by

$$-\dot{r} = \frac{2\gamma c_B^\alpha D_B^\alpha v^\beta}{\left(c_B^\beta - c_B^\alpha\right) RT} \frac{1}{r^2}. \tag{3.190}$$

In the further treatment, Wagner makes the assumption that the distribution function, $f(r,t)$, can be expressed in terms of a time-dependent function, $g_D(t)$, and a function $h_D(\xi)$, which depends on the ratio of radii

$$\xi = \frac{r}{r^*(t)}. \tag{3.191}$$

For the distribution function, the following *Ansatz* is made

$$f(r,t) = g_D(t)\xi^2 h_D(\xi). \tag{3.192}$$

With equation (3.187), one obtains the identity

$$\int_0^\infty h_D(\xi)\xi^2 \mathrm{d}\xi = \int_0^\infty h_D(\xi)\xi^3 \mathrm{d}\xi. \tag{3.193}$$

After insertion of these relations into equation (3.172) and elimination of $g_D(t)$, the evolution of the critical radius is found as

$$\dot{r}^* = \frac{2\gamma c_B^\alpha D_B^\alpha v^\beta}{3\left(c_B^\beta - c_B^\alpha\right)\gamma_D RT} \frac{1}{(r^*(t))^2}, \tag{3.194}$$

where the abbreviation

$$\gamma_D = \int_0^\infty h_D(\xi)\xi^2 \mathrm{d}\xi \tag{3.195}$$

has been used. This equation can be integrated and delivers

$$r^*(t) = r_0^* \left[1 + \frac{t}{\tau_D}\right]^{\frac{1}{3}}, \tag{3.196}$$

with the time constant, τ_D, given as

[20]The evolution of the mean precipitate properties is determined by the dissolution kinetics of the smallest precipitates.

$$\tau_D = \gamma_D \frac{\left(c_B^\beta - c_B^\alpha\right) RT}{2\gamma c_B^\alpha D_B^\alpha v^\beta} \left(r_0^*\right)^3. \tag{3.197}$$

The evolution of the number density of precipitates is obtained when considering that the mean precipitate volume is proportional to $\overline{r}^3$, which, for long times, t, is identical to $\overline{r}^3 = \left(r^*\right)^3$. As the time-independent volume fraction of particles is proportional to the product of particle density, m, and mean precipitate volume, m, is thus inversely proportional to $\overline{r}^3$ and, with equation (3.194), it is given as

$$m(t) = m_0 \left[1 + \frac{t}{\tau_D}\right]^{-1}. \tag{3.198}$$

In the further analysis of the stationary distribution function, and after some algebra (see ref. [74] for details), the differential equation

$$\left(\gamma_D^{-1}\xi^3 - 3\xi + 3\right)\frac{\mathrm{d}h_D}{\mathrm{d}\xi} + \left(6\gamma_D^{-1}\xi^2 - 3\right)h_D = 0 \tag{3.199}$$

is obtained. In equation (3.199), the time, t, does not occur explicitly any more, indicating that the *Ansatz* made in equation (3.192) is suitable for further analysis of the quasi-stationary evolution of the size distribution of coarsening particles. For evaluation of the distribution function, h_D, from equation (3.199), it is necessary to evaluate the value of γ_D determined by equation (3.193). After multiplication with ξ^3 and integration in parts between 0 and ξ, the following relation is obtained

$$\left[\left(\gamma_D^{-1}\xi^3 - 3\xi + 3\right)\xi^3 h_D(\xi)\right] + 9\left[\int_0^\xi h_D(\xi)\xi^3 \mathrm{d}\xi - \int_0^\xi h_D(\xi)\xi^2 \mathrm{d}\xi\right] = 0. \tag{3.200}$$

In this equation, the difference of integrals vanishes with the help of equation (3.193) in the limit of $\xi \to \infty$. Consequently, for evaluation of γ_D, the equivalent equation

$$\lim_{\xi\to\infty}\left[\left(\gamma_D^{-1}\xi^3 - 3\xi + 3\right)\xi^3 h_D(\xi)\right] = 0 \tag{3.201}$$

is obtained. The expression

$$\gamma_D^{-1}\xi^3 - 3\xi + 3 \tag{3.202}$$

has a minimum for $\xi = \gamma_D^{1/2}$ with a function value of $3 - 2\gamma_D^{1/2}$. When setting this expression equal to zero, Wagner finally obtained

$$\gamma_D = \frac{9}{4} \tag{3.203}$$

and, from equation (3.199), the distribution function, h_D, becomes

$$\frac{d\ln h_D}{d\xi} = -\frac{\frac{24}{9}\xi^2 - 3}{\frac{4}{9}\xi^3 - 3\xi + 3} = -\frac{\frac{7}{3}}{3+\xi} - \frac{\frac{11}{3}\xi - 4}{\left(\frac{3}{2} - \xi\right)^2}. \tag{3.204}$$

Integration with the norm condition

$$h_D(\xi = 0) = 1 \tag{3.205}$$

finally delivers the distribution function, h_D, as

$$h_D\left(\xi \le \frac{3}{2}\right) = \left(\frac{3}{3+\xi}\right)^{\frac{7}{3}} \left(\frac{\frac{3}{2}}{\frac{3}{2} - \xi}\right)^{\frac{11}{3}} \exp\left(-\frac{\xi}{\frac{3}{2} - \xi}\right). \tag{3.206}$$

$$h_D\left(\xi \ge \frac{3}{2}\right) = 0$$

The stationary size distribution is thus

$$f(r,t) = \frac{4}{9}\xi^2 \left(\frac{3}{3+\xi}\right)^{\frac{7}{3}} \left(\frac{\frac{3}{2}}{\frac{3}{2} - \xi}\right)^{\frac{11}{3}} \exp\left(-\frac{\xi}{\frac{3}{2} - \xi}\right). \tag{3.207}$$

The same result has been obtained independently by Lifshitz and Slyozov [75] shortly before. Equation (3.207) is known as the Lifshitz-Slyozov-Wagner (LSW) *size distribution* and it represents the stationary distribution function for precipitates that is obtained after long time of coarsening in the limit of small volume fraction of precipitates. The shape of the LSW size distribution is shown in Figure 3.33.

The maximum value of the LSW function is found from the equation $16\xi^3 + 27\xi - 54 = 0$, with the solution $\xi = 1.135$.

In summary, the time evolution of mean radius and number density obtained by LSW is given by

$$r^*(t) = r_0^* \left[1 + \frac{t}{\tau_D}\right]^{\frac{1}{3}}, \tag{3.208}$$

$$m(t) = m_0 \left[1 + \frac{t}{\tau_D}\right]^{-1}, \tag{3.209}$$

with the time constant, τ_D, given by equation (3.197). An equivalent version of these laws can be expressed as

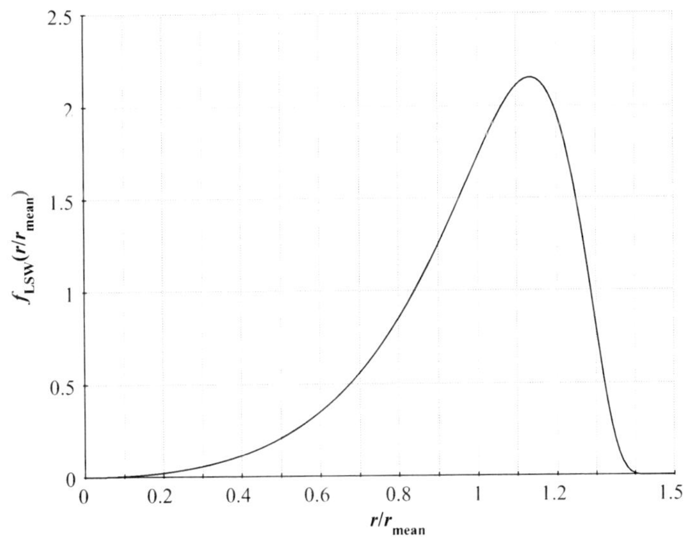

Figure 3.33. Stationary size distribution function given by classical LSW theory.

$$\bar{r}^3 = \bar{r}_0^3 + K_{LSW} \cdot t, \tag{3.210}$$

with

$$K_{LSW} = \frac{\left(r_0^*\right)^3}{\tau_D} = \frac{2\gamma c_B^\alpha D_B^\alpha v^\beta}{\gamma_D \left(c_B^\beta - c_B^\alpha\right) RT} = \frac{8\gamma c_B^\alpha D_B^\alpha v^\beta}{9\left(c_B^\beta - c_B^\alpha\right) RT}. \tag{3.211}$$

Note that the rate constant, K_{LSW}, is often written for the limit of $c_B^\beta - c_B^\alpha \approx 1/v^\beta$. K_{LSW} then becomes

$$K'_{LSW} = \frac{8\gamma D_B^\alpha c_B^\alpha \left(v^\beta\right)^2}{9RT}. \tag{3.212}$$

For a thourough analysis of classical LSW theory, the interested reader is also referred to, for instance, the review of P. Voorhees [76].

3.6.2 Extensions of LSW Theory for Finite Phase Fraction Effects

Following the original treatment of Lifshitz and Slyozov [75] and Wagner [74], several theories have been developed later to extend the approximation of negligible volume fraction of second phase particles to finite volume fractions. Most of these approaches have been motivated by the observation that experimentally determined size distributions of precipitates are commonly significantly broader than the stationary theoretical solution of LSW given by equation (3.207).

Interestingly, all of these later models preserve the principal time dependence of the mean radius evolution with an exponent of $t^{1/3}$. The major difference is found in the kinetic factor, K_{LSW}, in equations (3.210) and (3.212). For further analysis, the kinetic equation is written as

$$\bar{r}^3 = \bar{r}_0^3 + K\left(f^\beta\right)\cdot t, \tag{3.213}$$

where $K\left(f^\beta\right)$ is a monotonically increasing function of the volume fraction, f^β, of the precipitate [76]. The main developments going beyond the zero phase fraction limit of LSW are summarized below.

3.6.2.1 The Modified LSW Theory of Ardell [77]

Ardell's theoretical treatment directly addresses the main weakness of the original LSW theory, which is the assumption that the volume fraction of the precipitation phase is negligibly small. Ardell modified the assumptions made for the diffusion geometry by assuming that the mean concentration of the matrix surrounding the average particle is not given by the concentration at infinity but at a sphere with its center located in the center of the particle and a radius corresponding to half the mean distance between the precipitates. As this radius depends on the number density and size distribution of the ensemble of precipitates, the volume fraction effect is thus incorporated into the model.

The major results of the modified LSW theory is the observation that (*a*) the coarsening rate increases strongly as the volume fraction of precipitates increases and (*b*) the stationary size distribution becomes significantly broader with a less distinct peak.

The coarsening rate constant, $K\left(f^\beta\right)$, is obtained as

$$K\left(f^\beta\right) = \frac{6\gamma D_{\mathrm{B}}^{\alpha} c_{\mathrm{B}}^{\alpha}\left(v^\beta\right)^2}{RT}\frac{\xi_m^3}{v}, \tag{3.214}$$

where

$$v = \frac{3\xi_m^2}{1+2\beta\xi_m-\beta}, \tag{3.215}$$

$$\beta = \frac{6\left(f^\beta\right)^{\frac{1}{3}}}{e^{3f^\beta}\Gamma\left(f^\beta\right)}, \tag{3.216}$$

and

$$\xi_m = \frac{\left(\beta^2+\beta+1\right)^{\frac{1}{2}}-(1-\beta)}{\beta}. \tag{3.217}$$

The parameter ξ_m represents the theoretical relative maximum radius and the function, $\Gamma\left(f^\beta\right)$, is given as

$$\Gamma\left(f^\beta\right) = \int_{8f^\beta}^{\infty} x^{-2/3}e^{-x}dx. \tag{3.218}$$

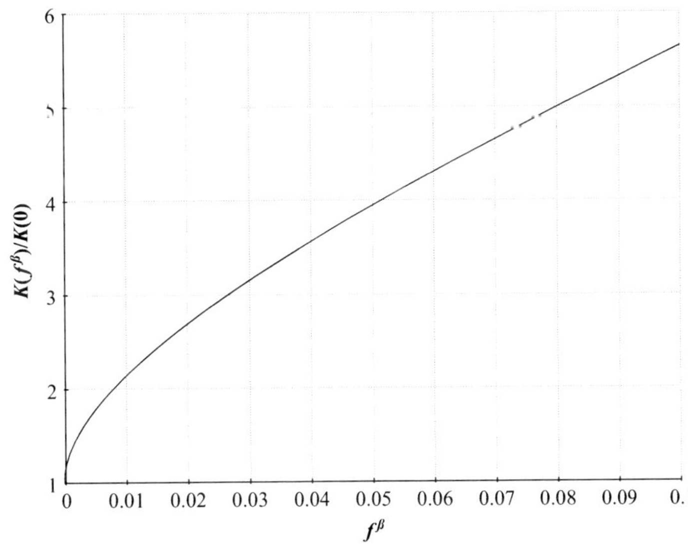

Figure 3.34. Relative coarsening rate obtained by Ardell [77] in dependence of the precipitate volume fraction, f^{β}.

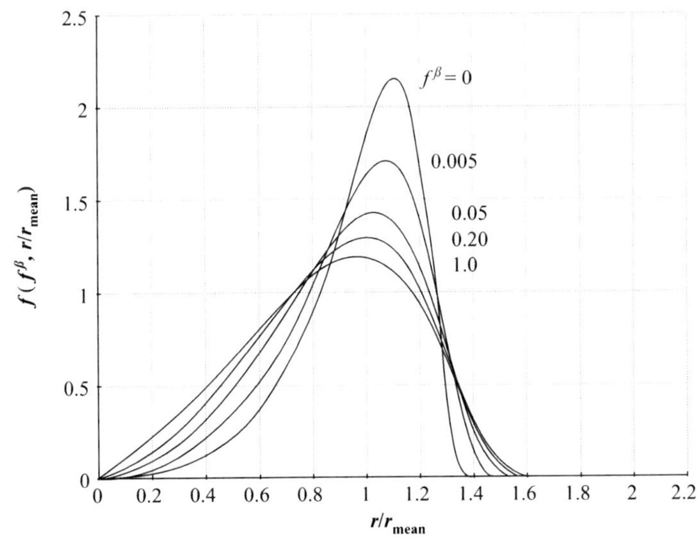

Figure 3.35. Coarsening size distribution function obtained by Ardell [77] in dependence of the precipitate volume fraction, f^{β}.

When evaluating these equations for increasing volume fraction of the precipitate phase, the dependence shown in Figure 3.34 is observed. The increase in growth rate is significant, thus emphasizing the importance of taking volume fraction into account by consideration of "local" conditions for the individual precipitates.

The size distribution of precipitates during coarsening at long times is shown in Figure 3.35. Apparently, even volume fractions clearly below 1% already led to a substantial broadening of the size distribution and significant deviations from the LSW limit of negligible volume fraction of the precipitate phase fraction.

3.6.2.2 The Brailsford and Wynblatt Theory [78]

The motivation for reinvestigation of the LSW theory performed by Brailsford and Wynblatt comes from the discrepancies between experimentally determined particle size distributions and both, LSW and Ardell's [77] theory. The analysis of the Ostwald ripening problem was carried out in two parts: (*a*) determination of the precipitate growth rate for a configuration of randomly arranged precipitates of the same type but different size, and (*b*) solution of the continuity equation for the precipitate distribution. The growth rate of the particles is then determined by the evaporation and adsorption of atoms from a homogeneous medium, which is determined by the entire precipitate population.

The results of the Brailsford and Wynblatt theory are somewhat between the LSW and Ardell theory. At a precipitate phase fraction of 0.5, the rate constant, $K\left(f^{\beta}\right)$, is increased by a factor of 3.5 compared to the LSW value.

3.6.2.3 The Davies, Nash, and Stevens (LSEM) Theory [79]

The basic idea behind the Lifshitz-Slyozov encounter modified (LSEM) theory is the possibility of "encounters," which is made possible if particles approach each other sufficiently close during growth. The probability for these encounters is higher, the higher the phase fraction of precipitates is. As a consequence of overlapping diffusion fields and narrow local arrangement, collisions between particles might occur leading to *particle coalescence*.[21] Davies, Nash, and Stevens emphasize that this possible growth mechanism must also be accounted for in the kinetic equation, although they find that its effect is "not great."

Davies, Nash, and Stevens demonstrate that the particle collisions increase the value of the rate constant, while at the same time leaving the principal time characteristics of the coarsening process unchanged with an exponent of 1/3. The theory predicts that the growth constant is increased by a factor of approximately three over the whole range of volume fractions. The steady-state distribution of particles predicted by this theory is broader than LSW and slightly more symmetric than the previous theories.

The LSEM theory was published shortly after the Brailsford and Wynblatt theory. In an additional note, Davies, Nash, and Stevens compare their approach with that of Brailsford and Wynblatt and conclude that the BW theory specifically excludes the effect of collisions/encounters from their treatment. The BW work is thus complementary to the LSEM theory, which explicitly takes these encounters into account.

3.6.2.4 The Tsumuraya and Miyata Theory [80]

In their work, Tsumuraya and Miyata develop six models for steady-state particle coarsening with the aim of explaining the experimentally observed shape of precipitate distributions. The coarsening process is driven by the reduction of interfacial area and volume diffusion of monomers.

---

[21] Precipitate coalescence and its influence on the evolution of precipitates are also dealt with in Section 6.3.

In the Tsumuraya and Miyata models, account is taken of the real spatial arrangement of particles during coarsening. For instance, one of the models represents a regular array of particles and it is used to explain the experimentally observed size distribution in Ni–Al with $f^{\beta} > 0.1$ and NiAl–Cr with $f^{\beta} > 0.2$. Another model is used to explain the size distribution in alloys with smaller phase fraction of precipitates.

3.6.2.5 The Marqusee and Ross Theory [81]

In their treatment of particle coarsening, Marqusee and Ross derive evolution equations for the precipitation parameters as well as the asymptotic particle distribution function for the late stages of coarsening. They start with writing down the diffusion equation for a single isolated particle in an infinite matrix. Then, they include the presence of a distribution of competing particles and add it to the general diffusion equations for the single particle.

Marqusee and Ross obtain a solution that conserves the principal $t^{\frac{1}{3}}$ dependence of the mean radius of the precipitate distribution. The rate constant is found to increase with increasing phase fraction of the minority phase as $\sqrt{f^{\beta}}$. The size distribution is broadened as a consequence of the competition between coalescence and monomer adsorption and desorption. The broadening is rather significant already at low volume fractions of the precipitate phase. Qualitatively, this theory is comparable to the Brailsford and Wynblatt theory [78], although they differ on a quantitative basis.

3.6.2.6 The Tokuyama and Kawasaki Theory [82]

Based on statistical mechanics, Tokuyama and Kawasaki employ the scaling expansion method to perform a spatial coarse graining to derive kinetic equations for the particle coarsening process. In their treatment, they try to address three major weaknesses of LSW theory: (*a*) the assumption of negligibly small volume fraction of the precipitation phase, where it is not clear how small f^{β} must be to conform with the term "negligibly small," (*b*) the application of mean field theory, which ignores the cooperative effects among the precipitates, and (*c*) the lack of consideration of statistical fluctuations in the size distribution caused by collision processes among the particles.

Tokuyama and Kawasaki show that there exist two distinct regions of coarsening, depending on the space–time scale: an intermediate stage and a late stage. The kinetic equations derived by this method are found to scale with $\sqrt{f^{\beta}}$ and they contain two terms: a collisionless drift term and a collision term. The collision term is found to be different from the conventional "encounter" integral of Lifshitz and Slyozov [75], as the former is of order $\sqrt{f^{\beta}}$ and describes distant "soft" collisions, whereas the latter is of order f^{β} and represents close "hard" collisions.[22]

---

[22]The term "soft" collision refers to the overlap and interaction of diffusion fields of neighboring particles, whereas the term "hard" collision denotes the actual process of coalescence where two particles collide and merge into a single larger particle. With soft collisions, the interaction between particles differs from the interaction given by a mean-field theory accounting for the diffusion field at infinity.

3.6.2.7 The Voorhees and Glicksman Theory [83,84]

Voorhees and Glicksman treat the multiparticle diffusion problem in the framework of an embedding technique, where the growing and shrinking precipitates are represented by point sources and sinks for the diffusing atoms. The diffusion problem outside the particle diffusion field is constructed using a potential theoretic technique, which is an adaptation of the Ewald method for calculation of lattice sums. The Ostwald ripening problem is reformulated in terms of an interaction matrix, the elements of which are Ewald sums. With this method, Voorhees and Glicksman discuss the influence of local diffusional interactions on the general coarsening characteristics.

In the first paper [83], Voorhees and Glicksman discuss the impact of their *Ansatz* on the coarsening behavior on the basis of their analytical model. In a first step, all governing equations are translated into a form where the space variables are dimensionless quantities, which scale with a characteristic diffusion length, l_c, and which remains to be specified by the problem under investigation. The value of the diffusion field is denoted by $\phi(\mathbf{r})$, where, in general, $\phi(\mathbf{r})$ represents a dimensionless potential such as, for instance, temperature, pressure, or solute concentration.

In a medium with m point sources or sinks at fixed positions, and for slowly changing diffusion fields, the evolution of the generalized potential can be described by a Poisson-type equation as

$$\nabla\phi(\mathbf{r}) = \sum_{i=1}^{m} -4\pi B_i \delta(\mathbf{r} - \mathbf{r}_i), \tag{3.219}$$

where B_i is the strength of the ith source or sink, δ is the Dirac function, $\mathbf{r}$ is the dimensionless vector locating the field point, and $\mathbf{r}_i$ is the vector locating the point source or sink.

A general solution of equation (3.219) is given by the linear combination

$$\phi(\mathbf{r}) = B_0 + \sum_{i=1}^{m} \frac{B_i}{|\mathbf{r} - \mathbf{r}_i|}, \tag{3.220}$$

with B_0 representing some constant reference potential, which, in general, can be set to zero. This formulation indicates clearly that each point source or sink in the system must contribute to the concentration field. Convergence of the summation is guaranteed only if the following condition holds:

$$\lim_{m\to\infty}\left(\sum_{i=1}^{m} B_i\right) = 0. \tag{3.221}$$

In order to solve the problem, Voorhees and Glicksman simplify the problem by assuming that the point sources and sinks are arranged in a somehow periodic manner. They use the concept of a Bravais lattice and a suitable basis. The Bravais lattice represents the spatial periodicity and the basis can approximate the spatial randomness. The sources and sinks can then be placed within a unit cell to form the basis, and the basis is then translated according to the chosen Bravais lattice to fill all space. Voorhees and Glicksman argue that this procedure can produce approximately spatially

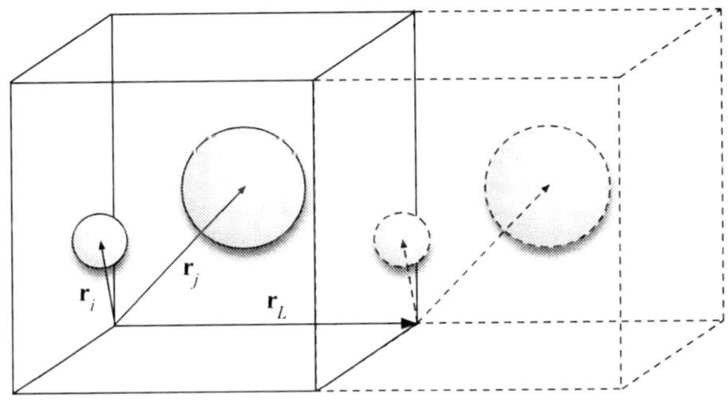

Figure 3.36. Construction of translational Bravais cells with periodicity vector, $\mathbf{r}_L$, by Voorhees and Glicksman [83]. $\mathbf{r}_i$ and $\mathbf{r}_j$ are vectors locating the centers of the point sources and sinks.

random structures, yet allowing for an analytical solution to the multiparticle diffusion problem. A sketch of this procedure is presented in Figure 3.36.

Based on this *Ansatz*, Voorhees and Glicksman obtain an analytical solution for the multiparticle problem, which allows taking into account the full effect of particle interactions. The solution is shown to be consistent with LSW theory in the limit of $f^\beta \to 0$. Moreover, the coarsening rates of small particles are not strongly affected by diffusional interactions. Small particles are expected to coarsen at a rate close to the one predicted by LSW, even at high volume fractions.

In the second paper [84], Voorhees and Glicksman apply their model to a statistically useful large number of particles on basis of a numerical implementation of their theory. The basic idea is to increase the number of sources and sinks in the single Bravais cell and to solve the diffusion problem numerically. For details of the analysis, the interested reader is referred to the original publication. In summary, the steady-state distribution of particles in the asymptotic limit are represented in Figure 3.37. In accordance with the previously outlined models, Voorhees and Glicksman find that the distribution broadens significantly at higher volume fractions of the precipitate phase, which was analyzed up to a volume fraction of 0.5.

3.6.2.8 The Enomoto, Tokuyama, and Kawasaki Theory [85]

In this approach, the finite volume fraction of particles is taken into account by a statistical theory that has been developed by these authors earlier. The competitive growth, as well as soft collisions, is considered on the basis of statistical correlations between the different particles. Enomoto, Tokuyama, and Kawasaki conclude that their theory is in excellent agreement with the numerical simulations performed by Voorhees and Glicksman [84]. This agreement is not observed if the effect of soft collisions is neglected.

3.6.2.9 The Marder Theory [86]

In 1987, Marder presented yet another approach to explain the experimentally observed broadening of size distributions compared to the mean-field solution of LSW. Marder tries to solve the

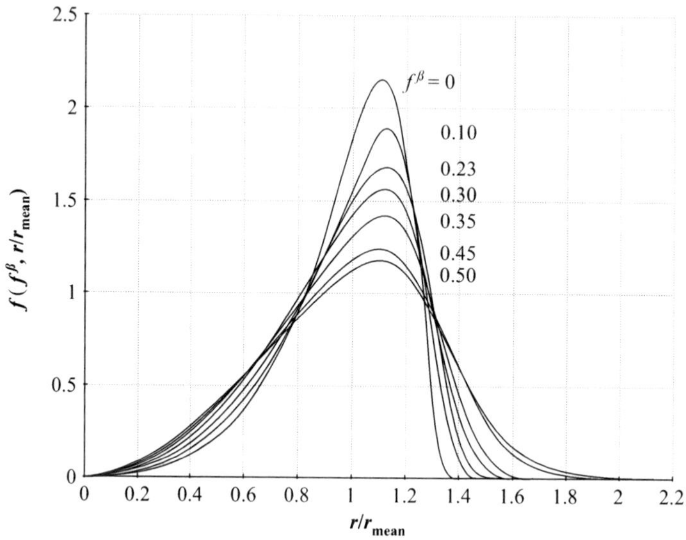

Figure 3.37. Coarsening size distribution function obtained by Voorhees and Glicksman [84] in dependence of the precipitate volume fraction, f^β.

problem of both, (i) predicting vastly broader size distributions than LSW, however, (ii) without predicting wildely accelerated coarsening rates. In Marder's theory, this goal is achieved by using a new model system from which a hierarchy of evolution equations is derived, which is expanded in the square root of the volume fraction of precipitates. The time evolution of correlations, which are identified as primary factors in determining a realistic solution to the problem, are analyzed through an "unusual" perturbation theory. The resulting differential equations are solved numerically and found to be in excellent agreement with experiment at a volume fraction of 5%.

3.6.3 Comparison of Theories

After the initial works of Lifshitz and Slyozov [75] and Wagner [74], several improved theories have been developed, mainly with the goal of explaining the experimentally observed broadening of the size distribution of coarsening particles compared to the original LSW mean-field theory. These theories are well described in, for example, the review articles by Voorhees [76] and by Tsumuraya and Miyata [80].

Interestingly, all of these theories conserve the principal time dependence of the mean radius and particle density evolution with $t^{1/3}$ and t^{-1}, respectively. The main differences in all these theories is the coarsening rate constant, $K(f^\beta)$, which is displayed as a function of volume fraction, f^β, in Figure 3.38.

The interested reader might finally be referred to some additional literature on coarsening [87–89].

3.6.4 Coarsening in Multicomponent Alloys

The coarsening rates and size distributions that have been developed in the previous section have been derived with knowledge of the equilibrium composition of the solute element at the

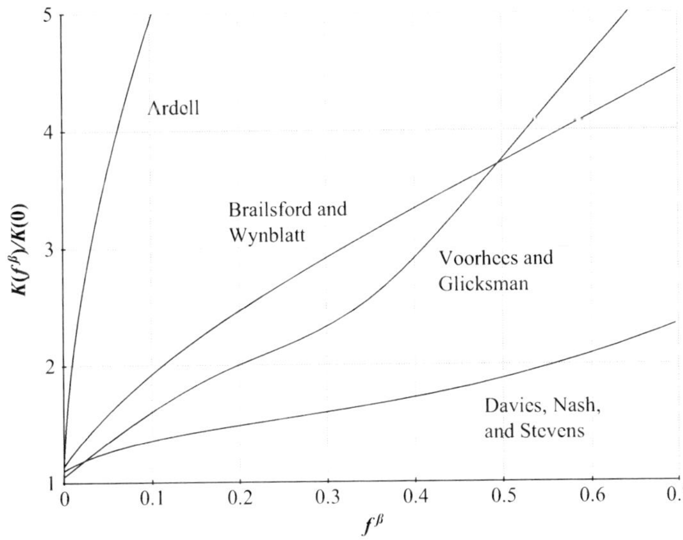

Figure 3.38. Dependence of the coarsening rate constant, $K(f^\beta)$, from volume fraction for different theories [84].

precipitate surface. In binary systems, this quantity is directly read from the phase diagram and modified for interfacial curvature effects using the Gibbs–Thomson equation. In multicomponent system, the interfacial compositions are not directly given (see Sections 3.3.1 and 3.3.3) but must be obtained from solution of the corresponding flux and mass balance equations.

One of the first works dealing with multicomponent effects in coarsening is that of Umantsev and Olson [90]. These authors also review some earlier work in this field. Umantsev and Olson present an approach for systems with a diagonal diffusivity matrix and they discuss how multicomponent effects, particularly nonideality of solutions, impact the coarsening behavior. Morral and Purdy [91] extended this work by taking into account also off-diagonal terms in the diffusivity matrix. They show that coarsening is independent of the solid solution thermodynamics. Solution of the multicomponent coarsening problem does, however, involve some matrix algebra.

Another similar approach was presented by Ågren et al. [92] at the 1997 Schloss Ringberg symposium. Accordingly, for a stoichiometric compound in a multicomponent solution and for a stationary size distribution, the coarsening rate constant is obtained as

$$K_{\mathrm{LSW}} = \frac{8\gamma v^\beta}{9\sum_{i=1}^{n} \frac{\left(X_i^\beta - X_i^\alpha\right)^2}{X_i^\alpha D_i^\alpha} RT}. \tag{3.222}$$

For evaluation of the equilibrium compositions, X_i^β and X_i^β, a thermodynamic equilibrium computation can be performed before the rate constant is evaluated. In the derivation, it is implicitly assumed that the displacements in local equilibrium content are small and the sum expression can be approximated by a constant value.

Additional information might be found in, for example, refs. [88,93–95].

3.7 Summary

Once a precipitate has nucleated and reached supercritical size, its further evolution can be described by the processes of attachment and detachment of monomers to the precipitate surface. Growth and dissolution of a precipitate are *deterministic* processes, in the sense that they describe the evolution of each individual particle in the thermodynamic system by unique and unambiguous kinetic laws. This behavior is in strong contrast to the laws controlling the nucleation process. The nucleation rate for second phase particles can only be quantitatively formulated in terms of statistical evidences and probabilities, but no information is available on a single nucleation event occurring on a particular place and at a particular time.

The transport of monomers through the matrix, which is a necessary condition for precipitate growth, is described by diffusion, with the flux of atoms toward the precipitate surface being controlled by the element gradients prevailing at the phase interface. A most effective solution to the binary diffusion-controlled growth problem has been developed by Zener, who proposed a simple approximation of the diffusion profiles surrounding the precipitate for planar as well as spherical geometry (see Section 3.2.2). Accordingly, the radius evolution of the precipitate scales with the square root of time. The corresponding kinetics is said to obey a *parabolic growth law*.

Similar expressions to the one obtained by Zener are presented for the quasi-stationary approach in Section 3.2.3. For the binary case, an implicit exact solution is also presented, which covers cases from low to high supersaturation equally. For simulation of diffusion-controlled precipitate growth, the equations summarized in this chapter are widely used in precipitation modeling.

The laws describing the diffusion-controlled evolution of precipitates in multicomponent environment are substantially more complex. The local equilibrium tie-lines, which determine the interfacial compositions, are no longer directly accessible from the phase diagram but they must be obtained as the solution of the multicomponent long-range diffusion problem in combination with appropriate mass balance conditions at the moving phase boundary. This issue is discussed in Section 3.3. Analytical solutions for the multicomponent case are no longer available and integration of the governing differential equations must be performed numerically on the computer. Luckily, powerful software is nowadays accessible that can carry out this task for the user.

In many cases, solution of the full moving boundary problem involving exact diffusion profiles in the matrix surrounding the precipitate becomes computationally expensive and approximations are introduced to reduce the complexity of the problem. Two corresponding models describing multicomponent precipitate growth are increasingly often used: (*a*) an approximation based on the local equilibrium concept (Section 3.3.4) and (*b*) a mean-field approach of the precipitation problem based on the thermodynamic extremal principle (Section 3.5). Both models are implemented in numerical precipitation kinetics codes and are available to the user in the form of comprehensive software packages. These are outlined in Chapter 9.

Once the matrix becomes depleted from solute precipitate-forming elements, particle growth stagnates and Ostwald ripening takes over controlling the further evolution of the precipitation system. This issue is the focus of Section 3.6, where several analytical and numerical approaches for coarsening are introduced and discussed. Analytical formulae describing the evolution of mean radius and number density of the precipitates as well as matrix supersaturation are derived. Although widely different shapes of the stationary size distribution are predicted by the various approaches, the principal time dependence of the mean radius evolution with the cube root of time ($t^{1/3}$) and the number density with the inverse of time (t^{-1}) is conserved by all coarsening models described in Section 3.6. These models are often used in practical evaluation of the long-time stability of precipitation stabilized microstructures for high-temperature applications.

In advanced precipitation kinetics studies involving nonisothermal temperature histories and complex multicomponent environments, the growth laws developed in this chapter are utilized in combination with appropriate theory for precipitate nucleation to set up numerical algorithms for precipitation kinetics simulations. These approaches are discussed in Chapter 5.

CHAPTER 4

INTERFACIAL ENERGY

When two initially isolated phases are brought into contact, the atoms constituting the two phases interact with each other across the new phase boundary. The part of the total energy of this system which is related to the interfacial region between two phases in contact is denoted as the *interfacial energy*, γ. Its value is commonly referred to in units of J/m^2.

The interfacial energy is a quantity which plays an important role in many continuums-mechanical approaches. The basic concept of interfacial energies root back to William Thomson [15] and Josiah Willard Gibbs [16] in their classical treatment of liquid droplets being in equilibrium with saturated vapor. Although originally considered as a surface energy, the concepts developed there can be applied straightforwardly to solid-state precipitation processes. Accordingly, the total free energy of a system with precipitates can be expressed as a superposition of volumetric and interfacial energy contributions (see Section 2.3 on solid-state nucleation theory). The equilibrium properties of systems of this kind can be obtained by minimization of the total system free energy.

The interfacial energy is a quantity that occurs as an important parameter in various materials-related simulation models. Unfortunately, its absolute value cannot be observed *directly* based on key experiments. The interfacial energy value is only accessible *indirectly*, that is, by measuring an interfacial energy-dependent quantity and subsequent extraction of the interfacial energy value from the corresponding theoretical model.

In the context of solid-state precipitation, interfacial energies are typically measured based on either nucleation experiments, where classical nucleation theory is employed to evaluate experimentally observed nucleation rates (see Section 2.3), or from coarsening experiments, where the LSW growth coefficient is extracted from the experimentally obtained mean radius and number density evolution (see Section 3.6). Whereas the former approach suffers from the difficulty of obtaining the experimental data, the latter approach carries the inherent problem that coarsening experiments can only deliver the product of diffusion coefficient and interfacial energy. As the diffusivities have large experimental scatter, it is difficult to obtain accurate values for the interfacial energy directly. If the evolution of supersaturation is measured simultaneously, this problem can be avoided. This issue has been reviewed by A. Ardell [96].

In this chapter, a theoretical concept is outlined which allows for the evaluation and prediction of interfacial energies from thermodynamic data that is stored, for example, in

CALPHAD-type[1] databases. In the treatment, advantage is taken of the close relation between interfacial energy and the enthalpy of solution.

4.1 The Nearest-Neighbor Broken-Bond Model

In 1938, R. Becker [97] introduced a thought experiment that provided a simple and powerful framework for calculation of interfacial energies. Consider two separated blocks of homogeneous phases, α_1 and α_2. Cut these blocks in half and assemble the unlike pieces again: Attach the two α_1 half-blocks to the two α_2 half-blocks. The difference in total energy of the two configurations represents the interfacial energy, γ, of the α_1 and α_2 contact surfaces. Figure 4.1 shows a sketch of this procedure.

When looking at the interface on the atomic scale, ideally, the contact area between the two phases is geometrically flat and the lattice planes of the two crystalline phases match perfectly. The latter feature is characteristic for *coherent* interfaces. In the simple case of primitive, cubic unit cells, interfaces between two phases could look like the ones shown in Figure 4.2.

In the first case (Figure 4.2[a]), the interface between the two phases is located along a low-index plane; in the second case (Figure 4.2[b]), the interface separates the two phases along a higher-index plane. Both interfaces are characterized by a sharp transition from one phase to the other, in which all atoms are perfectly separated on either of the two sides. No mixing of atoms between the phases and across the interface is present: The interface can be viewed as being ideally thin. This special case of phase boundary is called a *sharp, planar* interface. We will soon see that this configuration allows for a closed analytical solution for interfacial energy calculation as the sharp interface approximation does not involve entropic terms in the evaluation of its thermodynamic properties. Before elaborating this, however, the Becker model will be discussed further.

Consider the energy of the two blocks of matter in the starting configuration (state I in Figure 4.1). Each block is considered to be a homogeneous crystalline solid solution, with α_1 consisting only of atoms A and α_2 consisting only of atoms B. No interactions between the two separate blocks shall be present in state I. The total energy of the initial system is given by the superposition of the sum of all atomic interactions in each of the two blocks. For the sake of simplicity, we assume that both phases have the same crystal structure, that the atomic interaction between two atoms of type *i* and *j* is independent of its local chemical environment, and that the interactions are restricted to nearest-neighbor lattice positions. Figure 4.3 shows a sketch of the effectively broken nearest-neighbor bonds across two types of lattices.

As the only difference between states I and II in Figure 4.1 is the newly created interface between the unlike blocks of matter, the interfacial energy is given as

$$\gamma = E_{\alpha_{12}}^{\text{new}} - E_{\alpha_1}^{\text{broken}} - E_{\alpha_2}^{\text{broken}}, \tag{4.1}$$

with γ representing the interfacial energy, $E_{\alpha_{12}}^{\text{new}}$ being the newly created bonds across the interface between the unlike blocks, and $E_{\alpha_1}^{\text{broken}}$ and $E_{\alpha_2}^{\text{broken}}$ representing the energy of the broken bonds in the initial configuration.

[1] Compare Section 1.3.3.

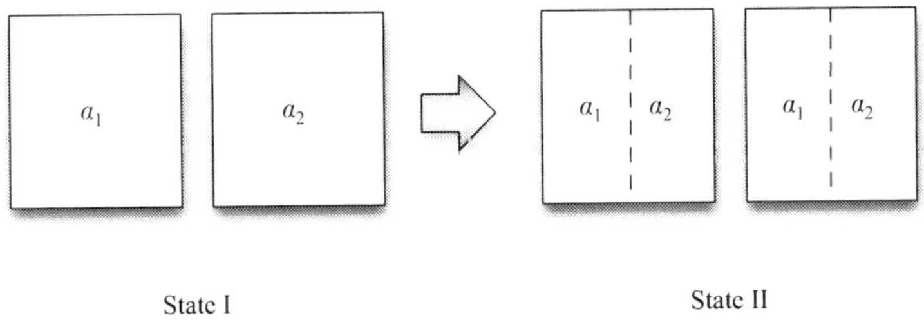

Figure 4.1. Scheme for calculation of the interfacial energy as proposed by R. Becker [97]. The dashed lines indicate the newly formed interfaces.

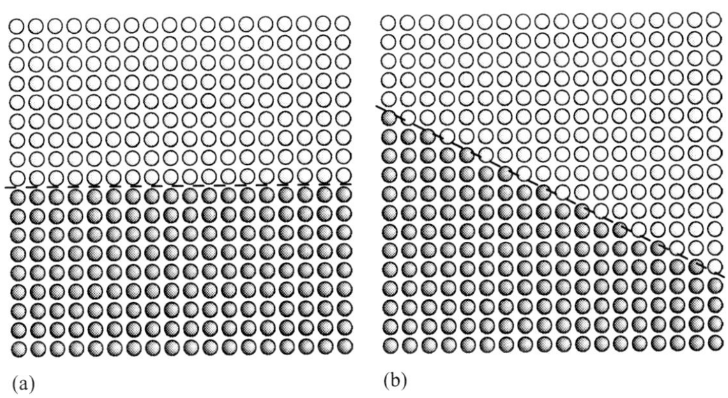

Figure 4.2. Sharp, planar, and coherent interfaces between two pure substances with different orientations in a two-dimensional simple cubic lattice.

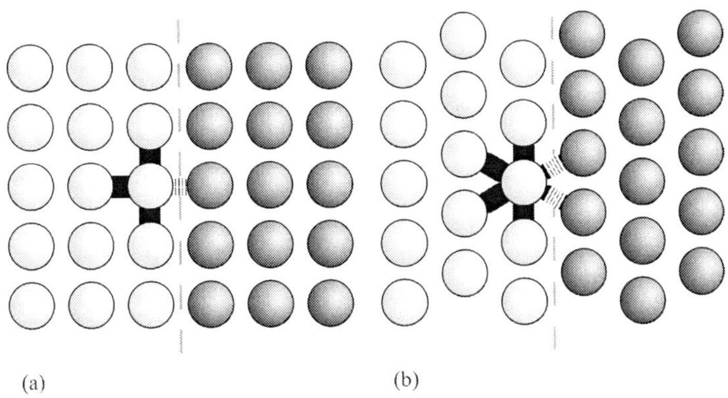

Figure 4.3. Broken atomic bonds across an interface in the primitive (a) and close-packed (b) system on a two-dimensional cubic lattice.

With the bond energy, ε_{ij}, between atoms i and j,[2] the energy related to the broken atomic bonds in the blocks with pure substances is given as

$$E_{\alpha_1}^{\text{broken}} = \frac{n_S z_S}{2} \cdot \varepsilon_{AA}$$
$$E_{\alpha_2}^{\text{broken}} = \frac{n_S z_S}{2} \cdot \varepsilon_{BB}, \tag{4.2}$$

where n_S is the number of atoms per unit area of interface and z_S is the number of broken bonds across the interface counted per interface atom. The energy of the newly formed interfaces[3] is given as

$$E_{\alpha_{12}}^{\text{new}} = n_S z_S \cdot \varepsilon_{AB}, \tag{4.3}$$

and the interfacial energy follows as

$$\gamma = n_S z_S \left[\varepsilon_{AB} - \frac{1}{2}\left(\varepsilon_{AA} + \varepsilon_{BB}\right) \right]. \tag{4.4}$$

When analyzing this result, we observe that the interfacial energy, γ, is only a function of the number of atoms per unit area of interface, the number of broken atomic bonds across the interface, and the difference in bond energy between the like and unlike atoms. Although the interfacial energy, γ, thus appears to be a readily accessible quantity, practical evaluation based on equation (4.4) shows a few weaknesses, which are mainly due to the assumptions that

- for calculation of γ, one must know the values of the interatomic bond energies. These values are not readily available, although modern computational materials science methods, such as Density Function Theory (DFT) or Molecular Dynamics (MD), show some potential for delivering these quantities from first principles calculations.
- the product, $n_S z_S$, is originally defined in terms of nearest-neighbor interactions. This is a convenient approximation, but, in most cases, it is unrealistic. Atoms interact with each other over considerably farther distances than only first-nearest neighbor.

An important contribution to solving the first problem has been made by David Turnbull (1915–2007), an American materials physicist and professor at Harvard University in Cambridge, USA, in 1955 [98]. This is outlined here. The solution to the second problem is pursued in Section 4.3.

Consider two homogeneous phases α_1 and α_2, consisting purely of atoms A and B, respectively. Using the formalism introduced in Section 1.3.2, the energy, E^{SS}, of the two blocks with one mole of atoms each is given as

[2] See also the treatment of regular solutions in Section 1.3.2.

[3] Since we obtain two identical blocks of dislike phases, the corresponding energy must be counted twice.

$$E_{\alpha_1}^{\mathrm{SS}} = \frac{N_A z_L}{2} \cdot \varepsilon_{AA}$$
$$E_{\alpha_2}^{\mathrm{SS}} = \frac{N_A z_L}{2} \cdot \varepsilon_{BB}, \tag{4.5}$$

with z_L being the coordination number (number of nearest-neighbor positions) and N_A being Avogadro's number.

After breaking all bonds between the atoms in the two blocks and merging them into a homogeneous solid solution, the total energy of the solid solution, $E_{\alpha_{12}}^{\mathrm{SS}}$, is obtained by bond counting as

$$E_{\alpha_{12}}^{\mathrm{SS}} = N_A z_L \cdot \varepsilon_{AB}. \tag{4.6}$$

The energy difference, ΔH_{sol}, between the two pure substances and the solid solution containing the mixture of A and B is known as the *enthalpy of solution*. It is given by

$$\Delta H_{\mathrm{sol}} = E_{\alpha_{12}}^{\mathrm{SS}} - \frac{1}{2}\left(E_{\alpha_1}^{\mathrm{SS}} + E_{\alpha_2}^{\mathrm{SS}}\right) = N_A z_L \cdot \left[\varepsilon_{AB} - \frac{1}{2}\left(\varepsilon_{AA} + \varepsilon_{BB}\right)\right], \tag{4.7}$$

and it has been introduced earlier in Section 1.3.2. With the effective interaction potential, ε^*, defined as

$$\varepsilon^* = \varepsilon_{AB} - \frac{1}{2}\left(\varepsilon_{AA} + \varepsilon_{BB}\right), \tag{4.8}$$

equations (4.4) and (4.7) can be rewritten as

$$\Delta H_{\mathrm{sol}} = N_A z_L \cdot \varepsilon^* \tag{4.9}$$

and

$$\gamma = n_S z_S \cdot \varepsilon^*. \tag{4.10}$$

Comparison of equations (4.9) and (4.10) directly delivers the governing equation for the interfacial energy as

$$\gamma = \frac{n_S z_S}{N_A z_L} \cdot \Delta H_{\mathrm{sol}}. \tag{4.11}$$

Equation (4.11) is the most important relation that directly connects the interfacial energy, γ, to the enthalpy of solution, ΔH_{sol}, and a structural factor, $n_S z_S / N_A z_L$. The interfacial energy is thus a quantity that can be evaluated uniquely and straightforwardly once the thermodynamics of the system of interest is known.

The treatment has so far been based on binary systems AB. It will be demonstrated, next, that equation (4.11) is equally valid for multicomponent alloys. The corresponding procedure has been suggested in ref. [99] and it is reproduced subsequently.

Consider a thermodynamic system consisting of a matrix phase, α_1, and a precipitate phase, α_2. Let $X_i^{\alpha_1}$ be the composition vector of the matrix and $X_i^{\alpha_2}$ be the same quantity of the precipitate. Both phases are assumed to be homogeneous solid solutions. The enthalpy of solution, ΔH_{sol}, of one mole of precipitate in the matrix can be calculated in three steps:

1. Dissolve the precipitate and count all bonds that are broken in the disassembly procedure. Each broken bond in the precipitate is of type *i-j* with a probability of $X_i^{\alpha_2} X_j^{\alpha_2}$.[4] For each atom, $z_L/2$ bonds are broken.
2. Create one mole of vacant lattice sites in the matrix phase by randomly selecting an atom in the bulk volume and depositing it at a random position on the surface of the matrix phase. Again, $z_L/2$ bonds are broken, with each bond being of *i-j* type with a probability of $X_i^{\alpha_1} X_j^{\alpha_1}$.
3. Insert the precipitate atoms into the matrix phase by filling up the empty site positions created in step (ii). In this procedure, z_L new bonds are created per precipitate atom. Under the assumption of an infinitely large matrix phase, the newly formed bonds are of type *i-j* with a probability of $X_i^{\alpha_1} X_j^{\alpha_2}$.

The enthalpy of solution of one mole of precipitate atoms is then given in analogy to equation (4.9) with

$$\Delta H_{\mathrm{sol}}^{\alpha_1\alpha_2} = N_A z_L \cdot \varepsilon^* \qquad (4.12)$$

and the effective interaction energy, ε^*, given by

$$\varepsilon^* = \sum_i \sum_j \varepsilon_{ij} \cdot \left[X_i^{\alpha_1} X_j^{\alpha_2} - \frac{1}{2}\left(X_i^{\alpha_1} X_j^{\alpha_1} + X_i^{\alpha_2} X_j^{\alpha_2} \right) \right]. \qquad (4.13)$$

The interfacial energy between a precipitate and the matrix in multicomponent environment can analogously be calculated in two steps based on Becker's thought experiment introduced earlier:

1. Split the two blocks in half and count the broken bonds per unit area. With z_S being the number of broken bonds per interface atom, and n_S the number of atoms per unit area of interface, $z_S n_S/2$ bonds are broken in the precipitate as well as matrix phase. The probabilities for a single bond being of *i-j* type are $X_i^{\alpha_1} X_j^{\alpha_1}$ and $X_i^{\alpha_2} X_j^{\alpha_2}$, respectively.
2. After recombining the unlike blocks, $z_S n_S$ new bonds are created with the probability for a single bond to be of *i-j* type being $X_i^{\alpha_1} X_j^{\alpha_2}$.

The interface energy is then

$$\gamma^{\alpha_1\alpha_2} = n_S z_S \cdot \varepsilon^*, \qquad (4.14)$$

[4]The interfacial energy occurs in cubic form in the exponent of the nucleation rate expression, see Sections 2.2.1 and 2.3.3.

with

$$\varepsilon^* = \sum_i \sum_j \varepsilon_{ij} \cdot \left[X_i^{\alpha_1} X_j^{\alpha_2} - \frac{1}{2}\left(X_i^{\alpha_1} X_j^{\alpha_1} + X_i^{\alpha_2} X_j^{\alpha_2} \right) \right]. \tag{4.15}$$

As identical terms for the effective interaction energy, ε^*, are observed in the enthalpy of solution as well as the interfacial energy expression, the validity of Turnbull's equation (4.11) in general multicomponent solutions is demonstrated.

4.2 Composition Dependence of the Precipitate–Matrix Interfacial Energy

In the previous section, the functional dependence of the interfacial energy, γ, from a structural factor, $n_S z_S / N_A z_L$, and the enthalpy of solution, ΔH_{sol}, have been derived. In the present section, the dependence of the interfacial energy from the chemical composition of precipitate and matrix is investigated further. Subsequently, the superscript "P" is used for the precipitate and "M" is used for the matrix phase.

Consider a binary system AB. Let the composition of the matrix be given in terms of the mole fraction variables, X_A^M and X_B^M, and the composition of the precipitate be X_A^P and X_B^P. Under the assumption of random alloys and using the identities $X_A^P = 1 - X_B^P$ and $X_A^M = 1 - X_B^M$, equation (4.2) becomes

$$\begin{aligned} E_M^{\text{broken}} &= \frac{n_S z_S}{2}\left(\left(x_A^M\right)^2 \varepsilon_{AA} + 2x_A^M x_B^M \varepsilon_{AB} + \left(x_B^M\right)^2 \varepsilon_{BB} \right) \\ &= \frac{n_S z_S}{2}\left(\varepsilon_{AA} + 2x_B^M \left(\varepsilon_{AB} - \varepsilon_{AA}\right) + \left(x_B^M\right)^2 \left(\varepsilon_{AA} + \varepsilon_{BB} - 2\varepsilon_{AB}\right) \right) \end{aligned} \tag{4.16}$$

$$E_P^{\text{broken}} = \frac{n_S z_S}{2}\left(\varepsilon_{AA} + 2x_B^P \left(\varepsilon_{AB} - \varepsilon_{AA}\right) + \left(x_B^P\right)^2 \left(\varepsilon_{AA} + \varepsilon_{BB} - 2\varepsilon_{AB}\right) \right). \tag{4.17}$$

The expression in bracket represents the sum of the products of bond energy and the probability of its occurrence (the latter is valid for the random alloy assumption). In the present system, equation (4.3) becomes

$$\begin{aligned} E_{PM}^{\text{new}} &= n_S z_S \left(x_A^M x_A^P \varepsilon_{AA} + \left(x_A^M x_B^P + x_A^P x_B^M \right) \varepsilon_{AB} + x_B^M x_B^P \varepsilon_{BB} \right) \\ &= n_S z_S \left(\varepsilon_{AA} + \left(x_B^M + x_B^P \right)\left(\varepsilon_{AB} - \varepsilon_{AA} \right) + x_B^M x_B^P \left(\varepsilon_{AA} + \varepsilon_{BB} - 2\varepsilon_{AB} \right) \right). \end{aligned} \tag{4.18}$$

The interfacial energy, which is given by the difference between the two configurations in the Becker model, then reads

$$\begin{aligned} \gamma &= n_S z_S \left[\left(x_B^M x_B^P - \frac{1}{2}\left(\left(x_B^M\right)^2 + \left(x_B^P\right)^2 \right) \right) \left(\varepsilon_{AA} + \varepsilon_{BB} - 2\varepsilon_{AB} \right) \right] \\ &= n_S z_S \left(x_B^M - x_B^P \right)^2 \left(\varepsilon_{AB} - \frac{1}{2}\left(\varepsilon_{AA} + \varepsilon_{BB} \right) \right). \end{aligned} \tag{4.19}$$

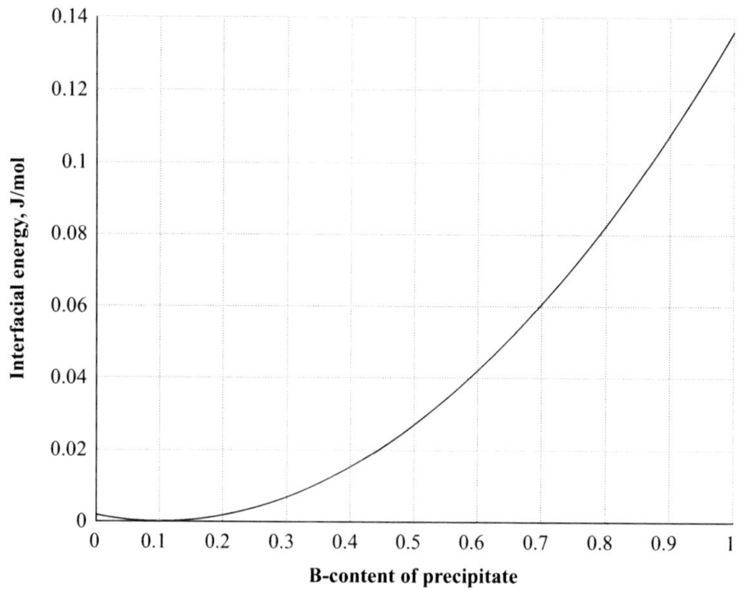

Figure 4.4. Quadratic composition dependence of the interfacial energy from the composition difference between matrix and precipitate evaluated for a regular solution with $L_1 = 20$ kJ/mol at 473 K and a matrix composition of $X_B = 0.1$.

With the effective interaction energy, ε^*, and the observation that the interaction term in brackets is independent of chemical composition in a first approximation, the composition dependence of the interfacial energy finally reads as

$$\gamma = n_S z_S \left(x_B^M - x_B^P\right)^2 \varepsilon^*. \tag{4.20}$$

Equation (4.20) manifests an important result, namely, that the interfacial energy depends quadratically on the difference in chemical composition between matrix and precipitate. This relation is shown exemplarily for a binary AB alloy with 10% B in Figure 4.4. The precipitate composition is varied on the X-axis, while the Y-axis displays the interfacial energy evaluated from equation (4.20). Exactly at $X_B = 0.1$, the matrix composition coincides with the precipitate composition and the interfacial energy is zero.

As the nucleation rate of precipitates is a strong function of the interfacial energy,[5] the quadratic dependences of the interfacial energy from the composition difference between matrix and precipitate has an important impact on the evolution of precipitation sequences involving several metastable phases: Precipitates with chemical compositions closer to the matrix phase are often kinetically favored and appear first in the sequence. The finally stable precipitate phases are often characterized by the largest differences in chemical composition between precipitate and matrix.

[5] The interfacial energy occurs in cubic form in the exponent of the nucleation rate expression, see Sections 2.2.1 and 2.3.3.

The effect of nonideality of solutions on the interfacial energy has been investigated by Borchers and Borman [100] using a pair interaction model. These authors considered interfaces between disordered phases as well as interfaces between a stoichiometric and a disordered phase. The interested reader is referred to this work for more information. When evaluating the interfacial energy from equation (4.11), that is, using the enthalpy of solution, the atomic bond effects are already incorporated in the thermodynamic description of the system.

Several researchers have utilized the theoretical treatment presented above to predict interfacial energies. The values have been used mainly for quantitative evaluation of coarsening rates in Ostwald ripening experiments. Interestingly, practically all of these works involve coherent precipitation in fcc crystals. Exemplarily, the analysis of Li, Saunders, and Miodownik [101] might be mentioned. These authors investigated interfacial energies for γ'-precipitates in Ni-base alloys, based on equation (4.11), to apply these values in a self-consistent manner to predict coarsening rates in multicomponent alloys.

The prediction of interfacial energies based on the nearest-neighbor broken-bond model appears to be well suited for fcc crystals. For bcc crystals, the situation is different. To the knowledge of the author, there are no reports available in literature that can act as a reference for successful application of equation (4.11). A major reason for this observation might be the fact that second-nearest neighbor interactions play a vital role in bcc crystals, whereas they are of lower relevance in fcc crystals.[6] Consequently, the original nearest-neighbor treatment of interfacial energies, as introduced by R. Becker, seems to be a good approximation for fcc structures, while they might be too severe a simplification for bcc crystals. This issue is pursued in the next section.

4.3 Generalization of the NNBB Approach—The GBB Model

Although the Nearest-Neighbor Broken-Bond (NNBB) approach of Becker [97] and Turnbull [98], equation (4.11), has been successfully applied to the prediction of interfacial energies in fcc crystals, application to other crystal structures, for example, bcc, remains critical due to the simplification of only nearest-neighbor interactions. Sonderegger and Kozeschnik [99] have taken this issue as motivation for development of a generalization, which takes into account also *n*-nearest neighbor interaction across the interface. Their approach is subsequently denoted as the Generalized Broken-Bond (GBB) model.

4.3.1 Effective Bond Energies and Broken Bonds

The main idea of the GBB approach is to replace the number of nearest-neighbor broken-bonds across the interface, z_S, and the nearest-neighbor coordination number, z_L, by effective quantities $z_{S,\mathrm{eff}}$ and $z_{L,\mathrm{eff}}$, which also take into account the interactions with atoms in higher-order coordination spheres. Equation (4.11) can then be written as

[6] A center atom in fcc crystals is surrounded by 12 nearest-neighbor atoms. In bcc crystals, only eight nearest-neighbors are present. The difference between first-nearest neighbor and second-nearest neighbor distances in fcc crystals is significantly larger than in bcc crystals.

$$\gamma = \frac{n_S z_{S,eff}}{N_A z_{L,eff}} \cdot \Delta H_{sol}. \tag{4.21}$$

In the GBB model, the weakening of interaction energies with distance is described by a weighting function, $f(r_k/r_1)$, where k represents the index of the coordination shell. The first-nearest neighbor shell is denoted by the index 1. Accordingly, the effective interaction energy, $\varepsilon^{*,k}$, of the kth shell reads

$$\varepsilon^{*,k} = \varepsilon^{*,1} \cdot f(r_k/r_1). \tag{4.22}$$

The enthalpy of solution and the interfacial energy are obtained by summation over all coordination shells, which delivers

$$\Delta H_{sol} = N_A z_{L,eff} \cdot \varepsilon^{*,1}, \tag{4.23}$$

and

$$\gamma = n_S z_{S,eff} \cdot \varepsilon^{*,1}. \tag{4.24}$$

The effective quantities for the number of broken bonds across the interface and the coordination number are

$$z_{S,eff} = \sum_{k=1}^{\infty} z_{S,k} \cdot f(r_k/r_1), \tag{4.25}$$

and

$$z_{L,eff} = \sum_{k=1}^{\infty} z_{L,k} \cdot f(r_k/r_1). \tag{4.26}$$

As the number of interface atoms per unit interfacial area can straightforwardly be obtained from the volumetric packing density, n_V, with $n_S = n_V^{2/3}$, the main task in evaluation of the GBB interfacial energy is now the determination of values for the structural parameters, $n_{S,eff}$ and $z_{L,eff}$, for each crystal lattice under consideration and for various different interface orientations.

The analysis of the effective number of broken bonds is performed in three steps:

1. Define the crystal structure and the orientation of the interface.
2. Count the effective number of broken bonds across the interface area A. Take into account the weighting factor, $f(r_k/r_1)$ for each bond.
3. Refer the number of broken bonds to the mean interfacial area of a single atom.

The procedure of bond counting across the interface is depicted exemplarily in Figure 4.5. The reference interfacial area, which is the shaded area, is denoted with the letter A. The dashed lines represent typical bond types, which must either be included in the bond counting process, or excluded. Accordingly, bond 1 represents a nearest-neighbor bond, which crosses the interface and finds one partner atom inside the reference interface area. This bond is to be counted. The same applies to bonds 2 and 3, although they extend further into the crystal lattice and thus belong to higher-order coordination spheres. Bond 4 is not counted since both atoms are located outside the reference area.

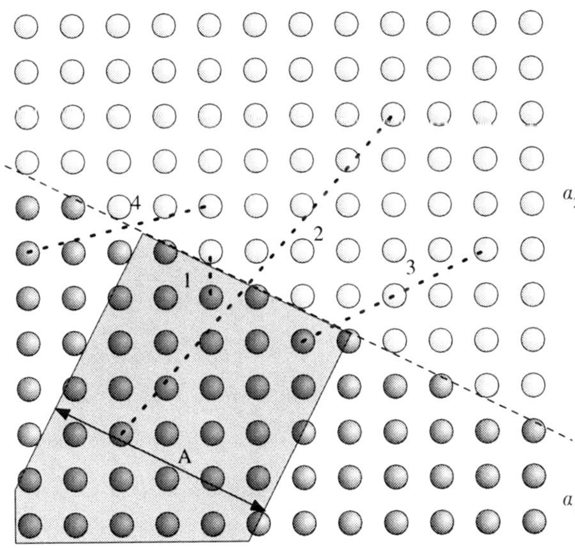

Figure 4.5. Broken bonds across a randomly oriented interface.

The actual analysis of Sonderegger and Kozeschnik [99] is carried out with periodic translational elements for bcc and fcc lattice and based on numerical integration. For the weighting function, the distance dependence has been assumed with values of r^{-6}, r^{-7}, and r^{-8}. The r^{-6} dependence corresponds to the classical Lennard-Jones potential for free atoms [102], while the higher exponents are representative for strong screening of the potentials as expected for metallic bonding.

When performing the summation of bond energies up to a shell index of k, all interactions above this limit are neglected. Consequently, the accuracy of the analysis increases with increasing number of interaction potentials that are taken into account. In the Figures 4.6 and 4.7, the analysis with k interaction shells is compared to the analysis using 500 interaction shells. As expected, the relative error decreases continuously, confirming that a sufficiently high number of interaction shells must be taken into account for accurate values of the structural factor.

For fcc and bcc structure, analytical correction functions are derived in ref. [99], which allow for a consideration of all bond interactions in shells higher than the index, k, to the structural factor. The relative errors are considerably smaller if this correction function is included. The corresponding functions are

$$\Delta z_{\mathrm{L,corr}} = n_V \frac{4\pi}{3} r_1^6 r_k^{-3}, \tag{4.27}$$

and

$$\Delta z_{\mathrm{S,corr}} = \frac{\pi}{2} n_V^{4/3} r_1^6 r_k^{-2}. \tag{4.28}$$

Table 4.1 summarizes the values obtained in the GBB approach for the structural factor $z_{\mathrm{S,eff}}/z_{\mathrm{L,eff}}$ for various interface orientations, interaction potentials, and bcc/fcc structure. The last line represents an average of the structural factor, which can be utilized in simulations where the orientation dependence of the interfacial energy is not explicitly incorporated.

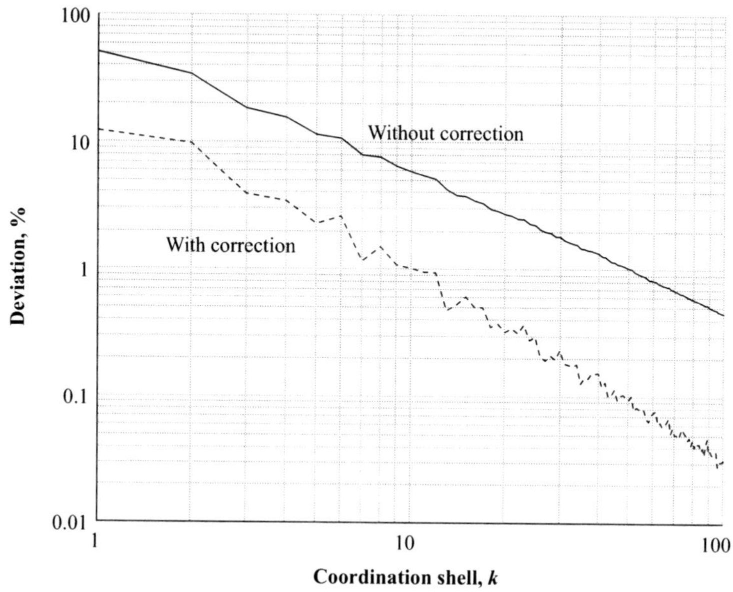

Figure 4.6. Maximum deviation of the structural factor, $z_{S,eff}/z_{L,eff}$, relative to the benchmark of 500 nearest-neighbor shells in the GBB model [99] for fcc structure.

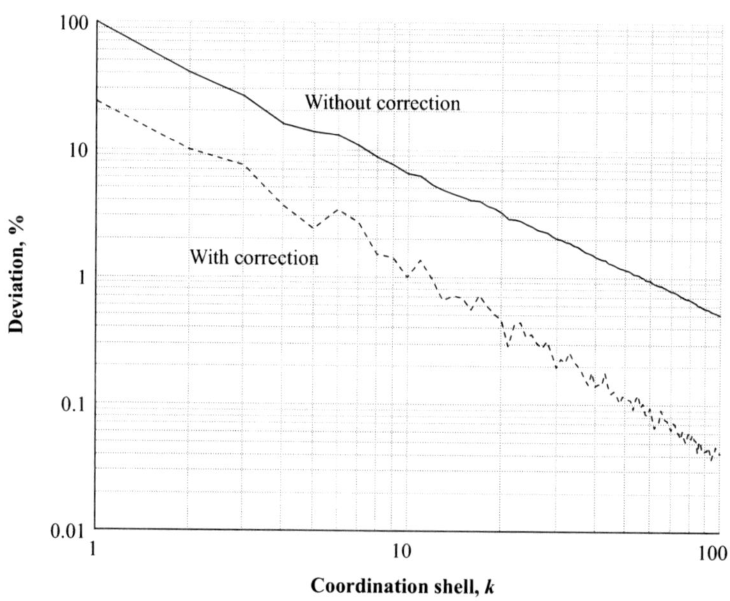

Figure 4.7. Maximum deviation of the structural factor, $z_{S,eff}/z_{L,eff}$, relative to the benchmark of 500 nearest-neighbor shells in the GBB model [99] for bcc structure.

Table 4.1. Structural factor, $z_{S,eff}/z_{L,eff}$, in the GBB approach for various interface orientations, interatomic potentials, and crystal structure.

	r^{-6}		r^{-7}		r^{-8}	
Orientation	**fcc**	**bcc**	**fcc**	**bcc**	**fcc**	**bcc**
[1,0,0]	0.3105	0.3332	0.2848	0.3113	0.2746	0.3041
[1,1,0]	0.3271	0.2905	0.3019	0.2617	0.2918	0.2488
[1,1,1]	0.2871	0.3351	0.2581	0.3096	0.2454	0.2986
[2,1,0]	0.3394	0.3283	0.3152	0.3029	0.3056	0.2925
[2,1,1]	0.3222	0.3224	0.2953	0.2956	0.2839	0.2837
[2,2,1]	0.3182	0.3281	0.2907	0.3012	0.2790	0.2893
[3,1,0]	0.3372	0.3355	0.3126	0.3116	0.3028	0.3025
[3,2,0]	0.3384	0.3174	0.3138	0.2904	0.3039	0.2790
[3,1,1]	0.3275	0.3361	0.3019	0.3111	0.2913	0.3007
[3,2,1]	0.3322	0.3198	0.3063	0.2921	0.2955	0.2796
[3,2,2]	0.3118	0.3319	0.2838	0.3056	0.2717	0.2939
[3,3,1]	0.3244	0.3198	0.2977	0.2921	0.2866	0.2798
[3,3,2]	0.3094	0.3328	0.2812	0.3066	0.2691	0.2951
[4,1,0]	0.3337	0.3379	0.3088	0.3144	0.2988	0.3057
[4,3,0]	0.3365	0.3110	0.3116	0.2835	0.3017	0.2716
[4,1,1]	0.3288	0.3398	0.3030	0.3158	0.2924	0.3064
[4,2,1]	0.3360	0.3314	0.3107	0.3053	0.3004	0.2941
[4,3,1]	0.3340	0.3150	0.3083	0.2869	0.2977	0.2742
[4,3,2]	0.3228	0.3290	0.2957	0.3021	0.2842	0.2901
[4,3,3]	0.3057	0.3340	0.2773	0.3079	0.2651	0.2965
[4,4,1]	0.3266	0.3141	0.3002	0.2861	0.2893	0.2736
[4,4,3]	0.3043	0.3344	0.2758	0.3083	0.2635	0.3297
Mean	**0.3291**	**0.3280**	**0.3020**	**0.3032**	**0.2909**	**0.2926**

From ref. [99].

Interestingly, the interfacial energy is only a relatively weak function of orientation, potential function, as well as crystal structure. This is indicated also in the Figures 4.8 and 4.9, where the structural factor is plotted in dependence of misorientation for fcc and bcc structures.[7]

The effective coordination numbers for fcc and bcc crystal structures are evaluated with $z_{L,eff,bcc} = 12.234$ and $z_{L,eff,fcc} = 14.442$. These values differ significantly from the traditional first-nearest neighbor values of $z_{L,bcc} = 8$ and $z_{L,fcc} = 12$.

[7] Note that two linearly independent lattice vectors with integer indices must be defined for three-dimensional interfaces.

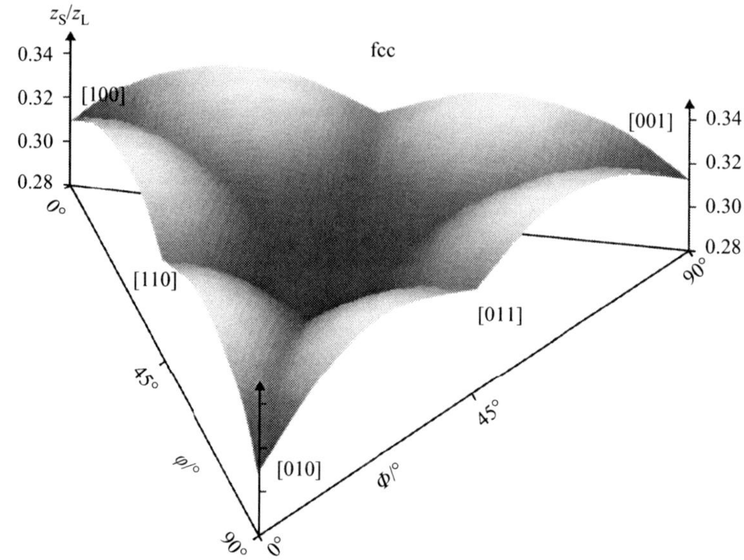

Figure 4.8. Structural factor, z_S/z_L, according to the GBB model [99] for fcc structure.

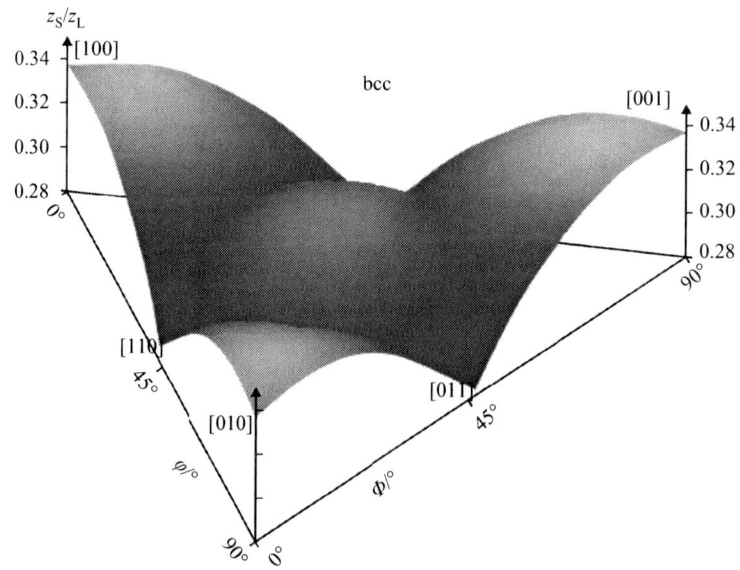

Figure 4.9. Structural factor, z_S/z_L, according to the GBB model [99] for bcc structure.

A rather surprising result obtained with the GBB model, which has already been indicated earlier, is that the values of the structural factors for fcc and bcc crystal structure are rather close and, therefore, the dependence of interfacial energy from crystal structure is weak. As the atomic interactions are found to be wide-range,[8] a sufficiently high number of neighbor shells must be included in the summation in order to obtain accurate interfacial energy values. This result could also be taken as an indication toward the assumption that the GBB model can also be a suitable means for evaluation of incoherent interfacial energies if only the chemical part of the interfacial energy is considered.

It has already been indicated that the choice of distance function according to Lennard-Jones with $f = (r_k / r_1)^{-6}$ is somehow arbitrary. For this reason, a comparison has been made by Sonderegger and Kozeschnik [99], where they compared the results obtained for the structural factor based on the Lennard-Jones potential with a number of *embedded atom potentials* that have been optimized to closely represent the bond energies in metals. In nine out of ten cases the deviation of the final results from the r^{-6}-dependence was below 1%.

The interfacial energy value obtained under the assumption of sharp, planar interfaces takes into account only the bond energies, which are broken across the interface. As the interface is assumed to be infinitely thin and, therefore, no mixing of atoms with entropic contributions to the interfacial energy is present, this value represents the upper limit of the (chemical contribution to the) interfacial energy. Diffuse interfaces, as well as interfacial curvature, are both effects that reduce the interface energy and will be dealt with in Section 4.4.

4.3.2 Comparison Between Theory and Experiment

The interfacial energy, γ, is an essential physical quantity in numerous theories of materials science. Interfacial energies play a major role in precipitation processes as well as microstructure evolution involving recrystallization and grain growth. The latter processes are concerned with interfaces between identical phases, and they are dealt with briefly in Section 2.4.5.

The dominant fields of application for interfacial energies in solid-state phase transformation simulation involve interfacial reactions between different phases. The interfacial energy is a key quantity in this context for evaluation of second-phase nucleation rates as well as precipitate coarsening rates. Unfortunately, only the latter process provides some potential for an experimental determination of interfacial energy values. Precipitates in coarsening reactions are usually large enough to exhibit a nearly planar interface shape, whereas nucleation of second phases mainly involves very small particle sizes in the order of nanometers. In this situation, interfacial curvature has a strong influence on the interface energy and adds a considerable amount of complexity to the experimental evaluation process.

Another source of information on interfacial energies is first-principles calculations. A huge amount of information derived from Density Functional Theory (DFT) (electronic band structure calculations) or Molecular Dynamics (MD) based on Embedded Atom (EAM)

[8]Although the interaction potential decreases with the weighting function, $f(r_k/r_1)$, the number of atoms belonging to a given coordination sphere increases and, partially, compensates for this effect. For exponents of 3 and less, the weighting function diverges.

potentials is available in literature. Unfortunately, only a minority of these calculations can be used for comparison with the results obtained in this section. For numerous analyses, the calculations are concerned with materials systems where no enthalpy of solution data are available from thermodynamic databases. More importantly, the analyses often involve special interface orientations and/or include coherency stress contributions which cannot directly be taken for comparison. For this reason, only a limited comparison is reproduced here as given in ref. [99] together with recent DFT values reported for Mg_5Si_6-β'' in the Al–Mg–Si system. The data are summarized in Table 4.2.

An important result of the comparison between interfacial energies obtained with the GBB model and first-principles density functional theory calculations is shown in the last line of Table 4.2. The values presented from ab-initio simulation show a spread of 100–449 J/m^2. This roots in the fact that the authors of this study [122] investigated numerous possible interface orientations between the Mg_5Si_6-β''-precipitate and the fcc Al matrix. The GBB value of 260 J/m^2 falls well inside the ab-initio range of values and might be seen as an average quantity including various interface orientations over the surface of the precipitate.

In multicomponent systems, a comparison of interfacial energies is inherently difficult because the interfacial energy is a strong function of the precipitate chemical composition. In Section 4.2, a quadratic dependence from the difference in element content between precipitate and matrix has been identified. In precipitate nucleation, this situation is even more involved because the nucleus chemical composition often deviates from the thermodynamic equilibrium composition due to the observation that the experimentally most successful nucleus will be the one with highest nucleation rate. This can be achieved with compositions yielding a minimum nucleation barrier (see Section 2.5.2). The appendix, nevertheless, contains a selection of calculated interfacial energies, which have been evaluated mainly under full equilibrium condition and which, for this reason, represent an

Table 4.2. Comparison between calculated and experimental interface energy data.

System	Precipitate	T / °C	γ_{Lit} / mJ/m^2	γ_{GBB} / mJ/m^2	Refs.
Ni–Al	Ni_3Al (γ')	500–715	14–80	36.2	88, 101, 103–106
Ni–Al	Ni_3Al (γ')	–273–1300	12–63	36.2	107–110
Ni–Cr (fcc)	Cr (bcc)	950	200	188	111
Fe–Cu	Cu (bcc)	600–750	520–610	570–630	112
Cu–Co	Co (fcc)	450–1150	108–1008	640	77, 113–118
Fe–Ni–Al	NiAl (B2-bcc)	700	28–40	35.7	119
Fe–Ni–Al–Mo	NiAl (B2-bcc)	700	29–42	33.8	119
Nimonic PE16	Ni_3Al (γ')	750	90.9	116.4	101, 120
Nimonic 80A	Ni_3Al (γ')	945	140	99.8	121
Al–Mg–Si	Mg_5Si_6-β''	0K*	100–449*	260	122

* Value obtained from first-principles calculations at 0 K.
Part of the data from ref. [99].

upper limit of possible values. This table can be used for reference purpose, however, keeping in mind that some of the phases[9] might have significantly lower interfacial energy values during nucleation due to the size effect and compositional constraints such as the minimum nucleation barrier criterion.

4.4 Interface Energy Correction for Small Precipitates

In the previous sections, it was demonstrated that the interface energy, γ, can be evaluated from the superposition of discrete broken atomic bonds, which are counted across unit interfacial area. In the case of a planar interface, the bond counting procedure can be performed conveniently on the basis of periodic translational volume elements and discrete atomic lattice positions. A similar *Ansatz* can, in principle, also be applied to curved interfaces. However, some basic modifications of the model are necessary to obtain a closed solution. It will be shown that the interfacial energy of a curved phase boundary can be expressed in terms of the energy of a planar boundary, γ_0, multiplied with a correction function denoted as $\alpha(\rho)$ with

$$\gamma(\rho) = \alpha(\rho) \cdot \gamma_0. \tag{4.29}$$

This is elaborated subsequently following the treatment suggested in ref. [123].

4.4.1 The Interface Energy Size Correction Function

In the integration procedure for the GBB bond energy of the planar interface, advantage has been taken from symmetric conditions across the interface: The number of atoms on both sides of the coherent interface is identical. Figure 4.10 indicates clearly that this situation does no longer apply to curved interfaces, which we are, however, concerned with in precipitation modeling.

Figure 4.11 shows a sketch of the areas (volumes) involved in the bond counting procedure for a planar (a) and a curved (b) interface. The reference volume elements located in the precipitate are denoted by a dark gray color and the symbol "A." The counterpart of the counting area in the matrix is shown in light gray. In the case of a planar interface, the reference area is rectangular and the construction of periodic volume elements is straightforward: In both, the matrix and precipitate phases, the areas extend into the bulk volume to a distance, r_k.

The case of a curved interface is only depicted exemplarily for the special case where the interaction radius corresponds exactly to the size of the precipitate. It is rather obvious from the sketch that for arbitrary curvature, that is, precipitate radius, and arbitrary interaction distance, r_k, the summation of broken bonds cannot easily be carried out using the same *Ansatz* of periodic reference elements as in the planar case. The problem can be solved, however, if the discrete atomic positions are replaced by the atomic charge density and the summation is replaced by an integral formulation.

[9]This statement is particularly important for solution phase, which can have a wide range of possible chemical compositions. A typical example for precipitate of a solution phase is bcc-Cu in ferritic steel.

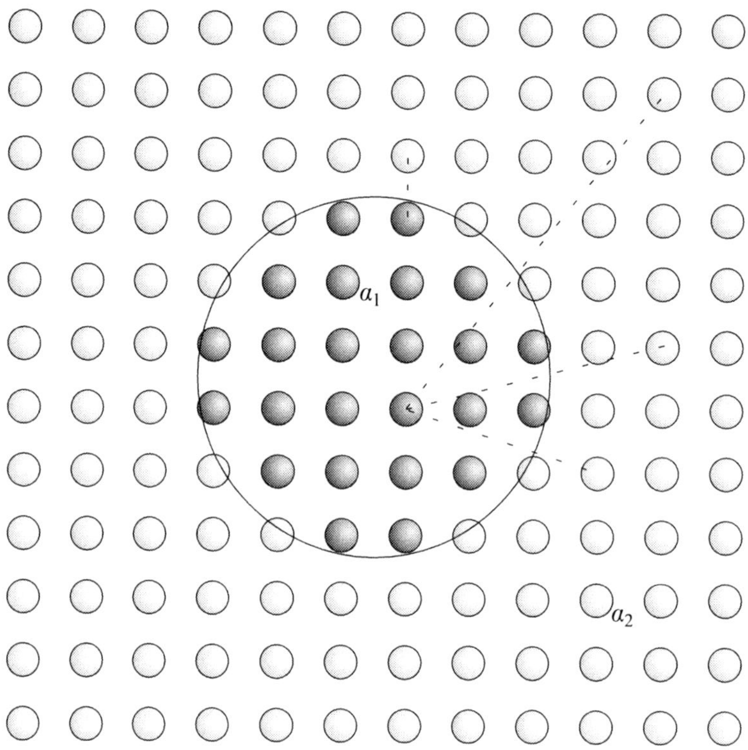

Figure 4.10. Small precipitate a_1 with some exemplary bonds. It is obvious that the total number of bonds coming from inside the precipitate can be substantially lower compared to planar interfaces.

In the planar case, the following steps have been performed to obtain the number of broken bonds across unit interfacial area:

I. Sum up all broken bonds extending from the reference area A of the precipitate into the matrix area up to a distance, r_k.
II. Normalize the result with respect to unit interfacial area.

In the spherical case, this procedure becomes rather involved when seeking periodic reference elements that take into account all possible crystallographic structures and, particularly, all possible combinations of precipitate radius and interaction distance, r_k. Therefore, the two basic steps, I and II, are generalized such that

Ia. *all possible* bonds inside the precipitate, z_P, are counted first;
Ib. this quantity is normalized to the number of bonds per atom in the precipitate;
Ic. this quantity is subtracted from the total number of bonds per atom (which is identical to z_L!).
II. steps Ia to Ic yield the number of broken bonds per atom in the precipitate which is finally normalized to the number of atoms per unit interfacial area.

Let $b(\rho)$ be the number of bonds within the sphere, n_A and n_V be the number of atoms per unit area and unit volume, and V_P and A_P be the volume and surface area of the precipitate. Steps Ib and Ic can be expressed as

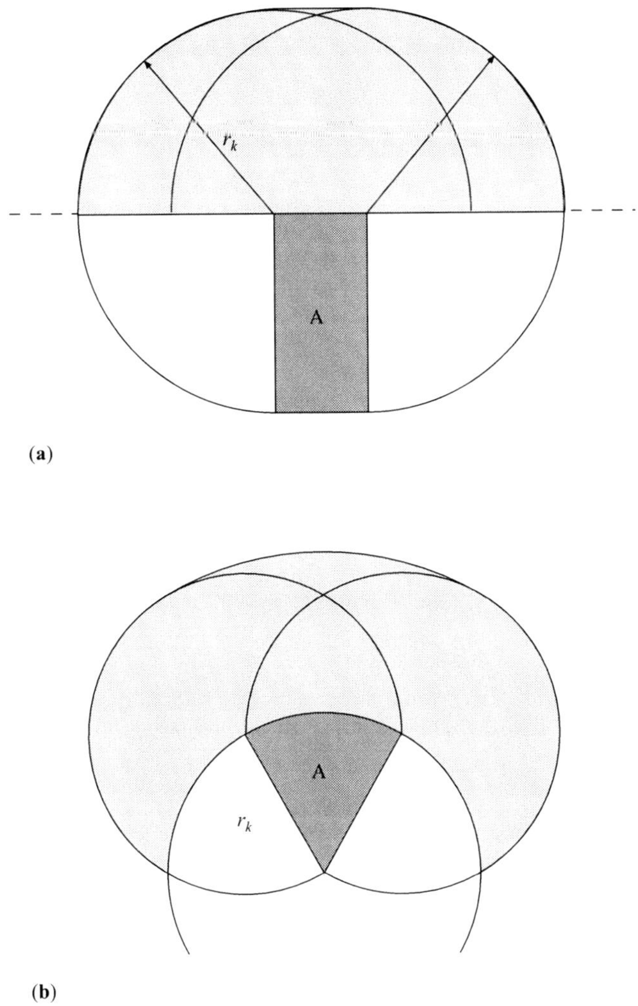

Figure 4.11. Sketch of model for bond counting with (a) planar interface and (b) curved interface. See text for explanation.

$$z_P = z_L - \frac{b(\rho)}{n_V V_P}. \tag{4.30}$$

The normalization, step II, delivers

$$z_S(\rho) = z_P \frac{n_V V_P}{n_A A_P}, \tag{4.31}$$

and with the expression for z_P we finally have

$$z_S(\rho) = \frac{\rho}{3} n_V^{1/3} z_L - \frac{b(\rho)}{n_V^{2/3} 4\pi\rho^2}. \tag{4.32}$$

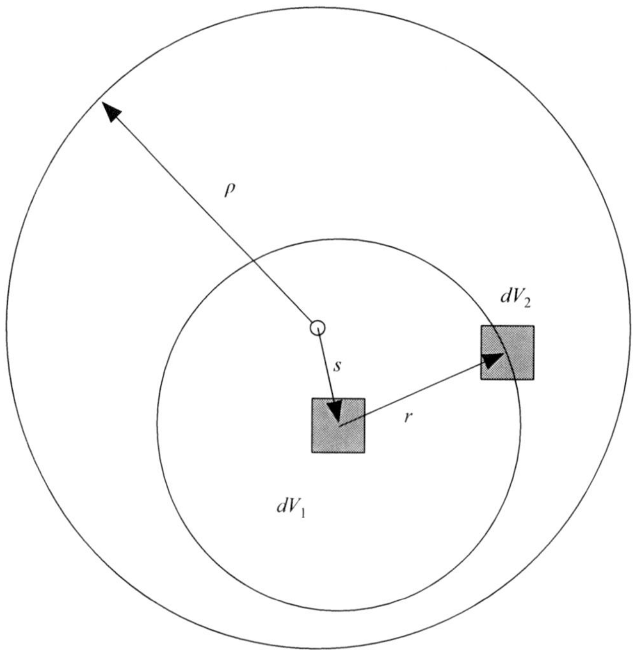

Figure 4.12. Two volume elements used for calculation of the number of broken atomic bonds inside the precipitate.

Apparently, the key for finding the size dependence of the interfacial energy is the calculation of the number of bonds inside the precipitate, $b(\rho)$. As indicated earlier, this task can be performed when replacing the discrete number of bonds in the atomistic lattice-oriented picture of the original GBB model, Section 4.3, by a continuum approach. Accordingly, for a single, infinitesimal bond we have

$$db = n_V dV_1 \cdot n_V dV_2 \cdot f(r), \tag{4.33}$$

where the coordinate variable, r, represents the distance between two volume elements as depicted in Figure 4.12. The function $f(r)$ is equivalent to the weighting function $f(r_k / r_1)$ of the original GBB model, taking into account that the strength of the atomic bonds decreases with distance. Since the classical Lennard-Jones potential with $f = (r_k / r_1)^{-6}$ has shown to be a realistic model potential for crystalline matter,[10] this dependence is utilized in the further treatment. Moreover, it is assumed that in the integration procedure, the two volume elements, dV_1 and dV_2, can only come to a minimum distance of r_c.

The integration of bond energies has been outlined in ref. [99] for the correction function to the discrete bond counting procedure. Inserting the minimum distance between two volume elements, r_c, for the radius, r_k, of the shell with index k in the corresponding equations (4.27) and (4.28) delivers

[10] See the corresponding comment at the end of Section 4.3.

$$z_{\mathrm{L,eff}} = n_{\mathrm{V}} \frac{4\pi}{3} r_1^6 r_{\mathrm{c}}^{-3}, \tag{4.34}$$

and

$$z_{\mathrm{S,eff}} = \frac{\pi}{2} n_{\mathrm{V}}^{4/3} r_1^6 r_{\mathrm{c}}^{-2}. \tag{4.35}$$

In a next step, the differential form for the infinitesimal bond, equation (4.33), is written in integral form as

$$b(\rho) = n_{\mathrm{V}}^2 \int_{V_1}\int_{V_2} f(r)dV_1 dV_2, \tag{4.36}$$

with the constraint $r > r_c$.

In spherical geometry, the volume integrals can be simplified for evaluation. Consider a spherical precipitate with radius, ρ. According to Figure 4.12, with the vector, **s**, connecting the center of the sphere with volume element, dV_1, and the vector, r, connecting dV_1 with dV_2, all volume elements with the same distance, s, from the center share the same geometrical properties and can be treated identically. With the transformations

$$dV_1 = 4\pi s^2 ds; \quad dV_2 = A dr, \tag{4.37}$$

and assuming Lennard-Jones potential, the expression for $b(\rho)$ becomes

$$b(\rho) = 4\pi n_{\mathrm{V}}^2 r_1^6 \int_s \int_r s^2 r^{-6} A dr ds. \tag{4.38}$$

The actual integration cannot be performed in a single step, but must be carried out for three different cases:

- *Case 1:* If the area A corresponds to the surface of a sphere, which is the case if the inner sphere in Figure 4.12 is located entirely inside the precipitate, integration is carried out between $0 \le s \le \rho - r_c$ and $r_c \le r \le \rho - s$. With $A = 4\pi r^2$, we obtain

$$b_1(\rho) = 16\pi n_{\mathrm{V}}^2 r_1^6 \int_{s=0}^{\rho - r_c} \int_{r=r_c}^{\rho - s} s^2 r^{-4} dr ds. \tag{4.39}$$

- *Case 2:* If the inner sphere intersects with the outer sphere, the integration sphere changes to a calotte. This case is shown in Figure 4.13. Integration is carried out between $0 \le s \le \rho - r_c$ and $\rho - s \le r \le \rho + s$. With

$$A = \pi r \left(\frac{\rho^2 - (s-r)^2}{s} \right), \tag{4.40}$$

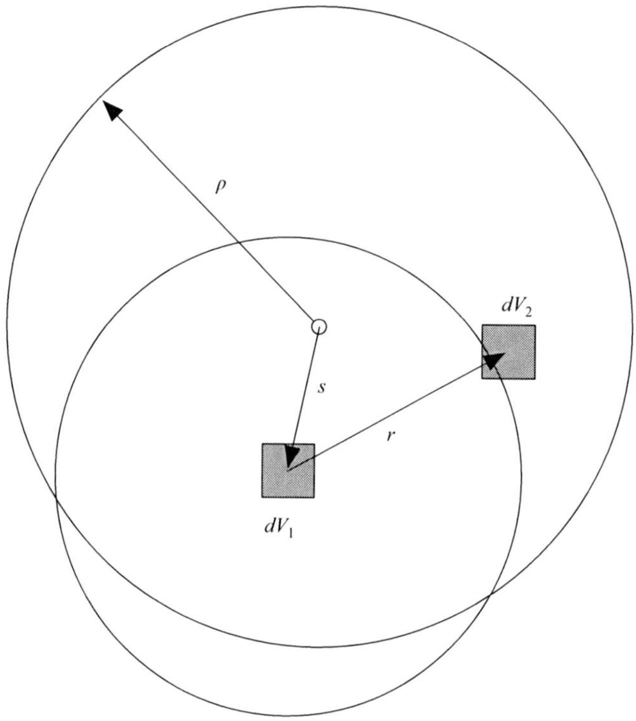

Figure 4.13. Integration geometry for calculation of the number of broken atomic bonds inside the precipitate.

we obtain
$$b_2(\rho) = 4\pi^2 n_V^2 r_1^6 \int_{s=0}^{\rho-r_c} \int_{r=\rho-s}^{\rho+s} sr^{-5}\left[\rho^2 - (s-r)^2\right] drds. \tag{4.41}$$

- *Case 3:* If the center of the sphere with the vector, **s**, is located outside the precipitate, the two spheres intersect always, as we have $r \geq r_c$. Integration is carried out between $\rho - r_c \leq s \leq \rho$ and $r_c \leq r \leq \rho + s$, in analogy to case 2. For the integral, we obtain

$$b_3(\rho) = 4\pi^2 n_V^2 r_1^6 \int_{s=\rho-r_c}^{\rho} \int_{r=r_c}^{\rho+s} sr^{-5}\left[\rho^2 - (s-r)^2\right] drds. \tag{4.42}$$

Finally, for the number of bonds inside the precipitate, one obtains after superposition of cases 1–3,

$$b(\rho) = \pi^2 n_V^2 r_1^6 \left\{ \frac{16}{9}\frac{\rho^3}{r_c^3} - \frac{22}{3}\frac{\rho^2}{r_c^2} + \frac{40}{3}\frac{\rho}{r_c} - \frac{15}{2} - \frac{11}{3}\ln\left(\frac{\rho}{r_c}\right) + \frac{1}{3}\ln(2) \right\}. \tag{4.43}$$

Insertion of this result into equation (4.32) delivers the number of broken bonds across the interface with

$$z_S(\rho) = \frac{\pi}{4} n_V^{4/3} r_1^6 \left\{ \frac{22}{3} r_c^{-2} - \frac{40}{3} \rho^{-1} r_c^{-1} + \rho^{-2} \left[\frac{15}{2} + \frac{11}{3} \ln\left(\frac{\rho}{r_c}\right) - \frac{1}{3} \ln(2) \right] \right\}. \tag{4.44}$$

These results are finally used to derive the size correction function, $\alpha(\rho)$, as

$$\alpha(\rho) = 1 - \frac{20}{11} \frac{r_c}{\rho} + \left(\frac{45}{44} - \frac{1}{22} \ln(2) \right) \cdot \frac{r_c^2}{\rho^2} + \frac{1}{2} \left(\ln \frac{\rho}{r_c} \right) \cdot \frac{r_c^2}{\rho^2}. \tag{4.45}$$

By comparing the number of broken bonds across the interface obtained from discrete bond counting with the values obtained from the continuum description, the lower integration radius, r_c, can be assigned an approximate value. This task has been carried out by Sonderegger and Kozeschnik [123], delivering $r_{c,\text{fcc}} = 0.2957$ and $r_{c,\text{bcc}} = 0.3036$. As these values are again very similar for both, the fcc and bcc crystal structures, these authors suggest using the general approximation

$$r_c = 0.3 \cdot r_1. \tag{4.46}$$

The size correction function, $\alpha(\rho)$, can finally be written[11] in terms of the nearest-neighbor distance, r_1, and the precipitate radius, ρ, with

$$\alpha(\rho) = 1 - \frac{6}{11} \frac{r_1}{\rho} + 0.08921 \frac{r_1^2}{\rho^2} + 0.045 \cdot \ln\left(\frac{10}{3} \frac{\rho}{r_1} \right) \frac{r_1^2}{\rho^2}. \tag{4.47}$$

4.4.2 Comparison with Size Correction in Vapor-Droplet Systems

In Chapter 2, where the nucleation of precipitates has been investigated, the equivalence between solid-state interfacial energies and surface energies of solid–liquid phase boundaries has been utilized in deriving expressions for the nucleation rate of precipitates in solid-state systems. The main result of the present section, so far, is that the interfacial energy of solid-state phase boundaries can be related to a structural factor and the enthalpy of solution. Consequently, the results obtained here for discrete atomic bonding on a crystal lattice can be compared directly to classical thermodynamic approaches to the size-dependence of the surface energy in vapor-droplet systems.

The curvature dependence of the interfacial energy has been treated in the context of surface tension several times. These models are all based on the fundamental relation for the pressure of a curved interface on the droplet with

$$p^\alpha - p^\beta = \frac{2\sigma}{\rho}, \tag{4.48}$$

[11] Note that a typing error exists in the original paper, [123], for the factor 0.08921, which is written correctly in the formula here.

where P is the pressure in phases α and β, σ is the surface tension, and ρ is the droplet radius. This relation can be combined with the Gibbs adsorption theorem [16]

$$n \cdot d\mu_\alpha + A \cdot d\sigma = 0. \tag{4.49}$$

n is the number of atoms in the droplet, μ is the chemical potential, and A is the surface area of the droplet. In addition, an interfacial region with some given thickness must also be taken into account, which is also in thermodynamic equilibrium. Following the treatment of Tolman [124], the following differential equation is obtained

$$\delta \cdot d\left(\frac{2\sigma}{\rho}\right) + d\sigma = 0, \tag{4.50}$$

with the Tolman length, δ, defined as

$$\delta = \frac{\Gamma}{\xi_\alpha - \xi_\beta}. \tag{4.51}$$

ξ is the density of the two phases. The quantity, Γ, represents a superficial density. Under the assumption that the interface thickness, δ, is independent of the droplet radius, $\delta = \delta_0$, and it is sufficiently small compared to the droplet radius, $\delta_0 << \rho$, Tolman developed the well-known expression for the radius dependence of the surface tension with

$$\sigma(\rho) = \frac{\sigma_0}{1 + \frac{2\delta_0}{\rho}}. \tag{4.52}$$

Without the limiting condition, $\delta_0 << \rho$, Khashchiev [125] derived the relation

$$\sigma(\rho) = \frac{\sigma_0}{\left(1 + \frac{\delta_0}{\rho}\right)^2}. \tag{4.53}$$

In both cases, the surface energy goes to zero with decreasing precipitate radius. Schmelzer, Gutzow, and Schmelzer [126] have reviewed the theoretical concepts for the size-dependence of the surface tension and they concluded that all of these models depend on the value of the Tolman length, δ. Moreover, Schmelzer, Gutzow, and Schmelzer deduct that "from thermodynamic arguments alone, no decision can be made as to what particular expression for $\delta(\rho)$ is the most appropriate one for a given substance at given thermodynamic constraints." In this sense, the solution presented in the previous section represents a significant contribution, as it does not depend on any kind of undetermined interface thickness and, therefore, provides a unique solution to the size-dependence of the surface (interfacial) energy of spherical objects being in equilibrium with an embedding substance.

The values of $\delta_0 = 0.25 \cdot r_1$ and $\delta_0 = 0.60 \cdot r_1$ in the Tolman approach are commonly considered as the lower and upper bound estimates for the critical interface thickness. The critical

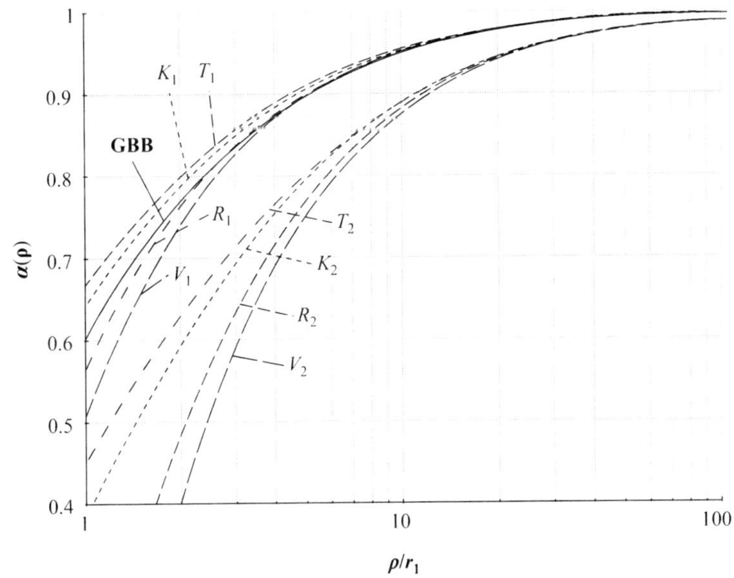

Figure 4.14. Comparison of the size correction function for the interfacial energy [123] with results obtained from thermodynamic models. T. Tolman [124]; K. Kashchiev [125]; R. Rasmussen [127]; V. Vogelsberger and Marx [128]. The subscripts "1" represent results for $\delta_0 = 0.25 \cdot r_1$, subscripts "2" for $\delta_0 = 0.60 \cdot r_1$. The symbol "GBB" denotes the results of the generalized broken-bond treatment.

radius obtained in the previous section with $r_c = 0.3 \cdot r_1$, therefore, agrees rather well with the value of the lower bound estimate of four different models depicted in Figure 4.14.

Finally, it is important again to emphasize that the decrease of the interfacial energy with size plays a particularly important role in nucleation theory. The typical sizes of nuclei in solid-state precipitation are in the order of 1 to 5 times r_1, which delivers correction values to the interfacial energy from approximately $\alpha(\rho) = 0.6$ to 0.9. These values enter in cubic form into the critical nucleation energy, which occurs in the exponent of the nucleation rate expression.

4.5 Energy of Diffuse Interfaces

At temperatures close to absolute zero, interfaces between two phases are *sharp*. This means that all atoms belonging to one phase are located on one side of the interface while all other atoms belonging to the other phase are located on the other side of the interface. At temperatures above zero, atoms from one phase can intermix with atoms of the other phase with a certain nonzero probability. The higher the temperature is, the higher is the probability for this mixing process and the more atoms are, on average, located on the "wrong" side of the phase boundary. Interfaces are considered as infinitely thin if no mixing of atoms occurs across the phase boundary. Interfaces that show a certain degree of mixing of atoms are denoted as *diffuse interfaces*. These are characterized by a certain finite interface thickness, as shown in Figure 4.15.

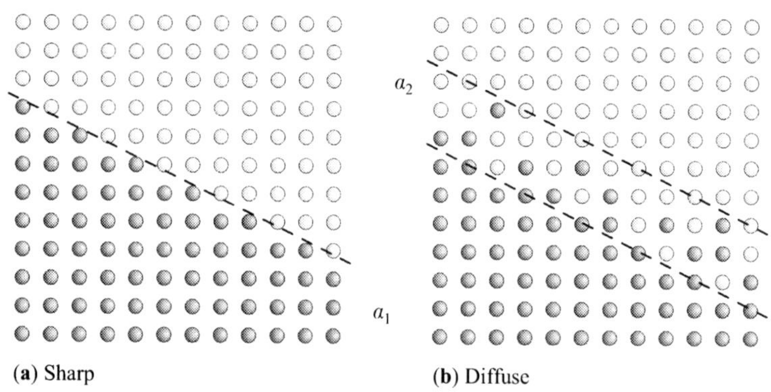

Figure 4.15. Schematic representation of a diffuse interface (a) at finite temperature compared to a sharp interface (b) at 0 K.

The fact that interphase boundaries can be diffuse and extend over some given distance is well known. It has been treated extensively in the framework of the *phase field* approach, which goes back to seminal work of John W. Cahn and John E. Hilliard [129–131]. The basic idea in this approach is to treat the energy of a phase separating system on the basis of an order parameter, η, which is unity in one phase and zero in the other. Between the two phases, across the interface, the properties of the phases change gradually from that of the phase α to that of the phase β. The degree of transition from α to β is determined by the value of the order parameter, η.

The idea of a continuous order parameter, as utilized in the phase field approach, has been successfully applied to numerous topics in materials science (see, e.g., the review of Steinbach [132], which also cites several other sources of reference on the phase field approach), demonstrating its huge potential in modeling of micro, meso, and macroscopic processes.

When going to atomic dimensions, as we have to in the treatment of phase boundaries, the continuous order parameter of the phase field model can alternatively be replaced by a discrete order parameter. For the description of interfaces that have a thickness of only a few atomic layers, this approach seems be a natural choice as the transition from one phase into the other occurs in discrete steps where the (minimum) thickness of each layer is given by the lattice parameter of the phases under consideration. Sonderegger and Kozeschnik [133] have recently adopted this concept in the development of a diffuse interface model based on the GBB concept.[12] Their approach is presented below.

4.5.1 Free Energy of a Diffuse Interface

In contrast to the sharp-interface model discussed earlier, where the free energy of the phase boundary can be obtained by mere counting of broken bonds across the interface, the energy of a diffuse interface must be evaluated taking into account (*a*) the *interface* energy between the two bulk phases and each individual (atomic) layer constituting the phase boundary, as well as (*b*) the *volume* energy of the interface layers. The second contribution arises from the

[12] The GBB approach is described in Section 4.3.

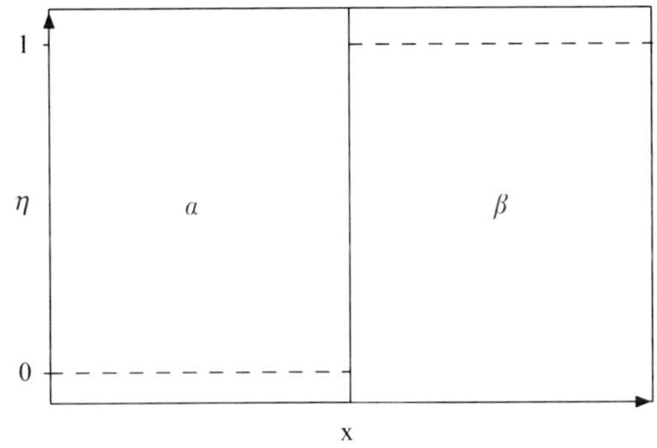

(**a**) Sharp interface

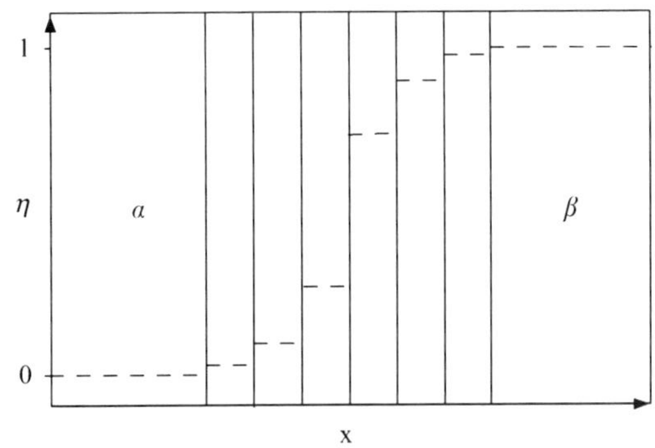

(**b**) Diffuse interface

Figure 4.16. Schematic representation of the discrete evolution of the order parameter η across the phase boundaries between two phases.

observation that the interface layers have a composition that deviates from both, the composition of phases α and β. The second contribution, (*b*), therefore, does not cancel out in the diffuse interface approximation when calculating the difference in total energy of the initial case with two separate blocks of phases α and β, and the final state where two unlike half-blocks are assembled.[13] Figure 4.16 shows a sketch of concentration profiles across a sharp and a diffuse interface expressed in terms of the order parameter, η.

For calculation of the total energy of the interface, the volume free energy of each interface layer contributes with an amount that is given by the difference in bulk free energy of the layer and the same amount of either pure phase α or pure phase β. These energy contributions are assumed to be available in the form of composition and temperature-dependent specific Gibbs energies. In the following treatment, it is further assumed that the chemical composition of each layer can be parameterized by only a single order parameter, η.

The energy contribution of the interfaces between the individual layers can be determined by superposition. Figure 4.16(b) indicates already that four contributions of interactions must be considered: (*a*) the interaction energy between bulk phases α and β, (*b*) the interaction energy

[13] This thought experiment goes back to R. Becker [97] and is described in Section 4.1.

between bulk phase a with each of the interface layers, (c) the interaction energy between bulk phase β with each of the interface layers, and (d) the interaction energy between each of the interface layers with all other interface layers. For the total free energy of the interface, we thus have

$$\gamma = \gamma_{\alpha,\beta} + \sum_{q=1}^{n} \gamma_{\alpha,q} + \sum_{p=1}^{n} \gamma_{\beta,p} + \sum_{p=1}^{n-1} \sum_{q=p+1}^{n} \gamma_{pq} + \sum_{p=1}^{n} \gamma_{\text{vol},p}, \tag{4.54}$$

where the indices refer to the interactions and $\gamma_{\text{vol},p}$ represents the volume free energy of the layer with index p. The free energies of the pure phases can be neglected as these terms cancel out in the interface energy evaluation.

Each individual interaction energy in equation (4.54) can now be treated on the basis of the results obtained in Section 4.3 for the sharp-interface energy in the framework of the GBB approach as given by

$$\gamma_{\text{sharp}} = \frac{n_S z_{S,\text{eff}}}{N_A z_{L,\text{eff}}} \cdot \Delta H_{\text{sol}}. \tag{4.55}$$

As the specific number of atoms in the interface, n_S, the coordination number, $z_{L,\text{eff}}$, and N_A do not depend on the properties of the interface, all contributions can be expressed in terms of the effective number of broken bonds, z_S, and the effective enthalpy of solution, ΔH_{sol}. Accordingly, we have

$$\gamma = \gamma_0 \cdot \frac{z_S}{z_{S,0}} \cdot \frac{\Delta H_{\text{sol}}}{\Delta H_{\text{sol},0}} = \gamma_0 \cdot \alpha\left(z_S\right) \cdot \beta\left(\Delta H_{\text{sol}}\right), \tag{4.56}$$

where γ_0, $z_{S,0}$, and $\Delta H_{\text{sol},0}$ represent the corresponding quantities of the pure phases. In this expression, the function $\alpha\left(z_S\right)$ accounts for the spatial arrangement of the system, whereas $\beta\left(\Delta H_{\text{sol}}\right)$ incorporates the influence of chemical composition.

When developing an expression for the parameter a, first, two neighboring layers of given thickness are considered. The thickness of the outer layers representing the undisturbed bulk phases is assumed to be infinite. Let the layers be separated by a sharp interface and let the left layer contain a volume element, dV_α, and the right layer contain a volume element, dV_β. With the distances a and b defined according to Figure 4.17, the individual bond contribution, d_b, of the interaction between the two volume elements reads

$$d_b = n_V dV_\alpha \cdot n_V dV_\beta \cdot f\left(r / r_1\right), \tag{4.57}$$

where n_V is the number of atoms per unit volume. For planar symmetry, the volume element, dV_α, can be substituted by $A \cdot da$, where A is the size of the interfacial area. d_b can then be directly transformed into dz_S reading[14]

[14] The analysis followed here is equivalent to the procedure applied in Sections 4.3 and 4.4.

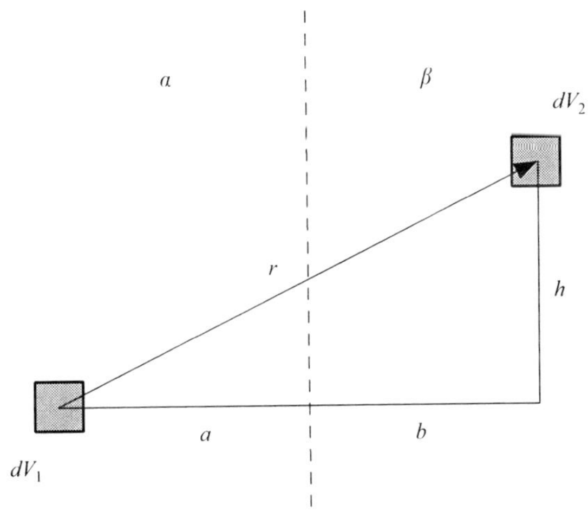

Figure 4.17. Definition of volume elements for integration of effective bond interactions.

$$dz_S = n_V^{4/3} \cdot f\left(r / r_1\right) \cdot da \cdot dV_\beta. \tag{4.58}$$

The volume element, dV_β, is next replaced by the infinitesimal variables db and dh, which delivers

$$dz_S = n_V^{4/3} \cdot f\left(r / r_1\right) \cdot 2h\pi \cdot da \cdot db \cdot dh. \tag{4.59}$$

With the Lennard-Jones potential, this equation can be integrated with respect to the weighting function, leading to

$$dz_S = 2\pi n_V^{4/3} r_1^6 \cdot \left[\left(a+b\right)^2 + h^2\right]^{-3} \cdot h \cdot da \cdot db \cdot dh, \tag{4.60}$$

where the weighting function has been taken as

$$f\left(r / r_1\right) = \left(r / r_1\right)^{-6} = \left[\left(a+b\right)^2 + h^2\right]^{-3} \cdot r_1^6. \tag{4.61}$$

Integration for the spatial coordinate, h, delivers the number of bonds between two parallel planes located in the distance, $a+b$, as

$$dz_S = \frac{\pi}{2} n_V^{4/3} r_1^6 \cdot \left(a+b+r_c\right)^{-4} \cdot da \cdot db. \tag{4.62}$$

In order to capture all bonds in plane b, integration for h is carried out between the limits of $-\infty$ to $+\infty$. The minimum bond length, r_c, has been introduced earlier in equation (4.46) with $r_c = 0.3 \cdot r_1$, where r_1 represents the atomic distance, and it must be accounted for in the integration procedure as outlined in refs. [99] and [123].

Equation (4.62) is the starting point for integration of the individual contributions to the interfacial energy as defined in equation (4.54). Before carrying out this task, however, this relation is utilized to evaluate the number of broken bonds, $z_{S,0}$, across a sharp planar interface, which will finally be used in the evaluation of the interface energy of a diffuse phase boundary as suggested in equation (4.56). This quantity can be obtained by integration of equation (4.62) for the case of two semi-infinite layers that are in immediate contact ($c = 0$). Accordingly, we get

$$z_{S,0} = \frac{\pi}{2} n_V^{4/3} r_1^6 \cdot \int_{b=0}^{\infty} \int_{a=0}^{\infty} \left(a + b + r_c\right)^{-4} \cdot da \cdot db, \tag{4.63}$$

with the solution

$$z_{S,0} = \frac{\pi}{2} n_V^{4/3} r_1^6 r_c^{-2}. \tag{4.64}$$

This result exactly reproduces the energy of the sharp interface, when inserted into equation (4.55) and using the GBB values of $z_{L,bcc} = 12.234$ and $z_{L,fcc} = 14.442$, ref. [99].

Next, equation (4.62) is generalized to the case where the two layers are separated by a gap of width, c. The number of bonds across the interface is then given as

$$z_S = \frac{\pi}{2} n_V^{4/3} r_1^6 \cdot \int_{b=\frac{c}{2}}^{b_0+\frac{c}{2}} \int_{a=\frac{c}{2}}^{a_0+\frac{c}{2}} \left(a + b + r_c\right)^{-4} \cdot da \cdot db. \tag{4.65}$$

The variables a_0 and b_0 represent the thickness of the two layers. From this equation, the function $\alpha(z_S)$ in equation (4.56) can be evaluated, which finally delivers

$$\begin{aligned}
\alpha_{pq} &= r_c^2 \cdot \left[\left(a_0 + b_0 + c + r_c\right)^{-2} + \left(r_c + c\right)^{-2} - \left(a_0 + c + r_c\right)^{-2} - \left(b_0 + c + r_c\right)^{-2}\right] \\
\alpha_{\alpha q} &= r_c^2 \cdot \left[\left(c + r_c\right)^{-2} - \left(b_0 + c + r_c\right)^{-2}\right] \\
\alpha_{\beta p} &= r_c^2 \cdot \left[\left(c + r_c\right)^{-2} - \left(a_0 + c + r_c\right)^{-2}\right] \\
\alpha_{\alpha\beta} &= r_c^2 \cdot \left(c + r_c\right)^{-2}.
\end{aligned} \tag{4.66}$$

The subscript indices denote the individual interactions introduced in equation (4.54).

In the next step, the factor $\beta(\Delta H_{sol})$ in equation (4.56) is evaluated. With the approximation that the composition profile of all elements can be parameterized by only a single order parameter, η,[15] a solution for β can be obtained, which allows for an estimation of the diffuse interface composition profile. Let X_i^p and X_i^q be the composition of layers p and q, and X_i^a

[15] This assumption means that segregation of individual atomic species into the interface cannot be treated explicitly in this approximation. However, the single-order parameter model delivers an upper-bound estimate of the interfacial energy, as any individual segregation would only occur under the constraint of interfacial free energy reduction.

and X_i^p be the corresponding quantities in the two bulk phases α and β. The enthalpy of solution of phase β in α is the given as[16]

$$\Delta H_{\text{sol}} = N_{\text{A}} z_{\text{L}} \varepsilon_0, \tag{4.67}$$

with the interaction energy, ε_0, defined as

$$\varepsilon_0 = \sum_{i,j} \varepsilon_{ij} \left[X_i^p X_j^q - \frac{1}{2}\left(X_i^p X_j^p + X_i^q X_j^q \right) \right]. \tag{4.68}$$

The composition of layers p and q can now be written in terms of the corresponding order parameters, η_p and η_q, as

$$\begin{aligned} X_i^p &= \eta_p X_i^\alpha + \left(1-\eta_p\right) X_i^\beta \\ X_j^q &= \eta_q X_j^\alpha + \left(1-\eta_q\right) X_j^\beta . \end{aligned} \tag{4.69}$$

With this parameterization, the functional expressions for β are obtained as

$$\begin{aligned} \beta_{pq} &= \left(\eta_p - \eta_q\right)^2 \\ \beta_{\alpha q} &= \eta_q^2 \\ \beta_{\beta p} &= \left(1-\eta_p\right)^2 . \end{aligned} \tag{4.70}$$

In the last step of this treatment, the volume free energies corresponding to the volume elements (layers), p, of the interface are evaluated. Assuming that half of the interface volume corresponds to phase α and the other half to phase β, the excess Gibbs energy of all volume elements in the interface is given as

$$\Delta G_{\text{vol}} = \frac{A}{v^{\alpha\beta}} \left[\sum_{p=1}^{n} g_p\left(\eta_p\right) \cdot a_{0,p} - \frac{d}{2}\left(g_\alpha + g_\beta\right) \right], \tag{4.71}$$

where A is the interfacial area, $v^{\alpha\beta}$ is the molar volume assumed to be identical for both phases α and β, $a_{0,p}$ is the thickness of layer p, and d is the thickness of the interface region. g_α and g_β are the specific volume free energies of the matrix α and precipitate phase β. The volume free energy can alternatively be expressed in terms of unit area of interface, which finally yields

$$\gamma_{\text{vol},p} = \left[g_p\left(\eta_p\right) - \frac{1}{2}\left(g_\alpha + g_\beta\right) \right] \cdot \frac{a_{0,p}}{v^{\alpha\beta}} . \tag{4.72}$$

[16]Compare Section 4.3.

With equations (4.54), (4.66), (4.70), and (4.72), the interfacial energy of a diffuse planar phase boundary is uniquely defined. The actual constitution of the interfacial composition profile can be evaluated by minimization of γ as given by equation (4.54). In the general case, this task must be achieved by numerical methods. In the special case of a regular solution, a closed analytical solution can be obtained. This is elaborated in the following section. The regular solution will also allow for a comparison with corresponding models developed on thermodynamic grounds.

4.5.2 Regular Solution Approximation for Diffuse Interfaces

Consider a binary[17] system with two interfacial layers between phases α and β. Let the composition of phase α be given by X^α and the composition of layer 1 by X. The regular solution model shall describe the thermodynamics of the system.[18] Due to the symmetry of the regular solution potential, the compositions of layer 2 and phase β are then given as

$$\begin{aligned} X^\beta &= 1 - X^\alpha \\ X_2 &= 1 - X. \end{aligned} \tag{4.73}$$

If X^α and X^β are fixed to their equilibrium values according to the phase diagram, the parameter X is the only independent variable in the system and its value can be obtained by minimization of the interfacial free energy.

The specific volume free energy of a binary regular solution is given by

$$\begin{aligned} g = {} & N_A \frac{z_L}{2}\left[\varepsilon_{11} X^2 + 2\varepsilon_{12}(1-X) + \varepsilon_{22}(1-X)^2\right] \\ & + RT\left[X \ln X + (1-X)\ln(1-X)\right]. \end{aligned} \tag{4.74}$$

This expression can be simplified having in mind that the concentration profile for regular solutions is symmetric. For evaluation of the volume free energy contribution to the interfacial energy of a phase boundary, one can, therefore, equivalently use the mean values of the two interfacial layers as well as the two bulk phases [133]. For the two interfacial layers, we get

$$\begin{aligned} g_{I1} = {} & N_A \frac{z_L}{2}\left[\varepsilon_{11} X^2 + 2\varepsilon_{12}(1-X) + \varepsilon_{22}(1-X)^2\right] \\ & + RT\left[X \ln X + (1-X)\ln(1-X)\right] \\ g_{I2} = {} & N_A \frac{z_L}{2}\left[\varepsilon_{11}(1-X)^2 + 2\varepsilon_{12}(1-X) + \varepsilon_{22} X^2\right] \\ & + RT\left[(1-X)\ln(1-X) + X \ln X\right], \end{aligned} \tag{4.75}$$

[17] It can be shown that equivalent equations are obtained for multicomponent systems if the parameterization of the composition profiles is performed with only a single-order parameter.

[18] See Section 1.3.2.

and the mean value of the free energy of the two interfacial layers is then given as

$$g_{I,\mathrm{mean}} = \frac{g_{I1}+g_{I2}}{2} = N_\mathrm{A}\frac{z_\mathrm{L}}{2}\left[X^2\left(\varepsilon_{11}+\varepsilon_{22}-\varepsilon_{12}\right)-X\left(\varepsilon_{11}+\varepsilon_{22}-\varepsilon_{12}\right)+\frac{\varepsilon_{11}+\varepsilon_{22}}{2}\right]$$
$$+RT\left[X\ln X+\left(1-X\right)\ln\left(1-X\right)\right]. \tag{4.76}$$

With the substitution

$$\varepsilon^* = \varepsilon_{12}-\frac{1}{2}\left(\varepsilon_{11}+\varepsilon_{22}\right), \tag{4.77}$$

the following expression is obtained

$$g^*_{I,\mathrm{mean}} = N_\mathrm{A}z_\mathrm{L}\varepsilon^* X\left(1-X\right)+RT\left[X\ln X+\left(1-X\right)\ln\left(1-X\right)\right]+K, \tag{4.78}$$

with

$$K = N_\mathrm{A}z_\mathrm{L}\frac{1}{2}\left(\varepsilon_{11}+\varepsilon_{22}\right). \tag{4.79}$$

The equivalent procedure can be performed for the mean energy of the two matrix phases, where a similar expression is obtained. Finally, when evaluating the difference in volume free energy, it is apparent that the parameter K vanishes, as it occurs in both, the energy expression for the interfacial layers as well as the mean volume free energy of the matrix phases. This parameter is, therefore, neglected in the further treatment.

Using the enthalpy expression for a regular solution, equation (1.31), at 0 K, the modified Gibbs energy becomes

$$g^* = \Delta H_{\mathrm{sol},0K}\cdot X\left(1-X\right)+RT\left[X\ln X+\left(1-X\right)\ln\left(1-X\right)\right], \tag{4.80}$$

with

$$\Delta H_{\mathrm{sol},0K} = N_\mathrm{A}z_\mathrm{L}\varepsilon^*. \tag{4.81}$$

For comparison with alternative models for diffuse interface energies, the solution enthalpy can be expressed in terms of the regular solution critical temperature, T_C, given as

$$T_\mathrm{C} = \Delta H_{\mathrm{sol},0K}/2R. \tag{4.82}$$

For the volumetric contribution to the interfacial free energy, we finally arrive at

$$\gamma_\mathrm{vol} = \frac{2RZd}{a^3N_\mathrm{A}}\cdot\left\{\begin{array}{l}2T_\mathrm{C}\cdot\left[\left(X-X^2\right)-\left(X_\alpha-X_\alpha^2\right)\right]\\ +T\left[\begin{array}{l}X\ln X+\left(1-X\right)\ln\left(1-X\right)\\ -X_\alpha\ln X_\alpha-\left(1-X_\alpha\right)\ln\left(1-X_\alpha\right)\end{array}\right]\end{array}\right\}, \tag{4.83}$$

where Z is the number of atoms per unit cell, d is the interface thickness, and a is the atomic distance. The width of the interface layers is assumed to be constant and identical to the lattice spacing.

In analogy to the general treatment of diffuse interfaces discussed in Section 4.5.1, the volumetric contribution is defined in terms of the sharp interface energy and a function, f_{vol}, according to

$$\gamma_{\mathrm{vol}} = f_{\mathrm{vol}} \cdot \gamma_0. \tag{4.84}$$

With

$$\gamma_0 = \frac{z_{\mathrm{L}}}{z_{\mathrm{S}}} \frac{Z^{\frac{2}{3}}}{a^2 N_{\mathrm{A}}} \cdot 2RT_{\mathrm{C}}, \tag{4.85}$$

we finally obtain

$$f_{\mathrm{vol}} = Z^{\frac{1}{3}} \frac{d}{a} \frac{z_{\mathrm{L}}}{z_{\mathrm{S}}} \cdot \left\{ \begin{array}{l} 2 \cdot \left[\left(X - X^2\right) - \left(X_\alpha - X_\alpha^2\right) \right] \\ + \dfrac{T}{T_{\mathrm{C}}} \left[\begin{array}{l} X \ln X + (1-X)\ln(1-X) \\ -X_\alpha \ln X_\alpha - (1-X_\alpha)\ln(1-X_\alpha) \end{array} \right] \end{array} \right\}. \tag{4.86}$$

The interface energy contribution is obtained by summation of the individual interface energies. The corresponding function, f_{intf}, is given as

$$f_{\mathrm{intf}} = \alpha_{\alpha,\beta}\left(1 - 2X_\alpha\right)^2 + 2\alpha_{\alpha,1}\left(X - X_\alpha\right)^2 + 2\alpha_{\alpha,2}\left(1 - X - X_\alpha\right)^2 + \alpha_{1,2}\left(1 - 2X\right)^2. \tag{4.87}$$

The unknown composition of layer 1, X, is finally obtained by minimization of the superposition of functions, f_{vol} and f_{intf}. This procedure delivers an implicit equation for X as

$$\begin{aligned} &2(1-2X) + 4Z^{-\frac{1}{3}} \frac{z_{\mathrm{L}}}{z_{\mathrm{S}}} \frac{a}{d} \left[\alpha_{\alpha,1}\left(X - X_\alpha\right) - \alpha_{\alpha,2}\left(1 - X - X_\alpha\right) - \alpha_{1,2}\left(1 - 2X\right) \right] \\ &- \frac{T}{T_{\mathrm{C}}} \ln\left(\frac{1}{X} - 1\right) = 0. \end{aligned} \tag{4.88}$$

In the limit of low and medium temperatures, the target composition, X, is similar to the matrix composition, X_α, and equation (4.88) can be approximately solved for X with

$$X = \left\{ 1 + \exp\left[\tilde{K} \frac{T_{\mathrm{C}}}{T} \left(1 - 2X_\alpha\right) \right] \right\}^{-1}, \tag{4.89}$$

and

$$\tilde{K} = 2 - 4Z^{-\frac{1}{3}} \frac{z_{\mathrm{S}}}{z_{\mathrm{L}}} \frac{a}{d} \left(\alpha_{\alpha,2} + \alpha_{1,2} \right). \tag{4.90}$$

4.5.3 Comparison with Other Models

In ref. [133], Sonderegger and Kozeschnik compare their GBB diffuse interface model to computational results obtained with the Cluster Variation Method (CVM) and the Cluster/Site Approximation (CSA), as well as experimental data obtained from Ostwald ripening studies.

In CVM [134,135], atoms are placed on various different positions of a rigid parent lattice. Systematic variations of atomic site occupations deliver expressions for the free energy of the system under consideration, the internal energy as well as the configurational entropy. CVM is considered to be a rigorous method, which means that the results of the computations do not depend on unknown parameters that are obtained by calibration to experimental results. When applying the CVM to the analysis of interfaces, the local constitution of the interface, which is its thickness and layer compositions, are obtained as a result of the energy minimization procedure utilized by this method.

The CSA represents a simplified method for evaluation of configurational entropies and energies (see, e.g., [136]). It is computationally less expensive than the CVM and it has been shown that it is an important technique for calculations in multicomponent systems where CVM becomes very cumbersome due to the high number of interaction terms.

For comparison, the results obtained in the previous section can be compared to results that have been computed for the same generic system, that is, regular solution and fcc structure. These data are available from Lee and Aaronson [137] using the NNBB model; from Cao et al. [138] using CVA, and from Asta [139] using CVM.

In Figure 4.18, the interfacial energy of a (100) phase boundary is compared for various computational techniques. The *y*-axis is normalized with respect to $k_B T_C / a^2$ for convenience. Apparently, with increasing temperature the interfacial energy decreases due to interfacial mixing and finally approaches zero when the critical temperature is reached. The various approaches deliver rather similar results. Particularly interesting from a practical view is that

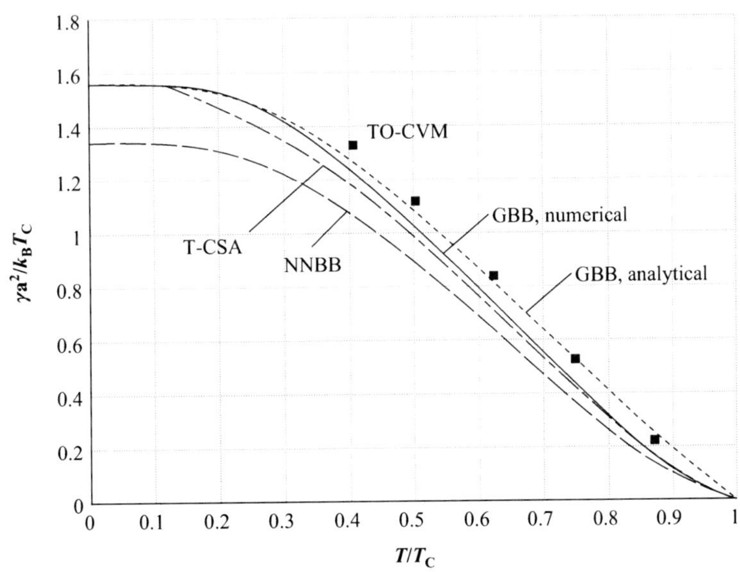

Figure 4.18. Energy of an (100) interface in fcc systems obtained with various computational methods: TO-CVM [139], T-CSA [138], NNBB [137].

the (thermodynamic) GBB approximation also corresponds well with the techniques based on rigorous computational procedures.

Finally, in ref. [133], the computational results for the interfacial energy of a diffuse phase boundary are compared with effective interfacial energies obtained from Ostwald ripening experiments [101]. Within the experimental uncertainty, the predicted values are found to be in good agreement with the experiments.

4.6 Summary

The concept of interfacial energy is well established and extremely useful in continuums-modeling of phase transformations (precipitation, condensation, recrystallization, etc.). Experimentally, this quantity is rather difficult to obtain because it is not possible to measure interfacial energies directly. In Ostwald ripening experiments, for instance, where the evolution of the mean precipitate radius and the number density are observed, one can only obtain the product of interfacial energy with diffusion coefficient. Only in a combination of measurements, where the evolution of supersaturation in a coarsening system is determined simultaneously with the evolution of mean radius and number density, a unique value for the interfacial energy can be obtained.

The determination of interfacial energies from nucleation and growth experiments also represents a widely utilized method. In this approach, the nucleation rate expression is used to calibrate the effective interfacial energy on precipitate densities or the evolution of the mean precipitate radius/phase fraction. Unfortunately, the values obtained from this methodology are strongly dependent on the size of the precipitates, as demonstrated in Section 4.4, as well as other important facets of early stage precipitation, such as crystallographic constraints or metastable precursor phases. As typical sizes of critical nuclei in precipitation systems are in the order of nanometers and below, the interfacial energy obtained by these experiments is typically 20–30% lower than the interfacial energy of the same phase boundary in later stages, where interfacial curvature can be neglected. This phenomenon is often ignored. Values reported in literature must, therefore, be considered with care. In addition, interfacial energy values reported from nucleation and growth experiments are also evaluated only indirectly and, therefore, inherit any possible shortcoming or deficiency of the theoretical concept used in the evaluation procedure.

The observation of interfacial energies based on computational methods represents a valuable alternative to the experimental approach. On the atomic scale, the value of the interfacial energy can be straightforwardly evaluated in the case of *sharp, planar* interfaces, where the phase boundary is assumed to be infinitely thin. In this special case, evaluation of the interfacial energy can be reduced to counting (broken) atomic bonds across the interface and its value can be determined with knowledge of the bond energies between the atoms involved or the enthalpy of solution of the precipitate in the multicomponent matrix. R. Becker and D. Turnbull have developed the corresponding theory several decades ago.

As the original theory is based on nearest-neighbor interactions only, its applicability to practical and quantitative evaluation of interfacial energies is limited. In the framework of the GBB model, Section 4.3, effective structural factors have been developed that avoid this deficiency. Moreover, correction factors for precipitate size (interfacial curvature), Section 4.4, and atomic mixing across the interface, Section 4.5, are now available that allow for the quantitative prediction of interfacial energies in arbitrary thermodynamic systems. Provided, of course, that accurate thermodynamic data for the involved solution enthalpies are available.

Despite the apparent potential of interfacial energy models, applications demonstrating the usage of "calculated" interfacial energy values are rare in precipitation modeling. Software, such as PrecipiCalc, PanPrecipitation, or TC-Prisma, collect and maintain interfacial energies in so-called *property databases*. The only exception in this aspect is the thermokinetic software *MatCalc*, where interface energies are routinely utilized as calculated from the GBB model.

CHAPTER 5

NUMERICAL MODELING OF PRECIPITATION

With appropriate theory for nucleation of precipitates at hand (which was elaborated in Chapter 2), and with a choice of different growth models for various types of applications (as discussed in Chapter 3), it is an obvious task, next, to combine and link a set of suitable models for the individual physical mechanisms to obtain a comprehensive simulation tool for the entire precipitation process. The purpose of this chapter is to provide an introduction and a discussion of features of the presently most popular approaches.

The numerical methods outlined below comprise of three approaches of fundamentally different methodology. In the first section, the well-known and widely appreciated *Kolmogorov–Johnson–Mehl–Avrami* (KJMA) model is discussed, which provides a closed analytical solution for the evolution of a phase transformation/precipitation process. Nowadays, this approach is often used to give a simple and easy computational representation of the experimentally obtained phase fraction evolution. By inspecting the values of the major kinetic parameters of the models, even some limited introspection into the metallurgical mechanisms governing the phase transformation process is possible.

The next approach, which is described in Section 5.2, is the *Langer–Schwartz* (LS) model, which, in contrast to a representation of the overall transformation progress in terms of a single progress parameter, provides evolution equations for the major precipitation parameters, mean radius, and number density. Due to several simplifying assumptions, the numerical complexity of this model is kept low and solutions to precipitation processes can be obtained with high computational efficiency.

The most versatile approach in this chapter, which is outlined in Section 5.3, is the *numerical Kampmann–Wagner* (NKW) model. With this simulation technique, the size distribution of precipitates is discretized into classes, which, in the beginning, is established from application of classical nucleation theory. For each of these classes, the evolution in time is traced individually utilizing an appropriate growth model. The NKW model is most widely used in advanced simulation tools for precipitation due to its conceptual simplicity and flexibility in practical use.

The three models elaborated in this chapter are certainly not the only ones available for precipitation simulation; they are certainly the most popular models, though. There exist interesting alternative approaches based on, for example, the Monte Carlo technique [140] or the particle-size-grouping method [141], which cannot be discussed here and the reader is referred to the original publications.

5.1 Kolmogorov–Johnson–Mehl–Avrami (KJMA) Model

Consider a phase transformation where the rate, $\dot{m}$, at which new particles of a second phase are formed, and the growth rate, $\dot{\rho}$, by which these particles grow, are known. Let the parameter, ξ, represent the progress of the phase transformation with a number between zero (no transformation has yet occurred) and one (the transformation is fully completed).

Also, take into account that regions of the transforming phases can come into contact in the course of the reaction and hinder themselves in their further evolution. The process, where growing regions of a phase (or grains) touch and block their further spatial extension is denoted as *hard impingement*. In precipitation processes, these events are rather rare and restricted to situations where the phase fraction of the second phase is appreciably high. However, precipitate growth is governed by diffusion of elements toward and away from the particle. A diffusion field of characteristic extension, thus defining the potential interaction range with other particles, surrounds each growing precipitate. In precipitation reactions, interaction between precipitates usually occurs if the corresponding diffusion fields of different particles overlap and precipitate growth is interfered by the competition for available solute atoms. The overlap of diffusion fields of growing precipitates is denoted as *soft impingement*.

5.1.1 Derivation of the KJMA Equation

A rather general theoretical framework describing the progress of a phase transformation as a function of time was developed between 1937 and 1941 by Andrei Nikolajewitsch Kolmogorov [142], a Russian mathematician (1903–1987), W. A. Johnson and Robert Franklin Mehl [143], an American metallurgist (1898–1976), as well as Melvin Avrami [144–146], a professor at the School of Mines, Columbia University, New York.[1] The concept is well known as the KJMA-model,[2] being applicable to a wide range of problems of phase transformation, such as precipitation, matrix transformation,[3] or recrystallization. A review of the KJMA model has been presented by M. Fanfoni and M. Tomellini [147].

Consider a phase α, which transforms into a phase β after having been quenched from the α one-phase field into the β one-phase field.[4] Starting at time $t = 0$ and evolving the system for a small time interval, Δt, a certain number of nuclei of the β phase will form in the supersaturated α matrix. This is shown in Figure 5.1(a), where three nuclei of the β phase are schematically drawn. Let the nucleation step occur under the constraint that the volume of the newly formed phase is sufficiently small compared to the volume of the matrix. Position the new nuclei randomly in the volume of the parent phase. The number of nuclei shall be given by $\Delta m = \dot{m} \cdot \Delta t$.

---

[1] Interestingly, I could not find any biography of Melvin Avrami. There exists a note by Ken Russell (http://improbable.com/airchives/miniair/2007/mini2007-01.htm) claiming that Melvin Avrami was of Palestinian origin, who later changed his name to Melvin Avrami Melvin.

[2] The order of authors is often also written as JMAK. In this context, we apply the order as given by the year of publication of the respective manuscripts.

[3] Typical examples of matrix transformations are the austenite decomposition into ferrite/pearlite in steel or the alpha/beta phase transformation in Ti-alloys.

[4] Equivalently, one may picture the case of a precipitate phase β forming in the $\alpha + \beta$ two-phase field.

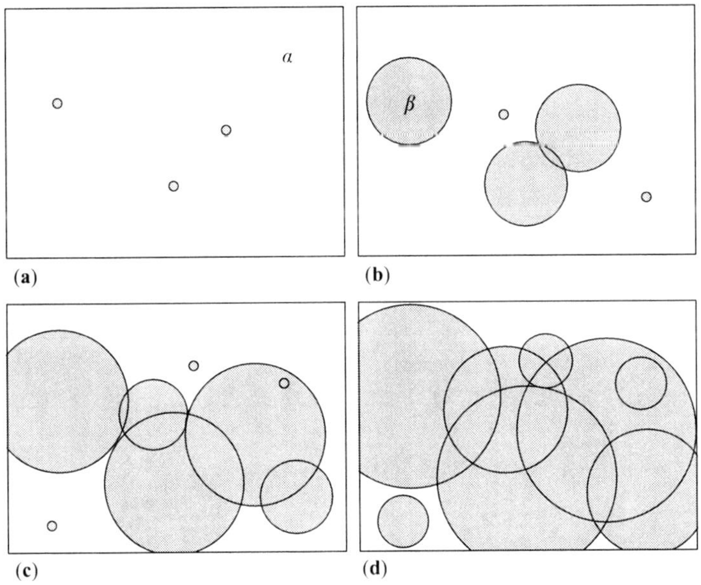

Figure 5.1. Illustration of the extended volume approach used to derive the KJMA model.

In the next time interval, let the existing precipitates grow with $\Delta\rho = \dot{\rho}\cdot\Delta t$. Add new nuclei to the system as determined by the nucleation rate, $\dot{m}$. Again, the position of the new precipitates is chosen at random and no account is taken of whether the new seeds become located in virgin α volume or in an already transformed region. A possible configuration after two steps is shown in Figure 5.1(b).

In Figures 5.1(c) and (d), two inherent problems of this approach become apparent:

- The continuous growth of the new phase leads to collisions of the growth fronts or to an overlap of the diffusion fields of the growing precipitates.
- The random placement of newly formed seeds can lead to nucleation in already transformed regions.

Both difficulties can be overcome, if two additional assumptions are made: (*a*) let the new phase nucleate and grow regardless of whether it involves regions of the parent phase or already transformed volume. Denote all volume that is created by either nucleation or growth as *extended volume*. (*b*) Relate the probability of a nucleation event occurring in untransformed volume and the probability of a growth event to involve untransformed regions only to the ratio between actual and extended volume.

For the given constant nucleation rate, $\dot{m}$, growth rate, $\dot{\rho}$, and keeping temperature constant, the extended volume, V^{ext}, created by a single growing precipitate is given by

$$V^{\text{ext}} = \frac{4\pi}{3}\dot{\rho}^3(t-\tau)^3, \tag{5.1}$$

where τ is the time at which the precipitate has nucleated. The superscript "ext" indicates that the volume of the sphere belongs to the extended volume and that impingement effects are ignored in this description.

For the total number, $\mathrm{d}m_\tau$, of nucleation events occurring between times τ and $\tau+\Delta\tau$, we obtain

$$\mathrm{d}m_\tau = \dot{m}_\tau V^{\mathrm{tot}}\mathrm{d}\tau, \tag{5.2}$$

where V^{tot} refers to the total volume of the transforming sample. The increase in extended volume within the time interval, $\mathrm{d}\tau$, is

$$\mathrm{d}V^{\mathrm{ext}} = V^{\mathrm{ext}}\cdot \mathrm{d}m_\tau = \frac{4\pi}{3}\dot{\rho}^3(t-\tau)^3\dot{m}_\tau V^{\mathrm{tot}}\mathrm{d}\tau. \tag{5.3}$$

The total extended volume, which is created by all nucleation events, is obtained after integration of equation (5.3). For constant nucleation rate, $\dot{m}$, and growth rate, $\dot{\rho}$, we get

$$V^{\mathrm{ext}} = \frac{4\pi}{3}\dot{\rho}^3\dot{m}V^{\mathrm{tot}}\int_0^t(t-\tau)^3\mathrm{d}\tau = \frac{\pi}{3}\dot{\rho}^3\dot{m}V^{\mathrm{tot}}t^4. \tag{5.4}$$

On comparison with the total volume, the total extended volume fraction, ξ^{ext}, is found with

$$\xi^{\mathrm{ext}} = \frac{V^{\mathrm{ext}}}{V^{\mathrm{tot}}} = \frac{\pi}{3}\dot{\rho}^3\dot{m}t^4. \tag{5.5}$$

For an analysis of the overall transformation progress, the extended volume fraction, ξ^{ext}, is related to the actual volume, ξ, by the assumption that the available volume for transformation is determined by the probability, P^{ut}, for finding untransformed regions. This is accomplished by multiplying the extended volume, V^{ext}, with P^{ut}. Consequently, the increase in real volume can be written as

$$\mathrm{d}V = \left(1-\frac{V}{V^{\mathrm{tot}}}\right)\mathrm{d}V^{\mathrm{ext}}, \tag{5.6}$$

with the probability term given by

$$P^{\mathrm{ut}} = \left(1-\frac{V}{V^{\mathrm{tot}}}\right). \tag{5.7}$$

Equation (5.6) can be straightforwardly integrated after separation of variables, yielding

$$\frac{V}{V^{\mathrm{tot}}} = 1-\exp\left(-\frac{V^{\mathrm{ext}}}{V^{\mathrm{tot}}}\right), \tag{5.8}$$

and

$$\xi = 1-\exp\left(-\xi^{\mathrm{ext}}\right). \tag{5.9}$$

After insertion for the extended volume fraction, equation (5.5), we finally obtain

$$\xi = 1-\exp\left(-\frac{\pi}{3}\dot{\rho}^3\dot{m}t^4\right). \tag{5.10}$$

Equation (5.10) describes the general evolution of a phase transformation involving both the nucleation stage as well as growth of the second phase until it comes to a stop due to impingement.

Although this expression has been derived for rather special conditions, that is, constant nucleation rate and constant growth rate, it can be shown that equations derived for alternative nucleation and growth conditions adopt similar forms. Consequently, the *KJMA equation* can be written in a generic form as

$$\xi = 1 - \exp(-k \cdot t^n)\,. \tag{5.11}$$

In this equation, k is generally denoted as the *Avrami coefficient* and n as the *Avrami exponent*. In most cases, the value of n is between 1 and 4. The Avrami coefficient contains the nucleation and growth rates and it is, therefore, a function of temperature as well as the thermokinetic boundary conditions of the transformation under investigation. The Avrami exponent, n, is closely related to the transformation mechanism. The characteristics of the KJMA equation is further explored in the following section.

5.1.2 Analysis of KJMA Parameters

In a first step, let the value of the Avrami coefficient be held at $k = 0.001$ and the shape of the KJMA equation is investigated for typical values of the Avrami exponent, n. The corresponding plots are shown in Figure 5.2(a) for linear time axis and in Figure 5.2(b) for log scale. Obviously, the Avrami exponent is strongly influencing the general shape of the KJMA kinetic curves. Low values around 1 deliver relatively flat curves, indicating slower transformation kinetics, and values of 3–4 lead to a sharp increase in the transformation parameter, ξ, with time. Higher Avrami exponents generally correspond to pronounced influence of nucleation events.

The same analysis is presented in Figure 5.3 for $k = 1$. High values of the Avrami coefficient are typical for transformations with high growth rates. The shift in time is much less pronounced compared to the case of $k = 0.001$. However, the shape of the KJMA kinetic curves is significantly altered by the value of the Avrami exponent.

Figure 5.4 presents an analysis of the influence of variations in the Avrami coefficient, k, with constant Avrami exponent $n = 2$. For values of $k = 0.1$ and less, the kinetic curves are mainly shifted to longer times, the general shape of the curves is essentially unaffected.

It has already been indicated earlier that the Avrami exponent, n, is closely related to the transformation mechanism. In essence, n can be viewed as a measure for the relative importance of nucleation and growth mechanisms. In cases where the Avrami exponent adopts values in the order of 1–2, nucleation rates are low to medium and the evolution of the transformation progress is predominantly controlled by growth. Higher exponents of 3–4 are often indication for pronounced nucleation in highly supersaturated systems. Values of $n = 2$ to 3 are often observed in precipitation systems with saturation of nucleation sites. In Table 5.1, some values for n are summarized in relation to possible phase transformation mechanism as compiled by J. W. Christian in his textbook on phase transformations [59].

Finally, for analysis of experimental data and evaluation of the Avrami exponent, the generic form of the KJMA equation, equation (5.11), can be further transformed. On applying the natural logarithm twice, we obtain

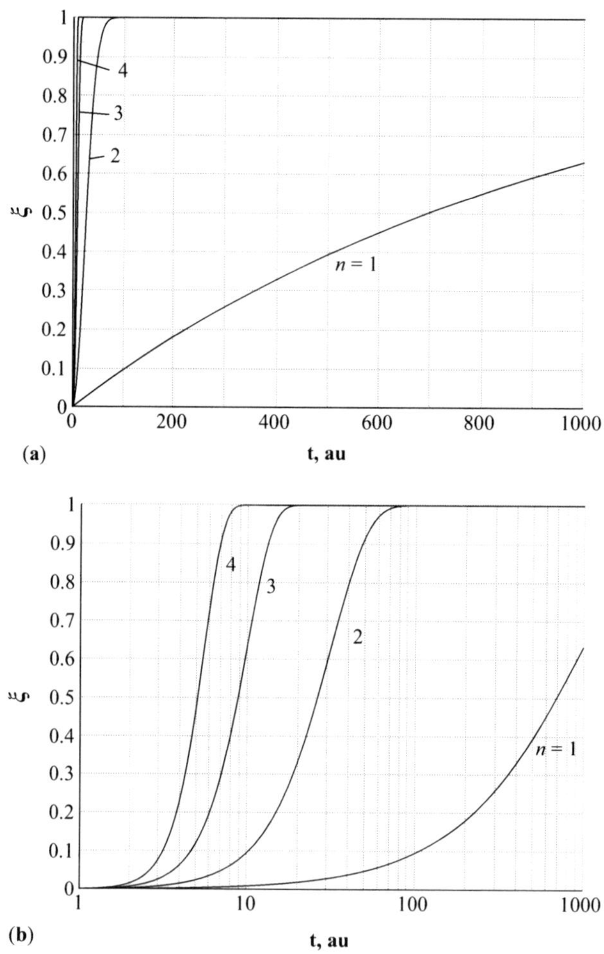

Figure 5.2. KJMA function for different values of the Avrami exponent, n, with $k = 0.001$.

$$\ln(-\ln(1-\xi)) = \ln(k) + n\ln(t). \tag{5.12}$$

On plotting $\ln(-\ln(1-\xi))$ versus $\ln(t)$, the Avrami exponent, n, can be directly read from the slope of the curve. The Avrami coefficient, k, can be observed at the intercept with the y-axis at $\ln(t) = 0$. This is exemplarily shown for the KJMA equation with $n = 2$ and $k = 10^{-3}$ ($\ln(k) = -6.91$) in Figure 5.5.

5.1.3 Multiphase KJMA Kinetics

The KJMA analysis has been utilized for description of phase transformations and precipitation reactions countless times in literature. One application is briefly outlined, next, because it is of particular interest for precipitation modeling. In this work, the KJMA model is used to evaluate the simultaneous evolution of multiple precipitates in typical 2 1/4wt%–Cr steel [148] and various different power plant steels [149] until it is finally formulated in general form with n precipitate types.

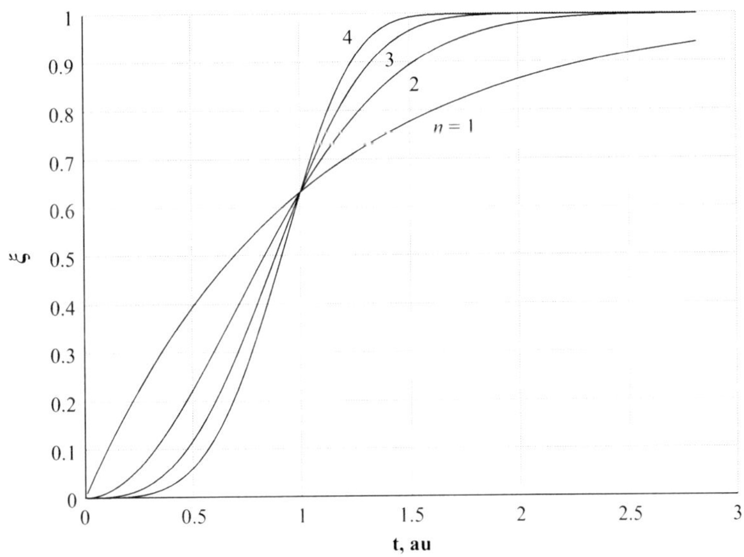

Figure 5.3. KJMA function for different values of the Avrami exponent, n, with $k = 1$.

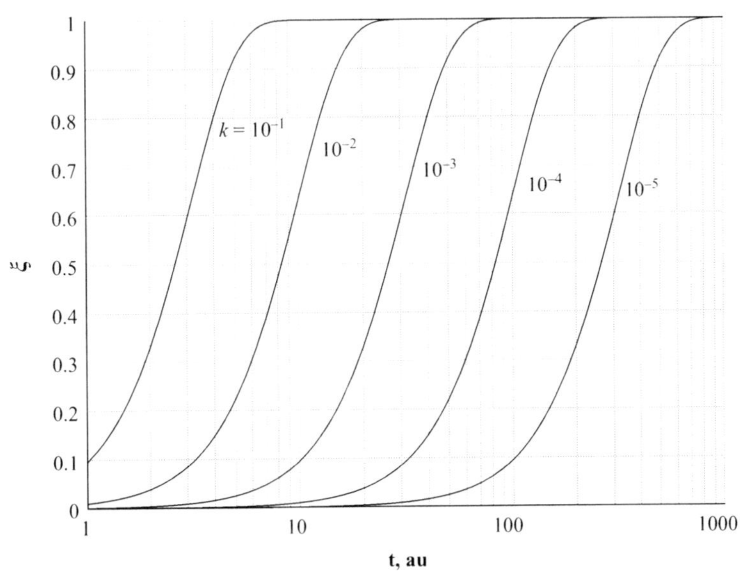

Figure 5.4. KJMA function for different values of the Avrami coefficient, k, with the Avrami exponent $n = 2$.

Consider two phases, first, which are denoted as β and τ. Let these phases be in direct competition for the precipitate forming elements. Then, the total volume, V^{tot}, that is available for precipitation is shared by the two phases according to

$$V^{\text{tot}} = V_{\beta}^{\text{tot}} + V_{\theta}^{\text{tot}}. \tag{5.13}$$

Table 5.1. Typical values of the Avrami exponent, n, in relation to phase transformation mechanism [59].

n	Interpretation
	Polymorphic changes, discontinuous precipitation, eutectoid reactions, interface-controlled growth, and so forth
1	Grain boundary nucleation
2	Grain edge nucleation
3	Zero nucleation rate
3–4	Grain edge nucleation
4	Constant nucleation rate
>4	Increasing nucleation rate
	Diffusion-controlled growth
>2.5	Small dimensions, increasing nucleation rate
2.5	Small dimensions, constant nucleation rate
1.5–2.5	Small dimensions, decreasing nucleation rate
1.5	Small dimensions, zero nucleation rate
1–1.5	Growth of particles of large volume
1	Needles and plates, large separation
1	Thickening of long cylinders
0.5	Thickening of large plates
2/3	Cottrell atmosphere formation

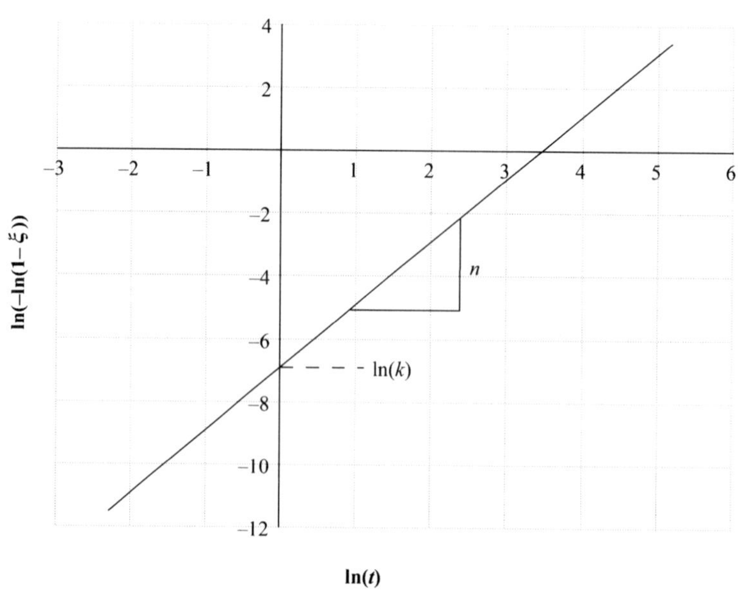

Figure 5.5. Evaluation of the Avrami exponent, n, and the Avrami coefficient, k.

The total increase in extended volume, equation (5.6), associated with nucleation and growth of the two phases is given in terms of the actual volumes, V_β and V_θ, as

$$\mathrm{d}V_\beta = \left(1 - \frac{V_\beta + V_\theta}{V^{\mathrm{tot}}}\right)\mathrm{d}V_\beta^{\mathrm{ext}}, \tag{5.14}$$

and

$$\mathrm{d}V_\theta = \left(1 - \frac{V_\beta + V_\theta}{V^{\mathrm{tot}}}\right)\mathrm{d}V_\theta^{\mathrm{ext}}. \tag{5.15}$$

Robson and Bhadeshia argue that the actual volumes, V_β and V_θ, are closely related, however, V_β will, in general, be a complex function of V_θ. It is, therefore, not possible to integrate equations (5.14) and (5.15) directly. To overcome this problem, a simplified relation

$$V_\theta = KV_\beta, \tag{5.16}$$

can be substituted, where K is a constant. For constant growth rate and constant nucleation rate, the value of K is given as

$$K = \frac{V_\theta}{V_\beta} = \frac{\dot{m}_\theta \dot{\rho}_\theta^3}{\dot{m}_\beta \dot{\rho}_\beta^3}. \tag{5.17}$$

With the constant K, we obtain the equations relating the increment of the real volume to that of the extended volume as

$$\mathrm{d}V_\beta = \left(1 - \frac{V_\beta + KV_\beta}{V^{\mathrm{tot}}}\right)\mathrm{d}V_\beta^{\mathrm{ext}}, \tag{5.18}$$

and

$$\mathrm{d}V_\theta = \left(1 - \frac{V_\theta + KV_\theta}{V^{\mathrm{tot}}}\right)\mathrm{d}V_\theta^{\mathrm{ext}}. \tag{5.19}$$

Integration of these equations delivers the relation between the real volume of the phase to that of the extended volume with

$$\frac{V_\beta^{\mathrm{ext}}}{V} = \frac{-1}{1+K}\ln\left(1 - \frac{V_\beta}{V}(1+K)\right), \tag{5.20}$$

and

$$\frac{V_\theta^{\mathrm{ext}}}{V} = \frac{-K}{1+K}\ln\left(1 - \frac{V_\theta}{V}\left(\frac{1+K}{K}\right)\right). \tag{5.21}$$

The evolution equations for the phase fractions are finally obtained as [148]

$$\xi_\beta = \left(\frac{1}{1+K}\right)\left\{1-\exp\left[-\frac{1}{3}(1+K)\pi\dot{\rho}_\beta^3\dot{m}_\beta t^4\right]\right\}, \tag{5.22}$$

and

$$\xi_\theta = \left(\frac{K}{1+K}\right)\left\{1-\exp\left[-\frac{1}{3}\left(\frac{1+K}{K}\right)\pi\dot{\rho}_\theta^3\dot{m}_\theta t^4\right]\right\}. \tag{5.23}$$

These equations resemble that of the one-phase KJMA equation with some extra factors accounting for the multiparticle nature of the present model. In the limit of small phase fractions, the equations are almost identical to the expressions for the evolution of each individual phase. The reason is that in the early stages of the reactions, almost all extended volume lies in the actual volume of the untransformed phase. At later stages, the volume fraction becomes less than in the single-phase case as the two competing phases share the same totally available volume.

For the volume increment of a third phase, γ, Robson and Bhadeshia obtain

$$\Delta V_\gamma = \left(1-\frac{\sum_{i=1}^{n} V_i}{V^{\mathrm{tot}}}\right)\mathrm{d}V_\gamma^{\mathrm{ext}}. \tag{5.24}$$

Unfortunately, there exist no general relations between phases in multiphase precipitation reactions such as to determine unique constants, K_i, for i phases. For instance, the precipitates interact with each other by removing or adding solute to the matrix and, thus, change the chemical composition of the matrix. The nucleation rates as well as the growth rates are functions of time, consequently, and are not constant as assumed in the simple model above. Therefore, the actual implementation of multiphase precipitation in the present multiparticle KJMA framework must be made in a more advanced model.

In their work, Robson and Bhadeshia introduce appropriate nucleation expressions for all precipitates of the system and set up growth and dissolution expressions for the radius evolution. The particular expressions take account of the special nucleation and growth characteristics of each precipitate type. The exact form of these equations is not reproduced here due to the fact that they are not of general nature and apply only for the very particular material and precipitate under investigation. The interested reader is referred to the original work for details [148].

A typical result obtained with the Robson and Bhadeshia approach is reproduced here. In ref. [149], these authors provide phase fraction curves for the simultaneous precipitation of M_3C, M_2X, and $M_{23}C_6$ phases in a 10CrMoV heat-resistant steel (Figure 5.6). The simulation is initiated with the equilibrium phase fraction of cementite (M_3C) as a starting condition. After 15 minutes, M_2X precipitates have nucleated and grown to their equilibrium phase fraction. At the same time, rapid dissolution of cementite sets in. $M_{23}C_6$ forms subsequently as the dominant carbide and dissolution of cementite continues until it almost dissolves at the end of the simulation.

The multiparticle KJMA model is certainly an interesting extension of the original KJMA approach. It is computationally efficient and offers the possibility for evaluation of time–temperature–precipitation diagrams from the individual simulations. However, it also shows a few drawbacks, among which there is the necessity for incorporation of each phase of the

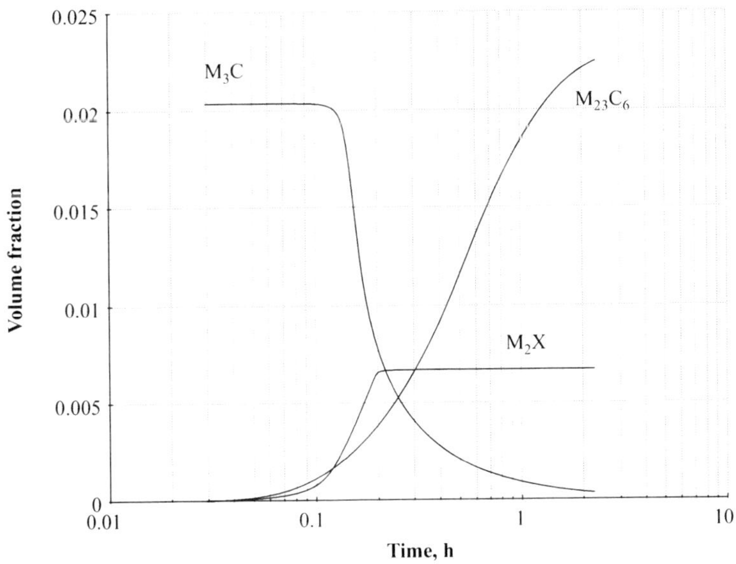

Figure 5.6. Precipitation sequence in a 10CrMoV heat-resistant steel as obtained with the multiparticle KJMA model [149].

system by special equations and mass balance relations. Moreover, one obtains only information on the evolution of the phase fraction while no results on radius and number density are delivered.

In the following sections, models allowing for a higher degree of detail on the physical nature of the precipitation reaction are introduced and discussed.

5.2 Langer–Schwartz Model

Motivated by ongoing discussions on the "apparent failure" of Classical Nucleation Theory (CNT) to quantitatively describe nucleation in precipitation systems, particularly with regard to droplet nucleation in supersaturated liquids, Langer and Schwartz (LS) developed a theoretical model [150] that takes into account simultaneously the nucleation process as well as the growth of already nucleated particles. The corresponding idea was formulated earlier by Binder and Stauffer [151]. LS argue that

> *the experimentally meaningful quantity is not the nucleation rate itself but the time required for the reaction to go to completion. To estimate a completion time, one must consider droplet growth as well as formation. Because the growth of droplets is controlled by diffusion—the experiments of interest have been performed in systems where the relevant order parameter is a locally conserved quantity—and diffusive processes over critical length scales become very slow near the critical point, the overall reaction rates may be appreciably different from estimates based on nucleation alone.*[5]

[5] This idea is considered by LS as a possibility to "*reconcile theory and experiment*" in nucleation.

5.2.1 The Original LS Model

In their treatment, Langer and Schwartz [150] make use of several—what they call "inelegant"—assumptions to be able to solve the problem. Their approach has shown, however, to deliver good predictions and it is, for this reason, utilized by many researchers for a fast and efficient numerical solution of precipitation problems. For details and motivation for the LS approximations, the interested reader is referred to the original work.

Consider the size distribution of precipitates, $f(r,t)$, with the radius, r, and time, t. The continuity equation describing the evolution of the size distribution is[6]

$$\frac{\partial f(r,t)}{\partial t} = \frac{\partial}{\partial r}(f(r,t)\dot{r}) + j(r,t), \tag{5.25}$$

where $j(r,t)$ is the distributed nucleation rate. For $j(r,t)$, which can be considered as a *distributed source of droplets*, the relation holds

$$J = \int_{r^*}^{\infty} j(r,t)\,\mathrm{d}r. \tag{5.26}$$

J is the nucleation rate. Instead of entering a specific expression for $j(r,t)$ and solving equation (5.26) directly, LS continue in a general way. A first major assumption is that they consider only particles with a size larger than the critical radius, r^*, to belong to the size distribution of precipitates. Consequently, the total number of precipitates, m_{LS}, is obtained from

$$m_{\mathrm{LS}} = \int_{r^*}^{\infty} f(r,t)\,\mathrm{d}r. \tag{5.27}$$

This assumption is eventually a good approximation in the early and medium stages of precipitation. However, it could be less accurate in the late stages during Ostwald ripening. LS chose this approximation for the merit of simplified mathematics in the further treatment.

LS then make an additional assumption about the shape of the distribution function, which is, in principle, artificial; however, it allows them to continue in their analyses without knowledge of the exact shape of the actual distribution function.

Consider the distribution of particles as schematically depicted in Figure 5.7. The particles belonging to the LS distribution are the ones inside the hashed area. In a rough approximation, LS introduce an *apparent* density of precipitates with an arbitrary function

$$f^*(r^*,t) = m_{\mathrm{LS}} \frac{b}{\bar{r}_{\mathrm{LS}} - r^*}, \tag{5.28}$$

where $\bar{r}_{\mathrm{LS}}$ is the mean radius of the LS distribution and b is an a priori undetermined constant. In their treatment, LS fix this value by comparison of their theory with the results of the LSW theory for $t \to \infty$ and choosing the value of $b = 0.317$ in such a way that the coarsening rate of the LS model becomes identical to the coarsening rate of LSW. Note that, in contrast to the LSW theory, the value of $\bar{r}_{\mathrm{LS}}$ remains always larger than r^* in all stages of the decomposition process.

[6] Compare also the starting point for derivation of the LSW theory, equation (3.170).

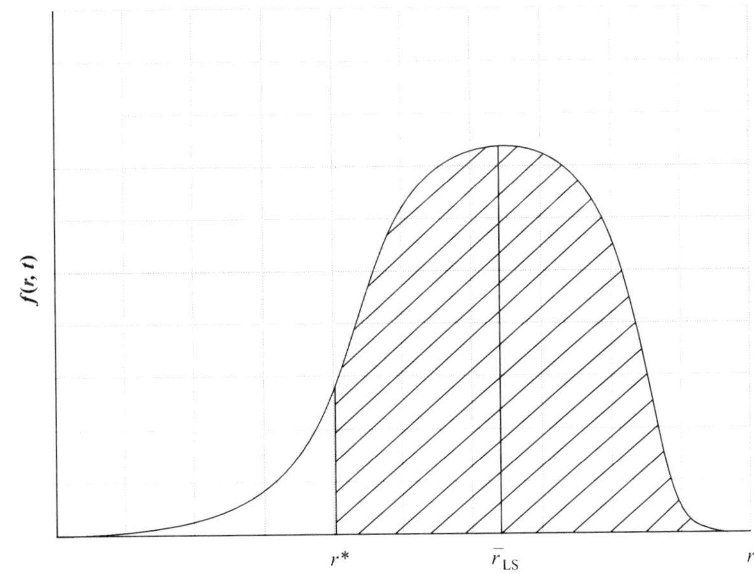

Figure 5.7. Schematic size distribution as assumed in the Langer–Schwartz theory.

Under these conditions, the evolution of the number density of particles, m_{LS}, is governed by nucleation as well as resolution and can be written as

$$\frac{dm_{LS}}{dt} = J - f^*(r^*, t)\frac{dr^*}{dt}. \tag{5.29}$$

With the mean radius of the distribution given by

$$\bar{r}_{LS} = \frac{1}{m_{LS}} \int_{r^*}^{\infty} f(r,t) \cdot r \cdot dr, \tag{5.30}$$

for the evolution of the mean radius of the distribution, LS find[7]

$$\frac{d\bar{r}_{LS}}{dt} = v(\bar{r}_{LS}) + (\bar{r}_{LS} - r^*)\frac{f^*(r^*, t)}{m_{LS}}\frac{dr^*}{dt} + \frac{1}{m_{LS}} J(r^*)(r^* + \delta r^* - \bar{r}_{LS}). \tag{5.31}$$

Equations (5.29) and (5.31) are the equations of motion describing the evolution of the LS particle distribution in time.

The first term on the right hand side of equation (5.31), $v(\bar{r}_{LS})$, represents the growth rate of precipitates with mean radius, $\bar{r}_{LS}$, in a supersaturated matrix. Typically, the classical expression

[7] Compare equation (12) in ref. [152].

for diffusion-controlled evolution of the radius of spherical precipitates, as obtained previously in equation (3.27), is inserted with

$$v(r) = \frac{D_B}{r} \cdot \frac{c_B^0 - c_B^{\alpha\beta}}{c_B^{\beta} - c_B^{\alpha\beta}}, \tag{5.32}$$

where D_B is the diffusion coefficient of the diffusing solute, c_B^0 is the concentration of B outside the diffusion field of the precipitate, c_B^{β} is the concentration inside the precipitate, and $c_B^{\alpha\beta}$ is the matrix concentration of B at the interface.

The second term on the right hand side of equation (5.31) accounts for the dissolution of particles of critical size, r^*, given by $f^*(r^*,t) \cdot dr^*$. The third term represents the change of the distribution function by nucleation of new particles with a size slightly larger than the critical radius, $r^* + \delta r^*$.[8]

Finally, for an analysis of precipitation, the equations of motion, (5.29) and (5.31), must be solved simultaneously with the mass conservation law

$$\frac{4\pi}{3} \bar{r}_{LS}^{3} \cdot m_{LS} \cdot \left(c_B^{\beta} - c_B^0\right) = \left(c_B^0 - \bar{c}_B\right), \tag{5.33}$$

where $\bar{c}_B$ is the mean concentration of B in the matrix.

In the further treatment of LS, the equations of motion, equations (5.29) and (5.31), are reformulated in terms of scaled variables. For details on this procedure and for an analysis of the time evolution of precipitate distributions based on the original LS model, the interested reader is invited to inspect equations (3.21) and (3.22) of the original work [150]. Instead of elaborating this task here, we proceed with an outline of the modified LS model as suggested by Kampmann and Wagner [152].

5.2.2 Modified Langer–Schwartz Model

One of the basic ingredients of the original LS theory, as outlined in the previous section, is the linearized version of the Gibbs–Thomson equation[9] with

$$X_B^{\alpha'} \approx X_B^{\alpha}\left(1 + \frac{2\gamma v^{\beta}}{\left(X_B^{\beta} - X_B^{\alpha}\right)RT} \cdot \frac{1}{\rho}\right), \tag{5.34}$$

with X_B^{α} and X_B^{β} being the equilibrium mole fractions of a diffusing element B in matrix and precipitate, respectively, γ being the interfacial energy, and ρ being the radius of the precipitate. $X_B^{\alpha'}$ is the local matrix composition at an interface with a precipitate of radius, ρ.

The early stages of a precipitation process are dominated by nucleation. Kampmann and Wagner (KW) [152] argue that application of the linearized Gibbs–Thomson equation can introduce significant error into the analysis, particulary at higher supersaturation, where the

[8] This condition is necessary because particles of exactly critical size have a growth rate of zero.

[9] Compare Section 1.5.1, as well as Section 3.6, equation (3.167).

critical radius of nuclei can be in the order of nanometers and less. These authors, therefore, suggest using the *nonlinearized* version of the Gibbs–Thomson equation, which is[10]

$$X_{\mathrm{B}}^{\alpha'} \approx X_{\mathrm{B}}^{u} \exp\left(\frac{2\gamma v^{\beta}}{RT} \cdot \frac{1}{\rho}\right). \tag{5.35}$$

Using this expression together with equation (5.32), the growth rate of the precipitate becomes

$$v(r) = \frac{D}{r} \cdot \frac{1}{c_{\mathrm{B}}^{\beta} - c_{\mathrm{B}}^{\alpha\beta}} \left[c_{\mathrm{B}}^{0} - c_{\mathrm{B}}^{\alpha\beta} \cdot \exp\left(\frac{2\gamma v^{\alpha}}{RT} \frac{1}{\rho}\right)\right]. \tag{5.36}$$

For further evaluation of the equations of motion, (5.29) and (5.31), scaling variables are introduced with:

$$\begin{aligned}
&r_{\mathrm{N}} = \frac{2v^{\alpha}\gamma_0}{RT}; \quad k_{\gamma} = \frac{\gamma}{\gamma_0}; \quad \xi_0 = \frac{c_{\mathrm{B}}^{0}}{c_{\mathrm{B}}^{\mathrm{eq}}}; \quad \xi = \frac{\bar{c}_{\mathrm{B}}}{c_{\mathrm{B}}^{\mathrm{eq}}};\\
&\xi_{\mathrm{B}}^{\beta} = \frac{c_{\mathrm{B}}^{\beta}}{c_{\mathrm{B}}^{\mathrm{eq}}}; \quad \rho = \frac{\bar{r}_{\mathrm{LS}}}{r_{\mathrm{N}}}; \quad \rho^{*} = \frac{r^{*}}{r_{\mathrm{N}}}; \quad \delta\rho^{*} = \delta r^{*} = \frac{\delta r^{*}}{r_{\mathrm{N}}};\\
&n = m_{\mathrm{LS}} \frac{4\pi}{3} r_{\mathrm{N}}^{3}; \quad \tau = \frac{D_{\mathrm{B}}}{r_{\mathrm{N}}^{2}} \cdot t; \quad \tilde{J} = J \frac{\mathrm{d}n}{\mathrm{d}N} \frac{\mathrm{d}t}{\mathrm{d}\tau}.
\end{aligned} \tag{5.37}$$

γ_{S} is a scaling parameter and it is taken in the following as $\gamma_{\mathrm{s}} = 0.1 J/m^2$. The continuity equation in the scaled version then reads

$$n = \frac{1}{\rho^{3}} \cdot \frac{\xi_0 - \xi}{\xi_{\mathrm{B}}^{\beta}}, \tag{5.38}$$

and the governing equations of the *modified Langer–Schwartz* (MLS) theory are finally obtained as

$$\frac{\mathrm{d}\rho}{\mathrm{d}\tau} + \frac{\rho}{3}\left\{\frac{1}{\xi_0 - \xi} + \frac{b \cdot k_{\gamma}}{\rho \ln \xi - k_{\gamma}} \cdot \frac{1}{\xi \ln \xi}\right\}\frac{\mathrm{d}\xi}{\mathrm{d}\tau} = -\tilde{J}\frac{\rho^{4}}{3}\frac{\xi_{\mathrm{B}}^{\beta} - 1}{\xi_0 - \xi}, \tag{5.39}$$

and

$$\frac{\mathrm{d}\rho}{\mathrm{d}\tau} + \frac{b \cdot k_{\gamma}}{\xi(\ln \xi)^{2}} \frac{\mathrm{d}\xi}{\mathrm{d}\tau} = \frac{1}{\xi_{\mathrm{B}}^{\beta} - 1} \cdot \frac{1}{\rho}\left\{\xi - \exp\left(\frac{k_{\gamma}}{\rho}\right)\right\} + \rho^{3}\frac{\xi_{\mathrm{B}}^{\beta} - 1}{\xi_0 - \xi} \cdot \tilde{J} \cdot \left(\frac{k_{\gamma}}{\ln \xi} + \delta\rho^{*} - \rho\right). \tag{5.40}$$

[10] This equation corresponds to approximation 1, equation (1.66), in Section 1.5.1. Note that it is a good approximation for the local B-content of the matrix only if the precipitate consists of almost pure atoms B.

In order to obtain the evolution of the precipitation system, equations (5.39) and (5.40) must be integrated numerically. It is important to note that, in the following evaluation of the MLS theory provided by Kampmann and Wagner, the nucleation rate expression is based on the time-dependent nucleation rate[11] as

$$\tilde{J} = NZ\beta^{*}\exp\left(-\frac{G^{*}}{k_{\mathrm{B}}T}\right)\exp\left(-\frac{\tau}{t}\right), \tag{5.41}$$

instead of the steady-state nucleation rate as proposed Langer and Schwartz in the original LS model. The parameters occurring in this equation are defined in Section 2.3.4.

5.3 Kampmann–Wagner Numerical Model

Based on the MLS theory, Kampmann and Wagner [152], in 1984, proposed a simulation framework for precipitation kinetics, which is now widely used and generally referred to as the *numerical Kampmann–Wagner* (NKW) model. In this approach, the following major assumptions are made over and above the MLS model:

1. The apparent density, $f^{*}(r^{*},t)$, equation (5.28), used in the original LS and in the MLS theory, is substituted by a discrete distribution of size classes, j, with a number of n_j precipitates with identical radius, r_j.
2. Dissolving particles below the critical radius, ρ^{*}, are assumed to belong to the size distribution.
3. The continuous time evolution of $f(r,t)$, or $f(\rho,\tau)$, is replaced by an appropriate time discretization with time steps Δt, or $\Delta\tau$, which must be chosen sufficiently small to maintain a smooth evolution of the precipitation system in the numerical integration procedure.

The first assumption (1) in the NKW model is essential, as it avoids the severe LS approximation of the real size distribution by the model function Figure 5.7 and represented by the apparent density, $f^{*}(\rho^{*},\tau)$, as given by equation (5.28). Accordingly, the real distribution function, $f(\rho,\tau)$, is subdivided into intervals, $[\rho_{j+1},\rho_j]$, with $|\rho_j-\rho_{j+1}|/\rho_j \ll 1$ and n_j precipitates in the nth interval. Applying simultaneously the NKW assumption (2), the precipitation parameters of the distribution are obtained as

$$n=\sum_{j=1}^{k_{\mathrm{sc}}} n_j; \qquad \bar{\rho}=\frac{1}{n}\sum_{j=1}^{k_{\mathrm{sc}}} n_j\bar{\rho}_j, \tag{5.42}$$

with k_{sc} representing the total number of size classes and

$$\bar{\rho}_j=\frac{1}{2}(\rho_j+\rho_{j+1}). \tag{5.43}$$

[11] Compare equation (2.90) in Section 2.3.4.

With these definitions, the continuity equation becomes

$$\left(\xi_{\mathrm{D}}^{\beta}-1\right)\sum_{j=1}^{k_{\mathrm{sc}}} n_j \bar{\rho}_j^3 = \xi_0 - \xi. \tag{5.44}$$

Item (3) in the above list, on the one hand, ensures numerical stability by limiting the size of the time interval Δt, or $\Delta\tau$, to values that ensure that the precipitate distribution evolves smoothly in time. On the other hand, this requirement also ensures that the precipitation parameters remain approximately constant within one time interval and the governing equations for nucleation and growth can be safely applied.

In the practical analysis, the width of each individual size class is monitored. If the width exceeds a given threshold value, the corresponding interval is subdivided into two smaller intervals under the constraint of conservation of both, distribution shape and mass. The first conservation requirement is achieved by adapting the relative number of precipitates in the subdivided classes according to the neighboring classes. If a class shrinks below a certain size, it is removed from the distribution, again, distributing the residual mass of these particles into the matrix under the constraint of mass conservation.

The fact that size classes with a mean radius, $\bar{\rho}_j < \rho^*$, also belong to the size distribution and are fully considered in the evolution algorithm allow for a convenient and implicit treatment of coarsening/Ostwald ripening inside the NKW model. Precipitates that are larger than the critical radius have positive growth rates, whereas precipitates that are smaller have negative values. Precipitate classes that shrink below a critical threshold are removed from the distribution. Consequently, after reaching the coarsening stage, size classes shrinking below the minimum size are continuously removed, thus reducing the total number of precipitates in the system.

Finally, the predictions of the MLS and the NKW model are compared in Figure 5.8. The simulations are reproduced from the work of Kampmann and Wagner [152]. Since the

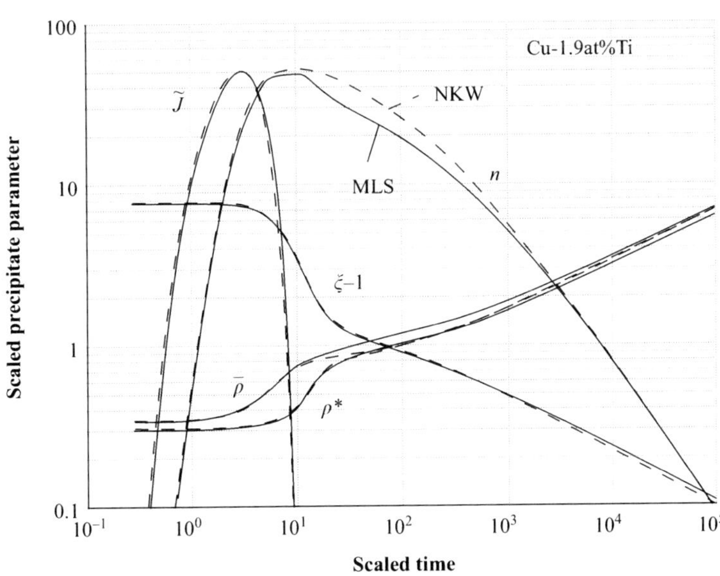

Figure 5.8. Comparison between MLS and NKW model [152].

analysis presented here shall only be of qualitative nature, for an exact description of the simulation parameters used for this comparison the interested reader is referred to the original publication.

In the simulation, both models are initialized with identical precipitate distribution and identical nucleation rate. Consequently, the results coincide almost perfectly for the early stages of the decomposition reaction. When nucleation comes to a stop, which is the point where the nucleation rate, $\tilde{J}$, goes to zero, the two models show some slight deviation. The reason for this observation originates in the fact that already in the late stages of growth, particles with a radius smaller than the critical radius, r^*, start to dissolve. At this stage, the MLS approach only counts particles with $r > r^*$ as belonging to the precipitate phase, whereas the NKW accounts for all particles with $r > 0$. As the apparent density, $f^*(\rho = \rho^*)$, in this stage, is proportional to $1/(\bar{\rho} - \rho^*)$, the MLS model provides a rather poor approximation and, therefore, the decrease in number density of precipitates is significantly more rapid than in the NKW model.

In the later stages of precipitation, the MLS and NKW model exhibit good accordance again, which is due to the particular choice of the constant b made by Langer and Schwartz in equation (5.28) such as to give identical results for the evolution of the mean precipitate parameters, radius, and number density, as predicted by LSW theory. Some differences are observed in the evolution of mean radius and critical radius. Whereas ρ^* approaches $\bar{\rho}$ in the NKW model, due to the assumptions on the shape of the apparent density in the LS approach, these quantities evolve with a clear spatial separation in the MLS model.

On comparing the two models, Kampmann and Wagner conclude that the MLS model gives a good representation of the overall phase transformation process, although it does not predict the course of precipitation with as much accuracy as the NKW approach due to the simplifying assumptions made by LS. Partially, these inaccuracies are, however, compensated by significantly less computational effort needed in evaluation of the MLS evolution equations compared to the NKW model.

5.4 General Course of a Phase Decomposition

In this section, precipitation processes are investigated in a generic sense, which means that hypothetical precipitation reactions are explored in terms of their characteristic stages. Moreover, a typical time–temperature profile in precipitation heat treatment is discussed as utilized in practical materials processing for precipitation strengthening.

5.4.1 Heat Treatments for Precipitation

In materials processing, precipitation of a dispersion of particles can have various different purposes. For instance, small precipitates in the nanometer range, which are present in fine dispersion, act as efficient obstacles for dislocation movement and, thus, represent an important means for materials strengthening. Medium-sized precipitates in the size range of 10 to several 100 nm can be efficient obstacles, which can pin grain boundaries and hinder grain boundary movement for the purpose of grain size control. Particles in the order of micrometers and more are metallurgically inert[12] but are often present in materials for the purpose of chemically binding unwanted or dangerous elements in the form of harmless compounds.[13]

The main types of precipitate morphologies found in a material's microstructure are

- *Primary precipitates*: Precipitation that occurs in the liquid state during solidification.
- *Grain boundary precipitates*: Precipitation that occurs along grain boundaries and grain boundary junctions. This type of precipitate morphology typically evolves during slow cooling with low to medium supersaturation.
- *Precipitation of a fine dispersion*: Closely spaced precipitates, which occur in high number density. This type of precipitate morphology is commonly wanted because it provides materials' strengthening by hindering dislocations from moving through the material.
- *Multimodal precipitate distribution*: Size distribution, in which larger and smaller precipitates of the same type occur. Often, smaller particles are responsible for precipitation strengthening whereas the larger particles are utilized for grain size control during thermal or thermomechanical treatment.

On the one hand, solid-state precipitation can occur as a basically uncontrolled by-product of materials manufacturing. For instance, during solidification of an alloy, microsegregation can lead to solute enrichment of interdendritic regions and precipitation of large particles inside these liquid domains. Primary precipitates usually grow to sizes in the order of a few micrometers and, as they are rather large, they are at the same time often widely spaced and occur in the microstructure with only small number. On the one hand, solid-state precipitation can occur as a basically uncontrolled by-product of materials manufacturing. For instance, during solidification of an alloy, microsegregation can lead to solute enrichment of interdendritic regions and precipitation of large particles inside these liquid domains. Primary precipitates usually grow to sizes in the order of a few micrometers and, as they are rather large, they are at the same time often widely spaced and occur in the microstructure with only small number density.[14] Figure 5.9(a) shows the typical appearance of primary precipitates in a polycrystalline material. Primary precipitates are often uncorrelated to the grain microstructure of the parent material, as grain growth at high temperatures after solidification can lead to substantial grain growth and, thus, detachment of primary solidification grain boundaries from the primary particles.

Figure 5.9(b) shows a typical precipitate microstructure that evolves during slow cooling processes. During slow cooling, precipitates tend to nucleate at grain boundaries and grain boundary junctions, as these heterogeneous nucleation sites offer very favorable conditions for nucleation (see Section 2.4). Once nucleated, these particles can grow fast due to the fast supply of solute along the grain boundaries. When the grain boundary precipitates continuously

[12] This means that they do not interact with the material's microstructure in terms of strengthening or grain boundary interactions. They can, however, be starting points for microcracks, which can be detrimental in applications involving dynamical mechanical loading.

[13] A typical example for this application is the MnS compound in steel, where Mn binds S to avoid grain boundary embrittlement caused by atomic S.

[14] Typical examples for primary precipitates are carbo-nitrides of Ti in microalloyed steel. Although the Ti content of this steel group is commonly very low, below 0.01 wt%, the high affinity of N and C to Ti, as well as the appreciable enrichment of Ti in the interdendritic regions during solidification, leads to formation of large Ti(N,C) particles of several μm size.

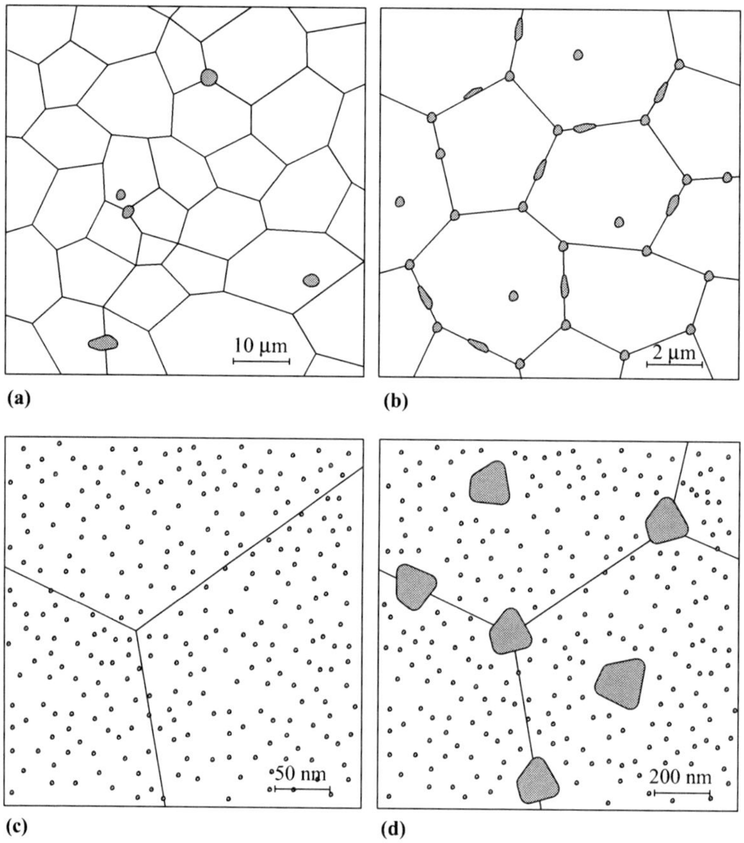

Figure 5.9. Morphology of precipitates for different applications in materials processing.

deplete the interior of grains from solute atoms, which is likely if the cooling rate is low, nucleation of precipitates inside the grain can be completely suppressed.[15]

On the other hand, precipitation can also occur on purpose as a consequence of controlled thermal and thermomechanical treatment. Depending on the desired precipitate morphology, which will define the final mechanical–technological properties, various types of precipitation heat treatments exist. Some of these involve plastic deformation; others are of purely thermal nature. The precipitate morphology shown in Figure 5.9(c) is a typical homogeneous distribution of a high number of fine particles to achieve maximum precipitation strengthening. Practically, this distribution can be achieved in alloys, which exhibit certain properties in terms of the phase diagram for the precipitation system. Typically, these systems are characterized by a one-phase region at high temperature where all elements connected to the precipitation process can be brought into homogeneous solid solution by annealing inside this region for sufficiently long time. A corresponding schematic binary A-B phase diagram is shown in Figure 5.10, where a B-rich phase β precipitates within an A-rich matrix α.

[15] Particles located at grain boundaries are often the reason for embrittlement which is why materials technologists commonly avoid this precipitate morphology in the final product.

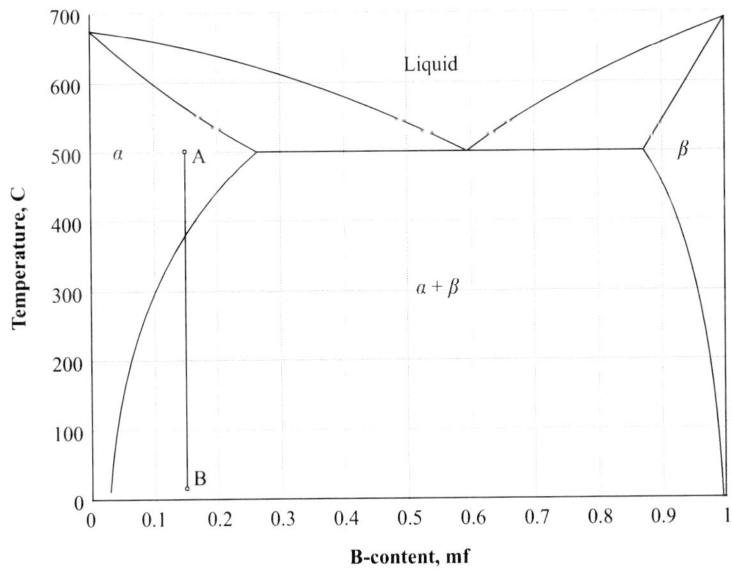

Figure 5.10. Archetype of a phase diagram for a precipitation system. After solution treatment at 500°C, point A, the alloy can be quenched to room temperature, point B, to bring the alloy into the supersaturated solid-solution condition.

At a hypothetical alloy composition of 0.15 mole fraction B, an alloy is heated up to a temperature of 500°C and held until a homogeneous solid solution (HSS) is established. Then, the alloy is rapidly cooled. After crossing the phase boundary between the α one-phase and $\alpha + \beta$ two-phase regions, the driving force for precipitation of β becomes positive. However, the quenching will be performed sufficiently fast, such that practically no long-range diffusion can occur to form precipitates of β. In essence, the B-atoms, which are equilibrated and homogeneously distributed in the α phase after solution treatment, become *frozen in* during the quench. The solution at room temperature practically inherits the distribution of B atoms from the solution treatment.

Precipitation heat treatments are typically realized in a sequence of three steps which are schematically depicted in Figure 5.11 and which are:

Step 1: Solution treatment
The alloy is heated up to a temperature that is above the solution temperature of the precipitate under consideration. Previously existing precipitates are dissolved and the elements constituting the precipitate are homogeneously distributed in the matrix. Typically, this step is carried out in the high-temperature one-phase region of the precipitation system. At the end of this step, the precipitation system is in the condition of a *homogeneous solid solution* (HSS).

Step 2: Quenching
From solution temperature, the alloy is quenched to a temperature, typically room temperature, where diffusion of elements is so slow that practically no precipitation can occur even after some time of rest before proceeding. The quench must be performed fast enough such

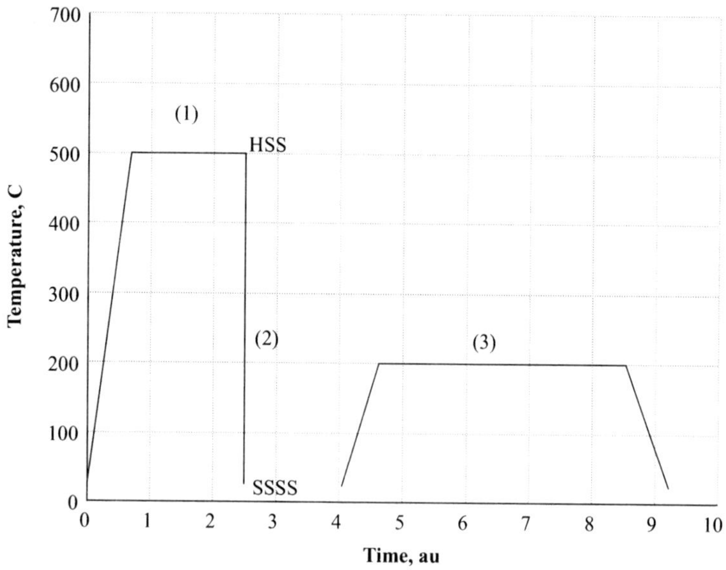

Figure 5.11. Classical heat treatment delivering a dense distribution of fine precipitates. Numbers in parenthesis correspond to the heat treatment steps described in the text.

that the high-temperature region below the $\alpha + \beta$ phase boundary is passed through safely avoiding formation of grain boundary as well as intragranular precipitates. As the solute atoms forming the precipitate are homogeneously distributed and the driving force for precipitation is at its maximum, this condition is denoted as supersaturated solid solution (SSSS).

In some alloy system, such as, for instance, Al, it is possible that a substantial amount of excess *quenched-in vacancies* is introduced into the material in this step. In this case, the acceleration effect of excess vacancies on all diffusive processes must be properly taken into account when performing simulations of the subsequent tempering step.[16]

Step 3: Tempering

In the final step of a typical heat-treatment procedure, the temperature is increased again to allow for accelerated diffusive processes and, thus, precipitation of a dense dispersion of small particles within reasonable times. The times are shorter, the higher the annealing temperature is chosen. However, increasing temperature simultaneously reduces the available chemical driving force and, thus, the nucleation rate. Usually, the tempering temperature is selected such that a compromise between short treatment time (T high) and maximum nucleation rate (T low) is found. At too low temperatures, the precipitate distribution will be very fine; however, annealing must be carried out for rather long times to support precipitation of sufficient amount of β to have optimum effect on mechanical properties. At too high temperature, the diffusion kinetics is fast; however, the nucleation rate decreases due to decreasing supersaturation and the precipitate distribution becomes increasingly coarse.

[16] See Section 7.4 for details on this topic.

Figure 5.9(d) sketches a typical multimodal size distribution of precipitates. These microstructures are typical for systems where the phase fraction of the precipitate phase is high, typically >30%. In the present case, there exist two distinct peaks in the size distribution at $r = 80\,\text{nm}$ and $r = 5\,\text{nm}$. This *bimodal* distribution is sometimes used in Ni-base alloy processing, where the larger particles are used for grain boundary pinning, while the small precipitates are responsible for precipitation strengthening. For the reason of making forming processes possible, the temperature must be raised so high that the smaller precipitates are all dissolved. Otherwise the Ni-base superalloy would simply be too hard to be deformed or it would fail in a brittle manner. Simultaneously, the larger precipitates must remain available for grain boundary pinning and grain size control. If the pinning effect of these particles gets lost, immediate grain growth would set in and the coarse grain microstructure would again become nondeformable.

The process of achieving this bimodal size distribution is commonly a two-step heat treatment, which is schematically shown in Figure 5.12. In the first part of the treatment, the alloy is heated up above the assumed solution temperature of the β phase, which is schematically shown in the phase fraction diagram Figure 5.13. Accordingly, the β phase dissolves entirely above 1100°C, which reasons a choice of $T_{sol} = 1.1150$°C. Afterward, the alloy is cooled to a temperature of $T_1 = 950$°C. From the phase fraction diagram, an equilibrium fraction of approximately 30% is read. As the alloy is cooled slowly to reaction temperature T_1, the driving force and the nucleation rate remain low. Consequently, only a small number of precipitates are formed, which can grow to substantial size, as the diffusional mobility of solute atoms is comparably high at this temperature. The reaction is stopped as soon as the phase fraction of β is close to its equilibrium value.

In the next step, the temperature is decreased to $T_2 = 750$°C. The cooling is done sufficiently fast such that no significant growth of the existing large β particles occurs. According to the phase fraction diagram, Figure 5.13, the equilibrium phase fraction at this temperature is approximately 45%, which makes a supersaturation of the system of about 15% phase fraction of β. This driving force might be sufficient to trigger an additional, second nucleation burst, this

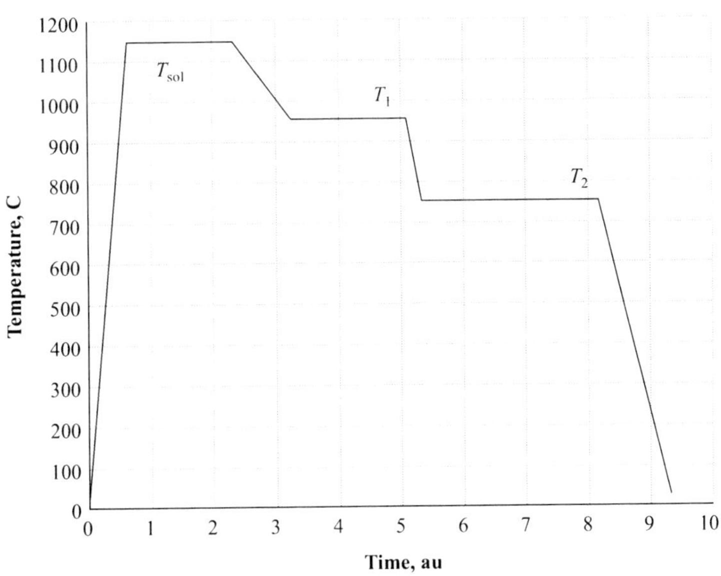

Figure 5.12. Stepwise heat treatment to create a bimodal size distribution of precipitates of the same type.

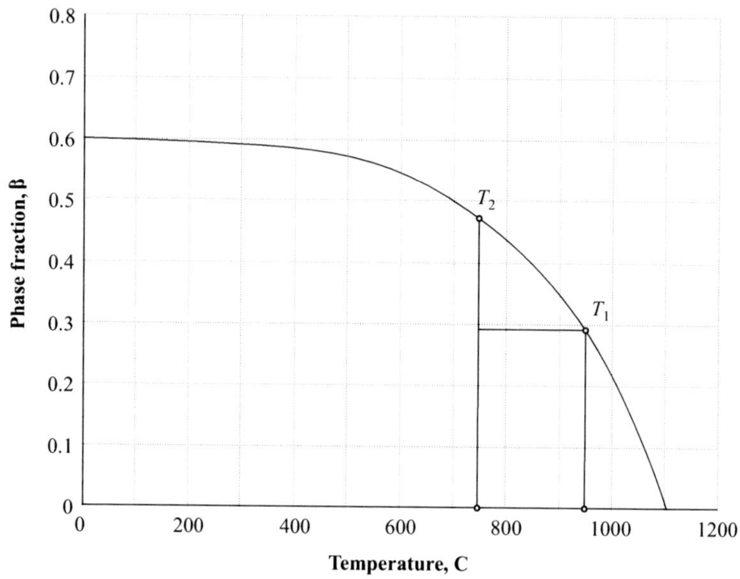

Figure 5.13. Equilibrium phase fraction diagram for the precipitate phase β for generation of multimodal size distributions.

time, however, at higher supersaturation due to faster cooling and at lower temperature, which brings about lower diffusional mobility and smaller diffusion fields around the particles. As a consequence, the nucleation rate is higher compared to the first nucleation wave and many new precipitates of β will form in between the corridors of the existing large precipitates from the first wave.

It is interesting to note, finally, that under suitable conditions, linear cooling can also produce the same effect of multimodal size distributions. Radis et al. [153] have investigated this issue experimentally as well as computationally in the Ni-base superalloy UDIMET 720 Li. The interested reader is pointed toward this publication.

5.4.2 Stages in Precipitate Life

The life of a solid-state precipitate starts with its birth process. Triggered by the precipitate nucleation event, the particle forms within the supersaturated solid solution. In Chapter 2, this topic is discussed in depth. We have found that nucleation of a second phase particle is driven by random compositional fluctuations. We have also concluded that it is impossible to predict the time and location of a single nucleation event by any means of physical model. However, the rate of nucleation can be formulated, provided that a sufficiently large ensemble of atoms is considered.

In Figure 5.14, typical snapshots are shown describing the individual stages of precipitate life. Consider a matrix α constituted by A atoms. Only the β precipitate-forming atoms B are displayed as spheres. In Figure 5.14(a), all B atoms are randomly distributed and this type of arrangement is denoted as *random solution*. It corresponds either to a situation where no chemical and mechanical interactions between A and B atoms exist, or where the sample is held at so high a temperature that entropy is entirely dominating the mixing process. Both situations are without practical interest because, in the first case, there will be no precipitate formed

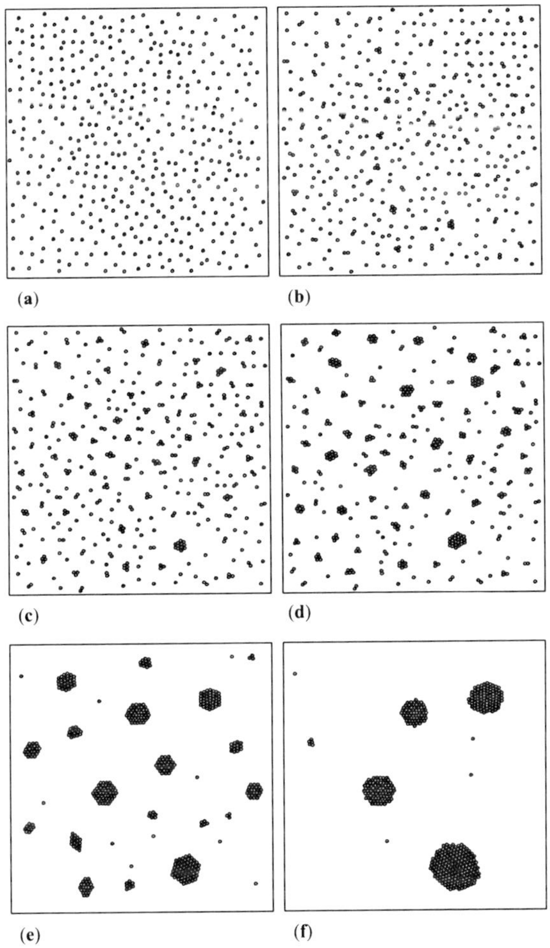

Figure 5.14. Stages in precipitation life.

after quenching due to complete miscibility of the two atomic species. In the second case, the temperature required for quasi-random mixing will be rather high and usually way above the melting point.

A typical snapshot of an *equilibrated solution* is shown in Figure 5.14(b) where several dimers, trimers, and multimers of B atoms are visible in the cloud of isolated B monomers. As already discussed in Section 2.3.4, a stationary size distribution is established after sufficiently long time annealing above the precipitate solution temperature. The shape of the size distribution is only dependent on the forward and backward rates of transition in size space. As long as the system is kept above the phase boundary, no stable nucleus can be established because the precipitates are still thermodynamically unstable.

Figure 5.14(c) represents a situation shortly after quenching into the α/β two-phase region. Due to positive driving force for precipitation below the phase boundary, precipitates reaching overcritical size are prone to stable continuous growth. This phenomenon is depicted schematically by the single larger particle in the lower right part of the figure. After some more time, several other nuclei have become overcritical (Figure 5.14[d]), while the existing nuclei have grown by continuous addition of other B atoms.

If the majority of nuclei have become overcritical, the process of growth controls the further life of the precipitates. The existing particles continue to increase their volume by incorporating more and more B atoms. The supersaturation of B decreases and nucleation is practically stopped. If all available B atoms are bound to the precipitates, the growth process is completed and the matrix surrounding the particles has reached a B concentration close to the equilibrium value, but not yet exactly. This situation is sketched in Figure 5.14(e) where most B atoms are tied up in the precipitates and only a small amount of B is still in solution. Remember that the exact amount of solute in the matrix is determined by the equilibrium value of the solubility corrected by the Gibbs–Thomson effect (see Section 1.5.1).

As the equilibrium value of the B solubility is not reached after growth, there exists a nonzero driving force to evolve the thermodynamic system to further decrease the total free energy. As discussed in Section 3.6, the Gibbs energy can be decreased by reducing the total precipitate–matrix interfacial area in the process of coarsening. In the course of Ostwald ripening, smaller particles with a high surface area to volume ratio dissolve and larger precipitates continue to grow. Thereby, the total interfacial area decreases with a simultaneous decrease of the total free energy. In Figure 5.14(f), only four larger particles have remained. As the interfacial area is reduced, simultaneously, the solubility of B atoms approaches its equilibrium value for the limit of no interfacial regions in the system.

5.4.3 Evolution of Precipitation Parameters

From the analysis of precipitate life as outlined in the previous section, three major stages in the evolution of a precipitate are identified:

- *Nucleation*: Starting in a supersaturated solid solution, overcritical nuclei of the new phase form. The nucleation process is of stochastic nature and it can only be quantitatively described on a statistical basis for large enough ensembles of atoms.
- *Growth*: Once a precipitate has reached overcritical size, it proceeds to growth by adsorption of additional precipitate-forming atoms. This process is deterministic in nature, which means that the evolution of every single precipitate can be unambiguously described and predicted by appropriate growth laws.
- *Coarsening*: Driven by the reduction of total interfacial area between precipitates and matrix, smaller precipitates dissolve and larger continue to grow. In the process of coarsening, the total number density of precipitates continuously decreases.

In the present section, a closer look is taken at the evolution of the main precipitation parameters: phase fraction, mean radius, and number density. Consider the simple case of a binary solution with a stoichiometric precipitate β. Let the specimen be solution-treated and quenched rapidly to a temperature, where the sample is held isothermally. In the supersaturated solid solution, precipitation of the β phase proceeds.

In Figure 5.15, the evolution of precipitation parameters is depicted exemplarily as obtained from a simulation for a precipitate constituted of almost pure B and a nominal B-composition of 0.03. Figure 5.15(a) shows the evolution of the phase fraction of precipitates, which is, qualitatively, similar to the phase transformation curves obtained with the KJMA model (Section 5.1.2). If the time axis is plotted in log scale, the phase fraction curve resembles an "S"-shape. Time is displayed in arbitrary units ("au"). The regions marked by vertical dashed and dotted lines are specified subsequently in the discussion of the other precipitation parameters.

While the phase fraction of the β precipitates continuously increases, the B-content of the matrix decreases. Figure 5.15b indicates that this process does not immediately bring the B-content to its full equilibrium value, but approaches it gradually after about 10^6 au of time. At this time, the matrix is in metastable equilibrium with 10^{24} precipitates with a mean radius of approx. 2 nm, which can be read from Figure 5.15(c) and (d). Due to the effect of interfacial curvature (Gibbs–Thomson effect), the matrix B content is higher compared to the case without interfaces. The steady decrease of the matrix B content proceeds with the progress of Ostwald ripening, leading to a continuous reduction of total interfacial area.

From Figure 5.15(d), we distinguish the region where nucleation occurs by dotted vertical lines. After about 10^3 au of time, the nucleation rate, designated by the dashed curve, increases sharply and reaches its maximum value between 10^4 and 10^5 au of time. Since the time-dependent

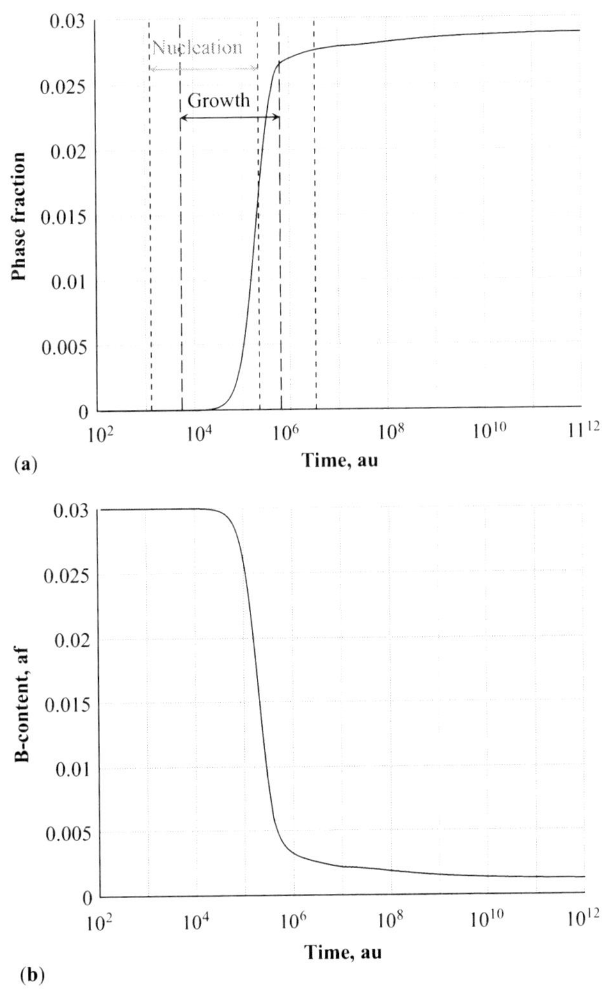

Figure 5.15. Evolution of precipitation parameters: Phase fraction and matrix B-content in atomic fraction; mean radius/critical radius and number density/nucleation rate. For position of vertical lines, see also plots (c) and (d). (*Continued*)

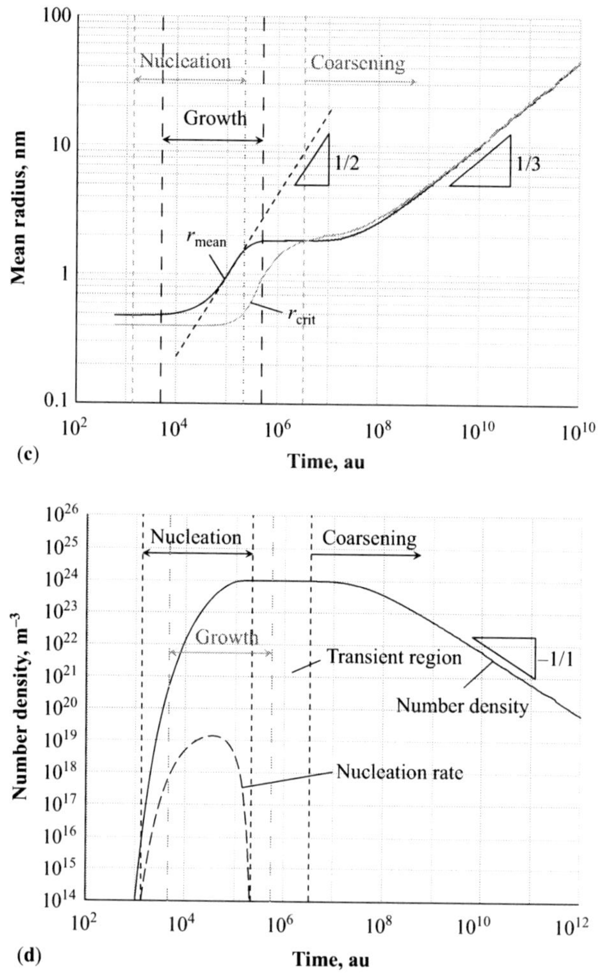

Figure 5.15. Evolution of precipitation parameters: Phase fraction and matrix B-content in atomic fraction; mean radius/ critical radius and number density/nucleation rate. For position of vertical lines, see also plots (c) and (d).

form of the nucleation rate expression, equation (2.90), is used in the simulations, the left shoulder of the nucleation rate curve is related to the transient nature of nucleation, which is associated with an incubation period[17] before, eventually, steady state nucleation is achieved. At around 10^5 au of time, Figure 5.15(b) shows a pronounced decrease of matrix supersaturation, which leads to a reduction of precipitation driving force and, consequently, leads to a cease of further nucleation. This observation is responsible for the steeply falling right shoulder of the nucleation rate curve.

The region associated with precipitate growth is identified from the mean radius plot in Figure 5.15(c). This image simultaneously displays the evolution of the critical radius, r^*. At early times, the critical radius is slightly below the mean radius, which is, however, rooted in the practical implementation of the nucleation process in the simulation algorithm. New

[17] See Section 2.3.4.

precipitates are added to the system with slightly overcritical radius, as particles with exactly the critical size have zero growth rates. While the mean radius increases due to precipitate growth, the critical radius initially remains almost constant. With decreasing supersaturation, r^* increases and crosses the mean radius line between 10^6 and 10^7 au of time. This might be seen as a first indication for the onset of Ostwald ripening. Another indication can be observed in the number density plot, Figure 5.15(d), where the beginning of Ostwald ripening can be related to the first dissolving precipitates.

An interesting region of precipitation is marked in Figure 5.15(d) as the transient region. In this period of time, the phase fraction, number density, as well as the mean radius remain almost constant. However, the supersaturation continues to decrease in this stage and the growth regime proceeds to the coarsening stage. Depending on the character of the transformation, this region can be very short or rather long. This will be discussed in the following section.

Finally, Figure 5.15 schematically shows triangles indicating the slopes of the mean radius and number density curves during growth and coarsening. In the growth stage, the tangent drawn to the mean radius curve has a slope of approximately 1/2. This is an indication for the parabolic kinetics of diffusion-controlled growth observed in this stage with a time dependence of $t^{\frac{1}{2}}$. In the coarsening stage, the kinetics of precipitation obey the time dependence of the LSW law which predicts that the mean radius of the size distribution evolves with $t^{1/3}$, while the number density decreases with t^{-1}. The characteristic features of the three stages in precipitate life are summarized briefly in Table 5.2.

Although each of these stages shows unique characteristics, in real precipitation reactions these stages exhibit more or less overlap or separation. It is shown, in the next section, that the nature of the entire precipitation reaction is controlled by the interplay of interfacial energy, driving force for precipitation, and phase fraction of the precipitate phase.

5.4.4 Overlap of Nucleation, Growth, and Coarsening

Consider the precipitation of a phase β in a supersaturated matrix α. Let us assume that the available driving force for the reaction is constant. For these conditions, investigate the progress of the phase transformation for different values of the interfacial energy, γ. Figure 5.15 summarizes typical results for the main precipitation parameters phase fraction, B-content of the matrix, mean radius, and number density of precipitates. Interestingly, the ratio between supersaturation/driving force and interfacial energy entirely determines the character of the transformation. For one limiting case, where the interfacial energy is high compared to the

Table 5.2. Evolution of precipitation parameters in different stages of precipitate life.

Stage	**Nature of region**	**Number density**	**Mean radius**
Nucleation	stochastic	++ new particles continuously nucleate	+ existing nuclei start growing
Growth	deterministic	o constant	++ continuous growth, parabolic with $t^{1/2}$
Coarsening	deterministic	– decreases with t^{-1}	+ increases slowly (LSW) with $t^{1/3}$

value of the driving force, the processes of nucleation, growth, and coarsening widely overlap, whereas for relatively high values of γ, these stages are clearly separated in time.

In Figure 5.16(a), the evolution of the phase fraction of β in the course of the precipitation reaction is displayed. For high values of γ, the phase fraction curves adopt the characteristic "S"-shape observed already earlier. With decreasing interfacial energy, the reaction stagnates after a steep increase and approaches the equilibrium phase fraction value at a significantly smaller rate. Figure 5.16(b), displaying the supersaturation of the thermodynamic system in terms of the B-content of the matrix, shows equivalent characteristics.

The reason for this behavior is found in Figures 5.16(c) and (d) showing the evolution of mean radius and number density. With decreasing interfacial energy, the reaction starts earlier with lower mean radius and higher number density of precipitates. As the radius of the particles

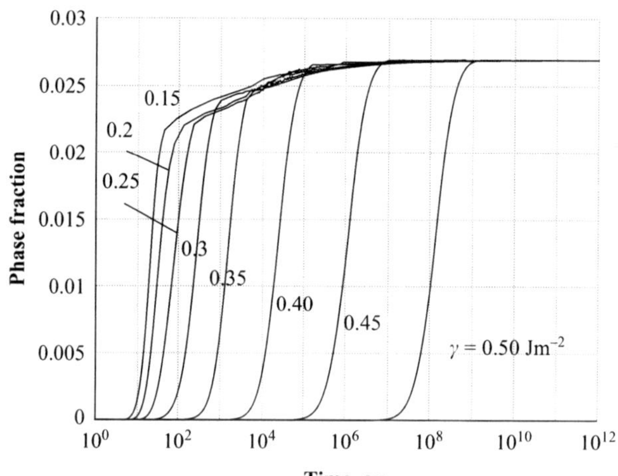

(a)

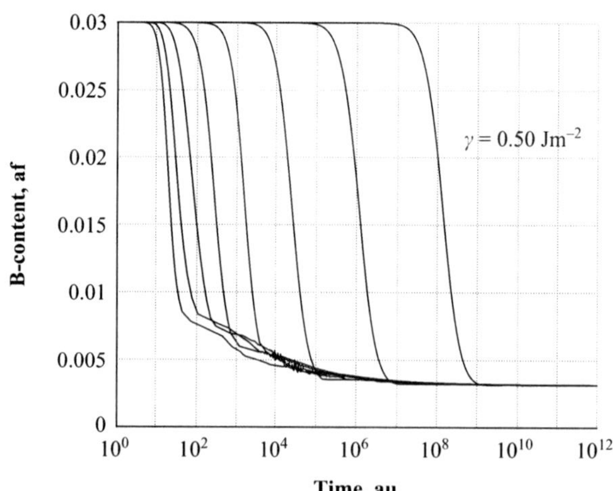

(b)

Figure 5.16. Evolution of precipitation parameters as a function of interfacial energy with constant driving force for precipitation (14 kJ/mol): Phase fraction and supersaturation; mean radius and number density. (*Continued*)

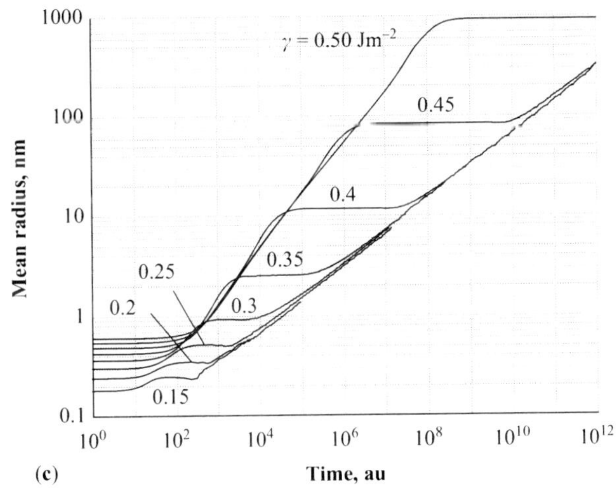

(c)

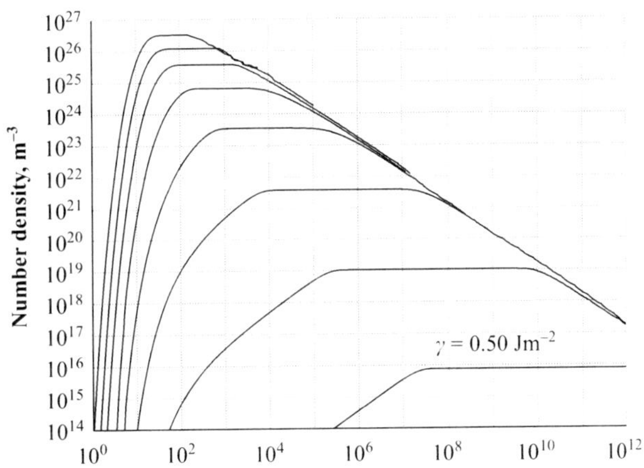

(d)

Figure 5.16. Evolution of precipitation parameters as a function of interfacial energy with constant driving force for precipitation (14 kJ/mol): Phase fraction and supersaturation; mean radius and number density.

and the corresponding diffusion fields are becoming smaller with lower interfacial energy, the reaction proceeds increasingly fast. However, with smaller precipitates, the Gibbs–Thomson effect becomes more prominent and the matrix content of B remains at a higher level. The reaction, thus, slows down earlier and the predominant physical mechanism evolving the system shifts from nucleation and subsequent growth to an overlapping reaction of nucleation, growth, and coarsening occurring at the same time. The evolution of the β phase fraction clearly reflects this behavior in the left shoulder of the phase fraction curve, which reaches the equilibrium value only progressively with increasing mean precipitate radius.

The variation of the mean radius of the precipitate populations and the number density of precipitates with interfacial energy is shown in Figures 5.16(c) and (d). With decreasing interfacial energy, the number density of precipitates increases sharply and the mean radius remains at a rather low level, while the width of the transient region of the precipitation reaction becomes

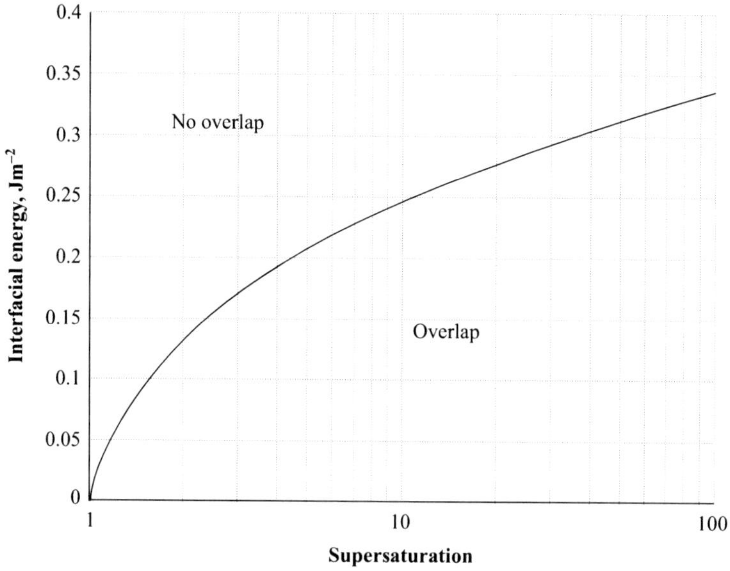

Figure 5.17 Regions with and without overlap of nucleation, growth, and coarsening according to Robson [154].

narrower with lower ratio of interfacial energy to driving force. At the lowest values, the nucleation and growth regions progress into the coarsening regime almost seamlessly and the plateau in the mean radius evolution becomes progressively smaller.

The plots of precipitation parameters with variation in γ demonstrate the importance of the interfacial energy in modeling precipitation. The major influence of this parameter is manifested in the nucleation rate expression, equation (2.76), where the interfacial energy appears in the exponent of the Boltzmann term in cubed form. In the present example, a variation of γ between 0.15 and 0.5 J/m^2 leads to precipitate densities between 10^{16} and 10^{27} particles/m^3. In the time evolution of the precipitate phase fraction, this variation causes a time shift over more than eight orders of magnitude. The mean radius of the precipitates adopts values between 1000 μm and less than 1 nm. Using a value of 0.6 J/m^2 entirely suppresses the precipitation reaction.[18]

Interestingly, the evolution of mean radius and number density for interfacial energy values below approximately 0.25 J/m^2 do not fall on the curves for the evolution of these parameters with higher interface energy values. The reason is that for low values of the interfacial energy, the nucleation, growth, and coarsening processes overlap and nucleation proceeds even if the first precipitates already dissolve in the Ostwald ripening process. As nucleation even occurs in the coarsening regime, the evolution of precipitation parameters deviates from the kinetics predicted by LSW theory.

[18] This behavior is the main reason, why the interfacial energy is a most popular fitting parameter. Small variations in this quantity can "switch on" or "switch off" the precipitation reaction and easily reproduce any intermediate state. In Chapter 4 of this book, a method is suggested for calculation of this quantity based on thermodynamic information.

For a thorough discussion of the interplay of nucleation, growth, and coarsening in precipitation reactions, the interested reader is referred to a paper by J. D. Robson [154]. This author has investigated this issue with the Numerical Kampmann–Wagner model. One of the main results of this work was the identification of a combination of parameters driving force and interfacial energy, where overlap of nucleation and coarsening occurs or not. The corresponding plot is reproduced in Figure 5.17.

In Figure 5.17, supersaturation is defined as the ratio of actual solute content to the corresponding equilibrium value. For small values of the interfacial energy, already small values of the supersaturation lead to an overlap of nucleation, growth, and coarsening. With increasing value of the interfacial energy, the supersaturation required to produce this overlap increases simultaneously.

5.5 Summary

The three most widely used approaches in precipitation modeling are the Kolmogorov–Johnson–Mehl–Avrami (KJMA), the Langer–Schwartz (LS), and the numerical Kampmann–Wagner (NKW) model.

The KJMA model is based on the extended volume approach and it dates back to the early 1940s. It basically solves the problem of describing the overall phase transformation kinetics in terms of a single transformation parameter located between 0 (no transformation) and 1 (transformation is completed). In the simplest version assuming constant nucleation and growth rates, the KJMA equation represents an exponential function with a kinetic coefficient and a time exponent. The KJMA approach has also been developed for simultaneous precipitation of multiple phases. The huge advantage of this model is its simplicity and computational efficiency: The evolution equations can be solved analytically and the final expression can be evaluated with a pocket calculator. However, as the two parameters of the model, the Avrami coefficient and the Avrami exponent, more or less represent fitting parameters, only limited introspection into the physical mechanisms governing the phase transformation is possible.

The LS model follows an entirely different modeling strategy. On assuming a simplified precipitate size distribution and introducing several simplifying assumptions, Langer and Schwartz developed a model that is capable of describing the evolution of the mean radius and the number density of precipitates based on suitable nucleation and growth rate expressions. The model allows for a detailed investigation of these important precipitation parameters as a function of alloy thermodynamics and diffusional mobility data.

The NKW model is the most detailed and versatile numerical precipitation kinetics approach. The evolution equations are derived from nucleation and growth rate expressions and, in contrast to the LS model, the size distribution of precipitates is represented in a discrete way. For each of the precipitate size classes, the growth conditions are evaluated and new precipitates are added to the precipitation system according to CNT. The NKW model is the approach with the highest degree of detail and possible introspection. It is most widely implemented in modern precipitation kinetics software.

CHAPTER 6

HETEROGENEOUS PRECIPITATION

In this chapter, models are introduced and discussed, which are useful in precipitation modeling and can, in one or the other, be coupled with the approaches discussed in the previous chapters. Many of the traditional models have been developed under simplifying geometrical assumptions for the benefit of simplicity. With the theory and concepts of this chapter, these models can be advanced and extended for situations where the original concepts do not apply.

6.1 Precipitation at Grain Boundaries

Nucleation at dislocations, grain boundaries, or other microstructural heterogeneities (see Section 2.4) is often preferred over homogeneous nucleation for particles with high volumetric misfit or crystallographic incommensurability with the parent lattice. Grain boundaries and grain boundary junctions provide energetically advantageous sites for precipitation where nucleation is substantially facilitated due to a reduction of grain boundary area[1] and the fact that grain boundaries can be seen as locally amorphous structures where crystal match is not a requirement for easy nucleation.

6.1.1 Problem Formulation

When precipitates grow at grain boundaries, the diffusion fields surrounding them are fundamentally different from the diffusion field around precipitates in an isotropic matrix. While the latter involves approximately spherical diffusion geometry, grain boundary precipitation involves both, fast grain boundary diffusion toward and away from the precipitate along the grain boundary, as well as slow bulk diffusion from the inside of the grain toward the grain boundary. The two diffusion geometries involved in random precipitation of spherical precipitates and grain boundary precipitation are shown schematically in Figure 6.1.

In the case of random precipitation, Figure 6.1(a), the diffusion fields are approximately spherical. The extension of the diffusion field is determined by the radially symmetrical composition profile (see Section 3.2.2) and mass conservation. In the case of grain boundary

[1] A theoretical model developed for this phenomenon by Clemm and Fisher is described in Section 2.4.4.

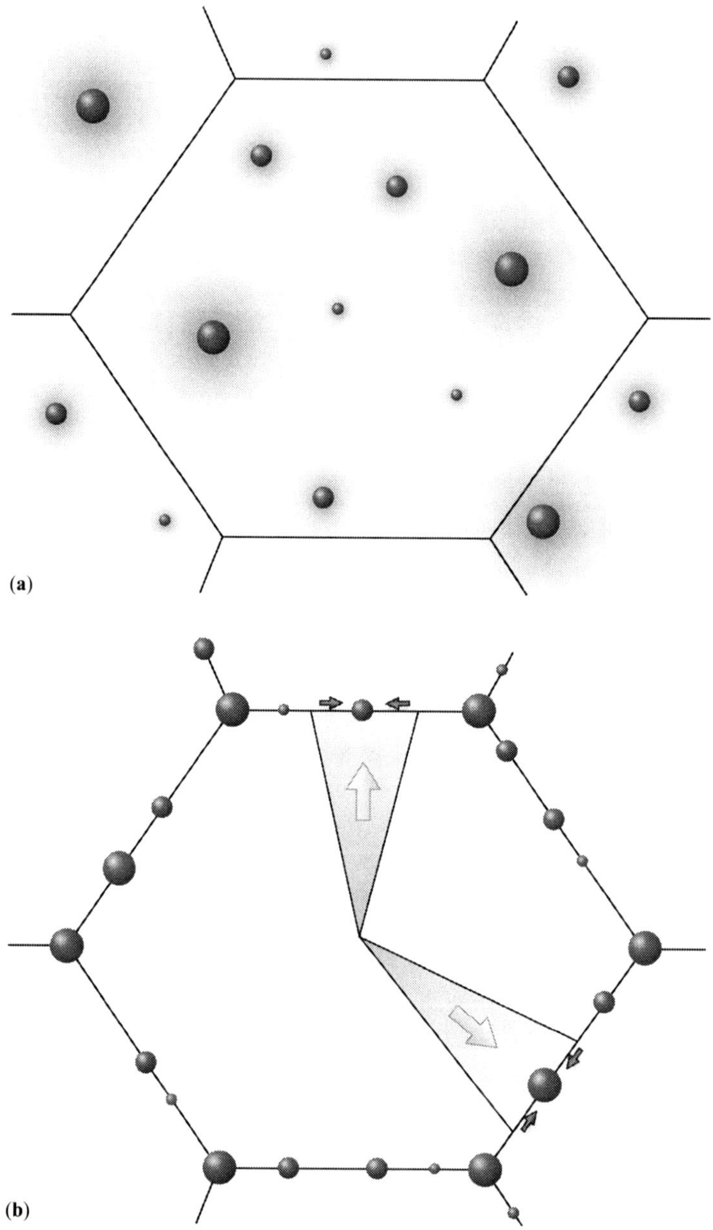

Figure 6.1. Schematic representation of the diffusion field around randomly nucleated spherical precipitates in an isotropic matrix (a) and diffusion fields for grain boundary precipitates (b).

precipitation, the diffusion field is nonsymmetrical because the grain boundary area acts as a high-velocity diffusion path where the mobility of atoms is significantly higher than the diffusional mobility in the undisturbed crystal bulk. This is indicated in Figure 6.1(b) by arrows, where the gray scale indicates fast and slow diffusion (darker color is faster diffusion).

Several researchers have studied the problem of grain boundary precipitation and various models exist in literature that deal with this issue. For the kinetics of nucleation see, for instance, the review of Lange, Enomoto, and Aaronson [155]. Interesting in the present context is the work of Aaron and Aaronson [156] who investigated the growth kinetics of precipitates in Al–4wt%Cu alloys at grain boundaries. These authors derived a model for the grain boundary precipitation process in planar geometry which was shortly after extended by Brailsford and Aaron [157].

The planar geometry is a reasonably good approximation for the early stages of the precipitation process. Aaron and Aaronson conclude, based on the analysis of experimental data using their model, that bulk diffusion of Cu through the Al crystal would give orders of magnitude too low for growth rates. Only when taking into account the "*collector-plate*" function of the grain boundaries, the growth rates can be reasonably well reproduced.

According to their model, the entire grain boundary is split into representative rectangular areas containing a single precipitate in their center. Each of these regions acts as a collector plate that can rapidly transport atoms moving into the collector-plate area to the precipitate absorbing these atoms. Diffusion inside the grain boundary is assumed to be significantly faster than the diffusion of atoms through the bulk. A corresponding sketch is shown in Figure 6.2.

The original model by Aaron and Aaronson [156] has been derived for planar geometry and a single diffusing species. Kozeschnik et al. [158] have introduced a model solving the problem for spherical geometry and general multicomponent systems using the thermodynamic extremal principle for derivation of suitable evolution equations.[2] In this model, the diffusion geometry is approximated by a spherical grain and a conical diffusion field extending from the grain boundary to the grain center. The cap of the cone represents the collector plate. The simplified geometry is shown in Figure 6.3 where the grain boundary precipitates are evenly distributed on the grain boundary area. The grayed region I depicts the grain volume, where the precipitate diffusion field extends and solute is collected or deposited for precipitate growth. The grain boundary area representing the collector plate is designated as region II.

Precipitate growth at a grain boundary is, initially, governed by the diffusion kinetics along the grain boundary area and diffusion into the grain boundary from the bulk crystal involving a thin region close to the grain boundary. The latter process slows down rather quickly, however, as the diffusion field extending into the grain interior becomes rapidly wider (parabolic growth law) and diffusion in the bulk is slow. Although the diffusive transport of solute atoms through the bulk volume is strictly limiting the growth of grain boundary precipitates, the particles are still interconnected with each other along the grain boundary. Diffusive "communication" of each precipitate with all others can easily occur along this short-circuit diffusion path and, although no growth can occur by this mechanism, the particles are subject to Ostwald ripening. Consequently, two mechanisms that determine the precipitate evolution at the grain boundary must be considered:

i. Growth of precipitates governed by diffusion inside the grain boundary (collector plate) and long-range diffusion of solute from the grain interior to the grain boundary.
ii. Coarsening of precipitates governed by fast diffusion along the grain boundaries.

Both mechanisms require separate treatment, which is elaborated subsequently.

[2]The thermodynamic extremal principle is described in Section 3.5.1. The method for deriving evolution equations for precipitate growth is outlined in Section 3.5.2.

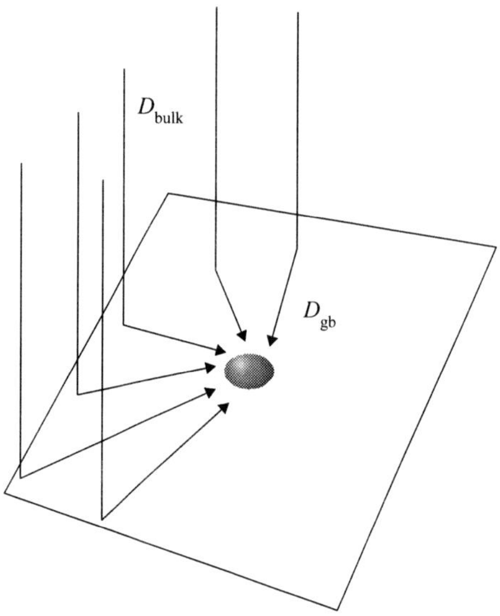

Figure 6.2. Sketch of the collector-plate mechanism suggested by Aaron and Aaronson [156]. Atoms arrive at the grain boundary at a velocity determined by the bulk diffusion coefficient, D_{bulk} (vertical lines). From there, they are transported toward the precipitate in the center at a rate determined by D_{gb}.

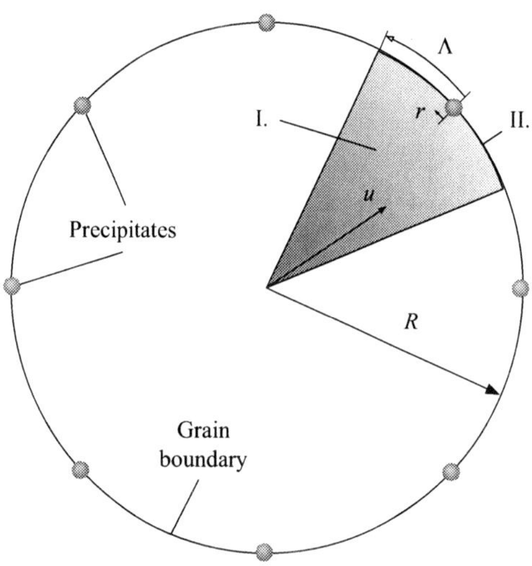

Figure 6.3. Schematic representation of the diffusion field assigned to a single grain boundary precipitate. The region designated as I is the diffusional zone in the bulk volume of the grain. Region II is the corresponding grain boundary area.

6.1.2 Diffusive Processes

As a starting point for the development of the grain boundary diffusion model, Kozeschnik et al. [158] assume that the diffusion zone of the collector plate is represented by a spherical cap with arc length 2Λ as shown in Figure 6.3. If the grain radius, R, is sufficiently large compared to Λ, the cap can be approximated by a circular disc and Λ is related to the number N of precipitates located at the grain boundary with

$$\pi\Lambda^2 N = 4\pi R^2, \tag{6.1}$$

and

$$\Lambda = \frac{2R}{\sqrt{N}}. \tag{6.2}$$

The precipitates are assumed to be stoichiometric and no diffusive flux is assumed to occur inside the precipitates.[3]

For the flux of component i outside the precipitate, the exchange of matter of a precipitate with the surface area $4\pi\rho_k^2$ is assumed to occur through a cylindrical mantle with radius $2\pi\rho_k\delta$, with δ being the thickness of the grain boundary. As the precipitate radius varies in time as ρ_k, mass balance at the interface requires

$$2\pi\rho_k\delta \cdot j_{ki}\left(\rho_k\right) = \dot{\rho}_k 4\pi\rho_k^2\left(c_{\mathrm{M}i}^{\mathrm{eq}} - c_{\mathrm{P}i}\right), \tag{6.3}$$

where $c_{\mathrm{M}i}^{\mathrm{eq}}$ is the equilibrium concentration of element i in the matrix and $c_{\mathrm{P}i}$ is the concentration of element i in the precipitate. For the radial diffusive flux, $j_{ki}(r)$, and the polar coordinate, r, defined in Figure 6.3, one gets

$$j_{ki}\left(r\right) = 2\dot{\rho}_k^2\left(c_{\mathrm{M}i}^{\mathrm{eq}} - c_{\mathrm{P}i}\right)\frac{\Lambda^2 - r^2}{\Lambda^2 - \rho_k^2} \cdot \frac{\rho_k^2}{r\delta}. \tag{6.4}$$

In analogy to the treatment outlined in Section 3.5, the dissipation due to diffusion in the grain boundary zone is given by

$$Q_{\mathrm{g}} = \frac{1}{2}\sum_{k=1}^{m_{\mathrm{g}}}\sum_{i=1}^{n}\int_{\rho_k}^{\Lambda}\frac{RT}{c_{\mathrm{M}i}^{\mathrm{eq}}D_{\mathrm{g}i}}\, j_{ki}^2\, 2\pi r\delta \cdot \mathrm{d}r, \tag{6.5}$$

where m_{g} is the number of precipitates located in one grain and n is the number of components. The factor 1/2 accounts for the fact that each precipitate corresponds to two grains. $D_{\mathrm{g}i}$ is the diffusion coefficient of component i inside the grain boundary. If the equilibrium concentration,[4] $c_{\mathrm{M}i}^{\mathrm{eq}}$, inside the grain is known, equation (6.5) can be solved and we get

$$Q_{\mathrm{g}} \approx 4\pi RT\sum_{k=1}^{m_{\mathrm{g}}}\alpha_k\rho_k^4\dot{\rho}_k^2\sum_{i=1}^{n}\frac{\left(c_{\mathrm{P}i} - c_{\mathrm{M}i}^{\mathrm{eq}}\right)^2}{c_{\mathrm{M}i}^{\mathrm{eq}}D_{\mathrm{g}i}}, \tag{6.6}$$

with the coefficient α_k given as

$$\alpha_k\left(\Lambda\right) = \frac{\left[\ln\left(\frac{\Lambda}{\rho_k}\right) - \frac{3}{4} + \left(\frac{\rho_k}{\Lambda}\right)^2\left(1 - \left(\frac{\rho_k}{\Lambda}\right)^2\right)\right]}{\left(1 - \left(\frac{\rho_k}{\Lambda}\right)^2\right)^2}. \tag{6.7}$$

[3]This assumption is a simplification compared to the model presented in Section 3.5. Solution of the coupled problem of simultaneous diffusion also inside the precipitate is rather involved and has not yet been carried out on a rigorous basis.

[4]The equilibrium concentration inside the grain is the same as the equilibrium concentration in the grain boundary as the present model neglects possible effects of grain boundary segregation.

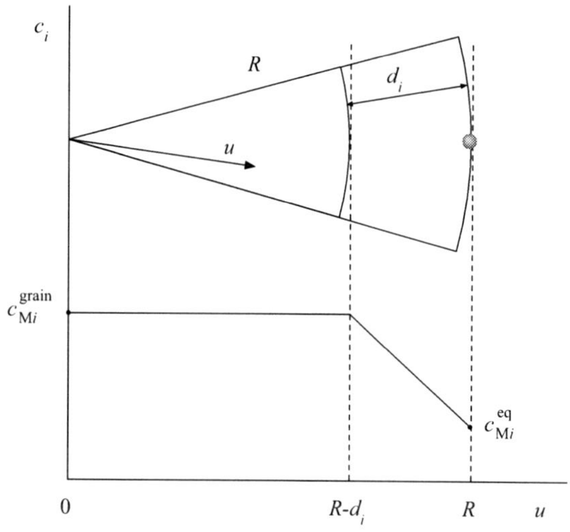

Figure 6.4. Simplified concentration profile for grain boundary precipitation in the model of Kozeschnik et al. [158].

With the dissipation term corresponding to the grain boundary as given by equation (6.6), in the next step, the dissipation of free energy inside the grain is considered. In order to solve this problem, a simplified composition profile is assumed in this model as shown in Figure 6.4. The initial composition is given by c_{Mi}^{grain} and the equilibrium composition is c_{Mi}^{eq}. As diffusion inside the grain boundary is fast compared to the bulk volume, the equilibrium value inside the grain boundary is reached instantaneously. In the course of precipitate growth, solute atoms from inside the grain are deposited inside the collector-plate area and a composition gradient develops. In the model of Kozeschnik et al. [158], this profile is approximated by a linear gradient in the interval $R - d_i \leq u \leq R$. The variable u is the distance from the center of the grain and d_i is the width of the interval (Figure 6.4) as

$$c_i = \frac{u + d_i - R}{d_i} c_{Mi}^{eq} + \frac{R - u}{d_i} c_{Mi}^{grain}. \tag{6.8}$$

The value of d_i is given by mass conservation with

$$\frac{1}{2} V \left(c_{Pi} - c_{Mi}^{eq} \right) = 4\pi \int_{R-d_i}^{R} \left(c_{Mi}^{grain} - c_i \right) u^2 \mathrm{d}u. \tag{6.9}$$

Performing the integral delivers a cubic equation for d_i with the solution

$$d_i = \frac{\beta_i}{3} - \frac{2R^2}{3\beta_i} + \frac{4R}{3}, \tag{6.10}$$

with the coefficients β_i given as

$$\beta_i = \frac{\sqrt[3]{\left[-44R^3 \left(c_{Mi}^{eq} - c_{Mi}^{grain} \right) - 162K_i + 18L_i \right] \cdot \left(c_{Mi}^{eq} - c_{Mi}^{grain} \right)^2}}{c_{Mi}^{eq} - c_{Mi}^{grain}}, \tag{6.11}$$

and

$$K_i = \frac{V}{8\pi}\left(c_{\mathrm{P}i} - c_{\mathrm{M}i}^{\mathrm{eq}}\right), \tag{6.12}$$

and

$$L_i = \sqrt{6R^6\left(c_{\mathrm{M}i}^{\mathrm{eq}} - c_{\mathrm{M}i}^{\mathrm{grain}}\right)^2 + 44R^3K_i\left(c_{\mathrm{M}i}^{\mathrm{eq}} - c_{\mathrm{M}i}^{\mathrm{grain}}\right) + 81K_i^2}\,. \tag{6.13}$$

For further details on the derivation procedure, the interested reader is referred to the original publication [158].

For the composition profile shown in Figure 6.4, the value of the diffusional flux, J_{ki}^*, at position $u = R$ is found as

$$\pi\Lambda^2 J_{ki}^* = -\frac{1}{2}4\pi\rho_k^2\dot{\rho}_k\left(c_{\mathrm{P}i} - c_{\mathrm{M}i}^{\mathrm{eq}}\right). \tag{6.14}$$

The flux, J_{ki}, in the region $R - d_i \le u \le R$ is approximated by the linear relation

$$J_{ki} = -\frac{2\rho_k^2\dot{\rho}_k\left(c_{\mathrm{P}i} - c_{\mathrm{M}i}^{\mathrm{eq}}\right)\left(u - R + d_i\right)}{\Lambda^2 d_i}. \tag{6.15}$$

The total dissipation due to diffusion in the grain volume is then given as

$$Q_{\mathrm{b}} = \sum_{k=1}^{m_{\mathrm{g}}}\sum_{i=1}^{n}\int_{R-d_i}^{R}\frac{RT}{c_i D_{\mathrm{b}i}}j_{ki}^2\frac{\pi\Lambda^2u^2}{R^2}\cdot\mathrm{d}u. \tag{6.16}$$

Entering equations (6.8) and (6.15) delivers

$$Q_{\mathrm{b}} = \sum_{k=1}^{m_{\mathrm{g}}}\sum_{i=1}^{n}\frac{\pi RT\rho_k^4\dot{\rho}_k^2 d_i\left(c_{\mathrm{P}i} - c_{\mathrm{M}i}^{\mathrm{eq}}\right)^2}{3D_{\mathrm{b}i}\Lambda^2R^2\left(c_{\mathrm{M}i}^{\mathrm{eq}} - c_{\mathrm{M}i}^{\mathrm{grain}}\right)}\Gamma_i, \tag{6.17}$$

with

$$\Gamma_i = \sum_{l=0}^{4}\Gamma_{il}\left(c_{\mathrm{M}i}^{\mathrm{grain}}\right)^{4-l}\left(c_{\mathrm{M}i}^{\mathrm{eq}}\right)^{l}. \tag{6.18}$$

The coefficients Γ_{il} and additional mass balance considerations for practical evaluation of the model parameters are given in Appendix C of ref. [158].

6.1.3 Evolution Equations for Precipitate Growth

For a rigorous evaluation of the total Gibbs energy and the corresponding driving forces in the system, the chemical potential profiles in the matrix and the respective integrals must be solved. In a first approximation, the total Gibbs energy, G, is expressed in terms of the mean concentrations, $c_{\mathrm{M}i}$, which are evaluated from the global mass balance as

$$\frac{4\pi}{3}R^3\left(c_{\mathrm{M}i}^{\mathrm{grain}} - c_{\mathrm{M}i}\right) = \frac{2\pi}{3}\sum_{k=1}^{m_{\mathrm{g}}}\rho_k^3\left(c_{\mathrm{P}i} - c_{\mathrm{M}i}\right), \tag{6.19}$$

delivering
$$c_{\mathrm{M}i} = \frac{2R^3 c_{\mathrm{M}i}^{\mathrm{grain}} - \sum_{k=1}^{m_g} \rho_k^3 c_{\mathrm{P}i}}{2R^3 - \sum_{k=1}^{m_g} \rho_k^3} \approx c_{\mathrm{M}i}^{\mathrm{grain}} + \left(c_{\mathrm{M}i}^{\mathrm{grain}} - c_{\mathrm{P}i}\right) \cdot \sum_{k=1}^{m_g} \frac{\rho_k^3}{2R^3}. \tag{6.20}$$

With m_i representing the number of moles of component i in the system, one obtains

$$\frac{4\pi}{3}\left(R^3 - \frac{1}{2}\sum_{k=1}^{m_g} \rho_k^3\right) c_{\mathrm{M}i} = m_i - \frac{2\pi}{3} c_{\mathrm{P}i} \sum_{k=1}^{m_g} \rho_k^3. \tag{6.21}$$

The total Gibbs energy, G, then follows as

$$G = \frac{2\pi g_{\mathrm{P}}}{3v^{\beta}} \sum_{k=1}^{m_g} \rho_k^3 + 2\pi\gamma \sum_{k=1}^{m_g} \rho_k^2 + \frac{4\pi}{3}\left(R^3 - \frac{1}{2}\sum_{k=1}^{m_g} \rho_k^3\right) \sum_{i=1}^{n} c_{\mathrm{M}i}\mu_i, \tag{6.22}$$

with g_{P} being the molar Gibbs energy of the precipitate, v^{β} being the molar volume of the precipitate, γ being the precipitate–matrix interfacial energy, and μ_i being the chemical potential of component i expressed in terms of $c_{\mathrm{M}i}$.

On substituting equation (6.21) into (6.22), the Gibbs energy is obtained as

$$G = \frac{2\pi g_{\mathrm{P}}}{3v^{\beta}} \sum_{k=1}^{m_g} \rho_k^3 + 2\pi\gamma \sum_{k=1}^{m_g} \rho_k^2 + \sum_{i=1}^{n}\left(m_i - \frac{2\pi}{3} c_{\mathrm{P}i}\right)\mu_i. \tag{6.23}$$

Based on the formalism derived from the thermodynamic extremal principle, Section 3.5.1, the evolution equations for the radius of the precipitate k can be obtained from

$$\frac{1}{2}\frac{\partial Q_g}{\partial \dot{\rho}_k} = -\frac{\partial G}{\partial \rho_k}, \quad k = 1, \ldots m_g, \tag{6.24}$$

and for the growth rate contribution from precipitate growth, $\dot{\rho}_k^{\mathrm{g}}$, we finally obtain

$$\dot{\rho}_k^{\mathrm{g}} = \frac{\sum_{i=1}^{n} c_{\mathrm{P}i}\mu_i - \frac{g_{\mathrm{P}}}{v^{\beta}} - \frac{2\gamma}{\rho_k}}{2RT\rho_k^2\left\{\alpha_k \sum_{i=1}^{n} \frac{\left(c_{\mathrm{P}i} - c_{\mathrm{M}i}^{\mathrm{eq}}\right)^2}{c_{\mathrm{M}i}^{\mathrm{eq}}\delta D_{gi}} + \frac{1}{12\Lambda^2 R^2}\sum_{i=1}^{n} \frac{d_i\Gamma_i\left(c_{\mathrm{P}i} - c_{\mathrm{M}i}^{\mathrm{eq}}\right)^2}{D_{\mathrm{b}i}\left(c_{\mathrm{M}i}^{\mathrm{eq}} - c_{\mathrm{M}i}^{\mathrm{grain}}\right)^5}\right\}}. \tag{6.25}$$

6.1.4 Evolution Equations for Precipitate Coarsening

For the precipitates located at grain boundaries, it is assumed that they communicate via diffusion along the grain boundaries, independent of the growth contribution derived in the previous section. For the coarsening process, the same diffusion geometry is assumed to be valid as used for precipitate growth, with the constraints

$$\frac{4\pi}{3}\sum_{k=1}^{m_g} \rho_k^3 = const., \tag{6.26}$$

or

$$4\pi \sum_{k=1}^{m_g} \rho_k^2 \dot{\rho}_k = 0. \tag{6.27}$$

In the evaluation of the dissipation term, it is assumed that the width of the collector plate is 2Λ instead of Λ to ensure sufficient overlap of the diffusion zones. With the coefficient $\beta_k = \alpha_k(2\Lambda)$, the dissipation in the grain boundary is found to be

$$Q_c \approx 4\pi RT \sum_{k=1}^{m_g} \beta_k \rho_k^4 \dot{\rho}_k^2 \sum_{i=1}^{n} \frac{\left(c_{Pi} - c_{Mi}^{eq}\right)^2}{c_{Mi}^{eq} \delta D_{gi}}. \tag{6.28}$$

During coarsening, only the second term in equation (6.23) contributes to the driving force and the governing equation is

$$\frac{1}{2}\frac{\partial Q_g}{\partial \dot{\rho}_k} + \lambda \cdot \left(4\pi\rho_k^2\right) = -\frac{\partial G}{\partial \rho_k}, \qquad k = 1, \ldots m_g, \tag{6.29}$$

with λ being a Lagrange multiplier. Partial differentiation of Q_c and G with equation (6.29) yields

$$\dot{\rho}_k^c = \frac{-\gamma - \lambda\rho_k}{RT\rho_k^3 \beta_k \sum_{i=1}^{n} \frac{\left(c_{Pi} - c_{Mi}^{eq}\right)^2}{c_{Mi}^{eq}\delta D_{gi}}}, \tag{6.30}$$

and the Lagrange multiplier, λ, follows from insertion of equation (6.30) into (6.27) as

$$\lambda = -\gamma \frac{\sum_{l=1}^{m_g} \left(\rho_l \beta_l\right)^{-1}}{\sum_{l=1}^{m_g} \left(\beta_l\right)^{-1}}. \tag{6.31}$$

Equation (6.30) can finally be rewritten as

$$\dot{\rho}_k^c = \frac{\gamma \left[-\frac{1}{\rho_k} + \frac{\sum_{l=1}^{m_g} \left(\rho_l \beta_l\right)^{-1}}{\sum_{l=1}^{m_g} \left(\beta_l\right)^{-1}} \right]}{RT\rho_k^2 \beta_k \sum_{i=1}^{n} \frac{\left(c_{Pi} - c_{Mi}^{eq}\right)^2}{c_{Mi}^{eq}\delta D_{gi}}}. \tag{6.32}$$

Both processes, growth and coarsening, are assumed to occur independently and the total rate of change in the size of each precipitate is given as

$$\dot{\rho}_k = \dot{\rho}_k^{\mathrm{g}} + \dot{\rho}_k^{\mathrm{c}}. \tag{6.33}$$

6.1.5 Growth Kinetics of Equisized Precipitates

The model developed in the previous sections is implemented in the software *MatCalc* (see Section 9.3). For an analysis of the behavior of the grain boundary precipitation model in comparison with the classical models involving spherical diffusion fields, the same simulation setup is used as before in Section 5.4.4, where the overlap of nucleation, growth, and coarsening has been investigated. To ensure a clear separation of the three precipitation stages, a constant interfacial energy value is used with $\gamma = 0.3\ J/m^2$. The grain boundary diffusion coefficient is assumed to be 10^4 times that of the bulk volume.

Figure 6.5 shows the evolution of phase fraction and radius of grain boundary precipitates over time (in arbitrary units). The four solid lines are evaluated for grain sizes (diameter) between $R = 1$ to 1000 μm, the dashed line represents the same calculation for spherical diffusion fields of randomly distributed precipitates. In the simulation, 10^{18} particles/m^3 are considered as being of identical size and identical chemical composition. At the beginning of the simulation, the precipitates are distributed over the grain boundary nucleation sites and the growth kinetics are evaluated based on equation (6.33). As no further nucleation is allowed to occur in this parameter study and all precipitates are identical in size and composition, this situation corresponds to a single particle growing in a periodic spatial arrangement. The simulation is completed after the matrix becomes depleted of solute atoms. As all precipitates have identical size, there is no driving force present for Ostwald ripening.

The evolution of the precipitate radius in Figure 6.5(b) exhibits several interesting features. For identical setup and considering the early stages, growth of grain boundary precipitates starts significantly earlier compared to the growth of randomly distributed precipitates with spherical diffusion fields. This observation is easily reasoned on the basis that the initial growth of grain boundary precipitates is governed by rapid grain boundary diffusion and the collector-plate mechanism that efficiently "collects" solute atoms from the close vicinity of the grain boundary.

After the immediate volume close to the grain boundary becomes depleted, the growth rate decreases because the diffusion distances from the grain interior to the grain boundary becomes increasingly wide and further growth is controlled by slow bulk diffusion. This effect is clearly visible for the largest grain sizes, where the radius versus time curve shows a bend after which the curves continue with different slope. The curves for smaller grain size show this effect at higher values of the radius, because more grain boundary are is available to support the collector-plate mechanism. Consequently, the fast growth regime can continue even at later stages.

When comparing the grain boundary precipitation kinetics to that of randomly distributed particles, the growth rates are generally lower, even in the regime where the collector-plate mechanism is most effective. For larger grain size, this effect becomes rather prominent and the growth rates are substantially lower. This observation is clearly visible also in Figure 6.5(a), showing the phase fraction versus time curves for the same set of simulations. Interestingly, the curve for the smallest grain size is rather similar in shape compared to the curve for random distribution. The main difference is a shift in time.

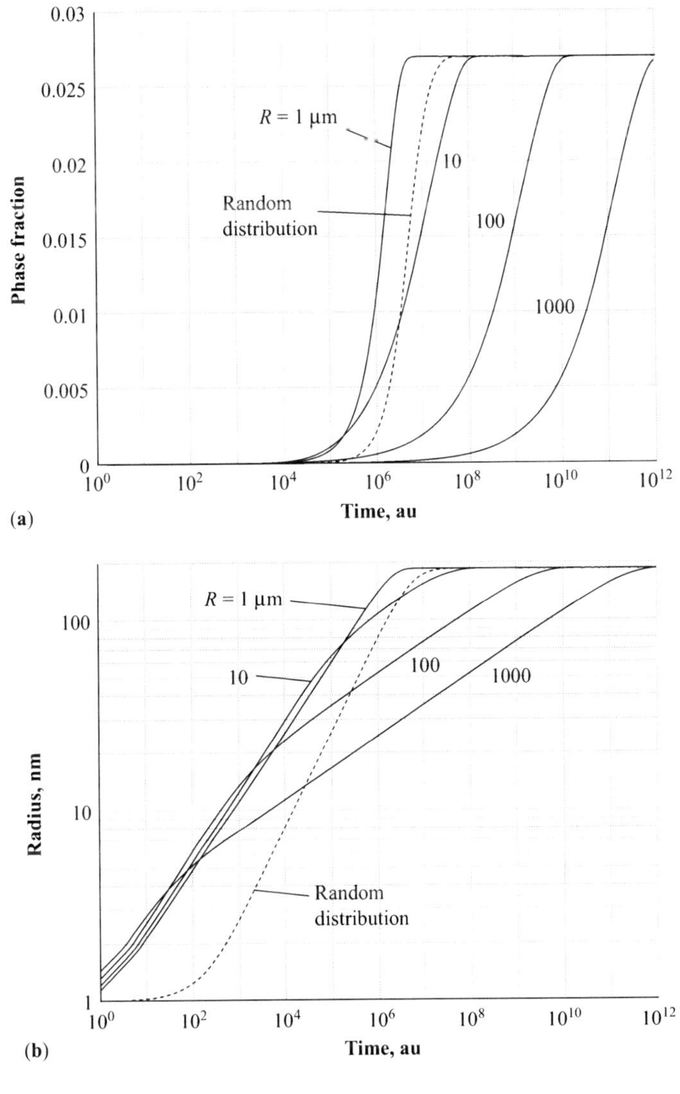

Figure 6.5. Evolution of phase fraction (a) and radius (b) for grain boundary precipitates compared to a random precipitate distribution with spherical diffusion fields.

6.1.6 Growth Kinetics of Nonequisized Precipitates

For an analysis of the evolution of an ensemble of nonequisized precipitates, the evolution equation (6.33) is applied together with a suitable nucleation model.[5] In ref. [158],

[5]Nucleation at grain boundary requires consideration of the particular diffusion geometry, which is different from randomly distributed precipitates. Nucleation at grain boundaries occurs in a rather small volume around the grain boundary. The compositional fluctuations occur mainly inside the grain boundary. As soon as a nucleus has formed, the depletion of this volume and the decrease of driving force must be taken into account separately from the depletion of the bulk volume inside the grain.

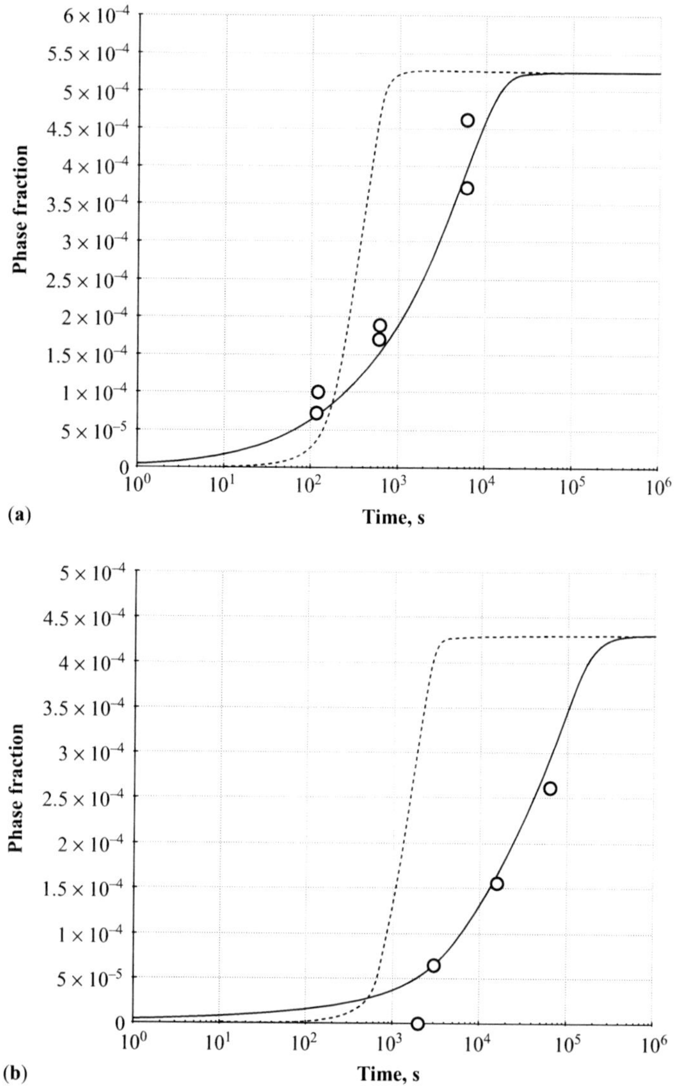

Figure 6.6. Evolution of phase fraction during grain boundary precipitation (solid lines) and comparison with experimental data [160] for AlN precipitation in microalloyed steel. (a) 0.058 wt% Al, 0.0058 wt% N, 927°C. (b) 0.079 wt% Al, 0.0072 wt% N, 982°C. Dashed line shows simulation for random precipitate distribution with spherical diffusion fields. Austenite grain diameters of 100 μm are assumed for alloy (a) and 50 μm for alloy (b).

Kozeschnik et al. apply their model to the case of AlN precipitation in microalloyed steel. This model system is adopted here because AlN precipitation in austenitic steel occurs predominantly at grain boundaries [159] due to considerable volumetric misfit of AlN compared to the Fe matrix.[6]

[6]Compare Section 2.4 on heterogeneous nucleation.

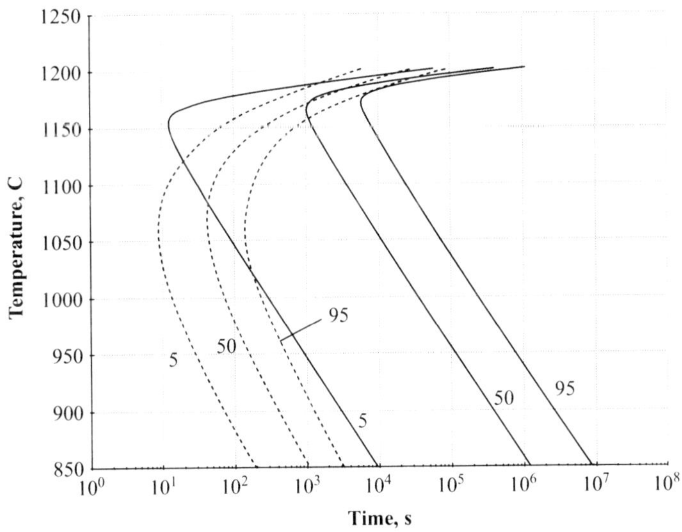

Figure 6.7. Calculated time–temperature–precipitation diagram for AlN in microalloyed steel. Solid lines are for grain boundary precipitation and dashed lines for random precipitation with spherical diffusion fields. The numbers represent the progress of transformation in percentage.

Figure 6.6 shows a comparison of simulation with experiment. The solid lines in this figure represent results of the precipitation simulation with the model for grain boundary geometry. The open circles are experimental phase fraction data from Leslie et al. [160] for alloys that have been solution treated for 15 minutes at 1260°C and then held at test temperature. Apparently, the shape of the phase fraction curve is very well reproduced by the grain boundary precipitation model, whereas the kinetics predicted for spherical diffusion geometry are clearly too fast and cannot describe the experimentally observed kinetics.

With the calibration for AlN precipitation in steel as performed in refs. [158,161], a time–temperature–precipitation (TTP) diagram is calculated and compared to the equivalent diagram for spherical diffusion geometry in Figure 6.7. The wider spacing of the 5, 50, and 95% phase fraction lines reflects the smaller growth gradients of grain boundary precipitates. At lower temperature, random precipitation succeeds grain boundary precipitation for the present setting of 10^{12} m^{-2} dislocation density and 100μm grain diameter. Interesting is the difference in shape. The nose temperature for grain boundary precipitation is almost 100 K above the nose temperature for spherical diffusion fields.

6.1.7 Coarsening Kinetics

In the formulation of the present grain boundary precipitation model, the growth and coarsening processes are considered explicitly in the evolution equations. This assumption is based on the rapid diffusional communication between precipitates, which are located on the grain boundary network. The coarsening behavior of grain boundary precipitates has been investigated by numerous researchers, for example, [162–164]. Exemplarily, the treatment of Ardell [162] is taken and compared to the results of the numerical grain boundary precipitation model, equation (6.33). For convenience, the functions describing the stationary size distribution during coarsening as obtained by Lifshitz, Slyozov and Wagner (LSW), equation (3.207), as well as Ardell, equation (24) in ref. [162], is reproduced below. Accordingly, the size distributions read

$$f_{\mathrm{LSW}}(r,t)=\frac{4}{9}\xi^2\left(\frac{3}{3+\xi}\right)^{\frac{7}{3}}\left(\frac{\frac{3}{2}}{\frac{3}{2}-\xi}\right)^{\frac{11}{3}}\exp\left(-\frac{\xi}{\frac{3}{2}-\xi}\right);\ \xi\leq\frac{3}{2}, \tag{6.34}$$

and

$$f_{\mathrm{Ardell}}(r,t)=\left(\frac{3}{4}\right)^4\xi^3\left(1-\frac{3\xi}{4}\right)^{-\frac{19}{6}}\left(\frac{3\xi^2}{16}+\frac{\xi}{2}+1\right)^{-\frac{23}{12}}\cdot$$

$$\exp\left(\frac{1}{2}-\frac{\frac{2}{3}}{\frac{4}{3}-\xi}-\frac{1}{6\sqrt{2}}\cdot\left(\tan^{-1}\left(\frac{1+\frac{3\xi}{4}}{\sqrt{2}}\right)-\tan^{-1}\left(\frac{1}{\sqrt{2}}\right)\right)\right);\ \xi\leq\frac{4}{3}. \tag{6.35}$$

In Figure 6.8, a representative size distribution as obtained from the numerical simulation with *MatCalc*, using the evolution equations derived in this section, is compared to an analytical solution provided by Ardell and the classical LSW model. Apparently, the numerical

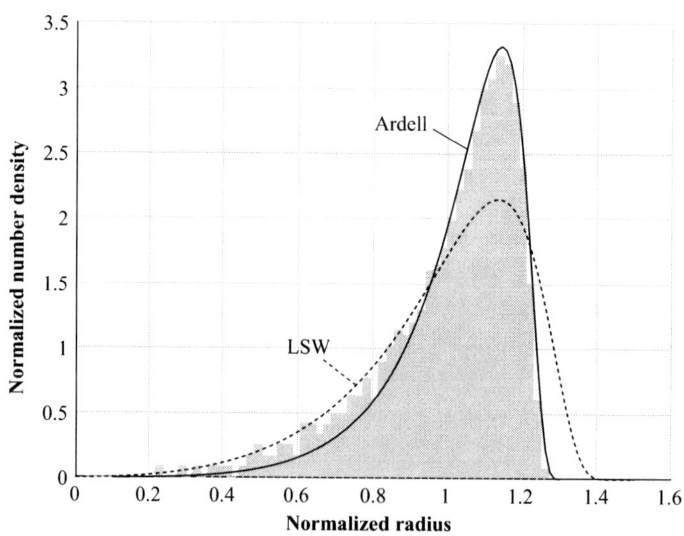

Figure 6.8. Comparison of size distributions as obtained with the numerical grain boundary precipitation model, equation (6.33), depicted as grey bars, the model of Ardell [162] (solid line) for grain boundary precipitation and the LSW model (dashed line) for random precipitation with spherical diffusion fields.

simulation is in good accordance with Ardells results. The LSW size distribution is comparably wider and has a less pronounced maximum.

In their analysis, Kozeschnik et al. [158] emphasize that the size distribution during coarsening is not a fully stationary function, as the geometric situation that applies to the evolution equations, is permanently changing due to the constant number of grains but decreasing number of precipitates. In their simulations, they find that good agreement between the numerically obtained and Ardell's size distribution is only achieved over two orders of magnitude in time. Outside this region, the distribution is wider with a less pronounced peak.

A major difference between random and grain boundary precipitation is also observed in the time dependence of the precipitation parameters. Whereas the mean radius of the LSW size distribution evolves with time as $t^{1/3}$, the dependence for grain boundary precipitates observed in the simulations is between $t^{1/4.2}$ and $t^{1/4}$. This is in good accordance with the theoretical value of $t^{1/4}$, which is predicted for purely grain boundary diffusion controlled coarsening. The time evolution of the number density is observed with approximately $t^{-1/1.4}$ compared to t^{-1} in the LSW model.

6.2 Anisotropy and Precipitate Shape

In practically all examples of diffusion-controlled phase transformations and precipitation reactions in this book, the shape of the product phase/precipitate has been approximated by either spherical or planar geometry. With this simplification, closed analytical expressions could be obtained describing the various reactions. In this section, the diffusion field around nonspherical precipitates is investigated in more detail. In the framework of the SFFK model for the evolution of precipitate radius and composition (see Section 3.5.2), correction functions have been developed [165], which allow for a convenient treatment of growth and composition evolution of elongated and disc-like precipitates, too.

6.2.1 Shape Parameter, *h*, and SFFK Evolution Equations

Prolate (cigar) or oblate (disc) shaped precipitates of different height to diameter ratio can conveniently be approximated by a family of isovolumetric cylinders as indicated in Figure 6.9. With the shape parameter, *h*, defined as

$$h = \frac{H}{D}, \tag{6.36}$$

where H is the height of the cylinder and D is its diameter, only a single parameter is needed to define the shape of the nonspherical precipitate. With this definition, small values of h represent discs, whereas large values represent needles. In the following treatment, h is assumed to remain constant. In practical simulation, h is considered as an input quantity that comes, for example, from experiment.

Starting point of the treatment is the SFFK model formulation for spherical precipitates, which is outlined in Section 3.5.2. Accordingly, the Gibbs energy of the precipitation system, G, and the dissipative processes, Q_1, Q_2, and Q_3, are formulated and linked together utilizing the thermodynamic extremal principle (TEP). Into the corresponding equations, a priori unknown

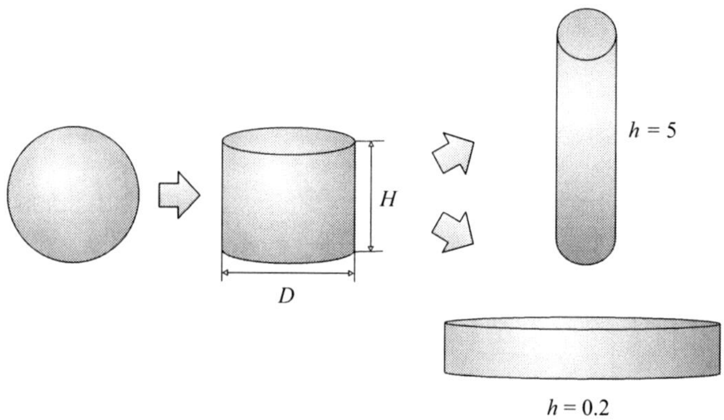

Figure 6.9. Definition of the shape parameter $h = H/D$.

shape factors, S_k, K_k, I_k, and O_k, are introduced, first, the value of which will be fixed in a second step. The value k denotes the index of a single particle in the ensemble of m precipitates.

Equations (3.104), (3.105), (3.108), and (3.114) are reformulated as

$$G = \sum_{i=1}^{n} N_{0i}\mu_{0i} + \sum_{k=1}^{m} \frac{4\pi\rho_k^3}{3}\left(\lambda_k + \sum_{i=1}^{n} c_{ki}\mu_{ki}\right) + \sum_{k=1}^{m} 4\pi S_k \rho_k^2 \gamma_k, \tag{6.37}$$

$$Q_1 = \sum_{k=1}^{m} \frac{4\pi K_k \rho_k^2}{M_k^{\mathrm{IF}}} \dot{\rho}_k^2, \tag{6.38}$$

$$Q_2 = \sum_{k=1}^{m} \sum_{i=1}^{n} \frac{4\pi I_k RT \rho_k^5}{45 c_{ki} D_{ki}} \dot{c}_{ki}^2, \tag{6.39}$$

$$Q_3 = \sum_{k=1}^{m} \sum_{i=1}^{n} \frac{4\pi O_k RT \rho_k^3 \left(\dot{\rho}_k \left(c_{ki} - c_{0i}\right) + \rho_k \dot{c}_{ki}/3\right)^2}{c_{0i} D_{0i}}. \tag{6.40}$$

The symbols are described in Section 3.5.2. Following the procedure outline there, the coefficients of the linear system of equations representing the evolution of the radius and composition of each precipitate are then obtained as

$$A_{11} = \frac{K_k}{M_k^{\mathrm{IF}}} + O_k RT \rho_k \sum_{i=1}^{n} \frac{\left(c_{ki} - c_{0i}\right)^2}{c_{0i} D_{0i}}, \tag{6.41}$$

$$A_{1i} = A_{i1} = \frac{O_k RT \rho_k^2}{3}\left(\frac{c_{ki} - c_{0i}}{c_{0i} D_{0i}} - \frac{c_{k1} - c_{01}}{c_{01} D_{01}}\right), \quad i = 2, \ldots, s, \tag{6.42}$$

$$A_{1i} = A_{i1} = \frac{O_k RT \rho_k^2}{3}\left(\frac{c_{ki+1} - c_{0i+1}}{c_{0i+1} D_{0i+1}} - \frac{c_{ks+1} - c_{0s+1}}{c_{0s+1} D_{0s+1}}\right), \quad i = s+1, \ldots, n-1, \tag{6.43}$$

$$A_{ij} = \frac{RT\rho_k^3}{45}\left[\left(\frac{I_k}{c_{ki}D_{ki}} + \frac{5O_k}{c_{0i}D_{0i}}\right)\delta_{ij} + \left(\frac{I_k}{c_{k1}D_{k1}} + \frac{5O_k}{c_{01}D_{01}}\right)\right], \quad i = 2,\ldots,s, j = 2,\ldots,s, \tag{6.44}$$

$$A_{ij} = \frac{RT\rho_k^3}{45}\left[\left(\frac{I_k}{c_{ki+1}D_{ki+1}} + \frac{5O_k}{c_{0i+1}D_{0i+1}}\right)\delta_{ij} + \left(\frac{I_k}{c_{ks+1}D_{ks+1}} + \frac{5O_k}{c_{0s+1}D_{0s+1}}\right)\right], \tag{6.45}$$

$$i = s+1,\ldots,n-1, \quad j = s+1,\ldots,n-1$$

$$A_{ij} = A_{ji} = 0, \quad i = 2,\ldots,s, \quad j = s+1,\ldots,n-1. \tag{6.46}$$

The symbol δ_{ij} stands for the Kronecker delta function. The right hand side of the equation system is given by

$$B_1 = -\frac{2S_k\gamma_k}{\rho_k} - \lambda_k + \frac{d_{k,\text{chem}}^{\beta}}{v^{\alpha}}, \tag{6.47}$$

$$B_i = -\frac{\rho_k}{3}\left(\mu_{ki} - \mu_{0i} - \mu_{k1} + \mu_{01}\right), \quad i = 2,\ldots,s, \tag{6.48}$$

$$B_i = -\frac{\rho_k}{3}\left(\mu_{ki+1} - \mu_{0i+1} - \mu_{ks+1} + \mu_{0s+1}\right), \quad i = s+1,\ldots,n-1. \tag{6.49}$$

With the evolution equations given, the functional dependence of the shape factors is elaborated next section.

6.2.2 Determination of Shape Factors

The shape factor, S_k, relates the surface area of the sphere to that of an isovolumetric cylinder. Setting the diameter of the precipitate to unity, the volume of the cylinder is $\pi h/4$, and its surface is $S_P = \pi(h + 1/2)$. A sphere of the same volume has a radius $\rho_k = (3h/16)^{1/3}$ and a surface $S_S = 4\pi(3h/16)^{2/3}$. The shape factor for the surface thus becomes

$$S_k = \frac{S_P}{S_S} = 0.7631 \cdot h^{\frac{1}{3}} + 0.3816 \cdot h^{-\frac{2}{3}}. \tag{6.50}$$

The shape factor, K_k, related to interface migration is obtained after assuming that the interface velocities $v_z = Ah/2$ for $z = h/2$, $v_z = -Ah/2$ for $z = -h/2$, and $v_r = A/2$ for $r = 1/2$. A designates a proper constant. This ensures that h remains constant during evolution of the precipitate. The dissipation due to interface migration corresponding to the cylindrical precipitate is then given as

$$Q_{1P} = \frac{1}{M_k}\left(\frac{\pi}{2}\left(\frac{Ah}{2}\right)^2 + \pi h\left(\frac{A}{2}\right)^2\right) = \frac{A^2\pi}{8M_k}\left(h^2 + 2h\right). \tag{6.51}$$

The rate of volume change for the precipitate based on these interface velocities is $3\pi Ah/4$. The interface velocity of the spherical precipitate of the same volume is $\dot{\rho}_k = A(3h/16)^{1/3}$. The dissipation due to interface migration of the spherical precipitate is

$$Q_{\mathrm{IS}} = \frac{4\pi\rho_k^2\dot{\rho}_k^2}{M_k} = \frac{4\pi A^2}{M_k}\left(\frac{3h}{16}\right)^{\frac{4}{3}}. \tag{6.52}$$

The shape factor related to interface migration is then

$$K_k = \frac{Q_{\mathrm{IP}}}{Q_{\mathrm{IS}}} = 0.2912 \cdot h^{\frac{2}{3}} + 0.5824 \cdot h^{-\frac{1}{3}}. \tag{6.53}$$

For the shape factors related to diffusion inside and outside the precipitate, the diffusive fluxes are determined numerically employing standard finite element software. On comparison of the numerical values with the known solution of the spherical case, the two missing shape factors are finally observed with

$$I_k \approx 0.4239 \cdot h^{\frac{4}{3}} + 0.6453 \cdot h^{-\frac{2}{3}}, \tag{6.54}$$

and

$$O_k \approx 1.0692 \cdot h^{\frac{2}{3}}. \tag{6.55}$$

The dependence of the four shape factors from the shape parameter, h, is shown in Figure 6.10. Whereas S_k, K_k, and I_k show a minimum in the vicinity of the spherical shape, the shape factor, O_k, varies continuously from low values for disc shape to high values for needle shape precipitates. This has interesting consequences if the growth rate of precipitates is evaluated as a function of precipitate shape, which is carried out in the next section.

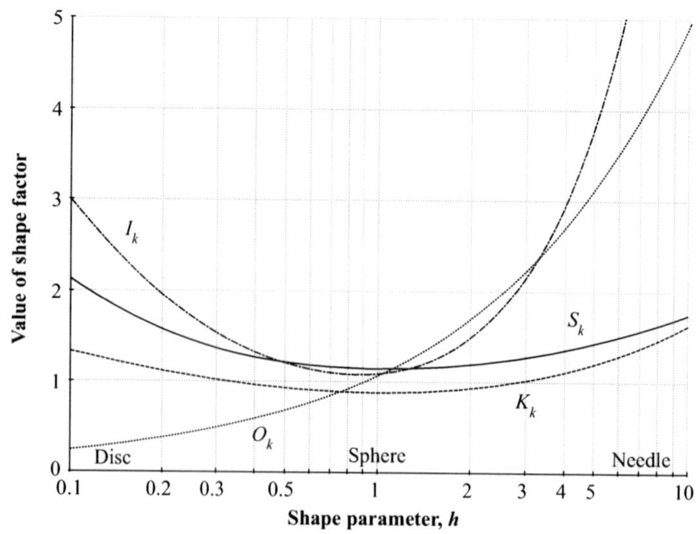

Figure 6.10. Shape factors as a function of the shape parameter, h.

6.2.3 Comparing Growth Kinetics

For the simplified case of precipitate growth with infinitely high interface mobility, M_k, which corresponds to a purely diffusion-controlled situation, ignoring mechanical stresses and a stoichiometric precipitate, the evolution of the precipitate radius, $\dot{\rho}_k$, is given by

$$O_k RT\rho_k \sum_{i=1}^{n} \frac{(c_{ki} - c_{0i})^2}{c_{0i} D_{0i}} \cdot \dot{\rho}_k = -\frac{2S_k \gamma_k}{\rho_k} + \frac{d^{\beta}_{k,\text{chem}}}{v^{\alpha}}. \tag{6.56}$$

Under these conditions, the growth rate is only dependent on the two shape factors, S_k and O_k. For a given set of values for driving force and interfacial energy, it is thus possible to directly compare the growth rates of precipitates with different shape parameter, h. In ref. [165], this task has been performed for $d^{\beta}_{k,\text{chem}} / v^{\alpha} = 10^8$ J/m^3 and $\gamma_k = 0.15$ J/m^2. These results are reproduced in Figure 6.11.

The solid line represents the scaled growth rate for a precipitate, which is large enough such that the influence of interfacial area on the growth rate becomes negligible. Under these circumstances, the growth rate continuously increases as the precipitate shape goes from needle to sphere to disc. The results obtained here thus indicate that the geometrical situation for diffusion-controlled growth is optimum for a disc. Obviously, the disc shape "collects" solute atoms most efficiently. The needle shape, in contrast, is least favorable in terms of equivalent volume increase. This result is due to the influence of the shape factor, O_k, which continuously increases with increasing shape parameter, h.

Another interesting feature is also shown in Figure 6.11 with regard to the influence of precipitate size. Apparently, the nonspherical shapes become increasingly unfavorable with decreasing precipitate size. The growth rates for the smallest particles are only positive in the close vicinity of $h = 1$. This is an indication that precipitates prefer spherical shape in the nucleation stage where the size of the precipitates is in the order of nm and less.

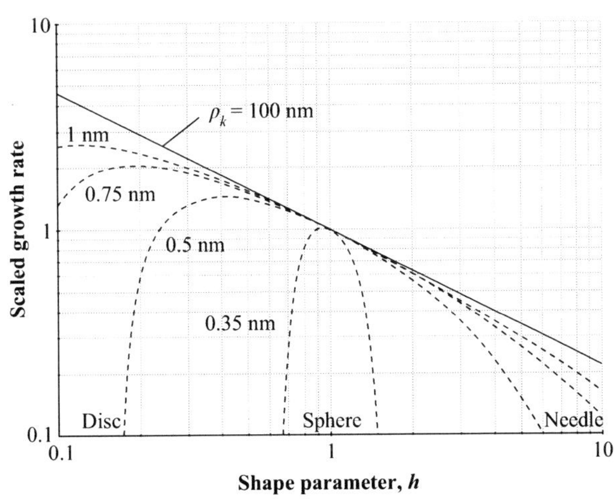

Figure 6.11. Growth rate of precipitates with variable radius as a function of the shape parameter, h.

6.3 Particle Coalescence

The size evolution of a precipitate in time is often controlled by the elementary processes of attachment and detachment of monomers (solute atoms). The individual adsorption and desorption events occur randomly with respect to the position on the precipitate surface. Each attachment and detachment event moves the center of gravity of the precipitate along the trajectory of the precipitate center and the location of the adsorption/desorption process. As a consequence, the precipitate virtually "moves" through the parent lattice and does not remain statically at the same spatial location.[7] The movement occurs randomly and it can formally be described by the same concepts as utilized in Section 7.1.3 for the diffusion of atoms as a consequence of the random walk of structural vacancies.

In precipitation systems with moving particles, the evolution of the system is not only determined by the elementary adsorption and evaporation processes of monomers, but also by the possibility of particle collisions with subsequent coalescence. The former mechanism forms the basis of practically all models for diffusion-controlled precipitate evolution that are treated in this book, being either a model for growth or a model for precipitate coarsening. In LSW theory,[8] for instance, small precipitates are assumed to evaporate monomers at a higher rate than large precipitates and the coarsening kinetics is described in terms of these two rates and the resulting diffusive flux between precipitates of different size. If precipitates can also move through the crystal, the possibility of collision and coalescence adds an additional evolution mechanism.

Once two particles have come into contact, they rapidly merge into one larger particle. This process is denoted as *coalescence*[9] and its kinetics is controlled by the minimization of the total surface area of the two particles. Various stages in a typical coalescence event are depicted in Figure 6.12. The sequence of images shows two precipitates coming into contact and merging quickly into a single larger precipitate as observed in a lattice Monte Carlo simulation[10] based on the vacancy exchange mechanism.[11] The spheres represent atoms of type B, which form precipitates inside the A-B miscibility gap. The simulation box has dimensions of $35 \times 35 \times 35$ bcc unit cells. The B-content is 3at%.

Interestingly, during the coalescence event, the neighboring smaller precipitate is practically unaffected, although the Monte Carlo snapshots clearly show that the smaller particle has considerable random drift.[12] The main reason for this observation is that the coalescence event for the two precipitates occurs on a smaller time scale compared to the cluster migration process

---

[7]This phenomenon is in a few aspects comparable to the phenomenon of Brownian motion, which has been observed and described in 1827 by the Scottish botanist Robert Brown (1773–1858). Brown observed pollen grains submerged in water in the microscope and found that these particles move randomly through the liquid. Later, this effect was explained by random collisions between the pollen grains and water molecules.

[8]See Section 3.6.

[9]In literature, the term "*coagulation*" is also often found for this phenomenon.

[10]Additional information on these simulations is given in ref. [166].

[11]See Section 7.1.

[12]For convenience, movies showing these processes are provided on the *MatCalc* web site http://matcalc.at -> Download.

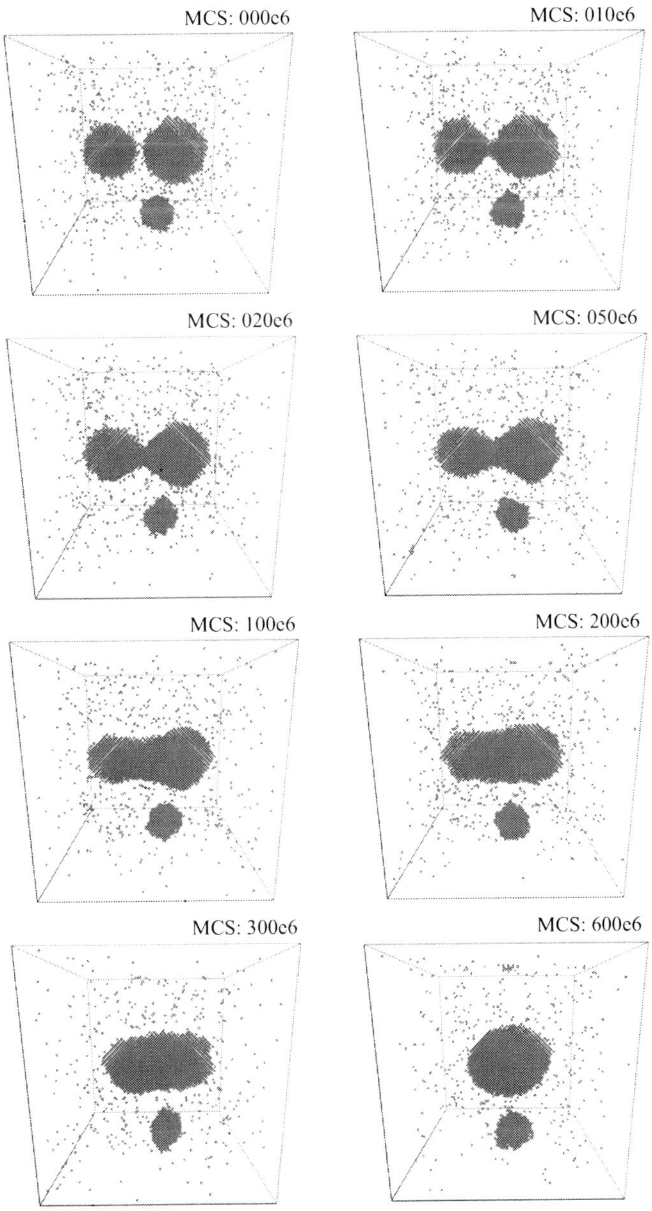

Figure 6.12. Various stages during a collision event involving two larger precipitates. The images are taken from a lattice Monte Simulation in the software *MatCalc*.[13] The simulation is continued in Figure 6.13.

through the crystal. Once the two precipitates form a stable atomic interconnection, the coalescence process proceeds rapidly and it can be even accelerated if the vacancy prefers to reside on positions inside the precipitate–matrix interface. In this case, the atoms are rapidly "shuffled" and transported along the precipitate surface.

[13] See Section 9.3.

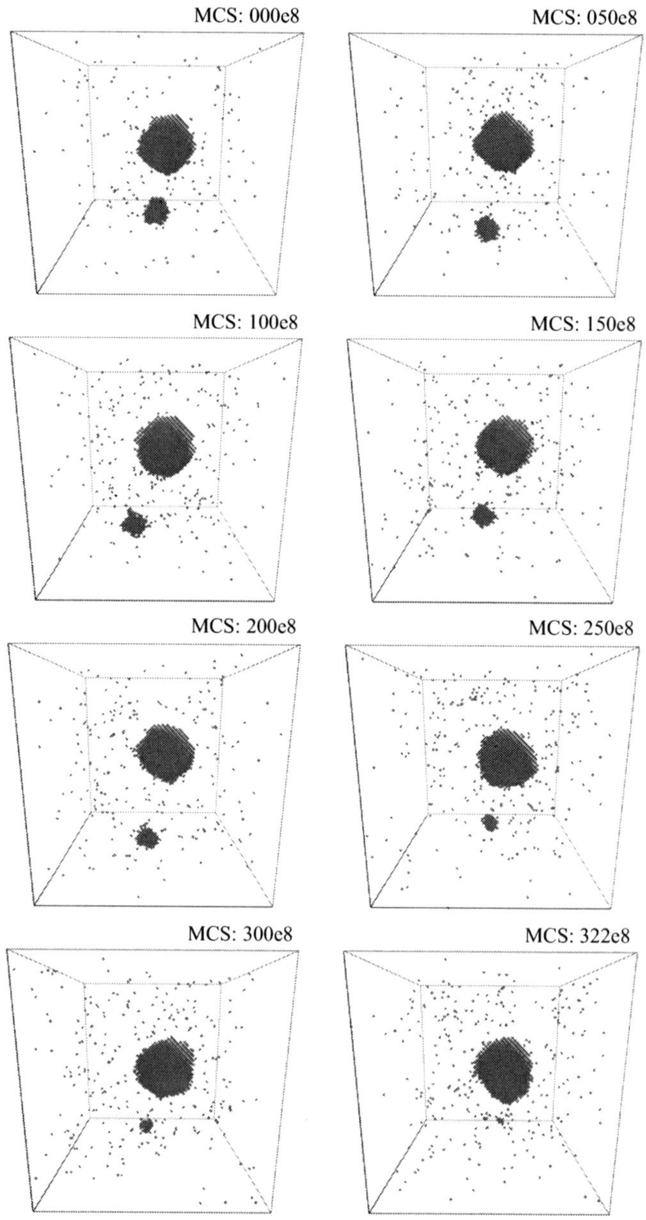

Figure 6.13. Various snapshots taken during a coarsening simulation with the lattice Monte Carlo technique. Note that the simulation time scale is 10^8 Monte Carlo steps compared to 10^6 in the previous coalescence simulation.

In Figure 6.13, the simulation is continued for the time after the coalescence event. It is important to note that the evolution now occurs only by adsorption and evaporation of monomers from the matrix and into it. The labels indicating the time scale in Monte Carlo steps (MCS) are two orders of magnitude larger. This observation indicates already that the precipitate evolution by coalescence is likely to play a major role in systems with high number density and phase fraction of precipitates and small distances between the surfaces of individual precipitates. In systems with low number density or low phase fraction, the distance between two random precipitates becomes large and, therefore, collision events are rather rare.

In the following subsections, the impact of precipitate/cluster migration is investigated with respect to its impact on the integral course of precipitation. The treatment is closely following the recent analysis by Warczok, Zenisek, and Kozeschnik [166] who utilize three different simulation methods to identify the dynamic behavior of precipitation systems subject to isolated particle coalescence as well as simultaneous adsorption/desorption and coalescence.

6.3.1 Diffusion Kinetics of Clusters

The average migration rate of an atom B can be described in terms of its diffusion coefficient, D_B, and the overall flux, J_B, is given approximately by Fick's first law, equation (7.16). The same concept of diffusion coefficients can be applied to dimers, trimers and, generally, n-mers. The issue of cluster diffusion has been treated in great detail be Binder and Stauffer [167] and Binder [168]. These authors suggest that the diffusion coefficient of clusters with size n, where n is the number of monomers/atoms in the cluster, is related to the diffusion coefficient of monomers as

$$D_n = D_1 \cdot n^{-d}, \tag{6.57}$$

with D_n denoting the diffusion coefficient of the cluster with size n and D_1 being the diffusivity of the monomer. The exponent d is typically between 4/3 and 1. These values have been confirmed by Fratzl and Penrose [169] using lattice Monte Carlo simulation with the vacancy–exchange mechanism. The dependency of the cluster diffusivities, D_n, is plotted for the limiting cases of the exponents $d = 4/3$ and $d = 1$ in Figure 6.14.

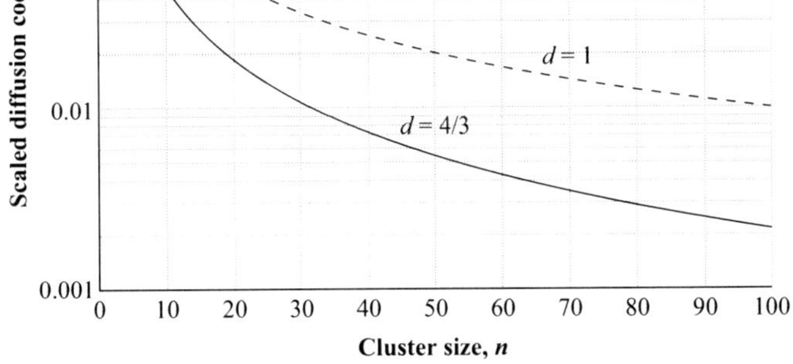

Figure 6.14. Scaled diffusion coefficient, D_n/D_1, of clusters of size n according to Binder–Stauffer kinetics, equation (6.57).

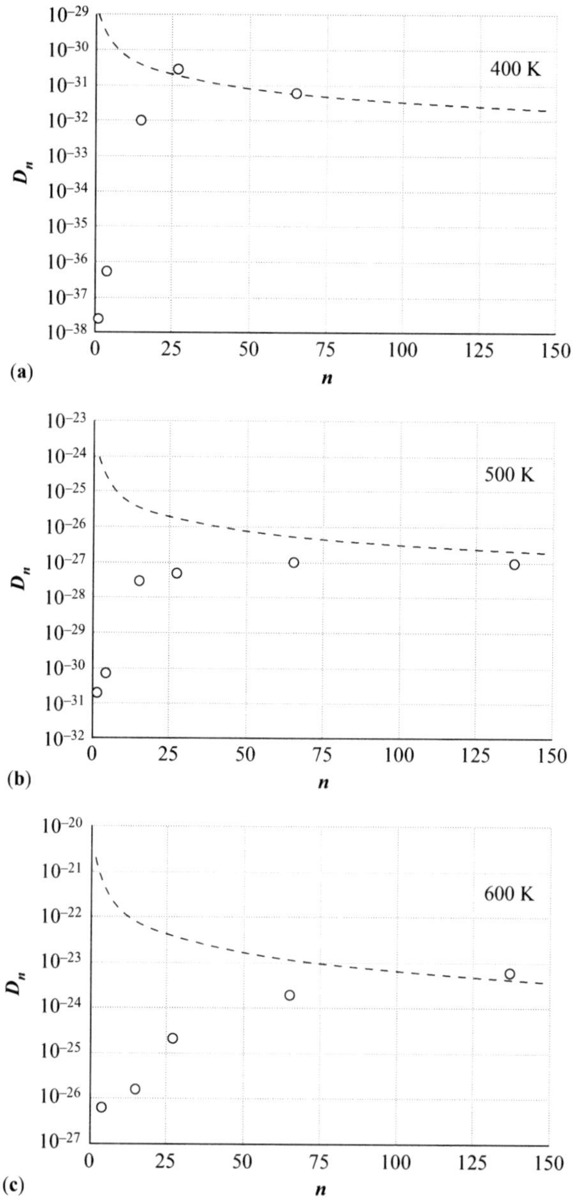

Figure 6.15. Diffusion coefficient of Cu-clusters in bcc Fe of size n according to Binder–Stauffer kinetics (dashed lines) and measured from kinetic Monte Carlo simulations of Soisson and Fu [170] (open circles).

According to equation (6.57), the diffusivity of a cluster with five atoms is lower by a factor of almost 10 compared to the diffusivity of a monomer when using an exponent of $d = 4/3$. Clusters containing approximately 30 atoms are slower already by a factor of 100. Using an exponent of $d = 1$ predicts a slightly weaker dependence of the diffusivity from cluster size. It is important, however, to recall that the assumptions made in the derivation of this formula are based on mere random adsorption and desorption arguments. The situation in real diffusion systems can be substantially more complex.

The Binder–Stauffer kinetics, as given in equation (6.57) and Figure 6.14, can be used to describe cluster diffusion in systems without pronounced vacancy preference to either the solute or solvent atoms. Using first-principles calculations of saddle point energies for diffusion and kinetic Monte Carlo simulations, Soisson and Fu [170] investigated the diffusion kinetics of bcc Cu clusters in a bcc Fe matrix. These authors placed Cu clusters of various sizes into the middle of a simulation box and measured the mean cluster displacement after a given number of Monte Carlo steps to extract diffusion coefficients. The simulations were carried out at rather low temperature to make sure that the clusters do not evaporate monomers into the surrounding matrix. The results of their simulations are shown in Figure 6.15.

The cluster diffusion coefficients observed in the Soisson and Fu simulations are largely different from the diffusion coefficients predicted by the Binder–Stauffer expression, equation (6.57), for small clusters. According to the simulations, at 600 K, a cluster containing 137 atoms is approximately 200 times more mobile than a Cu monomer. At 400 K, the difference is even more pronounced, however, the diffusion coefficients are already very small and, therefore, hardly of any practical relevance.

A major conclusion from these investigations may be drawn with respect to the governing physical mechanisms controlling precipitation in such systems: As Cu monomers in bcc Fe are found to be significantly slower than clusters of some intermediate size, the basic assumptions behind the precipitate growth and coarsening mechanisms as discussed in Chapter 3 are likely to be inappropriate. Consequently, application of the classical growth and coarsening models to these special precipitation systems is critical and the evaluation of the exact growth kinetics is certainly of higher complexity than suggested in many pieces of corresponding literature.

As the exchange of matter between matrix and precipitate, as well as between the precipitates themselves, also involves entire clusters, the effective diffusion coefficient controlling the diffusive fluxes are different from the diffusion coefficients of monomers. Moreover, the situation is even more complex when considering the vacancy–exchange mechanism as the governing process for diffusion. As the vacancy can reside preferentially inside the precipitate or at the precipitate–matrix interfaces, the mobility of the entire cluster can be substantially changed. As a consequence, the evolution of the precipitate system must also consider the possibility of cluster collision and coalescence. This is treated in the following subsection.

6.3.2 Evolution of Precipitation Systems by Coalescence

The evolution of multiparticle systems by coalescence has been treated several times in literature. The fundamental work of Smoluchowski [171] shows that the evolution of the precipitate number density, N_i, of a size class i can be described by an equation of the type

$$\frac{dN_i}{dt} = \frac{1}{2} \sum_{\substack{j=1 \\ k=i-j}}^{j=i-1} \beta(j,k) N_j N_k - \sum_{j=1}^{\infty} \beta(i,j) N_i N_j, \qquad (6.58)$$

where $\beta(i,j)$ is the collision rate of particles with sizes i and j. In some special cases, this set of equations can be solved analytically, depending on the functional form of $\beta(i,j)$, which, in turn, depends on the particle collisions being of either Stokes [172], turbulent [173], Brownian [174], or gradient type [175]. In many practical cases, numerical solution is required.

For the special case of particle migration due to Brownian motion, Smoluchowski derived an expression for $\beta(i,j)$, which is a function of the cluster diffusion coefficients, D_i and D_j, and the precipitate radii, r_i and r_j, as

$$\beta(i,j) = \frac{1}{2}\left(D_i + D_j\right)\left(r_i + r_j\right). \tag{6.59}$$

In the liquid phase, the cluster diffusivities are found to be approximately independent of the cluster size. In this case, closed analytical solutions are possible [171]. If the cluster mobility is a function of cluster size, the situation must be treated numerically.

Assume that the clusters moving through the crystal are represented by gas particles, which are characterized by their position with respect to their center of mass and their mobility, given as a function of precipitate size. Evolution of this precipitation system is assumed to occur exclusively by collision of particles. In the simulation, the particles are assigned discrete sizes, given in terms of the number of monomers in the particle. In each simulation time step, the particles are allowed to move along a randomly generated vector. As soon as two particles approach each other to less than an atomic distance, the particles are merged (coalescence event), relocated to their common center of mass, and, furthermore, treated as a single new particle. The mobility of each cluster is given by Binder–Stauffer kinetics, equation (6.57), with $d = 4/3$.

The results of these simulations are shown in the next section in comparison with results from a Monte Carlo analysis and continuums-mechanical modeling of the Fe–3at%Cu system.

6.3.3 Simultaneous Adsorption/Evaporation and Coalescence

Consider a bcc lattice occupied with 97% Fe and 3% Cu atoms. One single lattice site is assumed to be empty, that is, it represents a structural vacancy. Let the vacancy exchange its position with any of the eight nearest-neighbor atoms. In this way, the precipitation system can evolve.

Let the energy, E, of the system be given by the superposition of interatomic bonds, ε_{ij}, with

$$\begin{aligned} E = {} & n^{(1)}_{FeFe}\varepsilon^{(1)}_{FeFe} + n^{(1)}_{CuCu}\varepsilon^{(1)}_{CuCu} + n^{(1)}_{CuFe}\varepsilon^{(1)}_{CuFe} + n^{(1)}_{FeVa}\varepsilon^{(1)}_{FeVa} + n^{(1)}_{CuVa}\varepsilon^{(1)}_{CuVa} \\ & + n^{(2)}_{FeFe}\varepsilon^{(2)}_{FeFe} + n^{(2)}_{CuCu}\varepsilon^{(2)}_{CuCu} + n^{(2)}_{CuFe}\varepsilon^{(2)}_{CuFe} + n^{(2)}_{FeVa}\varepsilon^{(2)}_{FeVa} + n^{(2)}_{CuVa}\varepsilon^{(2)}_{CuVa}, \end{aligned} \tag{6.60}$$

where the superscript denotes either the nearest-neighbor or the second-nearest-neighbor shell and n_{ij} is the number of bonds of a certain type. The interaction parameters, ε_{ij}, between the individual atoms have been determined such that the experimental Fe-Cu phase boundaries are reproduced at the temperature of interest. In the present case, the simulations are carried out at 500°C. Details on the simulation procedure and the values of parameters are described in ref. [166].

Although the focus of this investigation is on the effect of vacancy–solute preference on the precipitation kinetics of the Fe-Cu system, several general observations can be drawn from the results. Let the analysis be carried out for an asymmetry parameter, a^*, given as

$$a^* = 2 \cdot \frac{\varepsilon_{FeVa} - \varepsilon_{CuVa}}{\varepsilon_{FeCu}}, \tag{6.61}$$

and describing the affinity of the vacancy to either the Fe or the Cu atoms. A value of a^* larger than zero is representative for a situation where the vacancy prefers to be surrounded by as many Cu atoms as possible. Negative values denote situations where the vacancy prefers the vicinity of Fe atoms and avoids being nearest-neighbor of a Cu atom. Figure 6.16 shows two simulations carried out for values of $a^* = +2$ and $a^* = -2$. The diagrams display symbols for each existing cluster size throughout the system evolution expressed in MCS. Interestingly, from these so-called *stair-fountain* diagrams, the character of the system in terms of the frequency of coalescence events can be nicely determined.

In Figure 6.16(a), the asymmetry parameter, a^*, has a value of –2, which means that the vacancy prefers to be surrounded by Fe atoms. In the course of the simulation, the vacancy stays mainly in the Fe-matrix and avoids the vicinity of Cu clusters. Consequently, the system evolution is determined mainly by the evaporation and adsorption of Cu monomers to the cluster, as material transport occurs mainly by vacancy–atom exchange in the matrix. The mobility of clusters is low, as this process represents only a weak motor for cluster migration.

In the other case, Figure 6.16(b) with $a^* = +2$, the vacancy prefers Cu-rich environment and, therefore, resides often in the vicinity or inside the Cu clusters. Consequently, Cu atoms are frequently "shuffled" over the cluster surface and the mobility of clusters increases rapidly.[14] As the entire clusters become rather mobile, the frequency of coalescence events also increases rapidly and the system evolves by adsorption/evaporation and particle collision simultaneously.

In the stair-fountain diagrams of Figure 6.16, these characteristics are clearly visible. In the case of $a^* = -2$, the individual symbols appear to be aligned along continuous kinetic paths. The "apparently" interconnected lines initially increase in size and, finally, decrease. This behavior resembles water falling down a fountain, thus giving these plots their name. In the case of $a^* = +2$, several lines seem to be abruptly ending or beginning. The interpretation of these points suggests that at each of these discontinuities two clusters merge into a single larger one by a coalescence event. The paths of two clusters suddenly end[15] while a single new path for a larger particle suddenly appears. In Figure 6.16(b), several of these collision events are marked with arrows for convenience.[16]

The results of these Monte Carlo simulations are displayed in Figure 6.17 in terms of the cluster density and the mean radius for values of a^* between +4 and –4. Again, pronounced differences can be observed for the evolution of the overall precipitation reaction. In Figure 6.17(a)

---

[14] If a certain interaction energy threshold is surpassed, the cluster mobility decreases again as the vacancy resides only inside the clusters. In this case, the vacancy cannot contribute to the redistribution of monomers over the cluster surface.

[15] The dissolution of two clusters is not always visible but hidden in the symbols of many other smaller particles.

[16] Movies demonstrating these effects can be viewed in the supplementary materials section of ref. [166] or they can be downloaded from the *MatCalc* software web site http://matcalc.at.

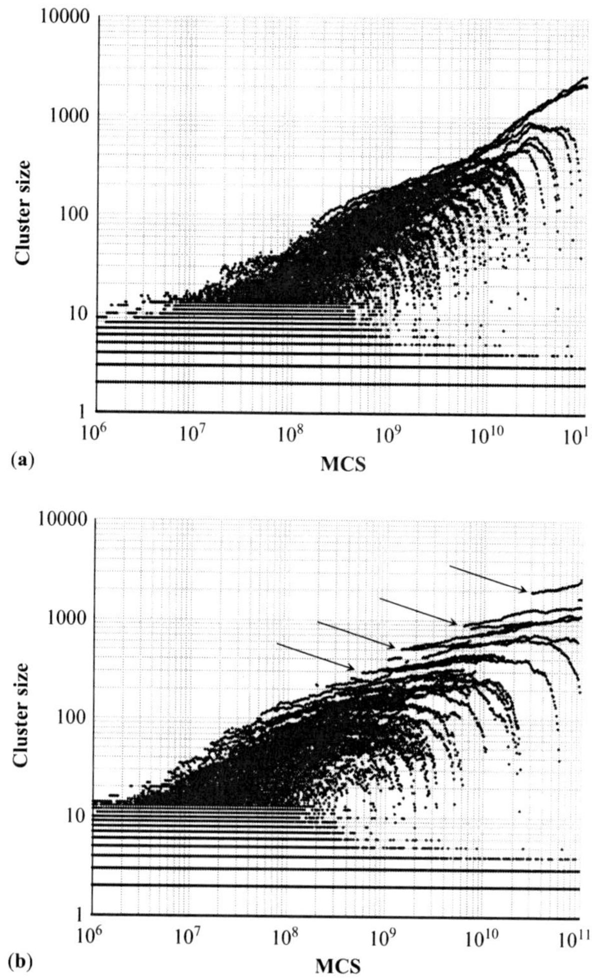

Figure 6.16. Stair-fountain diagrams from Monte Carlo simulation with (a) $a^* = -2$ and (b) $a^* = +2$. The arrows indicate the occurrence of coalescence events.

and (b), higher negative values of a^* clearly lead to a deceleration of the entire precipitation process. Both, the number density as well as the mean radius curves are shifted to longer times with increasing a^*. The principle shape of the curves remains conserved, though. The reason for this retardation is identified in the tendency of the vacancy to avoid the vicinity of Cu atoms. As the exchange frequency between vacancy and Cu decreases, so does the effective diffusivity of the Cu monomers.

The evolution of precipitation parameters for values of a^* between 0 and +2 shows similar behavior. Neither a pronounced shift nor a change of shape is observed. For a^* higher than +2, the situation changes drastically. With higher positive values of a^*, the vacancy tends to reside inside the Cu clusters and, consequently, is not available for Cu monomer diffusion outside the clusters. At intermediate positive values, the vacancy moves along the surface of the cluster, however, and facilitates cluster movement and cluster

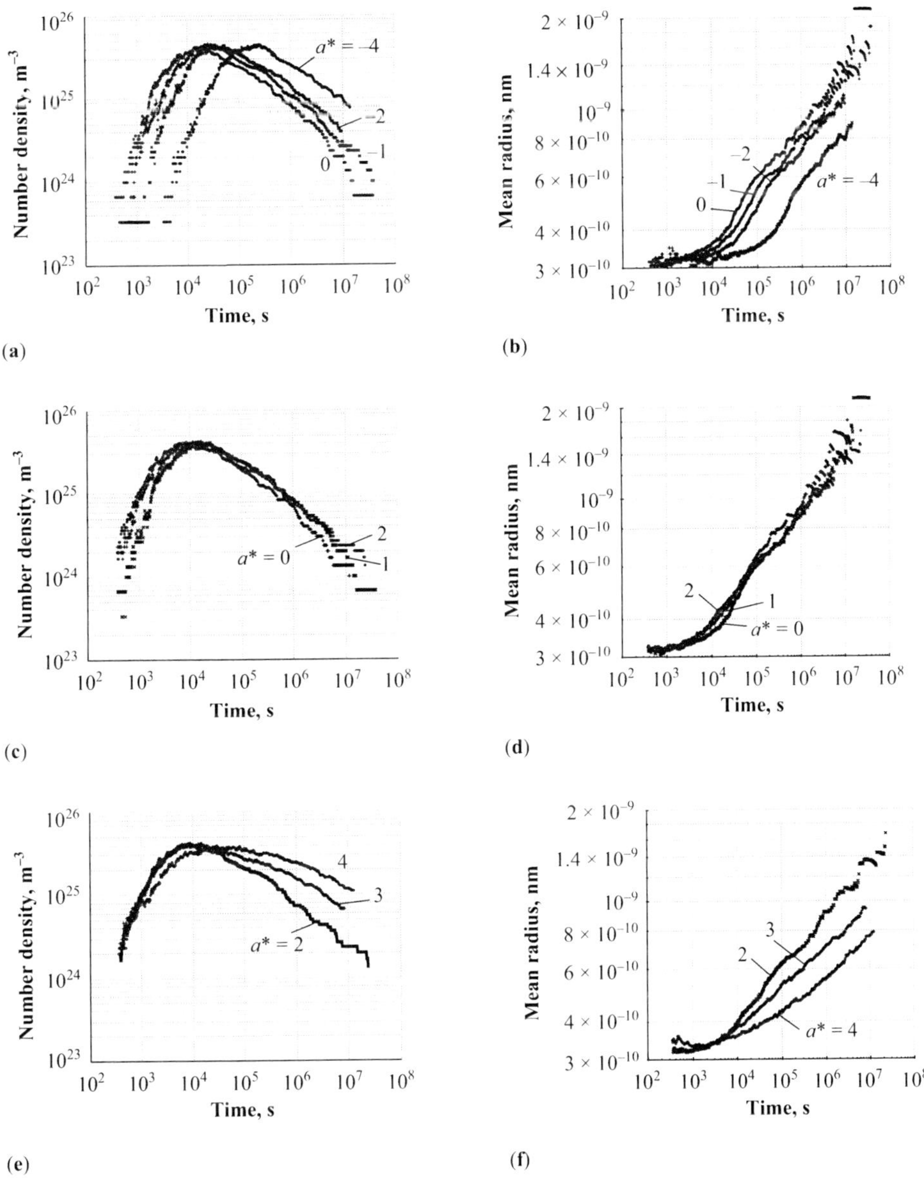

Figure 6.17. Evolution of precipitation parameters as a function of the asymmetry parameter, a^*, as obtained from lattice Monte Carlo simulation.

coalescence. The system, therefore, predominantly evolves by the coagulation mechanism. With high positive values of a^*, this effect reduces again and the precipitation process is slowed down.

Interestingly, the effect of positive a^* values on the initial stages of the precipitation reaction is weak. The preference of the vacancy to Cu atoms supports Cu diffusion. Only after sufficiently large Cu-clusters exist, which are potential vacancy traps, the further precipitation reaction becomes more sluggish.

6.3.4 Phenomenological Treatment of Particle Coalescence

The careful numerical study of systems evolving solely by particle coalescence as well as systems with simultaneous adsorption/evaporation of monomers provides a basis for the formulation of a continuums model of coagulation for straightforward implementation into NKW[17]-type models. A corresponding attempt has been undertaken in ref. [166], where the cluster coagulation process has been introduced into the evolution equation for the number density, N_i, of size classes i. In a simple, semiempirical approximation, the rate of N_i has been set proportional to the total cluster density, N, the mean cluster cross-sectional area represented by the square of the mean particle radius, $\bar{\rho}^2$, the diffusion coefficient, D_i, of clusters of size class i, and a dimensionless proportionality coefficient, k_{co}, with

$$\dot{N}_i = k_{\mathrm{co}} \cdot N\bar{\rho}^2 D_i . \tag{6.62}$$

With the corresponding implementation in the *MatCalc* software,[18] the effect of k_{co} on the evolution of precipitation parameters is investigated in Figure 6.18. Without consideration of coalescence, that is, $k_{\mathrm{co}} = 0$, the same shape of number density and mean radius curves is observed as already discussed in Section 5.4.4. The number density of precipitates initially increases sharply due to nucleation of particles in the supersaturated matrix. In the course of the reaction, supersaturation decreases, nucleation seizes, and the system continues to evolve by growth of existing precipitates until supersaturation goes to zero and all available solute is concentrated in the precipitates. The following transient region is characterized by a stasis in number density and mean radius until, finally, the system enters the coarsening regime.

If coalescence is accounted for in the simulation, the increase of number density of precipitates is limited by the relative weight of the coalescence mechanism compared to the adsorption/evaporation mechanism incorporated in the classical evolution equations. With increasing values of k_{co}, the value of the number density maximum decreases. Moreover, the number density decreases immediately after reaching the maximum value. The stasis in number density as well as mean radius disappears and both quantities adopt an increasingly continuous evolution with vanishing transient region. Additionally, the evolution of the mean precipitate radius is drastically accelerated compared to the curve without coalescence. Finally, all curves approach the adsorption/evaporation-controlled coarsening curves in the limit of long simulation times.

The precipitation characteristics of the combined process can be straightforwardly interpreted when comparing to the time evolution of purely coalescence-controlled or purely adsorption/evaporation-controlled systems. The corresponding results, as obtained from the numerical coalescence simulations described in Sections 6.3.2 and coarsening kinetics from classical Ostwald ripening as described in Section 3.6.1, are displayed by straight lines with slopes given by the corresponding models in Figure 6.19. In the absence of coalescence, the number density evolution follows the well-known t^{-1} law after entering the coarsening regime, whereas the mean radius follows the typical $t^{1/3}$ characteristics. In contrast, the coalescence kinetics can be well described by time exponents of approximately $t^{-1/2}$ for the number density and $t^{1/6}$ for the mean radius. For the present example, the lattice Monte Carlo results and the continuums

[17] Numerical Kampmann–Wagner approach, see Section 5.3.

[18] See Section 9.3.

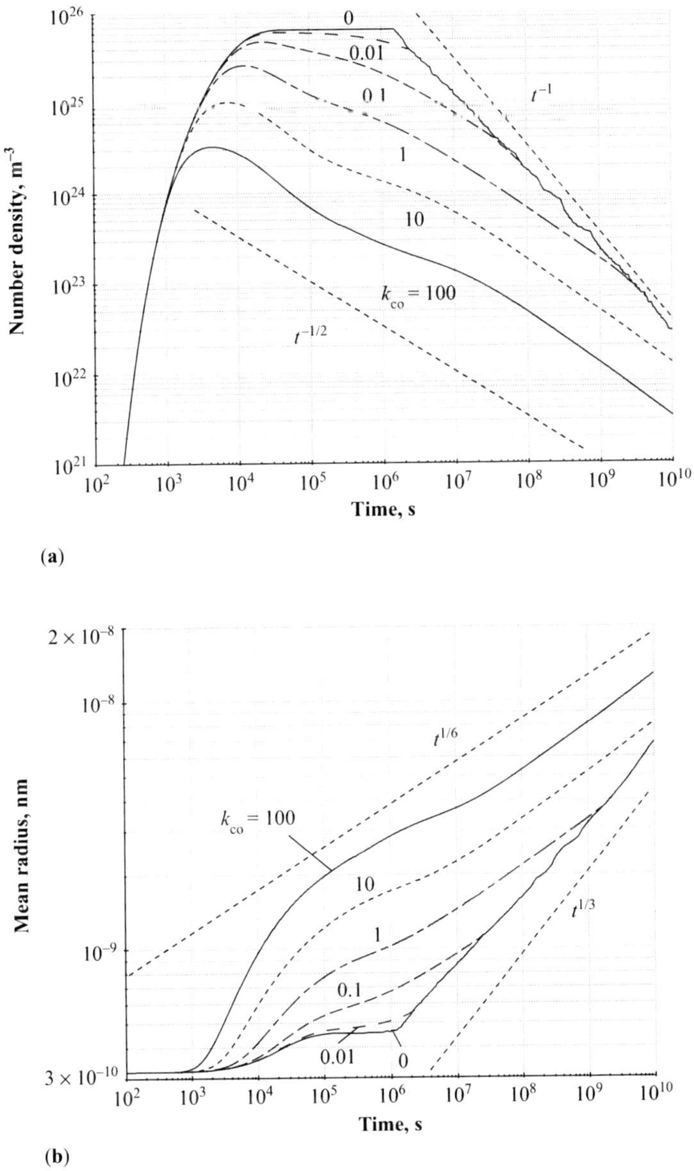

Figure 6.18. Evolution of precipitation parameters as a function of the coalescence coefficient, k_{co}.

modeling approach coincide well, thus demonstrating that the phenomenological algorithm can give a realistic representation of the combined process.

6.3.5 Comparison with Experiment

The course of a precipitation reaction with simultaneous adsorption/evaporation of monomers and particle coalescence in the Fe-1.5at%Cu system was investigated by Warczok et al. [176]. In this work, the results of lattice Monte Carlo simulations, continuums modeling, and

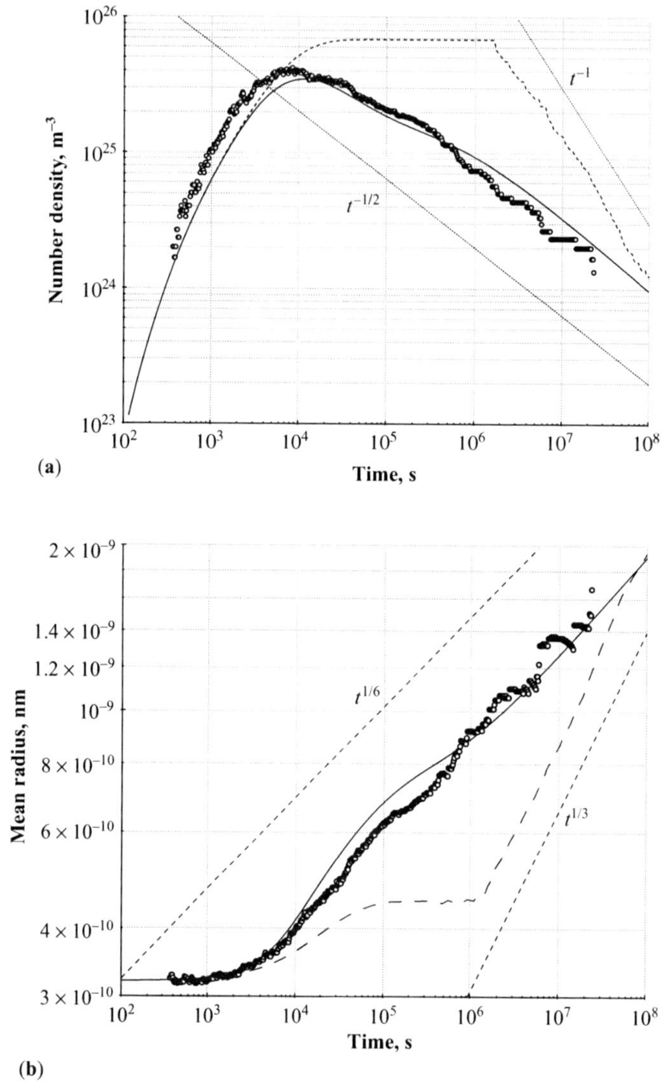

Figure 6.19. Comparison of lattice Monte Carlo simulations (symbols) with the continuums approach (solid lines). The evolution of precipitation parameters, number density, and mean radius without coalescence are shown as dashed lines. The theoretical limits from LSW theory and coalescence kinetics are shown as straight lines.

experimental analysis based on 3D-atom probe tomography and small angle neutron scattering are compared for an Fe-1.5at%Cu alloy heat treated at 500°C. The results of this analysis are compiled in Figure 6.20.

The experimental evidence displays a pronounced peak in number density between approximately 10^1 and 10^2 s without any plateau stemming from a transient region. The data from atom probe field ion microscopy (APFIM) and small angle neutron scattering (SANS) are consistent and coincide well with the numerical modeling.

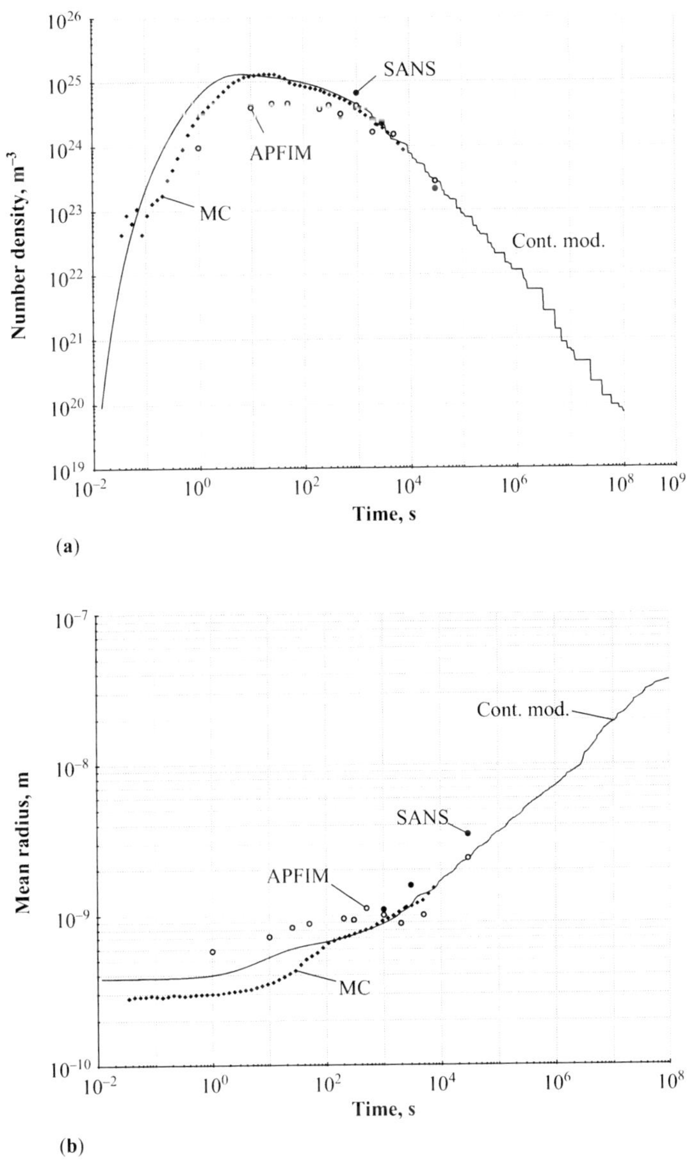

Figure 6.20. Comparison of lattice Monte Carlo simulation (MC) and continuums modeling (cont. mod.) with experimental data from atom probe field ion microscopy (APFIM) and small angle neutron scattering (SANS) for precipitation in the Fe–1.5at%Cu system.

6.4 Simultaneous Precipitation and Diffusion

Modeling solid-state precipitation and phase transformation can be performed at various levels of detail: On the one hand, the diffusion profiles around moving phase boundaries can be of interest to the researcher and the progress of the phase transformation is followed in terms of the position of the transformation interface and the evolution of the composition profiles. On the other hand, the evolution of individual precipitates or entire precipitate populations may be in the center of interest. In this case, the researcher is advised to use models describing the precipitation parameters in terms of evolution equations for the radius and composition of precipitates or precipitate distributions.

In some cases, the researcher is interested in the combined process: The evolution of the composition profiles determined by long-range diffusion as well as the simultaneous evolution of precipitation processes on various spatial locations in the sample. Typical examples for processes where long-range diffusion and precipitation occur simultaneously are the carburization of Cr-steel or Ni-base alloys, the post-weld heat treatment of dissimilar welds, the plasma-supported nitridation of complex high-strength tool steel, or the oxidation of materials during high-temperature processing. Potential simulation approaches for a combined treatment of long-range diffusion and precipitation are the focus of this section.

6.4.1 Numerical Treatment in the Local-Equilibrium Limit

The first numerical models addressing these problems are reported by Bongartz et al. [177] for carburization of a Ni-base alloy and Buchmayr and Kirkaldy [178] for diffusion in Fe–Cr–C alloys. Later, Engström, Höglund, and Ågren [179] and Helander and Ågren [180] demonstrated the general applicability of the DICTRA approach[19] to this kind of problems. Kozeschnik et al. [181,182] utilized the *MatCalc*[20] software to describe the evolution of carbide-free zones in dissimilar Cr-steel weldments.

Based on his previous work, Bongartz et al. [177], in 1989, suggested a general numerical model that is capable of dealing with simultaneous long-range diffusion and precipitation reactions in systems with an arbitrary number of diffusing elements and precipitating phases. In order to make the problem accessible to computational analysis, these authors make an important assumption: If the long-range diffusion problem occurs on sufficiently large length scale, which is the case if the interparticle spacing is much smaller than the long-range diffusion distance, then the precipitates will have sufficient time to come into thermodynamic equilibrium with the surrounding matrix at each time interval. Locally, the system will be in equilibrium, independent of the long-range diffusion processes. The basic features of this model are outlined below.

Consider volume diffusion of elements through a matrix of given structure. Subdivide space into appropriate intervals and solve the multicomponent diffusion equation with respect to the matrix content of each element by a suitable numerical technique.[21] A corresponding sketch is shown in Figure 6.21 for the example of diffusion of an atomic species B into a cylindrical sample. Vertical dashed lines symbolize the cells. Arrows indicate the diffusive interaction between cells for the cell with index *i*.

In each of the simulation cells, precipitation of an arbitrary number of phases can take place. The exact type and amount of each phase is given by the thermodynamic properties of the system and the overall chemical composition of the cell. As the precipitates are assumed to be in local equilibrium with the matrix, atoms from the matrix phase can be picked up after each

[19] See Section 9.1.

[20] See Section 9.3.

[21] A straightforward method to solve the diffusion problem is the Finite Differences technique.

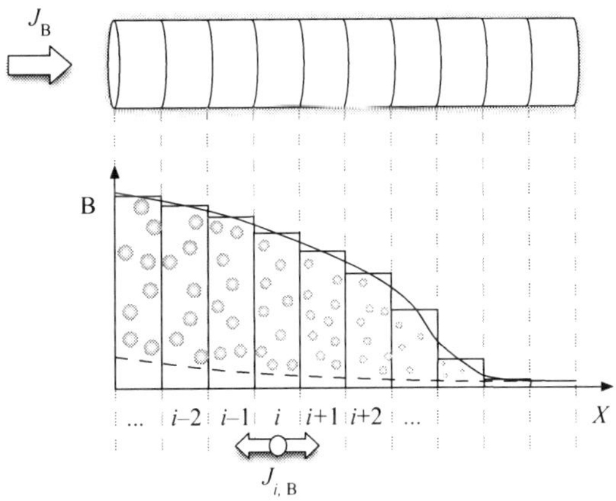

Figure 6.21. Spatial discretization of a diffusion sample. Top: Cylindrical sample with in-flux of atoms B into the volume. Bottom: Representation of the sample with discrete cells. The composition profile for the total amount of B (solid line) is replaced by a stepwise profile. The matrix B composition is indicated as a dashed line. The spheres represent precipitates.

diffusion step and be bound inside the precipitates. These solute atoms do not participate in the long-range diffusion process.[22]

The basic assumptions of Bongartz et al. [177] for the numerical treatment of simultaneous long-range diffusion and precipitation are that the entire process can be subdivided into two separate simulation steps with:

- *The diffusion step*: Evaluate the diffusive fluxes between the cells based on the amount of unbound elements (matrix composition). For this step, solve the multicomponent diffusion equation.[23] The precipitates do not contribute in this step except for the fact that they represent obstacles to the diffusion process. This is commonly taken into account in the form of a phenomenological labyrinth factor.
- *The equilibrium step*: After the diffusion step, for each simulation cell, evaluate thermodynamic equilibrium between the matrix phase and the precipitate phases. The precipitation status of each phase is described by its phase fraction only. The basis for this calculation is the total element content that is the sum of elements bound in the precipitates, the matrix composition from the previous simulation step, as well as the amount of solute entering or leaving the cell in the diffusion step.

The controlling algorithm selects simulation time increments such that the desired accuracy of the calculation is maintained. In each simulation step, the equilibrium amount of precipitate phases increases or decreases, depending on the amount of new solute entering or leaving the cell. Moreover, the mixture of different precipitate types can also change as a consequence of the evolution of overall composition determined by the long-range diffusion processes.

In practical carburization experiments, the size of precipitates in various distances from the interface will change with time. This observation is indicated in Figure 6.21 by the varying

---

[22] In real diffusion-precipitation systems, the precipitates represent obstacles that slow down the effective kinetics of the long-range diffusion process. This effect is commonly taken into account on a phenomenological basis by a so-called *labyrinth factor*. In the work of Engström, Höglund, and Ågren [179], this quantity is chosen to be proportional to the phase fraction square, f^2.

[23] This topic is dealt with in Section 7.2.5.

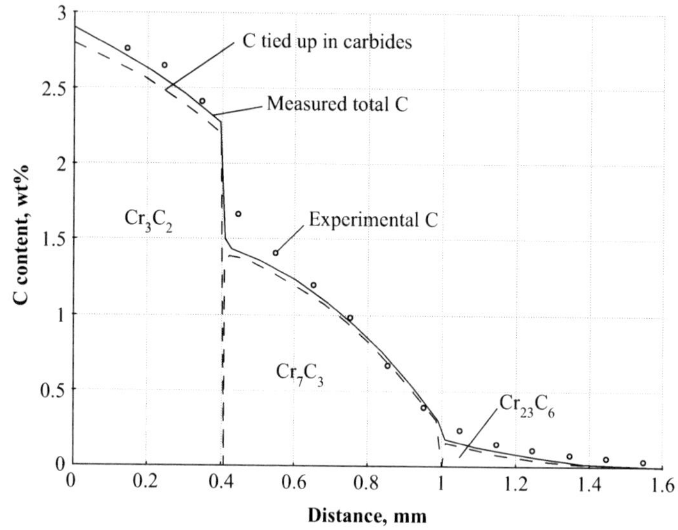

Figure 6.22. Experimental data and numerical simulation of the carburization treatment of a Ni-25wt%Cr alloy for 1000 h at 850°C as reported by Bongartz et al. [177].

diameter of the spheres. In the local-equilibrium model for simultaneous long-range diffusion and precipitation, this phenomenon is not tracked as only the total amount of solute in each cell enters an equilibrium simulation that, per definition, ignores any kinetic effects or size effects of precipitates (Gibbs–Thomson effect). The full long-range diffusion–precipitation problem is computationally significantly more challenging and it is dealt with briefly in Section 6.4.3.

6.4.2 Comparison of Local-Equilibrium Simulations with Experiment

The prototype process for simultaneous long-range diffusion and precipitation considered in the studies of Bongartz et al. [177] is the carburization of a Ni-25wt%Cr alloy in an Argon–methane gas mixture at 850°C for 1000 h. During carburization, three types of carbides are observed: Cr_3C_2, Cr_7C_3, and $Cr_{23}C_6$. After the carburization treatment, the authors measured the total amount of elements as a function of distance from the reaction surface, as well as the type of Cr-carbide. Typical results of the simulation and the experimental data are shown in Figure 6.22.

Apparently, the simulations of Bongartz et al. are in very good agreement with experiment. However, in these simulations, the authors have adjusted the thermodynamic parameters of the system, which are part of their analysis, and the diffusion coefficients such that this agreement could be achieved. Motivated by this weakness, Engström, Höglund, and Ågren [179] have performed simulations for the same set of experimental data, however, in their simulations, using independent thermodynamic and mobility data. These have become available shortly after the work of Bongartz et al. in the form of the software packages Thermo-Calc and DICTRA[24]

[24] See Section 9.1.

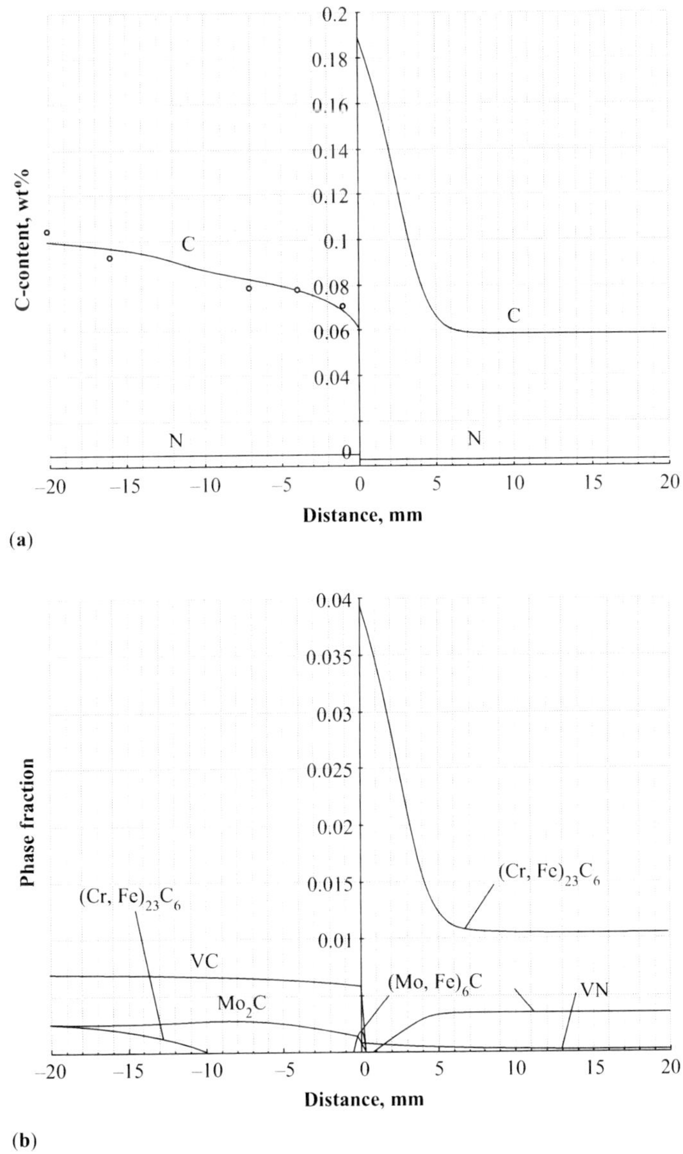

Figure 6.23. Experimental data and numerical simulation of dissimilar weld of 0.5CrMoV and 2Cr-steel heat treated at 565°C for 1.8×10^5 s as reported by Kozeschnik et al. [181,182]. Experimental points of the C-content have been determined from combustion analysis.

together with the comprehensive CALPHAD-type thermodynamic and mobility databases used by these software products.

In the new simulations carried out in DICTRA, the diffusional cross-terms, which have been neglected in the Bongartz et al. simulations, are fully accounted for and the thermodynamic information of the system is taken from the CALPHAD assessment of the Ni–Cr–C phase diagram. The corresponding simulations described in ref. [179] give an almost perfect reproduction of the experimental data, thus supporting the potential of computational thermodynamics and kinetics in application to technological processes.

In the same work, Engström, Höglund, and Ågren [179] also apply their model to the heat treatment of dissimilar welds of Cr-steels. They compare simulations for welds between a ferritic Fe–1wt%Cr and ferritic Fe–12wt%Cr alloy with experimental data and find good agreement with experiment in terms of the C-profile on the two sides of the weld, as well as the distribution of phases as a function of distance from the interface. Helander and Ågren [180] apply the DICTRA software to the combined diffusion–precipitation processes inside a weld between austenitic and ferritic steel. Again, the simulations give a good representation of the experimental data.

A similar investigation is reported by Kozeschnik et al. [181,182], who investigated the phase distribution and C profile across three dissimilar welds that have been in operation in a power plant for up to 1.8×10^5 s (23 years) at 565°C. Using the *MatCalc* software,[25] both, the C and N content in the steel are simulated. Again, the comparison between experiment and simulation gives good accordance (see Figure 6.23).

6.4.3 Coupled Diffusion and Precipitation Kinetics

One of the major advantages of the local-equilibrium based long-range diffusion and precipitation model described in the previous section is its comparably adequate requirement of computational resources. On standard desktop computers, corresponding simulations can be performed within minutes or hours. The simulation results provide information on the concentration profiles of elements as well as the distribution of phases (phase fractions!) as a function of distance from the interface.

If the model assumption of small interparticle spacing, compared to the dimensions of the long-range diffusion problem, is inappropriate or if the simulator is interested in the evolution of precipitation parameters as a function of time and space, the model must be extended to include also the time evolution of the system in each single simulation cell.

Unfortunately, corresponding simulations are not yet reported in literature. The presently available software packages are either capable of treating the kinetic evolution of a precipitation system in a single representative volume element (applies to all software packages described in Chapter 9), or they can solve the simultaneous long-range diffusion–precipitation problem in the local-equilibrium limit (DICTRA, *MatCalc*). However, the necessary theoretical concepts are available. The major step in development of a full kinetic diffusion–precipitation model will be to replace the equilibrium step in the Bongartz et al. [177] model (Section 6.4.1) by a full kinetic precipitation step. Instead of evaluating a single thermodynamic equilibrium, the full kinetic model should progress the precipitation system in time based on a set of adequate evolution equations.

Applications that are in need of such simulation algorithms are manifold:

- Nitridation of tool steel, where the kinetic evolution of carbides changing to nitrides is of interest for the final strengthening potential of the surface region
- Thermomechanical treatment of steel in the austenite/ferrite two-phase region (advances multiphase steel), where precipitation processes occur in a complex multiphase microstructure

[25] See Section 9.3.

- Simulation of the bainite transformation, where precipitation of carbides occurs either between two ferrite sheaves (upper bainite) or inside the ferrite sheaves (lower bainite)
- Evolution of precipitation-free zones around grain boundaries
- And many more …

The coupled diffusion–precipitation kinetics feature is not yet available in any of the present software packages. However, it is likely that future developments in this direction will include these capabilities. This shall hopefully be reported in future editions of this book.

CHAPTER 7

DIFFUSION

A large portion of phase transformation reactions in solid-state matter involves the long-range diffusive transport of atoms because the equilibrium chemical composition of the participating phases commonly differs from one phase to the other. If one phase grows on the expense of another phase, atoms are redistributed between the two phases and the atomic diffusive transport often represents the governing physical mechanism controlling the kinetics of the phase transformation progress.

In this chapter, we will first look at the features of diffusion on the atomic scale. The relation between atomic vibrations and long-range diffusion is explored and formulated. Then, diffusion will be investigated on the continuum scale where the macroscopic transport of mass is the focus of the analysis. The forces acting on the diffusing atoms are defined and the temperature-dependence of diffusion, as well as the impact of dislocations and grain boundaries, is explored. Finally, the role of excess structural vacancies on diffusion is investigated.

The treatment presented here is aimed at giving a brief overview of the topic as far as it is necessary to understand the role of diffusion in phase transformations. The structure of the first sections of this chapter is similar to the treatment given in Chapter 5 of ref. [11]. For details and further reading, the reader is referred to special textbooks, for example, refs. [14,183–186].

7.1 Mechanisms of Diffusion

Atoms in the solid state have close atomic bindings to the neighboring atoms. These bonds try to stabilize the atomic aggregate in certain configurations. At the same time, the atoms are thermally excited, which means that they vibrate around their equilibrium positions with an *average* energy corresponding to the temperature of the aggregate. The kinetic energy of individual atoms can be significantly higher, however, than the average value of the ensemble and, with a certain probability, these atoms can leave their original position in the material and move (jump) to another site.

As the probability, P, of a successful jump is determined by the height of the activation barrier and the thermal energy of the atom, we can write

$$P \approx \exp\left(-\frac{Q}{RT}\right), \tag{7.1}$$

where Q is the activation energy for diffusion, R is the universal gas constant, and T is temperature. The value of Q is typically constant for a given material, crystal structure, magnetic state, and diffusion mechanism.

7.1.1 Diffusion in Crystalline Materials

In crystalline solids, diffusion is made difficult by the periodic potential and arrangement of atoms in the lattice structure. In the vast majority of cases, even if the kinetic energy of the individual atom would be high enough to support a jump to a neighboring lattice site, other atoms will already occupy these positions and impede the jump. Consequently, various models have been proposed for the mechanism supporting diffusion in crystals. The most important ones are schematically shown in Figure 7.1.

In Figure 7.1(a), the atom indexed I directly changes place with atom II. As this procedure cannot be supported in a single step, this process occurs in several substeps: (*a*) atom I leaves its original site to jump into an interstitial position. To do so, the atoms surrounding the new position must be pushed aside and they become "squeezed" in the remaining space. (*b*) Atom II "presses through" the corridor to the original position of atom I. In this sequence, atom I can change place with atom II and both atoms are displaced by one atomic distance. One major problem persists, however, with this model for diffusion: The elastic energy that is required to displace the atoms in step (*a*) is rather high. Estimates on the effective jump frequencies that can be obtained with this model quickly showed that this mechanism is straightforward, however, its activation barrier is much too high to explain the fast diffusion rates observed experimentally.

From this dilemma, two separate "schools" of concept emerged in the 1940s suggesting alternative mechanisms to explain the experimentally observed faster exchange rates. The first group of researchers proposed the "ring mechanism" depicted in Figure 7.1(b): In a concerted action of four neighboring atoms, which "rotate" in the form of a ring, the activation barrier is significantly lower than in the direct exchange mechanism and the estimated diffusion rates also become orders of magnitude faster. The theoretical values approach the orders of magnitude of experimental diffusion rates. One of the main proponents of the *ring mechanism* was Robert Franklin Mehl, an American metallurgist (1898–1976).

The second model trying to explain the fast diffusion rates observed in crystals is the *vacancy–exchange* mechanism. This model is based on the observation that each crystal contains a certain amount of point defects that is typical for each material and temperature. In the context

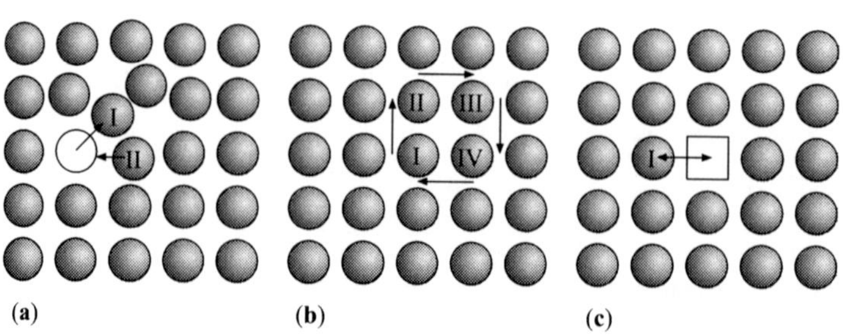

Figure 7.1. Mechanisms for atomic exchange in crystals.

of diffusion, the most relevant point defects are vacancies, that is, unoccupied lattice sites. In the vacancy–exchange mechanism, any of the neighboring atoms can change place with the vacancy, which apparently needs much less activation energy compared to the direct exchange mechanism. Similar to the ring mechanism, vacancy–exchange can deliver diffusion rates that are in basic agreement with experiment. The main proponent of vacancy–atom exchange as the dominant mechanism for diffusion in crystals has been Ernest Oliver Kirkendall, an American chemist and metallurgist (1914–2005).

In a series of three papers [187–189] that appeared within a span of eight years, Kirkendall was able to prove that atom–vacancy exchange represents the dominant diffusion mechanism in crystalline solids. His main argument, which could unambiguously distinguish between any of the two theories, was that in the case of the predominance of the ring mechanism, the diffusivity of all (substitutional) components in the system would be of the same magnitude. In the case of the predominance of the vacancy–exchange mechanism, the diffusivity of individual atomic species could be significantly different from the others, if the vacancy would feel attracted by any of the diffusing species.

On the example of the diffusion of Zn in alpha-brass, Kirkendall and coworkers showed that Zn diffuses significantly faster than Cu, the only explanation for this observation being that the vacancy–atom exchange is the operative mechanism for diffusion in crystalline systems.[1] This issue is utilized further in the next section where diffusion is investigated on the atomic scale.

7.1.2 The Principle of Microscopic Time Reversal

Consider an atom, which is located at a certain position in the periodic lattice of a crystalline material. Let us assume that the atom is somehow marked and that we are capable of following this atom on its journey through the material. The mobility of the atom shall be facilitated by exchange with vacancies.[2] On the microscopic scale, diffusion shows some interesting features that are not obtained when looking at this phenomenon from a macroscopic perspective.

Consider an atom or vacancy in two-dimensional space that travels through a crystal with 20 random jumps. If an observer records the sequence of jumps with some sort of camera and shows a playback of this sequence in forward or backward motion afterward, any viewer of this scene will not be able to distinguish unambiguously the direction of the *arrow of time* for this process. Even if we mark all involved atoms and play the sequence randomly in forward or backward direction, both directions will appear as plausible to the spectator as the other. The spectator observes that microscopic diffusion exhibits a feature that is denoted as the *principle of time reversal*. A typical sequence of 20 atomic jumps is shown in Figure 7.2.

Whereas it is apparently not possible to distinguish the arrow of time in microscopic diffusion, experience tells us that diffusion does show a directional behavior in time when watched on the macroscopic scale. This characteristic of "large-scale" macroscopic systems has already

[1] Interesting background information on this topic is provided in ref. [190].

[2] It is easy to imagine that a vacant lattice position "travels" through the crystalline material by repeated exchange with any of the neighboring atoms. In analogy, an atom can be considered as "travelling" through the material by repeated exchange with "bypassing" vacancies that exchange place with this particular atom. Every exchange event displaces the atom by one atomic position.

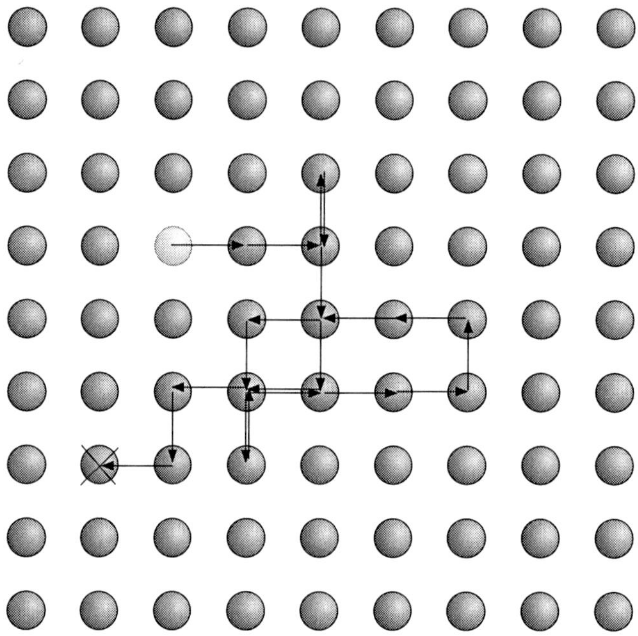

Figure 7.2. Sequence of 20 atomic jumps in solid-state diffusion picturing the principle of time reversal.

been emphasized in Figure 1.4 of Section 1.3.1, where the dissolution of a drop of ink has been reasoned on the basis of the configurational entropy of a thermodynamic system. In Figure 7.3, the temporal evolution of two blocks of matter consisting of either atoms A or B is depicted, which are brought into immediate contact at time $t = 0$.

When looking at the sequence of images Figure 7.3(a)–(d), any external observer will have no trouble to put the images into correct order. From experience, and common knowledge, we know that condition (a) represents the starting condition and condition (d) is close to the final state of the system. From experience, we know that unlike atoms tend to mix. We hardly ever experience situations where a state (d) transforms *spontaneously* into a state (a).[3] On a macroscopic scale, diffusion does not exhibit reversible character, although it is reversible on a microscopic scale. We conclude that macroscopic diffusion is of *irreversible* nature.[4]

[3] In a classical thought experiment, one half of a given volume is initially filled with a certain amount of gas molecules while the other half volume remains empty. A tight membrane separates the two volumes. As soon as the membrane is removed, the gas molecules distribute evenly and homogeneously in the entire volume. An external spectator hardly ever observes the reverse process, nor would he expect it, in which all gas atoms spontaneously decompose into two half volumes, one with and the other without gas molecules.

[4] The discrepancy between microscopic reversibility and macroscopic irreversibility has stimulated controversial discussion since it was introduced in 1867 by the Scottish physicist James Clerk Maxwell (1831–1879). A physical interpretation of this apparent dilemma has finally been given by Ludwig Boltzmann in his treatment of configurational entropy. See Section 1.3.1.

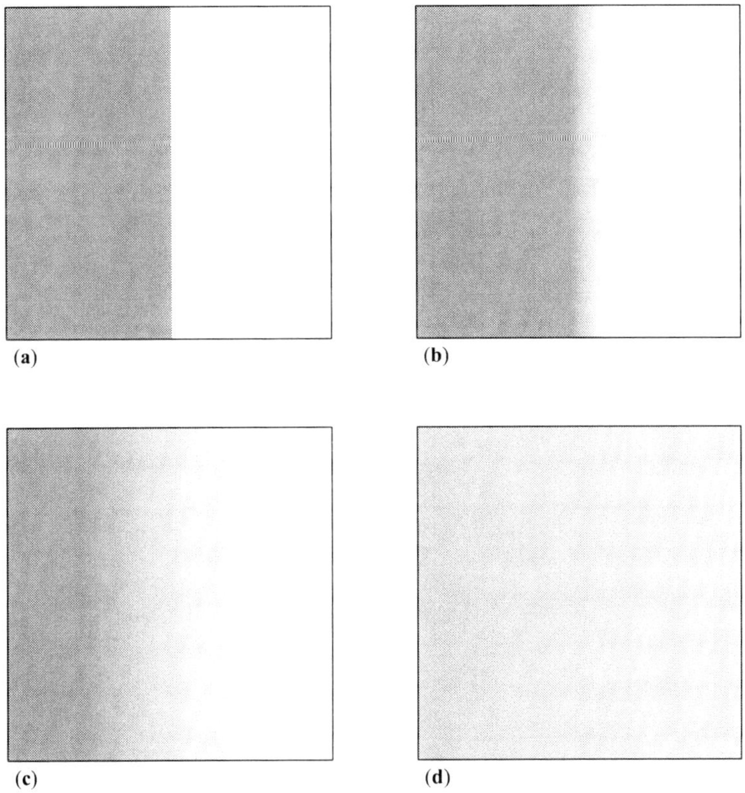

Figure 7.3. Macroscopic diffusion/mixing of atoms of a gray phase into atoms of a white phase.

7.1.3 Random Walk Treatment of Diffusion

Consider an ensemble of k "random walkers" that move on a rigid lattice in sequences of single atomic jumps in *random* direction. Let these jumps occur in only one dimension for the merit of simplicity, first. Each random walker, thus, moves along the x-direction in steps of $\Delta\lambda = \pm 1$. The possible sequence of up to four jumps is summarized in Table 7.1.

After a single step, the walker is found at a position given by the distance, $\Delta\lambda = \pm 1$. In the consecutive step, in 50% of all cases, the walker steps back in the same direction and the effective displacement is zero. In 50% of all cases, the walker continues in the same direction and the effective displacement is $\Delta\lambda = \pm 2$. If the experiment is repeated over and over again, the mean displacement $<\lambda_n>$ of k walkers going n steps is given as

$$<\lambda_n> = \frac{1}{k}\sum_{i=1}^{k}\lambda_i = 0. \tag{7.2}$$

Obviously, the mean displacement of all walkers is always zero and does, therefore, not represent a useful quantity to describe the movement of random walkers in space. However, Table 7.1 also indicates that although the mean displacement of walkers is zero, individual walkers can

Table 7.1. Random walk sequences in one dimension.

n	**Sequence**	$\Delta\lambda$	λ^2	$\sqrt{\frac{1}{k}\cdot\sum\lambda^2}$
1	+1	+1	1	$\sqrt{\left(\frac{1}{2}\cdot 2\right)}=(1)^{\frac{1}{2}}$
	−1	−1	1	
2	+1+1	+2	4	$\sqrt{\left(\frac{1}{4}\cdot 8\right)}=(2)^{\frac{1}{2}}$
	+1−1	0	0	
	−1+1	0	0	
	−1−1	−2	4	
3	+1+1+1	+3	9	$\sqrt{\left(\frac{1}{8}\cdot 24\right)}=(3)^{\frac{1}{2}}$
	+1+1−1	+1	1	
	+1−1+1	+1	1	
	−1+1+1	+1	1	
	+1−1−1	−1	1	
	−1+1−1	−1	1	
	−1−1+1	−1	1	
	−1−1−1	−3	9	
4	+1+1+1+1	+4	16	$\sqrt{\left(\frac{1}{16}\cdot 64\right)}=(4)^{\frac{1}{2}}$
	+1+1+1−1	+2	4	
	+1+1−1+1	+2	4	
	+1−1+1+1	+2	4	
	+1+1−1−1	0	0	
	+1−1+1−1	0	0	
	+1−1−1+1	0	0	
	+1−1−1−1	−2	4	
	−1+1+1+1	+2	4	
	−1+1+1−1	0	0	
	−1+1−1+1	0	0	
	−1−1+1+1	0	0	
	−1+1−1−1	−2	4	
	−1−1+1−1	−2	4	
	−1−1−1+1	−2	4	
	−1−1−1−1	−4	16	

be many steps away from the origin. To get a measure of the spreading behavior, the *root mean square* (RMS) value of the individual displacements can be employed. The RMS is defined as the square root of the average value of the *squared* displacements with

$$<\lambda_n^2> = \frac{1}{k}\sum_{i=1}^{k}\lambda_i^2. \tag{7.3}$$

The RMS displacement provides a measure of the average straight-line distance between the origin and the position of the walkers. For the RMS displacement, employing the results of Table 7.1, we observe

$$\sqrt{<\lambda_n^2>} = \sqrt{n}\cdot\lambda. \tag{7.4}$$

Equations (7.2) and (7.4) describe the observation that although the mean displacement of random walkers is zero, the walkers nevertheless spread out. It can be shown that these relations are also valid in two and three dimensions. Equation (7.4) describes the region where an observer can expect to find the walkers after n steps; and it also indicates that the region of expected positions extends with time, that is, the square root of steps n. In Figure 7.4,

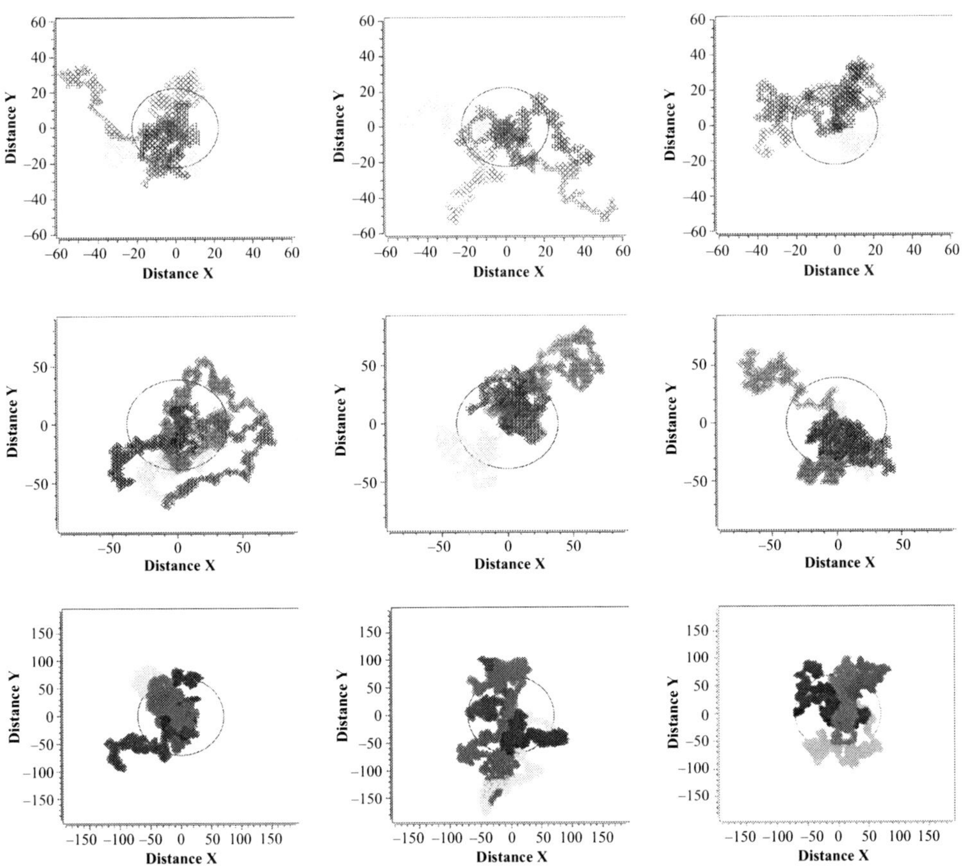

Figure 7.4. Series of random walkers going different number, n, of steps: first row: $n = 250$, second row: $n = 750$, third row: $n = 2500$. Each diagram contains the track of 5 walkers. The spherical region for the RMS displacement is evaluated from equation (7.4).

the pathway of random walkers[5] [191] is shown and compared to the prediction of equation (7.4).

For only low number of steps, a large variation of values is observed and the RMS displacement gives only a rough estimate for the region of most likely walker location. For the highest value of steps, 2500, the prediction given by equation (7.4) is more representative. It is important, therefore, to realize that the RMS displacement can only provide information on the *average* distance from the origin, but it gives no information on the expected position of a single walker after a given number of steps. The random walks presented in Figure 7.4 clearly show that individual walkers can go far beyond the RMS displacement value given by equation (7.4), while others are observed close to the origin.

7.1.4 The Einstein–Smoluchowski Equation

Diffusion in solid matter needs thermal activation, which manifests itself in oscillations of each atom around the spatial equilibrium position. The maximum frequency, ν, by which the atoms vibrate is known as the *Debye frequency*,[6] and it is in the order of 10^{12}–10^{13}/s for many metals (see, e.g., Table 2.4 in ref. [192]). With the Debye frequency (frequency of oscillations) denoted as ν and z_L representing the coordination number, that is, the number of nearest-neighbor sites, the attempt frequency of jumps is given by nz_L. Although the attempt rate for a diffusive jump is rather high, nevertheless, only a comparably small number of jumps will be successful. The probability of a single exchange event is proportional to the exponent of the ratio between the activation energy, Q_D, and the kinetic energy of the atom, $k_B T$. The effective jump frequency, Γ_D, is then given as

$$\Gamma_D = \nu z_L \exp\left(-\frac{Q_D}{RT}\right). \tag{7.5}$$

Note that Q_D is often given in units of J/mol, which is why we have utilized the relation $R = N_A k_B$.

Equation (7.5) can now be combined with the result of the random walk treatment of diffusion outlined in the previous section. Each successful jump will move the atom over one atomic distance. After substituting the number of jumps in equation (7.4) and substituting the radius coordinate r for x, the mean diffusion distance $<r^2>$ follows as

$$<r^2> = \Gamma_D \lambda^2 \cdot t. \tag{7.6}$$

On the other hand, the RMS displacement of walkers can equivalently be formulated in terms of the *second moment* of the walker distribution as

---

[5] The calculations were performed with the software *MatCalc*, Section 9.3, based on the random number generator *Mersenne Twister* [191].

[6] Named after the Dutch physicist and physical chemist Peter Joseph William Debye (1884–1966).

$$<r^2> = \int_0^\infty r^2 f(r,t)\,dr. \tag{7.7}$$

The function $f(r,t)$ is the probability density for finding a walker at a given position r and given time t. The solution to this problem is elaborated later in Section 7.2.2. On inserting equation (7.34) for $f(r,t)$, we get

$$<r^2> = \frac{1}{(4\pi Dt)^{d/2}} \int_0^\infty r^4 \exp\left(\frac{-r^2}{4Dt}\right) dr. \tag{7.8}$$

With the substitution

$$z = \frac{r}{\sqrt{4Dt}}, \tag{7.9}$$

the integral on the right-hand side can be carried out with

$$\int_0^\infty z^4 \exp\left(-z^2\right) dz = \frac{3\sqrt{\pi}}{8}. \tag{7.10}$$

For the RMS displacement of a spreading diffusant originating at $x = 0$, we finally arrive at

$$<r^2> = 2dDt, \tag{7.11}$$

where d denotes the dimensionality of the diffusion problem. The relation between the mean square displacement and the macroscopic diffusion coefficient is often denoted as the *Einstein–Smoluchowski equation*, as it was derived independently by A. Einstein [193] and M. Smoluchowski [194].

It is important to emphasize here that in the derivation of equation (7.11), we have made use of the "macroscopic" solution to Fick's second law for a spreading diffusion cloud. Because Fick's laws are phenomenological relations, equation (7.11) also represents a phenomenological relation.

Finally, it is interesting to combine this result with the RMS displacement of random walkers. Comparison of equations (7.4) and (7.11) leads to

$$\Gamma_D \lambda^2 = 2dD, \tag{7.12}$$

where Γ_D is the effective "microscopic" jump frequency, λ is the "microscopic" mean random walker (RMS) displacement, and D is the "macroscopic" diffusion coefficient. Accordingly, the frequency of successful diffusive jump attempts is

$$\Gamma_D = \frac{2dD}{\lambda^2}, \tag{7.13}$$

and, vice versa, the macroscopic diffusion coefficient reads in terms of the microscopic jump frequency and the observed RMS displacements as

$$D = \frac{1}{2d}\Gamma_{\mathrm{D}}\lambda^2. \tag{7.14}$$

This equation, thus, delivers a handy relation between microscopic and macroscopic quantities describing diffusion.

7.2 Macroscopic Models of Diffusion

Macroscopic diffusion is a phenomenon that we observe regularly, every day in our life. The majority of diffusion encounters occurs in the context of heat transport. In winter, we dress up with thick clothes to avoid too fast diffusion of heat from our body to the cold environment. In the kitchen, heat diffuses from the heat source through the cooking pots into the food. Moreover, even without the application of extra heat from ovens, salt and spices, applied onto the surface of goods, can penetrate into the bulk volume of the goods to makes them a delicious experience. In any case, either in the transport of energy or mass, diffusion governs the kinetics of the process. This observation is the focus of the following analysis.

7.2.1 Phenomenological Laws of Diffusion

The British chemist and physiologist Thomas Graham (1805–1869), was one of the first researchers to perform a systematic analysis of diffusion [195]. Graham investigated the flow of gas atoms through a small hole into vacuum as well as the diffusion of salt in aqueous solution. Graham was able to show that the diffusion rate, or mass flow, is proportional to the concentration difference between the two reservoirs of matter. However, Graham did not formulate the fundamental laws of diffusion. This was shortly afterward achieved by Adolf Eugen Fick, a German physiologist (1829–1901), who postulated that the laws of diffusion are equivalent to the laws of heat conduction, as observed earlier by Jean Baptiste Joseph Fourier, a French mathematician and physicist (1768–1830), and the laws of transport of current, as deducted by Georg Simon Ohm, a German physicist (1789–1854).

According to Fick, the flux, J, of matter is proportional to the gradient in concentration with

$$J = -D\frac{\partial c}{\partial x}, \tag{7.15}$$

where x is a spatial coordinate and D is a proportionality factor denoted as *diffusion coefficient*. Equation (7.15) is generally known as *Fick's first law*. Interestingly, this law was derived on purely phenomenological grounds and experimental observation. Adolf Fick published his research on diffusion in 1855 [196]. As outlined in the previous section, the rigorous derivation of the laws of diffusion is due to Einstein [193] and Smoluchowski [194].

In multiple dimensions, the spatial gradient in the variable x can be expressed using the Nabla operator, ∇. Fick's first law for the diffusive flux vector, $\mathbf{J}$, becomes

$$\mathbf{J} = -D\nabla c, \tag{7.16}$$

where

$$\nabla = \begin{pmatrix} \partial/\partial x \\ \partial/\partial y \\ \partial/\partial z \end{pmatrix}. \tag{7.17}$$

Fick's first law defines a unique relation for the stationary mass transport in the presence of a composition gradient.

When investigating diffusive processes further, a representative volume element (RVE), $\Delta x \Delta y \Delta z$, may be considered as depicted in Figure 7.5. Let the diffusive flux occur in the x direction, first. The change of total mass related to the RVE may be given in terms of the diffusive flux in the x direction J_x^{in} going into the RVE and the flux J_x^{out} leaving the RVE with

$$\Delta_x(\#\text{moles}) = -\left(J_x^{\text{out}} - J_x^{\text{in}}\right)\Delta y \Delta z \cdot \Delta t. \tag{7.18}$$

As the concentration c of atoms in the RVE is

$$c = \frac{\#\text{moles}}{\Delta x \Delta y \Delta z}, \tag{7.19}$$

the total change of moles of atoms related to the flux in x direction is

$$\Delta c = \frac{\Delta_x(\#\text{moles})}{\Delta x \Delta y \Delta z}. \tag{7.20}$$

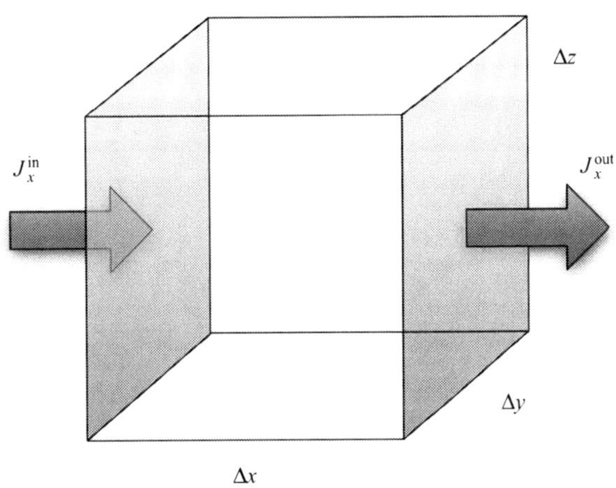

Figure 7.5. Mass conservation for a representative volume element.

On comparison of equations (7.18) and (7.20), we find

$$\frac{\Delta c}{\Delta t} = -\frac{J_x^{\text{out}} - J_x^{\text{in}}}{\Delta x}. \tag{7.21}$$

With the abbreviation $\Delta J = J_x^{\text{out}} - J_x^{\text{in}}$, generalization to three dimension and using differentials instead of the difference expressions, we finally arrive at

$$\frac{\partial c}{\partial t} = -\nabla \mathbf{J}. \tag{7.22}$$

Equation (7.22) represents a general relation for mass conservation in an RVE for diffusive mass transport in the absence of sinks and sources. When inserting Fick's first law for the diffusive flux, **J**, equation (7.16), we obtain the general relation

$$\frac{\partial c}{\partial t} = -\nabla\left(-D\nabla c\right). \tag{7.23}$$

If the diffusion coefficient, D, is independent of concentration, this relation simplifies to

$$\frac{\partial c}{\partial t} = D\nabla^2 c. \tag{7.24}$$

Equation (7.24) is generally known as *Fick's second law*. It is a second-order linear partial differential equation describing transient diffusion processes.

Using standard algebra, this equation can be expressed in terms of planar, cylindrical, and spherical coordinates by substituting the Laplace operator, ∇^2. With the symbols x, y, and z for Cartesian coordinates, r, θ, and z for cylindrical coordinates and r, θ, and ϕ for spherical coordinates, the transient diffusion equations become

$$\frac{\partial c}{\partial t} = D\left(\frac{\partial^2 c}{\partial x^2} + \frac{\partial^2 c}{\partial y^2} + \frac{\partial^2 c}{\partial z^2}\right), \tag{7.25}$$

$$\frac{\partial c}{\partial t} = \frac{D}{r}\left[\frac{\partial}{\partial r}\left(r\frac{\partial c}{\partial r}\right) + \frac{1}{r}\frac{\partial^2 c}{\partial \theta^2} + r\frac{\partial^2 c}{\partial z^2}\right], \tag{7.26}$$

$$\frac{\partial c}{\partial t} = \frac{D}{r^2}\left[\frac{\partial}{\partial r}\left(r^2\frac{\partial c}{\partial r}\right) + \frac{1}{\sin\theta}\frac{\partial}{\partial \theta}\left(\sin\theta\frac{\partial c}{\partial \theta}\right) + \frac{1}{\sin^2\theta}\frac{\partial^2 c}{\partial \phi^2}\right]. \tag{7.27}$$

If the diffusion problem can be reduced by symmetric boundary conditions, that is, linear diffusion with rotational symmetry, equations (7.25)–(7.27) simplify to

$$\frac{\partial c}{\partial t} = D\frac{\partial^2 c}{\partial r^2}, \tag{7.28}$$

$$\frac{\partial c}{\partial t} = \frac{D}{r}\cdot\frac{\partial}{\partial r}\left(r\frac{\partial c}{\partial r}\right), \tag{7.29}$$

$$\frac{\partial c}{\partial t}=\frac{D}{r^2}\cdot\frac{\partial}{\partial r}\left(r^2\frac{\partial c}{\partial r}\right), \tag{7.30}$$

where r denotes the spatial variable.

With the substitution $\lambda = r/\sqrt{t}$, equations (7.28)–(7.30) can be further transformed into the general form of the linear diffusion equation

$$-\frac{\lambda}{2}\frac{\partial c}{\partial \lambda}=\frac{D}{\lambda^{p-1}}\cdot\frac{\partial}{\partial \lambda}\left(\lambda^{p-1}\frac{\partial c}{\partial \lambda}\right), \tag{7.31}$$

where the index p has the values 1, 2, or 3 for planar, cylindrical, or spherical symmetry, respectively. Equation (7.31) is an ordinary parabolic differential equation in the variable λ. For the case of $p = 1$, this equation is also known as the *Boltzmann diffusion equation.*

7.2.2 Special Solutions of Fick's Second Law

The macroscopic laws of diffusion can be solved analytically for special cases of well-defined boundary conditions.

7.2.2.1 Spreading of a Diffusant From a Point Source

Consider a point mass, M, which is initially located at a position $x = 0$. With the concentration, c, of atoms and assuming one-dimensional diffusion, the entire mass can be viewed as being concentrated in a single plane. Applying the mass conservation law, we obtain

$$\int_{-\infty}^{+\infty} c(x,t)\,dx = 1. \tag{7.32}$$

The solution to equation (7.24) with the constraint (7.32) is elaborated in many textbooks, e.g., [14,184], and delivers

$$c(x,t)=\frac{M}{2\sqrt{\pi Dt}}\exp\left(\frac{-x^2}{4Dt}\right). \tag{7.33}$$

The second term in this equation indicates that the diffusing atoms will spread according to a Gaussian distribution. This situation is shown schematically in Figure 7.6 for different values of Dt.

The general solution to this problem in d dimensions and unit mass M is

$$c(r,t)=\frac{M}{(4\pi Dt)^{d/2}}\exp\left(\frac{-r^2}{4Dt}\right), \tag{7.34}$$

where the variable r denotes the distance from the origin. In this form, equation (7.34) represents the probability distribution of walkers spreading from the point source with a diffusion coefficient D and time t.

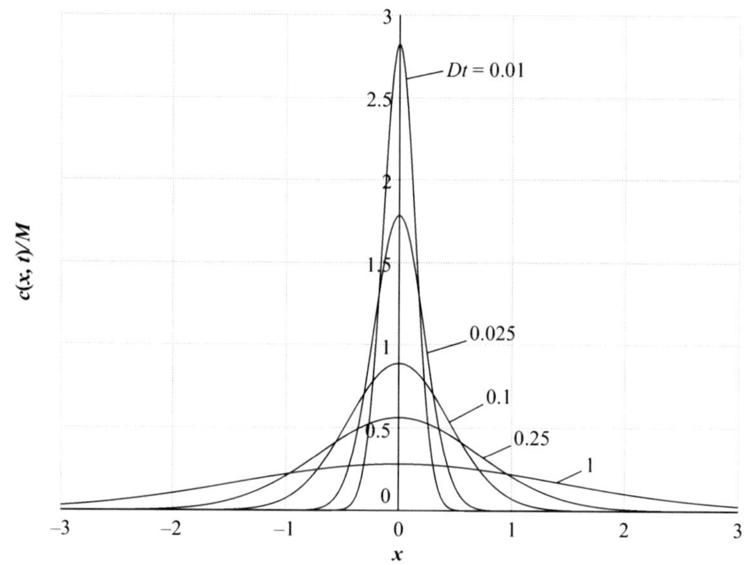

Figure 7.6. Spreading of diffusant from a source of mass, M, located at the origin according to equation (7.33).

7.2.2.2 Diffusion into a Semi-Infinite Sample

Consider one-dimensional diffusion into a semi-infinite sample along the x-direction. In contrast to the previous example, where the spreading of a diffusant was considered with all atoms initially concentrated in a mass M located at the center of the sample, we now deal with a situation where the open face of a semi-infinite sample is in contact with an infinite reservoir of atoms at a concentration, c_0. The atoms are allowed to diffuse into the sample with kinetics given by Fick's second law. The corresponding situation is typical for technological processes such as carburization and nitridation, or the reverse processes of decarburization and denitridation.

The boundary condition of the problem is a constant concentration at the boundary of the semi-infinite sample given as

$$c(0,t) = c_0. \tag{7.35}$$

The solution to Fick's second law is then given as

$$c(x,t) = c_0 \cdot \operatorname{erfc}\left(\frac{x}{\sqrt{4Dt}}\right), \tag{7.36}$$

with the *error function*, "erf(z)" and its complement, "erfc(z)," given as

$$\operatorname{erfc}(z) = 1 - \operatorname{erf}(z) = 1 - \frac{2}{\sqrt{\pi}} \int_0^z \exp\left(-\eta^2\right) d\eta. \tag{7.37}$$

In Figure 7.7, several diffusion profiles for values of $z = x/\sqrt{4Dt}$ are plotted. Note that these curves are self-similar for z = const.

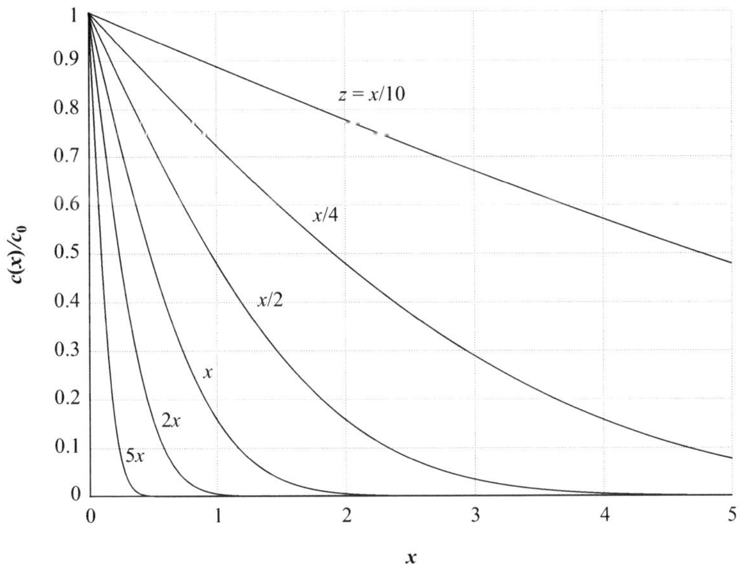

Figure 7.7. Composition profile for diffusion into a semi-infinite sample according to equation (7.36).

7.2.3 Numerical Solution

In many practical applications, either the diffusion problem cannot easily be reduced to simple one-dimensional geometry or the composition dependence of the diffusion coefficient[7] is too complex to be able to take advantage of analytical solutions of Fick's second law. In these cases, alternative techniques must be applied to obtain the solution of the transient diffusion equation.

For a numerical solution of the diffusion problem, continuous space is divided into discrete spatial cells, or volumes. Using standard numerical methods, such as the Finite Element, the Finite Volume, or Finite Differences method, the evolution of composition profile in time is obtained by integration of the ordinary differential equations (ODE) for transient diffusion in each of the cells. The ODE integration can be performed by Euler forward integration, implicit integration involving the solution of a linear system of equations, as well as advanced higher-order Runge–Kutta methods. Details for these solution procedures are described in numerous textbooks on differential equations. A simple example using the Finite Differences method is outlined in ref. [11], Section 5.4.

Alternatively, specialized software codes for diffusion simulation can be utilized, which often provide mobility or diffusion databases along with the program. One of these codes is the program *Profiler* [197] that is based on the square-root diffusivity method. Other programs, which involve CALPHAD-type thermodynamic and mobility databases, are the software packages DICTRA and *MatCalc*. The latter two products are described in Sections 9.1 and 9.3, respectively.

[7] More exactly, the *diffusion coefficient matrix*. See Section 7.2.5, later.

7.2.4 Diffusion Forces and Atomic Mobility

When Adolf Fick formulated the phenomenological relation between the mass flux and the gradient of chemical composition with the proportionality coefficient, D, his conclusions were based on the results of investigations on the diffusion of salt ions in aqueous solution. Strictly speaking, Fick's first law, equation (7.16), describes the special case where the salt ions do not interact with each other chemically and the diffusion process is governed by random walk of atoms/vacancies only. In a more rigorous picture, the generalized forces acting on the diffusing species must be considered as an additional component constituting the driving force for macroscopic diffusion.

If a force, F, acts on a body, this force accelerates the body according to Newtonian mechanics. The acceleration acts in the direction of the force. This is evident, for instance, in gravitational forces acting on a mass or electrical fields acting on a charged particle. As a force, F, is generally defined as the gradient of a potential, Φ, we can write

$$F \propto \nabla\Phi. \tag{7.38}$$

In diffusion, this force is given by the negative gradient of the chemical potential with

$$F = -\nabla\mu. \tag{7.39}$$

In crystalline materials, the force acting on the atoms is counteracted by the effect of the periodic potential of the lattice. The vast majority of atoms will remain at their original lattice positions. According to the discussion in Section 7.1.1, migration of a lattice atom is facilitated if the neighboring lattice site is empty, that is, it is occupied by a lattice vacancy. In the case of an external force, the probability for the atom–vacancy exchange in the direction of the force will be increased.

On a macroscopic level, the resistance of atoms against movement in the direction of the generalized force can be interpreted as *diffusional friction* characterized by a friction coefficient, f_D. With an effective drift velocity, u, of a large number of atoms caused by the force, F, we can write

$$F = f_D \cdot u. \tag{7.40}$$

On the other hand, the drift velocity, u, is given in terms of the generalized force, F, and the friction coefficient, f_D, with

$$u = \frac{1}{f_D} \cdot F = B \cdot F, \tag{7.41}$$

where B is the mobility of atoms. Note that mobility and friction coefficient are related as $B = 1/f_D$. The *diffusional mobility*, B, is an intrinsic property of an atomic species in a given material. It defines the physical response of an ensemble of atoms to a generalized force, F. In the case of diffusion, this force is given by the gradient of chemical potential, $\nabla\mu$. For the flux, J, of atoms passing through a plane of unit area in unit time, we thus obtain

$$J = cu = -cB\nabla\mu. \tag{7.42}$$

Equation (7.42) represents the fundamental relation between the diffusional flux of atoms, J, and the generalized force, $\nabla\mu$, with the proportionality factor, $-cB$. c is the concentration of the diffusing element.

This relation can be investigated further and compared to the *phenomenological* diffusion equation (7.22), Fick's first law, after inserting for the chemical potential gradient, $\nabla\mu$. In general solutions, the chemical potential of an element B can be written in terms of the chemical activity, a_B, with[8]

$$\mu_B = \mu_{B,0} + RT\ln(a_B). \tag{7.43}$$

With the activity coefficient, f_B, given as $a_B = f_B \cdot X_B$, the chemical potential gradient becomes

$$\nabla\mu_B = RT\left[\nabla\ln(f_B) + \nabla\ln(X_B)\right]. \tag{7.44}$$

This expression can be reformulated as

$$\nabla\mu_B = \frac{RT}{X_B}\left[\frac{X_B\nabla\ln(f_B)}{\nabla X_B} + 1\right]\nabla X_B, \tag{7.45}$$

or

$$\nabla\mu_B = \frac{RT}{X_B}\left[\frac{\nabla\ln(f_B)}{\nabla\ln(X_B)} + 1\right]\nabla X_B. \tag{7.46}$$

After inserting into equation (7.42), transforming into mole fraction variables, X_B, and substituting the Nabla operator by partial differentials in the square brackets, we obtain

$$J_B = -v^{\alpha} X_B B_B \nabla\mu_B = \frac{RT}{X_B}\left[\frac{\partial\ln(f_B)}{\partial\ln(X_B)} + 1\right]\nabla c_B. \tag{7.47}$$

This result can be compared with the definition of Fick's first law, equation (7.16), which delivers a relation between the diffusion coefficient, D_B, and the mobility, B_B, as

$$D_B = RT\cdot B_B\left(1 + \frac{\partial\ln(f_B)}{\partial\ln(X_B)}\right). \tag{7.48}$$

In systems without chemical interaction between atoms, for example, in ideal solutions, the term in round brackets becomes unity because the activity coefficient, f_B, becomes unity and the logarithm terms vanish. Then, equation (7.48) reduces to

$$D_B = RT\cdot B_B, \tag{7.49}$$

yielding a direct relation between the *microscopic* mobility of atoms, B_B, which was derived from purely atomistic considerations, and the *macroscopic* diffusion coefficient, D_B, which was developed earlier from the phenomenological considerations of Fick's equations.

---

[8] See also Section 1.3.2, equation (1.40).

Interestingly, the present treatment suggests that even in the absence of external forces, the atoms experience a force that causes a net flux of atoms. This force cannot be measured directly, which is why it is generally denoted as a *generalized* force. In ideal solutions, this force is directly proportional to the concentration gradient.

In general solutions, the effect of nonideality can be accounted for in the form of the *thermodynamic factor*, ϕ_B, as

$$D_B = RTB_B \cdot \phi_B, \tag{7.50}$$

where

$$\phi_B = 1 + \frac{\partial \ln(f_B)}{\partial \ln(X_B)}. \tag{7.51}$$

In evaluation of equation (7.51), care must be taken because the derivative is taken with respect to the mole fraction variable, X_B. In multicomponent systems, the thermodynamic factor, ϕ, is obtained as

$$\phi_i = \frac{X_i}{RT}\left(\frac{\partial \mu_i}{\partial X_i} - \frac{1}{1-X_i}\sum_{j \neq i} X_j \frac{\partial \mu_i}{\partial X_j}\right). \tag{7.52}$$

Derivation of equation (7.52) is elaborated in, for example, ref. [11].

7.2.5 Multicomponent Diffusion

Our treatment of diffusion has so far been concerned with a single diffusing species denoted as B, only. This is generalized now for multicomponent systems with n elements. For a component i, the diffusional flux is given as

$$J_i = -c_i B_i \nabla \mu_i. \tag{7.53}$$

In this relation, the diffusive flux, J_i, of element i is defined in terms of the concentration, c_i, that represents the number of moles of atoms that are available for carrying the net mass flow, the mobility, B_i, and the generalized force, $\nabla \mu_i$. As the chemical potential, $\mu = \mu(\mathbf{X})$, is a function of all composition variables, with $\mathbf{X}$ representing the chemical composition vector, the gradient of the chemical potential already accounts for all multicomponent effects that may come from solution nonideality.

From a practical point of view, the chemical potentials might not be readily accessible. It is convenient, then, to describe the diffusion process in terms of concentration gradients as done in Fick's laws. By application of the chain derivation rule and using the spatial coordinate, r, for the diffusive flux we find

$$J_i = -c_i B_i \sum_{j=1}^{n} \frac{\partial \mu_i}{\partial X_j} \frac{X_j}{\partial r} = -X_i B_i \sum_{j=1}^{n} \frac{\partial \mu_i}{\partial X_j} \frac{c_j}{\partial r}. \tag{7.54}$$

It is convenient, now, to introduce the *intrinsic* diffusion coefficient matrix, D_{ij}, with

$$D_{ij} = -X_i B_i \frac{\partial \mu_i}{\partial X_j}. \tag{7.55}$$

The intrinsic diffusion matrix forms an $n \times n$ symmetric matrix, which is

$$D_{ij} = \begin{bmatrix} D_{11} & D_{12} & \cdots & D_{1n} \\ D_{21} & D_{22} & & \\ \vdots & & \ddots & \\ D_{n1} & & & D_{nn} \end{bmatrix}. \tag{7.56}$$

The intrinsic flux of component i can be formulated in compact form as

$$J_i = D_{ij} \nabla c_j, \quad i, j = 1 \ldots n, \tag{7.57}$$

or

$$J_i = \sum_{j=1}^{n} D_{ij} \frac{\partial c_j}{\partial r}. \tag{7.58}$$

The summation in equation (7.58) emphasizes that in multicomponent diffusion, the impact of every single element on every other element must be summed up. Only then, effects, such as uphill diffusion of individual components against their own composition gradient [198], become tractable by computer modeling.

The individual fluxes, J_i, represented by equation (7.58) are not independent, however, neither are the concentration gradients, $\partial c_j / \partial r$. Based on mass conservation considerations, one of the n fluxes and one of the n concentration gradients can always be expressed in terms of the others. Since the sum over all n concentrations, c_i, is

$$\sum_{j=1}^{n} c_j = \frac{1}{v^\alpha}, \tag{7.59}$$

where v^α is the molar volume, for the derivative, we get

$$\sum_{j=1}^{n} \frac{\partial c_j}{\partial r} = 0. \tag{7.60}$$

When choosing the concentration gradient of component n as the dependent one, the concentration gradient for this element can be written as

$$\frac{\partial c_n}{\partial r} = -\sum_{j=1}^{n-1} \frac{\partial c_j}{\partial r}. \tag{7.61}$$

The constraint (7.61) can be used to reformulate the general form of Fick's first law, equation (7.58), in terms of independent gradients. The summation term of the reference element, n, can be written outside the summation to yield

$$J_i = \sum_{j=1}^{n-1} D_{ij} \frac{\partial c_j}{\partial r} + D_{in} \frac{\partial c_n}{\partial r} = \sum_{j=1}^{n-1} D_{ij} \frac{\partial c_j}{\partial r} - D_{in} \sum_{j=1}^{n-1} \frac{\partial c_j}{\partial r}$$
$$= \sum_{j=1}^{n-1} \left(D_{ij} - D_{in} \right) \frac{\partial c_j}{\partial r}. \tag{7.62}$$

Just like before, the diffusion coefficients in parenthesis can be lumped into a diffusion matrix, which is now of dimension $(n-1)\times(n-1)$, and reads

$$D'_{ij} = \begin{bmatrix} D_{11} - D_{1n} & D_{12} - D_{1n} & \cdots & D_{1(n-1)} - D_{1n} \\ D_{21} - D_{2n} & D_{22} - D_{2n} & & \\ \vdots & & \ddots & \\ D_{(n-1)1} - D_{(n-1)n} & & & D_{(n-1)(n-1)} - D_{(n-1)n} \end{bmatrix}. \tag{7.63}$$

The matrix, D'_{ij}, is denoted as the *reduced* intrinsic diffusion matrix. The $n - 1$ independent diffusional fluxes are then

$$J_i = D'_{ij} \nabla c_j, \qquad i, j = 1 \ldots (n-1), \tag{7.64}$$

or

$$J_i = \sum_{j=1}^{n-1} D'_{ij} \frac{\partial c_j}{\partial r}. \tag{7.65}$$

In the treatment of multicomponent diffusion, so far, we have assumed that the external observer moves synchronously with the lattice planes of the diffusion sample. The respective coordinate system is denoted as the *lattice frame of reference*. In his series of papers, Kirkendall [187–189] has shown that inert markers that are inserted into the contact plane of a diffusion couple, migrate with respect to the coordinate system of the sample if the diffusion coefficient of the diffusing species differs.[9] If the position of the external observer is stationary with respect to the position of the entire diffusion sample, the coordinate system is denoted as the *laboratory frame of reference.*

The intrinsic fluxes, J_i, are converted from the lattice frame of reference into the laboratory frame of reference with a Galilean transformation, that is, by the addition of a convective flux term. The drift velocity, v_D, by which the intrinsic coordinate system moves relative to the external observer, is given by

$$v_D = J_{Va} v^{\alpha}, \tag{7.66}$$

[9] Kirkendall has investigated the diffusion of Cu and Zn in alpha-brass. He observed that Zn atoms move faster into the Cu-rich side of a diffusion couple than vice versa. The flux of Zn-atoms is counteracted by a flux of vacancies in the opposite direction, as diffusion occurs by the vacancy–atom exchange mechanism. Across the interface, a nonbalancing net flux of mass exists, therefore, as mass (Zn atoms) is effectively transported from the Zn-rich side of the contact plane to the Cu-rich side, vacancies are moved in the opposite direction. For an external observer, the initial contact plane of the diffusion couple moves into the direction of the Zn-rich side.

with the net flux of vacancies across the interface defined as

$$J_{\text{Va}} = \sum_{j=1}^{n} J_j, \tag{7.67}$$

and v^{α} being the molar volume. The convective fluxes, J_i^{conv}, are finally determined as

$$J_i^{\text{conv}} = c_i v_{\text{D}} = -X_i \sum_{j=1}^{n} J_j. \tag{7.68}$$

With the intrinsic fluxes, J_i^{conv}, given by equation (7.58), and with the fluxes, $\bar{J}_i$, in the laboratory frame of reference given as the superposition of the intrinsic and convective fluxes as

$$\bar{J}_i = J_i^{\text{intr}} + J_i^{\text{conv}}, \tag{7.69}$$

the flux of component i in the laboratory frame of reference becomes

$$\bar{J}_i = -\sum_{j=1}^{n} D_{ij} \frac{\partial c_j}{\partial r} + X_i \sum_{k=1}^{n} \sum_{j=1}^{n} D_{kj} \frac{\partial c_j}{\partial r}. \tag{7.70}$$

This expression can be further evaluated with introduction of the Kronecker delta, δ_{ij}, where $\delta_{ij} = 1$ if $i = j$, and $\delta_{ij} = 0$ if $i \neq j$. Expansion of the first term and rearrangement delivers

$$\begin{aligned} \bar{J}_i &= -\sum_{k=1}^{n} \sum_{j=1}^{n} \delta_{ij} D_{kj} \frac{\partial c_j}{\partial r} + X_i \sum_{k=1}^{n} \sum_{j=1}^{n} D_{kj} \frac{\partial c_j}{\partial r} \\ &= -\sum_{k=1}^{n} \sum_{j=1}^{n} (\delta_{ik} - X_i) D_{kj} \frac{\partial c_j}{\partial r}. \end{aligned} \tag{7.71}$$

The summations in equation (7.71) are performed over all n diffusing elements. Again, one of the concentration gradients can be expressed in terms of the others, equation (7.61). After inserting into equation (7.70), we obtain

$$\bar{J}_i = -\sum_{j=1}^{n-1} \left(D_{ij} - D_{in}\right) \frac{\partial c_j}{\partial r} + X_i \sum_{k=1}^{n-1} \sum_{j=1}^{n-1} \left(D_{kj} - D_{kn}\right) \frac{\partial c_j}{\partial r}. \tag{7.72}$$

The *independent* fluxes in the laboratory frame of reference are finally obtained after some algebra as

$$\bar{J}_i = -\sum_{j=1}^{n-1} D'_{ij} \frac{\partial c_j}{\partial r}, \tag{7.73}$$

with the diffusion coefficient matrix given as

$$D'_{ij} = -\sum_{m=1}^{n-1}\sum_{s=1}^{n-1}\left(\delta_{js} - X_j\right)\left(\delta_{mi} - X_i\right)D_{ms}. \tag{7.74}$$

The relations presented in this section are used in most of the recent software packages for diffusion simulation. Commonly, databases are provided with the software containing the diffusional mobility of atoms in CALPHAD-type notation, that is, polynomials in T, P, and $\mathbf{X}$. The thermodynamic aspects are accounted for in thermodynamic databases storing the composition-dependent Gibbs energy of phases, from which the thermodynamic factor and, finally, the diffusion coefficients in either intrinsic or laboratory frame of reference are obtained.

7.3 Activation Energy for Diffusion

Diffusion in crystalline materials occurs predominantly by random jumps of atoms into empty neighboring lattice positions (substitutional diffusion) or into sites of the interstitial sublattice (interstitial diffusion). Between the individual sites an energy barrier exists as a consequence of the periodic potential of the crystal lattice, which the migrating atom must overcome before it can move from one lattice position to another. This potential barrier is denoted as the *activation energy for diffusion*.

Diffusion of an atomic species in a matrix of the same atomic sort is denoted as *self diffusion*. If the atoms under consideration migrate in an environment of only unlike atoms, the diffusion process is denoted as *impurity* or *tracer diffusion*. In both cases, it is assumed that the atoms do not chemically interact and that the diffusion process occurs by random walk only. If the solution is nonideal, which is commonly the case in concentrated alloys, the diffusion kinetics of atoms is concentration dependent and the diffusion problem must be described by *chemical diffusion*. The latter effect can be taken into account, for instance, in the form of the thermodynamic factor, equation (7.52), or in the way outlined in Section 7.2.5 for multicomponent diffusion. In this section, the focus is laid on the temperature dependence of the diffusion process.

7.3.1 Temperature Dependence of the Diffusion Coefficient

The rate at which thermally activated processes occur can be estimated from kinetic theory that describes the reaction rate, k, as being proportional to the fraction of atoms having a kinetic energy[10] of at least the activation energy, ΔH_R, with

[10] A physical explanation for the exponential dependence of the reaction rate comes from statistical mechanics, where the distribution of the kinetic energy of vibrating particles is described by Maxwell–Boltzmann statistics as $\frac{N_i}{N} = \frac{g_i \exp\left(-E_i/k_B T\right)}{\sum_j g_j \exp\left(-E_j/k_B T\right)}$, where N_i is the number of atoms in a given state, N is the total number of states, g_i designates the number of degenerate states, and E_i is the energy of the particular state.

$$k \approx \exp\left(-\frac{\Delta H_{\mathrm{R}}}{k_{\mathrm{B}}T}\right), \tag{7.75}$$

where k_{B} is the Boltzmann constant and T is temperature. This law was initially proposed by Jacobus Henricus van 't Hoff, Jr., a Dutch physical and organic chemist (1852–1911). A few years later, Svante Arrhenius, a Swedish chemist and physicist (1859–1927), provided a physical interpretation of this relation. Equation (7.75) is known as the *Arrhenius law*.

For the temperature dependence of the diffusion coefficient, D, we can thus write the proportionality

$$D \approx \exp\left(-\frac{\Delta H_{\mathrm{mig}}}{k_{\mathrm{B}}T}\right) = \exp\left(-\frac{\Delta h_{\mathrm{mig}}}{RT}\right), \tag{7.76}$$

where ΔH_{mig} and Δh_{mig} are the *migration enthalpy* and *molar migration enthalpy*, respectively. This quantity represents the height of the energy barrier between two neighboring lattice positions during diffusion.

On an interstitial sublattice, the migration enthalpy represents the major barrier that must be overcome by thermal excitation and, with the pre-exponential factor, D_0, we get

$$D = D_0 \exp\left(-\frac{q_{\mathrm{D}}^{\mathrm{interst}}}{RT}\right), \tag{7.77}$$

where $q_{\mathrm{D}}^{\mathrm{interst}}$ is the *activation energy for diffusion* on the interstitial sublattice. It is related to the migration enthalpy as

$$q_{\mathrm{D}}^{\mathrm{interst}} = \Delta h_{\mathrm{mig}}. \tag{7.78}$$

On the substitutional sublattice, atomic jumps can only occur into empty neighboring sites, that is, vacancies. The rate at which atomic jumps occur is, therefore, also proportional to the density of vacant sites. For the equilibrium site fraction of thermal vacancies, $X_{\mathrm{Va}}^{\mathrm{eq}}$, in Section 1.6, equation (1.78), we have found

$$X_{\mathrm{Va}}^{\mathrm{eq}} = \exp\left(-\frac{\Delta g_{\mathrm{Va}}^{\mathrm{f}}}{RT}\right) = \exp\left(-\frac{\Delta h_{\mathrm{Va}}^{\mathrm{f}}}{RT}\right)\exp\left(\frac{\Delta s_{\mathrm{Va}}^{\mathrm{f}}}{R}\right), \tag{7.79}$$

and the diffusion coefficient on the substitutional sublattice becomes

$$D \approx \exp\left(\frac{\Delta s_{\mathrm{Va}}^{\mathrm{f}}}{R}\right)\exp\left(-\frac{\Delta h_{\mathrm{Va}}^{\mathrm{f}}}{RT}\right)\cdot\exp\left(-\frac{\Delta h_{\mathrm{mig}}}{RT}\right). \tag{7.80}$$

The exponential term containing the vacancy formation entropy is independent of temperature and it is, therefore, commonly incorporated into the pre-exponential factor, D_0. On the substitutional sublattice, the diffusion coefficient becomes

$$D = D_0 \exp\left(-\frac{q_D^{\text{subst}}}{RT}\right) = D_0 \exp\left(-\frac{\Delta h_{\text{Va}}^{\text{f}}}{RT}\right) \exp\left(-\frac{\Delta h_{\text{mig}}}{RT}\right), \tag{7.81}$$

with

$$q_D^{\text{subst}} = \Delta h_{\text{Va}}^{\text{f}} + \Delta h_{\text{mig}}. \tag{7.82}$$

In Table 7.2, typical values for the activation energies for self diffusion as well as impurity diffusion are summarized for some elements. Note that the activation energies of substitutional elements are generally higher than the values for interstitials. Whereas the activation energy for diffusion of substitutional elements contains two contributions, $\Delta h_{\text{Va}}^{\text{f}} + \Delta h_{\text{mig}}$, the activation energy for diffusion of interstitial elements (e.g., C in Fe) contains only the migration part. This makes interstitial diffusion orders of magnitude faster than substitutional diffusion, particularly at lower temperatures.

The diffusion equation can be rearranged using the natural logarithm as

$$\ln D = -\frac{q_D}{RT} + \ln D_0. \tag{7.83}$$

Accordingly, the natural logarithm of the diffusion coefficient plotted against $-q_D/RT$ yields a straight line. This is shown exemplarily for the diffusion coefficient of Al in Figure 7.8 that is drawn for the values of q_D and D_0 from Table 7.2. Extrapolation of this line to $\ln(D) = 0$ delivers the value of $\ln(D_0)$.

Although the Arrhenius equation represents a simple closed mathematical form for the temperature dependence of thermally activated processes, diffusion characteristics are usually more complex and need advanced mathematical descriptions. The short list of coefficients summarized in Table 7.2 is, therefore, only an indication of the approximate value of diffusion coefficients.

Table 7.2. Typical values for the vacancy formation enthalpy, Δh_{Va}, and the activation energy for self/impurity diffusion, q_D, frequency factor, D_0, as well as the diffusion coefficient, D, at different temperatures for some metallic systems.

El	Matrix	Δh_{Va} kJ/mol	q_D kJ/mol	D_0 m²/s	D(298 K) m²/s	D(873 K) m²/s	D(1273 K) m²/s
Al	Al (fcc)	65	127.2	1.4×10^{-5}	7.1×10^{-28}	3.4×10^{-13}	—
C	Fe (fcc)	—	147.8	1.2×10^{-5}	1.5×10^{-31}	1.7×10^{-14}	1.0×10^{-11}
Fe	Fe (bcc)*	154	218.0	4.6×10^{-5}	—	—	3.6×10^{-14}
Fe	Fe (fcc)	135	286.0	7.0×10^{-5}	5.2×10^{-55}	5.4×10^{-22}	1.3×10^{-16}
Ni	Ni (fcc)	172	287.0	2.3×10^{-4}	1.1×10^{-54}	1.5×10^{-21}	3.9×10^{-16}
Mo	Mo (bcc)	289	437.5	1.3×10^{-5}	2.7×10^{-82}	8.7×10^{-32}	1.5×10^{-23}
W	W (bcc)	386	526.3	4.0×10^{-6}	2.2×10^{-98}	1.3×10^{-37}	1.0×10^{-27}

* The diffusion coefficient of bcc Fe cannot be described by a single activation energy value over the entire range of temperatures. See discussion in text.

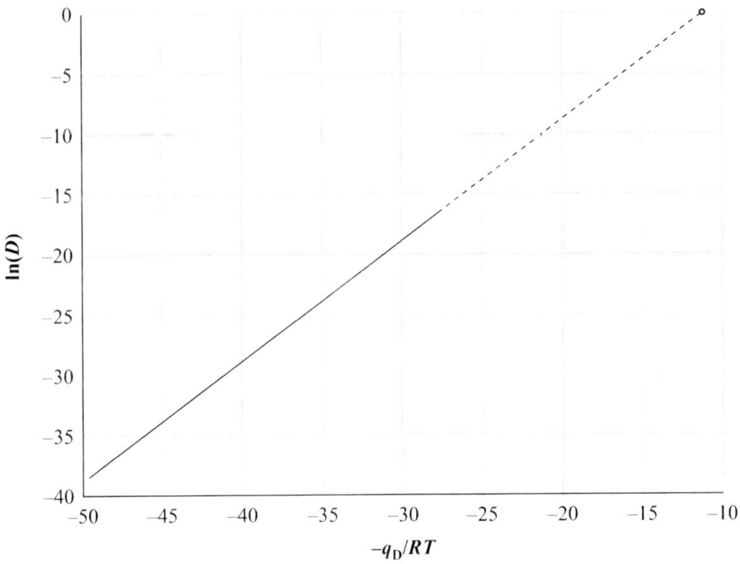

Figure 7.8. Self-diffusion coefficient of Al plotted as ln(D) versus $-q_D/RT$. Solid line corresponds to the temperature range from 125°C to 650°C. Dashed line is extrapolation to $T \to \infty$.

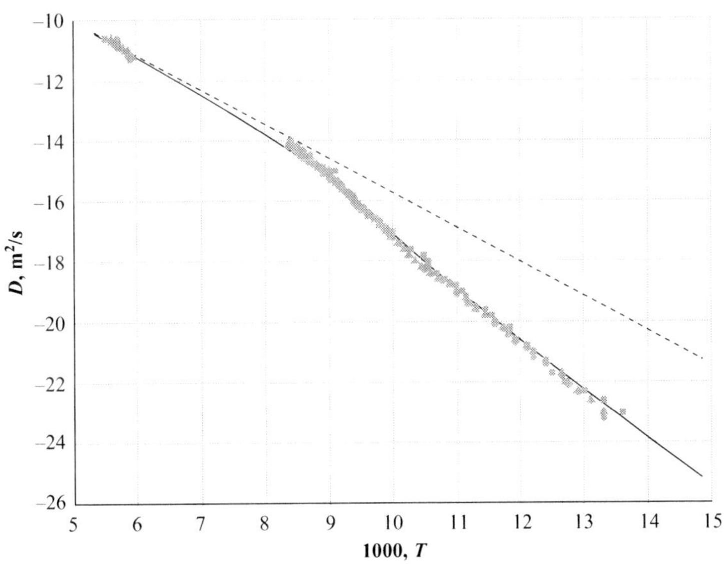

Figure 7.9. Self-diffusion coefficient of bcc Fe including the effect of ferromagnetism (solid line) and without magnetic contribution (dashed line). Calculations are performed according to the magnetic model of Jönsson [199].

In ferro-magnetic materials, for instance, the effect of magnetic ordering on diffusion cannot be neglected. This is shown exemplarily for the self-diffusion of bcc Fe, which has a Curie temperature of 1043 K (770°C). Figure 7.9 compares the self-diffusion coefficient of Fe as described by Jönsson [199] with experimental points measured by various researchers. For

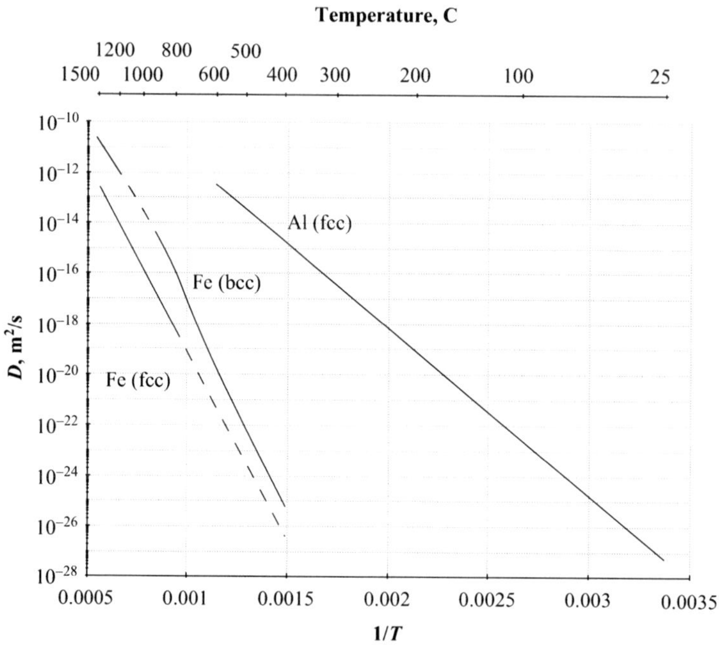

Figure 7.10. Self-diffusion coefficient of Al and Fe. Evaluated from the *MatCalc* mobility databases.

details on the evaluation and the underlying model, the interested reader is referred to Jönssons original work. According to these data, magnetic ordering decreases the diffusivity of Fe below the Curie temperature by several orders of magnitude.

In alloys exhibiting short-range ordering, such as NiTi shape memory alloys, or FeAl, or TiAl based alloys, the fact that nearest-neighbor vacancy exchange is energetically unfavorable compared to second-nearest neighbor jumps must be included in the modeling procedure for diffusivities. This issue has been dealt with several times in literature and details are beyond the scope of this book. The interested reader is pointed toward an article by Helander and Ågren [200], who suggest an efficient CALPHAD-type model for diffusion in alloys with B2-bcc short-range ordering.

Figure 7.10 compares self-diffusion coefficients of Al and Fe over a wide range of temperatures. Close to the melting points of these elements, both self-diffusion coefficients adopt comparable values around 10^{-13} m^2/s. This finding is in line with the observation that the equilibrium density of thermal vacancies close to the melting temperature is similar with values between 10^{-3} and 10^{-4} for most metallic systems.[11]

A comparison of self-diffusion coefficients of the same element with different crystal structure indicates that diffusion in bcc crystals is generally faster by a factor of roughly 10–100 compared to fcc crystals. This observation is depicted in Figure 7.10, comparing the diffusivities of Fe in fcc and bcc structure. The reason for this effect is found in the lower packing density of the bcc structure, which is only 0.68 compared to 0.74 of fcc. In the close-packed structure, diffusion is more difficult because the gaps between the individual atoms are generally smaller.

[11] See Section 1.6 on structural vacancies for details.

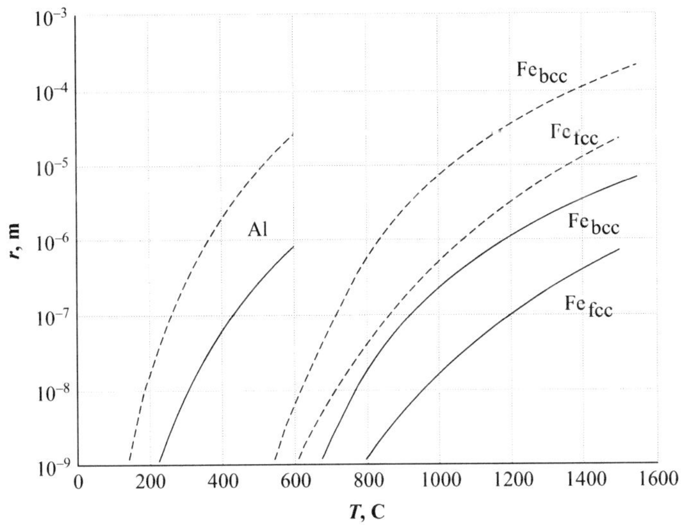

Figure 7.11. Mean diffusion distance, r, for Al and Fe self-diffusion. The solid lines are evaluated for a diffusion time of 1 s, the dashed lines for 1000 s.

Finally, some typical mean diffusion distances are evaluated in Al and Fe for diffusion times of 1 s and 1000 s in Figure 7.11. The evaluation is based on the Einstein–Smoluchowski relation, equation (7.11), which delivers

$$r = \sqrt{2Dt}, \tag{7.84}$$

for the one-dimensional case. Accordingly, this relation straightforwardly delivers an order of magnitude guess of the diffusion distance and, thus, allows for an estimate of the maximum interaction distance for diffusion-controlled reactions.

For Al alloys below approximately 200°C, the mean diffusion distance is in the order of a few atomic distances. The same is observed for bcc Fe below 700°C and fcc Fe below 800°C. Consequently, these temperatures represent threshold values for these alloys below which only insignificant diffusion can occur and diffusional phase transformations become rapidly sluggish. For diffusion times of 1000 s, these temperatures are roughly 100 K lower. Experimentally, phase transformations involving long-range diffusion of substitutional atoms are rarely observed in steel or Ni-base superalloys below 600 to 700°C.[12] At these temperatures and below, phase transformations in steel occur by alternative mechanisms such as the bainite or martensite reaction.[13]

In practical materials processing, the thermomechanical treatment of alloys is often carried out at temperatures where long-range diffusion-controlled phase transformations would be too

[12] Long-range diffusion is required, for instance, in the austenite to ferrite transformation in steel: The ferrite phase pushes austenite-stabilizing elements ahead of the interface into the parent austenite phase, which requires long-range diffusion of elements in the austenite.

[13] Precipitation of particles in the nm-range involves only medium-range diffusion with smaller diffusion distances. Precipitation of alloy carbides is thus still possible even at temperatures below the threshold, where long-range diffusion is already too sluggish to happen in noticeable time.

sluggish on the basis of the above treatment. Still, diffusional phase transformations are observed experimentally. The reason for this apparent discrepancy is twofold in most of these cases:

i. *Short-circuit diffusion paths*: Diffusion coefficients are commonly measured under conditions close to thermodynamic equilibrium and for microstructures where diffusion occurs predominantly through the bulk crystal. In real crystals, several types of lattice defects exist where diffusion occurs substantially faster, typically several orders of magnitude. The most prominent defects of this type are incoherent phase boundaries, dislocations, high-angle grain boundaries, and subgrain boundaries. If sufficient volume of defects exist, noticeable mass transport occurs also along these paths and the effective diffusion coefficient increases.
ii. *Thermal excess vacancies*: Substitutional diffusion occurs by atom–vacancy exchange. If a material is heated up to elevated temperature, held until thermal equilibrium is achieved, that is, the number of structural vacancies has reached a value close to equilibrium, and then quenched, the vacancies have insufficient time to diffuse to appropriate vacancy sinks to annihilate. The microstructure becomes supersaturated by structural excess vacancies, which can significantly speed up diffusion, sometimes by several orders of magnitude.

By any of these two mechanisms, long-range diffusional phase transformations and medium-range diffusional precipitation reactions can occur at noticeable diffusion rates even at temperatures where "equilibrium-near" diffusion would be much too slow. Typical technological processes making use of these mechanisms are quench and temper treatments or processes involving plastic deformation.

7.3.2 Diffusion Along Dislocations and Grain Boundaries

Dislocations are one-dimensional lattice defects that are characterized by a compressive and a tensile stress field around the dislocation core (see Figure 2.17 in Section 2.4.1). Inside the core region, diffusion is significantly faster than in the undisturbed lattice because the interatomic distances are larger, thus decreasing the potential barrier for diffusional jumps, and generation and annihilation of vacancies can occur easily on dislocation jogs (see also next Section 7.4). The same conclusion applies also to high-angle grain boundaries or incoherent phase boundaries, with the additional feature that these lattice defects represent two-dimensional objects, where diffusion is even more accelerated compared to the dislocation core.

The main effect of lattice defects on the diffusion coefficient is their impact on the activation energy for diffusion. Inside a grain boundary or dislocation, diffusion can occur without the mandatory requirement of atomic lattice position exchange with a vacancy that passes by in the course of its random walk through the crystal. Particularly, the core of the grain boundary resembles a liquid/amorphous structure and diffusion can occur also by alternative mechanisms such as atomic rotation or direct site exchange. Although the exact nature of diffusion in grain boundaries and dislocations will not be explored in detail here, it is evident that atomic migration inside these lattice defects is characterized by a significant reduction of the activation energy for diffusion.

In Figures 7.12 through 7.15, the diffusion data compiled in the Landolt–Börnstein materials database [201] has been evaluated [202] and homogenized to provide estimates about

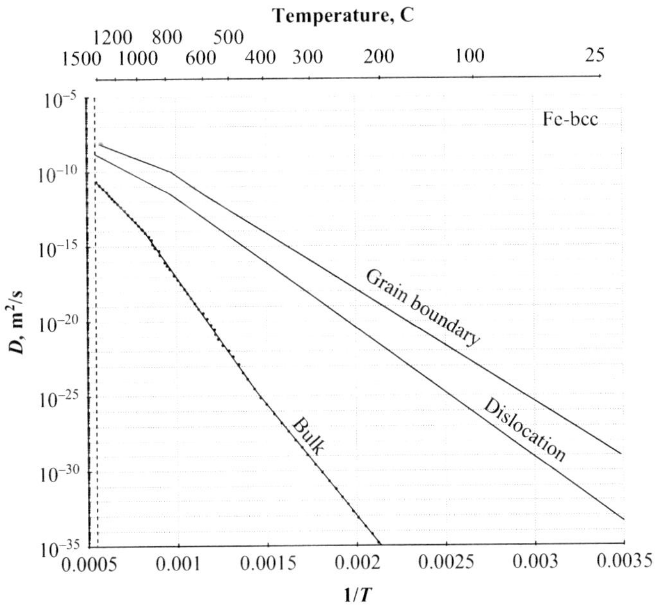

Figure 7.12. Self-diffusion of bcc-Fe in bulk, dislocation, and grain boundary. The dots at the bulk diffusivity line represent values obtained from the *MatCalc* mobility database, which take into account the effect of ferromagnetism below the Curie temperature. The solid bulk line is from the approximation with three linear segments.

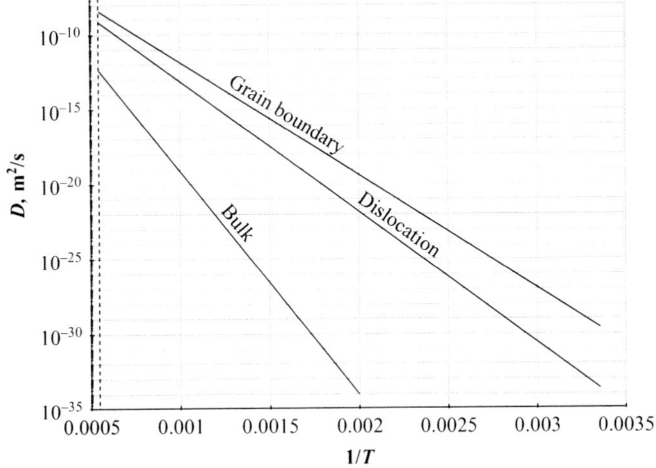

Figure 7.13. Self-diffusion of fcc-Fe in bulk, dislocation, and grain boundary.

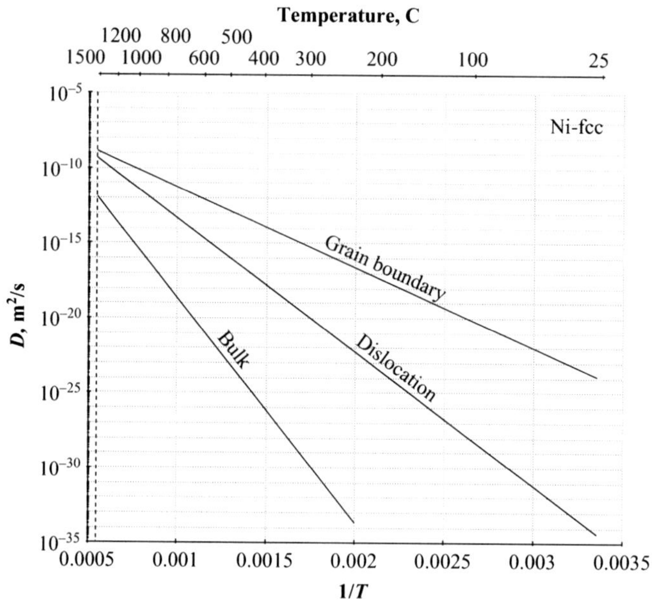

Figure 7.14. Self-diffusion of fcc-Ni in bulk, dislocation, and grain boundary.

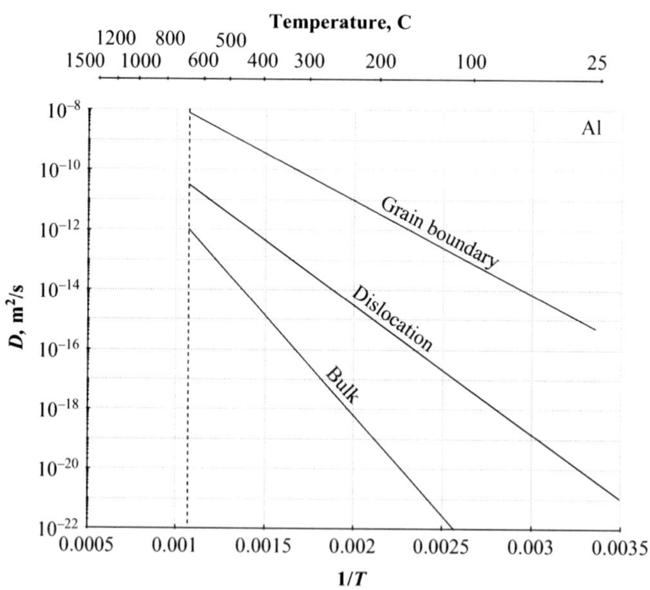

Figure 7.15. Self-diffusion of fcc-Al in bulk, dislocation, and grain boundary.

the kinetics of diffusion in dislocations and grain boundaries. These values are essential input parameters for models dealing with grain boundary precipitation (Section 6.1) or models for evaluation of effective diffusion coefficients in the presence of excess deformation-induced dislocations. The values displayed in the graphs are summarized in Table 7.3.

Table 7.3. Typical pre-exponential factors and activation energies for self-diffusion.

El.	Struct.	*T*, K	Type	D_0, m²/s	*Q*, kJ/mol
Al	fcc	298–933	bulk	1.4×10^{-5}	127.2
		298–933	dislocation	1.5×10^{-6}	83.2
		298–933	grain boundary	2.0×10^{-5}	60.2
Fe	bcc	298–693	bulk	6.0×10^{-4}*	285.0*
		693–1214	bulk	1.5×10^{0}*	330.0*
		1214–1811	bulk	2.0×10^{-4}*	240.0*
		298–693	dislocation	1.5×10^{-6}	150.0
		693–1214	dislocation	3.8×10^{-3}	195.0
		1214–1811	dislocation	5.0×10^{-7}	105.0
		298–693	grain boundary	6.0×10^{-7}	125.0
		693–1214	grain boundary	1.5×10^{-3}	170.0
		1214–1811	grain boundary	2.0×10^{-7}	80.0
Fe	fcc	298–1811	bulk	7.0×10^{-5}	286.0
		298–1811	dislocation	4.5×10^{-6}	167.0
		298–1811	grain boundary	5.5×10^{-5}	145.0
Ni	fcc	298–1728	bulk	2.3×10^{-4}	287.0
		298–1728	dislocation	4.0×10^{-5}	171.0
		298–1728	grain boundary	1.2×10^{-5}	122.0

* The self-diffusion values for bcc Fe cannot be described by a single equation due to the strong influence of the ferromagnetic–paramagnetic transition at 770 K. The values provided in this table give a reasonably close representation of the CALPHAD assessment by Jönsson [203], which is used in many mobility databases.

It should be noted, finally, that the values shown in Table 7.3 and the corresponding figures are aimed at giving an order of magnitude estimate of grain boundary and dislocation diffusion coefficients only. The available data on diffusion in dislocations and grain boundaries show considerable experimental scatter and uncertainty, therefore, these numbers should be considered with proper care.

In Table 7.4, the ratios between bulk diffusivity and diffusion along dislocations and grain boundaries are summarized. These values can be used to obtain estimates on the heterogeneous diffusion values in the absence of corresponding experimental data, but availability of multi-component values for bulk diffusivities. They are shown graphically in Figure 7.16.

The ratio between bulk and heterogeneous diffusion paths shows that dislocation and grain boundary diffusion will be significantly more important with decreasing temperature. The values at low temperature reach several ten orders of magnitude, which indicate that diffusion at

Table 7.4. Pre-exponential factors and activation energies for self diffusion describing the ratio between bulk ("b") and dislocation/grain boundary ("x") diffusivity.

El.	Struct.	*T*, K	Type	$D_{0,x}/D_{0,b}$	Q_x–Q_b, kJ/mol
Al	fcc	298–933	dislocation	1.1×10^{-1}	44.0
		298–933	grain boundary	1.4×10^{0}	67.0
Fe	bcc	298–1811	dislocation	2.5×10^{-3}	135.0
		298–1811	grain boundary	1.0×10^{-3}	160.0
Fe	fcc	298–1811	dislocation	6.4×10^{-2}	119.0
		298–1811	grain boundary	7.9×10^{-1}	141.0
Ni	fcc	298–1728	dislocation	1.7×10^{-1}	116.0
		298–1728	grain boundary	5.0×10^{-2}	165.0

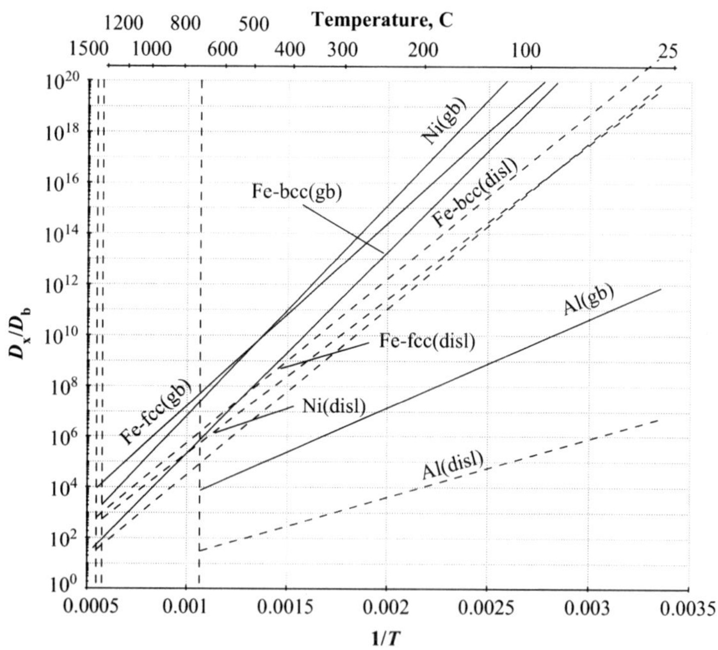

Figure 7.16. Ratio between bulk and grain boundary (gb)/dislocation (disl) self-diffusion for Al, Fe, and Ni.

low temperature will occur practically only along the high-diffusivity paths. Even at typical deformation temperatures for thermomechanical treatment of materials, the ratio is at least 10^6 to 10^9. This observation provides powerful means for the materials engineer in controlling microstructure evolution during thermomechanical processes by controlling the materials grain size and dislocation density with combinations of successive processes of deformation and recovery/recrystallization.

7.4 Excess Structural Vacancies

Diffusion in crystalline solid material is governed by the vacancy–atom exchange mechanism. Every time an atom changes place with a neighboring vacant lattice site, net transport of mass occurs in units of a single atom carried over one atomic distance. As, on average, the rate of successful diffusional jumps and, thus, the net transport of mass that can be attributed to a single vacancy is the same, an n-fold increase in vacancy concentration gives an n-fold increase in diffusion rate.

With this statement, the expression for the diffusion coefficient, equation (7.81), can straightforwardly be expanded to account for excess structural vacancies with the actual site fraction of vacancies, y_{Va}, on the substitutional sublattice and the equilibrium site fraction, $y_{\mathrm{Va}}^{\mathrm{eq}}$, as

$$D = D_0 \exp\left(-\frac{q_{\mathrm{D}}^{\mathrm{subst}}}{RT}\right) \cdot \frac{y_{\mathrm{Va}}}{y_{\mathrm{Va}}^{\mathrm{eq}}}. \qquad (7.85)$$

Obviously, if the actual vacancy concentration equals the equilibrium value, equation (7.85) reduces to the original form, equation (7.81).[14]

In this chapter, the process of vacancy generation and annihilation is investigated in more detail. In Section 7.4.2, evolution equations are derived for the vacancy site fraction in polycrystalline materials with sources and sinks for vacancies. Finally, the impact of excess vacancies in practical precipitation modeling is explored.

7.4.1 Vacancy Generation and Annihilation

Structural[15] vacancies are empty sites on the parent lattice of a crystal. They are equilibrium features of a crystal and stabilized by configurational entropy. It is important to recognize that vacancies cannot be created or annihilated easily inside a perfect crystal.[16] However, they can appear or disappear rather easily on heterogeneities of the polycrystalline microstructure such as

- free surfaces,
- grain boundaries,
- incoherent or semicoherent phase boundaries,
- subgrain boundaries,
- dislocation jogs, and so on.

[14] The explicit use of equation (7.81) is aimed at emphasizing that excess structural vacancies have substantial influence on the diffusivity of substitutional elements, while their effect on interstitial diffusion is generally negligible.
[15] The term *structural vacancies* is used to distinguish this type of lattice imperfection from, for example, *interstitial vacancies*, which are synonymous for empty lattice positions on the interstitial lattice.
[16] Structural vacancies can also be created by neutron bombardment, for instance, in nuclear reactors. This will not be treated here, however, as it is beyond the scope of our study.

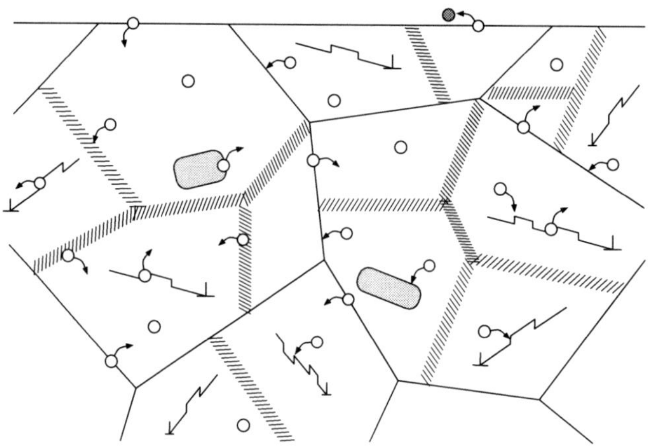

Figure 7.17. Polycrystalline microstructure with typical vacancy generation and annihilation sites: free surfaces, high-angle grain boundaries, incoherent phase boundaries, subgrain boundaries, and dislocation jogs.

The mechanism of vacancy creation on a free surface has already been illustrated in Figure 1.19 of Section 1.6, where an atom located initially in the topmost atomic layer of a crystal jumps onto the surface and leaves an empty lattice site behind. One of the atoms below replaces the missing atom on this site and thus migrates the vacancy into the interior of the crystal.

We now look at things from a more macroscopic point of view. Figure 7.17 shows a sketch of a typical polycrystalline microstructure together with prominent positions for structural vacancy generation and annihilation. The image sketches several grains with grain boundaries below a free surface, as well as dislocations with jogs and precipitates (gray) with assumed incoherent phase boundaries. Circles symbolize vacancies. The arrows indicate possible jump directions for vacancy annihilation or generation.

Several microstructural heterogeneities can be considered as perfect *vacancy sources* and *vacancy sinks*. For instance, free surfaces are excellent places for vacancy generation and annihilation. The same applies to grain boundaries and incoherent phase boundaries, which can be considered as layers of amorphous structure where the crystal periodicity is substantially disturbed. The energy barrier for vacancy generation is significantly reduced at these sites. The figure also shows subgrain boundaries that have eventually been introduced into the crystal by deformation and recovery processes. Subgrain boundaries represent structures that are composed of arrays of dislocations. They also provide a large number of vacancy sources and sinks.

Dislocations do not represent ideal sources and sinks for vacancies per se. An edge dislocation, for instance, can be viewed as a linear one-dimensional defect that is created by inserting an atomic half-plane into the crystal. The atoms along the dislocation line are arranged periodically, which means that removal or addition of a single atom or vacancy is difficult and associated with the creation of a dislocation jog. As this process is energetically expensive, it is more likely that vacancies attach to or detach from existing jogs. This is indicated schematically in Figure 7.17 where the arrows always point toward or originate from dislocation jogs.

If a material is held at some elevated temperature for a sufficiently long time, the continuous generation and annihilation of vacancies at the various sinks and sources of the microstructure reach a stationary state and the structural vacancy concentration equilibrates. If the temperature is now changed, the vacancy concentration adapts again to the new conditions by annihilation and generation of vacancies at the available sources and sinks. However, this process takes time due to the diffusive nature of this process. If time is insufficient—for example, at low temperatures—excess or shortage of vacancies can occur compared to the equilibrated state, which will influence the diffusivity of atoms according to equation (7.85) and must be taken into account in kinetic simulations.

Subsequently, a physical model is introduced that describes the evolution of nonequilibrium vacancies in microstructures containing ideal sources and sinks.

7.4.2 Modeling Excess Vacancy Evolution

Vacancies can migrate through the crystal by exchanging position with neighboring atoms. At ideal sources and sinks, vacancies are continuously created and annihilated. In the previous section, it has been concluded that this production and annihilation process leads to a stationary state after sufficient time and, thus, to an equilibration of the vacancy concentration in the material. Fischer et al. [204] have recently derived a model that describes the evolution of the structural vacancy site fraction based on vacancy generation and annihilation at

- dislocation jogs (DJ) with a jog density H_j;
- Frank loops (FL), with a density N and a mean radius r;
- grain boundaries (GB) with a grain size (diameter) d_g.

Let y_{Va} be the site fraction of vacancies on the substitutional sublattice. The generalized chemical potential, $\overset{*}{\mu}_{Va}$, of the vacancies can be written as

$$\overset{*}{\mu}_{Va} = \mu_{Va} - \bar{\Omega}\sigma_H, \tag{7.86}$$

where $\bar{\Omega}$ is the volume corresponding to one mol of lattice sites[17] and σ_H is the hydrostatic stress. The chemical potential of vacancies is given as

$$\mu_{Va} = RT \ln \frac{y_{Va}}{y_{Va}^{eq}} = RT \ln y_{Va} + \Delta h_{Va}^{f}, \tag{7.87}$$

where y_{Va}^{eq} is the equilibrium site fraction of vacancies and Δh_{Va}^{f} is the vacancy formation enthalpy. The generalized chemical potential then reads

$$\overset{*}{\mu}_{Va} = RT \ln y_{Va} + \Delta h_{Va}^{f} - \bar{\Omega}\sigma_H. \tag{7.88}$$

In a previous treatment of diffusion and creep in multicomponent alloys with nonideal sinks and sources for vacancies, Svoboda, Fischer, and Fratzl [205] derived an expression for the vacancy production/annihilation rate, $\dot{y}_{Va} = \partial y_{Va}/\partial t$, with

$$\dot{y}_{Va} = -\bar{\Omega}\,\mathrm{div}\,\vec{j}_{Va} + \alpha\left(1 - y_{Va}\right), \tag{7.89}$$

[17] Note that $\bar{\Omega}$ differs from the molar volume v^{α} because it refers to the number of lattice sites (including vacancies) instead of the number of atoms. In most cases, $\bar{\Omega}$ can be approximated by v^{α}, though.

where $\vec{j}_{\mathrm{Va}}$ is the macroscopic diffusive vacancy flux and α is an independent variable found to be (see equation [54] in ref. [205] and equation [3] in ref. [204])

$$\alpha = -\frac{\overset{*}{\mu}_{\mathrm{Va}}}{K\bar{\Omega}}. \tag{7.90}$$

The parameter K is the bulk viscosity. If we assume that $\operatorname{div}\vec{j}_{\mathrm{Va}} = 0$ and y_{Va} <<, the rate of vacancy generation/annihilation follows as

$$\dot{y}_{\mathrm{Va}} \approx \alpha = -\frac{\overset{*}{\mu}_{\mathrm{Va}}}{K\bar{\Omega}}. \tag{7.91}$$

As the generalized chemical potential, $\overset{*}{\mu}_{\mathrm{Va}}$, is only dependent on the actual values of y_{Va} and σ_{H}, and $\bar{\Omega}$ is practically constant, the value of $\dot{y}_{\mathrm{Va}}$ is only dependent on the structure and value of the bulk viscosity, K.

The total vacancy evolution determined by the three types of sources/sinks is given as

$$\dot{y}_{\mathrm{Va}} = \dot{y}_{\mathrm{Va}}^{\mathrm{DJ}} + \dot{y}_{\mathrm{Va}}^{\mathrm{FL}} + \dot{y}_{\mathrm{Va}}^{\mathrm{GB}}. \tag{7.92}$$

Subsequently, we derive the corresponding expressions for K, which will depend on the paths of the vacancies to their specific sinks and sources.

7.4.2.1 Annihilation at Dislocation Jogs

The dislocation jogs are assumed to represent ideal sources and sinks. The dissipation of free energy shall occur by diffusion of vacancies to and from the jogs. Let the jogs be homogeneously distributed in the bulk crystal with a density H_{j}. The volume corresponding to a single jog is $1/H_{\mathrm{j}}$ and the radius of a sphere where the vacancies are uniformly collected and transported to the jog by diffusion is given as

$$r_{\mathrm{s}} = \left(\frac{3}{4\pi H_{\mathrm{j}}}\right)^{\frac{1}{3}}. \tag{7.93}$$

The radial vacancy flux in a distance, ρ, from the center of the sphere is then

$$j_{\mathrm{Va}} = -\frac{r_{\mathrm{s}}^3 - \rho^3}{3\rho^2}\frac{\alpha}{\bar{\Omega}} \quad \left(a/2 < \rho \le r_{\mathrm{s}}\right). \tag{7.94}$$

a is the atomic distance and the radius of the core region of the jog is assumed to be $a/2$. The rate of Gibbs energy dissipation can be written as

$$Q_{\mathrm{s}} = \int_{a/2}^{r_{\mathrm{s}}} \frac{4\pi\rho^2 j_{\mathrm{Va}}\,\mathrm{d}\rho}{A_{\mathrm{Va}}} \approx \frac{\alpha^2}{2\pi a H_{\mathrm{j}}^2 A_{\mathrm{Va}}\bar{\Omega}^2}. \tag{7.95}$$

A_{Va} is a kinetic coefficient that has been derived by Svoboda et al. in ref. [206] with

$$A_{Va} = \frac{y_{Va}\tilde{D}_{eq}}{\bar{\Omega}RTy_{Va}^{eq}f}, \tag{7.96}$$

where the effective vacancy diffusion coefficient, $\tilde{D}_{eq}$, in a system with m substitutional elements is

$$\tilde{D}_{eq} = \sum_{k=1}^{m} y_k D_k^{eq}. \tag{7.97}$$

D_k^{eq} are the tracer diffusion coefficients of elements k for equilibrium vacancy concentration. The quantity f is the geometrical correlation factor ($f_{fcc} = 0.7815; f_{bcc} = 0.7272$).

On the other hand, the dissipation of Gibbs energy is also given by[18]

$$Q_s = \frac{\alpha^2 K}{H_j}. \tag{7.98}$$

Comparison of equations (7.95) and (7.98) provides the bulk viscosity, K_{DJ}, for vacancy annihilation/generation at homogeneously distributed dislocation jogs as

$$K_{DJ} = \frac{1}{2\pi a H_j A_{Va}\bar{\Omega}^2}. \tag{7.99}$$

After inserting for A_{Va}, the bulk viscosity becomes

$$K_{DJ} = \frac{RTy_{Va}^{eq}f}{2\pi a H_j y_{Va}\bar{\Omega}\tilde{D}_{eq}}. \tag{7.100}$$

With equations (7.90) and (7.86), the vacancy evolution in the presence of dislocation jogs finally becomes

$$\dot{y}_{Va}^{DJ} = \frac{2\pi a H_j y_{Va}\tilde{D}_{eq}}{y_{Va}^{eq}f}\left(\ln\frac{y_{Va}}{y_{Va}^{eq}} - \frac{\bar{\Omega}}{RT}\sigma_H\right). \tag{7.101}$$

7.4.2.2 Annihilation at Frank Loops

Frank loops are typical microstructural features of metallic materials, which are highly supersaturated in thermal vacancies after rapid quenching from elevated temperature. Frank loops are created when excess vacancies, agglomerated in a two-dimensional plate-like form, and the atomic layers above and below the vacancy plate collapse. In this process circular dislocation rings are formed, which grow continuously by condensation of additional vacancies.

[18] See ref. [205] for details.

Experimental evidence for Frank loop formation is often reported for pure metals. A general description of Frank loops is given, for example, in the textbook by Böhm [207]. Other sources are referenced in [204].

Let us consider N loops with an initial radius of r_s and an actual radius of r. Following the principles of classical nucleation theory,[19] the initial radius, r_s, must be larger than some critical radius, r_{crit}. The Gibbs energy change of Frank loop formation can be written as

$$\Delta G_{FL} = 2\pi r \gamma_{FL} - \frac{\pi r^2 a \mu_{Va}^*}{\bar{\Omega}} = 2\pi r \gamma_{FL} - \frac{\pi r^2 \mu_{Va}^*}{a^2 N_A}, \tag{7.102}$$

and derivation with respect to r and setting the derivative equal to zero delivers

$$r_{crit} = \gamma_{FL} \frac{N_A a^2}{\mu_{Va}^*}. \tag{7.103}$$

γ_{FL} is the Frank loop line energy. For circular loops, the conservation law for vacancies delivers

$$\frac{y_{Va,s} - y_{Va}}{a^3} = \frac{N\pi\left(r^2 - r_s^2\right)}{a^2}, \tag{7.104}$$

where the subscript "s" denotes the values of the initial quantities. The actual vacancy site fraction, y_{Va}, is then found as

$$y_{Va} = y_{Va,s} - N\pi a\left(r^2 - r_s^2\right). \tag{7.105}$$

The evolution of the Frank loop radius, $\dot{r}$, with respect to time is obtained after differentiation. It is related to the evolution of the vacancy site fraction, $\dot{y}_{Va}^{FL}$, with

$$\dot{r} = -\frac{\dot{y}_{Va}^{FL}}{2\pi N a r}, \tag{7.106}$$

and the jog density, H_j, which is expressed in terms of the distance, p, between jogs as

$$H_j = N\frac{2\pi r}{p}. \tag{7.107}$$

In ref. [204] geometrical consideration has delivered the approximation that roughly every fifth atom along a Frank loop can be considered as a jog yielding $p \approx 5a$.

On evaluation of the vacancy evolution rate due to Frank loop formation, the Gibbs energy of a perfect crystal without defects is denoted as G_0. The Gibbs energy of a crystal containing vacancies and N Frank loops is then

---

[19] See also Section 2.3.2.

$$G = G_0 + \frac{y_{\mathrm{Va}} \overset{*}{\mu}_{\mathrm{Va}}}{a^3 N_{\mathrm{A}}} + N 2\pi \gamma_{\mathrm{FL}} r. \tag{7.108}$$

The generalized chemical potential, $\overset{**}{\mu}_{\mathrm{Va}}$, driving the generation and annihilation of vacancies at Frank loops follows as

$$\overset{**}{\mu}_{\mathrm{Va}} = a^3 N_{\mathrm{A}} \frac{\partial G}{\partial y_{\mathrm{Va}}} = \overset{*}{\mu}_{\mathrm{Va}} + 2\pi a^3 N_{\mathrm{A}} N \gamma_{\mathrm{FL}} \frac{\mathrm{d}r}{\mathrm{d}y_{\mathrm{Va}}}. \tag{7.109}$$

With the vacancy balance, equation (7.106) and equation (7.87), we obtain

$$\overset{**}{\mu}_{\mathrm{Va}} = \overset{*}{\mu}_{\mathrm{Va}} - \gamma_{\mathrm{FL}} \frac{N_{\mathrm{A}} a^2}{r} = RT \ln \frac{y_{\mathrm{Va}}}{y_{\mathrm{Va}}^{\mathrm{eq}}} - N_{\mathrm{A}} a^3 \sigma_{\mathrm{H}} - \gamma_{\mathrm{FL}} \frac{N_{\mathrm{A}} a^2}{r}. \tag{7.110}$$

The differential equation for the vacancy evolution at Frank loops finally follows as

$$\dot{y}_{\mathrm{Va}}^{\mathrm{FL}} = -\frac{4\pi^2 N r y_{\mathrm{Va}} \tilde{D}_{\mathrm{eq}}}{y_{\mathrm{Va}}^{\mathrm{eq}} f p} \left[\ln \frac{y_{\mathrm{Va}}}{y_{\mathrm{Va}}^{\mathrm{eq}}} - \frac{N_{\mathrm{A}} a^3}{RT} \left(\sigma_{\mathrm{H}} + \frac{\gamma_{\mathrm{FL}}}{a r} \right) \right], \tag{7.111}$$

where equation (7.91) has been used as well as equation (7.110) for the generalized chemical potential, and equation (7.107) for the jog density, H_{j}.

The evolution of excess vacancies and Frank loops is determined by the two differential equations (7.106) and (7.111). Interestingly, Frank loops are unstable microstructural features, as the equilibrium vacancy concentration in the presence of Frank loops is higher than in its absence. Consequently, Frank loops are expected to decay again in favor of linear defects, such as dislocations, or planar defects, such as grain boundaries. If no other sources and sinks of vacancies exist, Frank loops exhibit coarsening characteristics, similar to precipitates or grains.

7.4.2.3 Annihilation at Grain Boundaries

Consider an isolated grain, without any internal sources or sinks for vacancies. For the merit of simplicity, assume that the grain can be represented by a sphere. If the vacancy concentration inside the grain deviates from the equilibrium value, the grain boundary will act as an ideal source or sink for vacancies and the diffusion of vacancies toward and from the grain boundary leads to an equilibration of the vacancy concentration.

The mass balance at the interface of the grain is given as

$$-\frac{(1 - y_{\mathrm{Va}})}{\bar{\Omega}} \dot{d}_{\mathrm{g}} = j_{\mathrm{Va}} (d_{\mathrm{g}}). \tag{7.112}$$

The quantity j_{Va} represents the radial diffusive flux of vacancies. The global mass balance reads

$$\frac{4\pi d_{\mathrm{g}}^3}{3} (1 - y_{\mathrm{Va}}) = const. \tag{7.113}$$

or

$$\dot{d}_\mathrm{g} = \frac{d_\mathrm{g}\dot{y}_\mathrm{Va}^\mathrm{GB}}{3(1-y_\mathrm{Va})}. \tag{7.114}$$

For the vacancy flux inside the grain, after some analysis, Fischer et al. [204] obtain

$$j_\mathrm{Va} = -\frac{\dot{y}_\mathrm{Va}^\mathrm{GB}\tilde{\rho}}{3\Omega}, \quad 0 < \tilde{\rho} < d_\mathrm{g}, \tag{7.115}$$

with $\tilde{\rho}$ being a spatial coordinate.

The vacancy generation and annihilation at sources and sinks at grain boundaries can now be evaluated by application of the thermodynamic extremal principle (TEP).[20] The Gibbs energy of a system consisting of a grain and half of a grain boundary (each grain boundary is shared by two grains) is given as

$$G = \frac{4\pi d_\mathrm{g}^3}{3\bar{\Omega}}\left[(1-y_\mathrm{Va})\mu_\mathrm{M} + y_\mathrm{Va}\mu_\mathrm{Va} - \bar{\Omega}\sigma_\mathrm{H}\right] + 2\pi d_\mathrm{g}^2\gamma_\mathrm{b}. \tag{7.116}$$

The subscript "M" denotes the atomic species of the matrix phase and γ_b is the grain boundary energy. If γ_b is assumed to be constant, and using the Gibbs–Duhem relation, the time derivative of G is obtained as

$$\dot{G} = \left\{\frac{4\pi d_\mathrm{g}^2}{\bar{\Omega}}\left[(1-y_\mathrm{Va})\mu_\mathrm{M} + y_\mathrm{Va}\mu_\mathrm{Va} - \bar{\Omega}\sigma_\mathrm{H}\right] + 4\pi d_\mathrm{g}\gamma_\mathrm{b}\right\}\dot{d}_\mathrm{g} + \frac{4\pi d_\mathrm{g}^3}{3\bar{\Omega}}(-\mu_\mathrm{M} + \mu_\mathrm{Va})\dot{y}_\mathrm{Va}^\mathrm{GB}. \tag{7.117}$$

The time derivative of the grain size can be eliminated with the help of equation (7.114), which delivers

$$\dot{G} = \frac{4\pi d_\mathrm{g}^3}{3(1-y_\mathrm{Va})}\left(\frac{\mu_\mathrm{Va}}{\bar{\Omega}} - \sigma_\mathrm{H} + \frac{\gamma_\mathrm{b}}{d_\mathrm{g}}\right)\dot{y}_\mathrm{Va}^\mathrm{GB}. \tag{7.118}$$

The dissipation term is evaluated as

$$Q = \frac{4\pi d_\mathrm{g}^5 f y_\mathrm{Va}^\mathrm{eq} RT}{45 y_\mathrm{Va}(1-y_\mathrm{Va})\bar{\Omega}\tilde{D}_\mathrm{eq}}\left(\dot{y}_\mathrm{Va}^\mathrm{GB}\right)^2, \tag{7.119}$$

and application of the TEP with

$$\dot{G} + Q = 0, \tag{7.120}$$

[20] See also Section 3.5.1.

directly delivers the result

$$\dot{y}_{\mathrm{Va}}^{\mathrm{GB}} = -\frac{15 y_{\mathrm{Va}} \tilde{D}_{\mathrm{eq}} f}{y_{\mathrm{Va}}^{\mathrm{eq}} f d_{\mathrm{g}}^{2}} \left(\ln \frac{y_{\mathrm{Va}}}{y_{\mathrm{Va}}^{\mathrm{eq}}} - \frac{\bar{\Omega}}{RT} \left(\sigma_{\mathrm{H}} - \frac{\gamma_{\mathrm{b}}}{d_{\mathrm{g}}} \right) \right). \tag{7.121}$$

Interestingly, the evolution equation for vacancies related to grain boundaries is formally identical to the equation for constant jog density at dislocations, where the quantity $15/d_{\mathrm{g}}^2$ appears in equation (7.121) compared to $2\pi a H_{\mathrm{j}}$ in equation (7.101).

Below, the vacancy evolution in polycrystalline materials is investigated using the evolution equations for dislocation jogs and grain boundaries. For an analysis of the evolution equations for Frank loops, the interested reader is referred to the original text by Fischer et al. [204].

7.4.3 Vacancy Evolution in Polycrystalline Microstructure

The important role of excess structural vacancies in materials processing is best known in Al-alloys. The standard processing routes for age-hardenable alloys commonly involve a solution treatment at temperatures between 500 and 600°C, followed by rapid quenching to room temperature, and subsequent annealing at temperatures between 150 and 250°C. Before the annealing step, the alloy is stored for some given time at room temperature.

When investigating the possible precipitation reactions at room temperature, one finds that the diffusion coefficient of alloying elements is too sluggish to lead to noticeable formation of precipitates. From Figure 7.10, the self-diffusion coefficient of Al is read to be in the order of 10^{-28}, which provides mean diffusion distances way below the atomic spacing even at long times. Figure 7.11, displaying the mean diffusion distances in time intervals of 1 s and 1000 s, shows that only above 200°C and for times longer than 1000 s the distances that can be covered by diffusion get of the order of 10 nm. Consequently, one would expect that Al-alloys are microstructurally stable at room temperature and no diffusional reactions would occur.

It has been realized, however, already 75 years ago, that significant precipitation of so-called Guinier–Preston zones (GP-zones) occurs when storing solution-treated and quenched Al-Cu alloys at room temperature for several days. In 1938, André Guinier, a French scientist and crystallographer (1911–2000), and George Dawson Preston, a British physicist (1896–1972), published two papers [208,209] in the same issue of *Nature* describing the "structure of age-hardened aluminium-copper alloys."[21] In both cases, nanometer-sized Cu platelets have been identified as the reason for strengthening even in the absence of elevated temperature treatment.

For practical investigation of the impact of excess structural vacancies on diffusion, consider a sample of pure Al that is heat-treated at 580 °C for sufficiently long time such that the concentration of vacancies can reach its equilibrium value. Let the sample be quenched to room temperature (25°C) at different rates and investigate the evolution of the vacancy concentration based on equations (7.101) for dislocation jogs and (7.121) for grain boundaries.[22] In Figure 7.18, the

[21] The history of this discovery has recently been reviewed [210].

[22] The following simulations are performed with the software *MatCalc* (see Section 9.3) and the corresponding thermodynamic and diffusion databases.

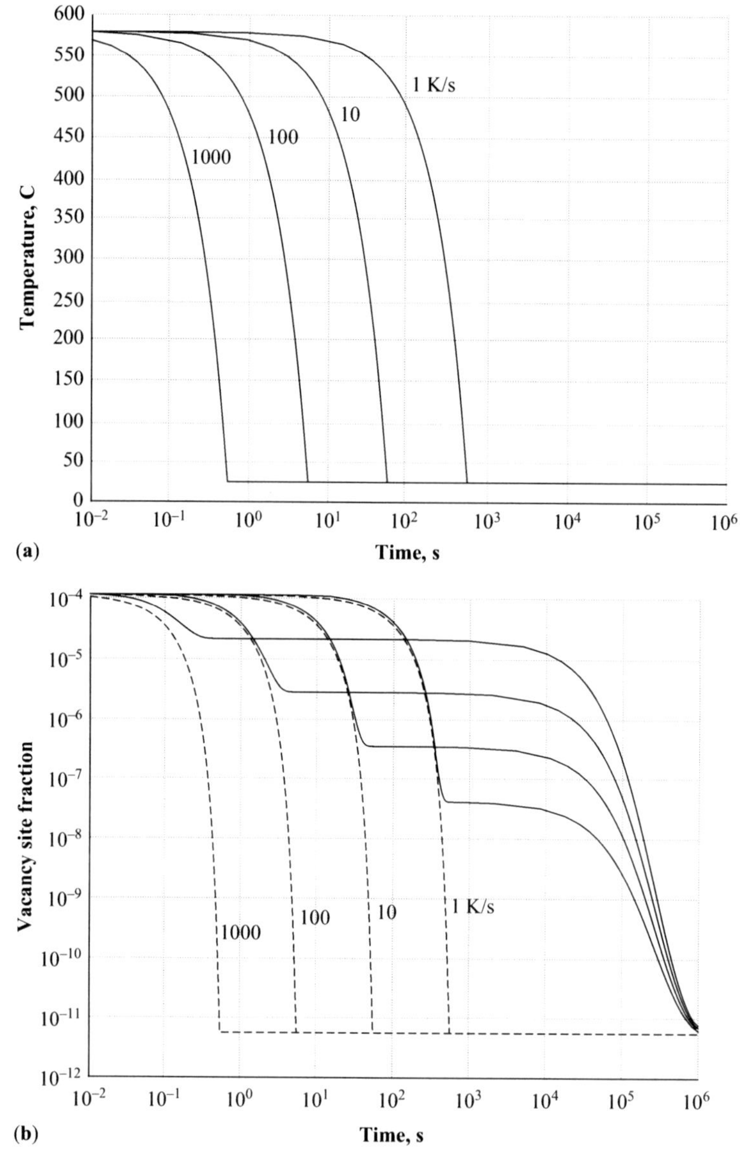

Figure 7.18. Thermal profile (a) and vacancy site fraction (b) during quenching of pure Al from 580°C to 25°C at different rates and holding for 10^6 s. Solid lines are drawn for y_{Va} and dashed lines for $y_{\text{Va}}^{\text{eq}}$.

temperature profiles and the corresponding vacancy concentrations are depicted for different quenching rates between 1 K/s and 1000 K/s. The concentrations are plotted for the actual and the equilibrium vacancy site fractions. The dislocation density is assumed to be $\rho = 10^{11}/\text{m}^2$ and the grain size is $d_{\text{g}} = 100\,\mu\text{m}$. The jog spacing on the dislocation line is taken as $p = 50a$, which means that every 50th atom along the dislocation line is assumed to be a jog.

The simulations start at 580°C in the fully equilibrated condition. During quenching, the equilibrium vacancy site fraction is described by the Arrhenius law (see Section 7.3.1). At low cooling rate, the actual vacancy concentration is found to closely follow this behavior of the equilibrium value down to rather low temperatures and vacancy site fractions.

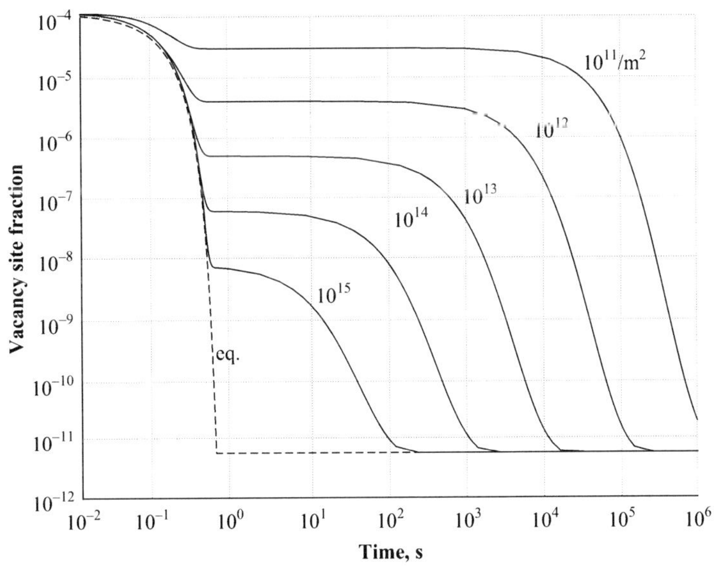

Figure 7.19. Vacancy evolution in Al after quenching from 580°C to 25°C at a rate of 1000 K/s. The numbers denote the dislocation density of the sample, "eq." designates the equilibrium vacancy site fraction.

At highest quenching rates, the vacancy site fraction starts to decrease in the very beginning, however, the annihilation process comes to a temporary seize with a plateau at a site fraction of approximately 10^{-5}. Apparently, the time that is required for the vacancies to reach the nearest sinks is insufficient and a substantial number of vacancies become "frozen-in" or "quenched-in." After reaching the temperature of 25°C, the actual vacancy site fraction is higher by a factor of more than 10^6 compared to the equilibrium value.

In view of the immense influence of excess vacancies on the diffusion kinetics in materials, which has been quantified previously in equation (7.85), this observation can straightforwardly explain the observed formation of GP-zones even at room temperature. A vacancy supersaturation of 10^6 leads to an acceleration of the diffusion kinetics by the same factor. Diffusive processes that would take 10^6 s under equilibrium conditions will then occur within only 1 s.

However, Figure 7.18 also shows that once the supersaturation is established by quenching, the vacancy concentration gradually decreases again with time until, in the present simulation, the equilibrium vacancy site fraction is reestablished after about 10^6 s. Consequently, the quenched-in vacancy effect is a transient effect only and will disappear after some characteristic time.

It is interesting to investigate the influence of microstructure on the evolution of quenched-in vacancies. In Figure 7.19, several curves for the structural vacancy concentration are plotted for different values of the dislocation density. The highest excess vacancy supersaturation is achieved at a dislocation density of $10^{11}/m^2$, which corresponds to the typical value of well-annealed fcc crystals. With a dislocation density of $10^{15}/m^2$, the maximum achievable supersaturation is 4 orders of magnitude smaller, due to the significantly shorter mean diffusion distances of the vacancies to the next sink.

In Figure 7.20, the same investigation is carried out for a variation of the grain size. The analysis indicates that, for the given conditions, a grain size of approximately 100 μm will

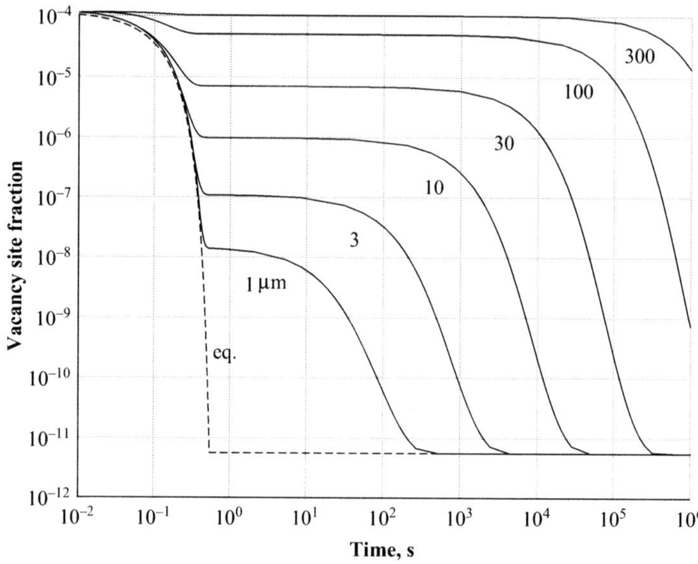

Figure 7.20. Vacancy evolution in Al after quenching from 580°C to 25°C at a rate of 1000 K/s. The numbers denote the grain size of the sample, "eq." designates the equilibrium vacancy site fraction.

freeze-in most of the thermal vacancies during the quench. Only for grain sizes below approximately 10 μm, substantial amount of the vacancies will have time enough to reach the next vacancy sinks to annihilate.

In view of these results, it is obvious that the evolution of structural vacancies can play a significant role in materials processing. Particularly in Al alloys, this effect is eminent and must not be neglected, see, for example, ref. [211]. In other alloy systems, the impact of excess vacancies seems to be less pronounced although there exists evidence of the interaction between NbC nucleation in microalloyed steel and quenched-in vacancies condensation [212].

7.5 Summary

In this chapter, some of the basic features of diffusion in crystalline solid materials are introduced and discussed. Although it is practically impossible to give a complete treatment of diffusion and diffusion modeling in a single chapter, the most relevant topics for precipitation modeling are outlined.

In the first part, the mechanism of diffusion in the periodic lattice of a metal is discussed. In the Einstein–Smoluchowski equation, the direct relation between microscopic and macroscopic diffusion is manifested and the atomic jump frequencies are correlated to continuums-mechanical, macroscopic diffusion coefficients.

In the section on macroscopic diffusion, the phenomenological laws of diffusion are investigated. Based on Fick's first and second laws, analytical solutions to some typical diffusion problems are derived. Then, the impact of chemical interaction of elements on diffusion is analyzed in the context of chemical diffusion coefficients involving either the thermodynamic factor or a full multicomponent framework for diffusion as nowadays used in advanced software for diffusion and phase transformation modeling.

The temperature-dependence of diffusion processes is discussed in the following section where the Arrhenius behavior of diffusion coefficients is analyzed in more detail as well as the physical basis of activation energies for diffusion. The latter quantity is found to be significantly lower at microstructural heterogeneities, such as dislocations and grain boundaries. Particularly, the effect of pipe diffusion along dislocations is found to be a major factor for precipitation modeling in the presence of deformation or in the context of thermomechanical treatment.

Finally, the impact of excess structural vacancies on diffusion and phase transformation kinetics is briefly investigated. A recent model describing the generation and annihilation of structural vacancies at dislocation jogs, grain boundaries, and Frank loops is introduced. In case studies, the relation between vacancy evolution and microstructural parameters, such as dislocation density and grain size, is explored. Although being an effect that is well known and practically utilized in Al-alloy processing, the impact of excess vacancies on diffusion and precipitation kinetics is less well investigated in other alloy systems and remains an open question.

CHAPTER 8

Design of Simulation

An experimenter is often faced with the problem that a large number of experimental parameters/variables should be assessed with minimum number of experiments, while, at the same time, achieving highest possible accuracy and significance. In experimental science, this topic is commonly addressed with a technique known as *Design of Experiment*. The aim of this method is to set up an optimum matrix of key experiments such that the identification of parameters is possible, unambiguously and accurately.

Similar to *Design of Experiment*, in this chapter, a corresponding technique in computational sciences is suggested: *Design of Simulation*,[1] with special emphasis on solid-state precipitation. The key question to be discussed here is, "How do I set up my computational framework such that I can achieve the results that I desire, and how can I obtain them with best quality and with adequate computational effort?"

The following section will discuss this issue with the example of fictitious precipitation kinetics simulations, which will serve as a template for the reader for his own type of problems. What comes next can readily be considered as a recipe for "How to design a numerical simulation" with special emphasis on the topics treated in this book.

8.1 General Considerations

A numerical simulation is aimed at reproducing a process occurring in nature, science, social life, economy, and so forth, on the researcher's computer. A computer simulation can have two purposes:

1. *Numbers*, "hard facts"
2. *Insight*, identification of mechanisms.

It is important to recognize that there are fundamental differences between these two intentions. In the first case, the simulation is performed to get accurate quantitative information

[1] *Design of Simulation* does not provide a recipe for minimum simulation effort with maximum output, but some considerations of how to design and setup simulations to achieve maximum benefit and introspection into a phase transformation process.

on a process that is commonly well known and for which the simulation technique is well developed. The art in performing this type of simulation is to set up the simulation framework with highest possible degree of detail and to feed the input parameters with the most precise input values available. The computer does the rest in a routine way. There are often well-known procedures available, recipes, for how to perform this type of "number crunching" simulation.

The situation is significantly different if computing is performed to gain insight into processes, which is, by the way, my personal favorite. In this case, ready-to-go recipes are generally missing and the "simulation experimenter" is often left alone in his/her search for a solution. Moreover, he/she can face a significant number of traps that will lead to erroneous results and conclusions. This issue is particularly dangerous if two or more errors compensate (fully or partially) and the simulation delivers results, which are plausible at first sight while the sequence of mechanisms and interactions describing the process can be substantially wrong.

When performing computer simulations in the field of solid-state precipitation, we are most likely confronted with simulations of the second type: identification of mechanisms and interactions governing the evolution of the precipitate microstructure. Precipitation kinetics simulations can be highly complex and involved. Unfortunately, the simulation experimenter is often not aware of this fact. Apart from a few positive and successful examples, where accurate *predictions* of number densities of precipitates, their distributions, heterogeneous nucleation sites, the evolution of the phase fractions or mean radii could be achieved, a standard procedure for prediction of general precipitation reactions is presently unavailable. The influences and interactions of chemical composition, interfacial energies, interface constitution, volume misfit strains, anisotropies in mechanical properties, coherency, heterogeneous nucleation sites, quenched-in vacancies, and so forth, on the precipitation process still need to be explored in more detail.

There exist, of course, quantitatively accurate approaches for precipitation phenomena, however, only for special aspects of solid-state precipitation. A successful example for predictive simulations is the evaluation of the diffusion-controlled movement of phase boundaries, either analytically (see Chapter 3) or numerically, using specialized software such as, for instance, DICTRA (Section 9.1). The DICTRA model is based on input parameters that come from independent kinetic experiments and the simulation procedure does not involve unknown fitting parameters.

As soon as precipitate nucleation must also be considered in the simulation setup, things are getting much more qualitative than quantitative. Accurate knowledge of interfacial energies is difficult, though reasonable estimates for this quantity are available (see Chapter 4). Nevertheless, the sensitivity of the results to small variations of the interfacial energy is immense; variations in this quantity in the percentage range can have a huge effect on the nucleation rate and, thus, the entire precipitation process.

When designing simulations, the researcher must be aware of these facts. The best we can currently achieve in terms of *predictive* precipitation kinetics simulations are *order-of-magnitude* forecasts. These are possible, however, and these should be the goal that we set to ourselves in our simulations. Once the designed simulation setup can reproduce the essential features of the precipitation reaction, which is the consistent evolution of all relevant precipitation parameters, the interactions among different types of precipitates, the correct heterogeneous nucleation sites, and so forth, in a last step, fine-tuning of input parameter can also bring quantitative agreement.

And we should restrict ourselves to *fine*-tuning. If we find that any of the input parameters must be modified to values outside its physically plausible range, most likely, we are on the wrong track. And this means, our conception of the phase transformation process is either incorrect or incomplete.

8.2 How to Design and Interpret a Solid-State Precipitation Simulation

When planning a precipitation simulation, the steps discussed in this section are aimed at giving a coarse outline that can help the simulation experimenter in setting up and designing his simulation in such a way that the results will be the ones that he is expecting.

Step 1. Phrase the problem

In order to design and plan a simulation, as a first step, the problem has to be accurately defined and phrased. The way, how you answer the question "*What do I want to simulate? What is the purpose of my simulation?*" will determine its outcome. In most cases, the problem is not well defined if you describe it as "*I want to calculate the phase transformation kinetics in alloy X.*" This statement is too general to allow for a useful definition of the problem. Instead, your phrasing should be "*I want to calculate the competitive nucleation and growth of Guinier–Preston zones and metastable precursor phases in the Al–Cu system subject to a solution treatment, quenching and annealing*" or "*I want to study the isothermal growth kinetics of allotriomorphic ferrite from austenite in duplex stainless steel.*" Although both questions are related to the simulation of phase transformations, the solutions to these problems are entirely different and the numerical simulation techniques to solve the problem are different, too. Whereas the first problem concerns the simulation of precipitation kinetics involving nucleation, growth, coarsening, and dissolution of multiple phases, the second problem is related to the multicomponent moving boundary problem in the area of interface kinetics. Different theoretical and numerical approaches are suitable for solution of these problems, and the appropriate choice will determine whether you will obtain the desired results or not.

The more precisely you phrase your problem, the easier you can proceed with the next steps. And the more accurate and useful will be the result of your simulation.

Step 2. Select degree of detail and abstraction

If a researcher designs a simulation, he/she would certainly desire to perform this simulation without unknown or badly defined input parameters, and to obtain all possible information with highest accuracy and relevance. Simultaneously, the simulation results should be accessible in reasonable time, which is commonly in the order of seconds, minutes, or hours, depending on the nature of the problem. Unfortunately, these two requirements act into opposite directions. The requirement of not using unknown input parameters in phase transformation modeling is best met by ab initio simulations, by calculations of the electronic structure of materials based on Schrödinger's equation. With the aid of useful approximations, such as the density functional theory (DFT), binding energies between atoms, elastic constants, enthalpies of formation, and even chemical potentials of arbitrary phases can be calculated and predicted on the computer; however, it is at the cost of exhaustive computation times in systems containing more than a few hundred atoms and temperatures above the absolute zero. Often, technically relevant system sizes are computationally accessible only if DFT is used in combination with Cluster Expansion (CE), Monte Carlo (MC), kinetic Monte Carlo (kMC), or Molecular Dynamics (MD). Still, even those methods require extensive computation and application to problems on the engineering scale is only feasible in a few selected cases.

Phase transformations in engineering materials are taking place on much larger length and time scales such that continuums-approaches are a convenient method of choice. Unfortunately, the various approaches differ in their degree of detail and abstraction and the user must carefully decide which approach is best suited for him/her. For instance, when dealing with interface kinetics, consideration of local concentration profiles and tracking of the interface position between the two phases is essential to obtain correct, predictive, and useful results. When dealing with precipitation kinetics problems involving multiple phases and multiple precipitates, tracking composition profiles is feasible only for a limited number of interacting particles, certainly too low in number to obtain precipitate size distributions or interactions between variably-sized precipitates in statistically significant number. In this case, the degree of abstraction must be increased and either some assumptions must be made on the shape of the composition profiles to avoid dealing explicitly with these profiles, or mean-field methods must be used to replace the computationally intensive front-tracking techniques.

This second step in design of simulation, the selection of the appropriate degree of detail and abstraction, is highly important because treating a problem on the wrong level of detail will not deliver the desired results. Either the results are obtained without the desired degree of detail and introspection, or the results are not delivered within the desired time, or they are not delivered at all. Phrasing the problem correctly in step 1 usually guides the simulation experimenter in the selection process for the appropriate degree of detail and abstraction.

Step 3. Place into proper context

Often, a problem appears to be very special and unique to the researcher and he/she assumes that he/she is the first and only one dealing with it. In fact, in most cases, it turns out that other research groups have treated similar problems, if not even the same, in one or the other way before. If you want to perform simulations, it is very important that you scan the available literature for corresponding evidence. The more information you can collect, the better your foundation for a successful solution of the problem will be. The experimental data is used to check and verify your calculations; they can be used to calibrate and fine-tune your input parameters and they will make the predictions in your own system more reliable, qualitatively as well as quantitatively.

For the problem stated in step 1, experiments could be helpful, which deal with precipitation in alloys, similar to the Al–Cu system, and which have been performed under well-defined experimental conditions. You can use these experiments to calibrate and verify your input parameters, your simulation setup, and then use these parameters and methods in your own problem. It is likely that your own simulation will deliver good results if your fine-tuning leads to a good reproduction of the available experiments in the similar material and for similar conditions.

Step 4. Determine dependent variables

In a simulation, the dependent variables are the variables that represent the outcome of the computer calculation. In an experiment, the dependent variables are the variables that are being measured. Commonly, the dependent variables are straightforwardly determined if step 1, phrasing of the problem, is performed well. For instance, the dependent variables in the first problem (precipitation kinetics in Al–Cu) are the number density, size distribution, and composition of precipitates of each phase type. These represent the basic set of dependent variables from which many other useful dependent variables, such as phase fraction and mean radius of the precipitates, can be derived. In the second simulation (ferrite interface kinetics in duplex steel), the basic set of dependent parameters comprises of the interface position and the composition profiles in ferrite and austenite.

Step 5. Determine independent variables

The independent variables are the variables that determine the pathway of the simulation, its progress in time, and the evolution of boundary conditions, such as temperature, pressure, chemical composition, dislocation density, grain size or carbon and nitrogen potentials during carburization, and nitridation treatment. Commonly, the set of independent variables is determined by the choice of simulation tool, that is, the software that is being used to tackle the problem. If the simulation experimenter works with alternative models (own or from literature) and self-written code, the set of independent variables is determined by the underlying model used in the simulation procedure.

Step 6. Define evolution of independent variables

The evolution of independent variables is closely related to the process that is being simulated. It can be a heat treatment for precipitation hardening, the thermomechanical treatment of high-strength structural steel, or diffusional and depositional processes in surface treatment of wear-resistant materials. In many cases, the evolution of independent variables is known at least to a degree, which is adequate for obtaining results at a sufficient level of accuracy to allow for sound conclusions about the mechanisms governing the process under investigation. If not, the efficiency of simulation is in serious question any way.

Still, experience shows that simulation experimenters tend to overestimate the accuracy of their knowledge of independent variables and, consequently, run simulations with partially inaccurate, incompatible, or inconsistent sets of input parameters. A sound assessment of the choice of values for the independent variables before drawing important conclusions is highly advisable.

Similarly important, a source of severe error arises if the focus of the simulation experimenter is directed too strongly at the *direct* problem and the *history* of the material before its *final* treatment is ignored. Typical examples for these potential traps in simulation routine are, for instance, seemingly isothermal precipitation processes that ignore the fact that the material has most likely undergone a solution and quenching treatment before, with subsequent heating to isothermal annealing temperature. Very often, nucleation processes of metastable precursor phases or segregation processes to lattice defects have occurred during this "pretreatment," which can heavily affect the evolution of the target phases. It is even possible that the target phases themselves nucleate before the isothermal heat treatment segment is reached. Using ill-defined evolution paths for independent variables can lead to severe misinterpretation of phase transformation mechanisms. In literature, numerous examples of simulations can be found, which ignore the history of the material and compensate this ignorance by (mis-) adjusted input parameters and physical quantities.[2] To avoid this kind of mistake, always start your simulation

[2] Remember the sensitivity of the results to variations of the interfacial energy. Small changes often cause severe differences in the precipitation behavior. This makes, unfortunately, the interfacial energy the most prominent fitting parameter in precipitation modeling.

in a well-defined condition, for instance, the solution-treated state. And then account for all individual processing steps that might propagate effects into the *direct* problem. If you start your simulation from an ill-defined state, don't be surprised if you need to tune your input parameters strongly to get accordance between simulation and experiment. Remember: "*Garbage in = garbage out*," particularly, if you expect your simulation to deliver useful results from an erroneous set of independent variables.

Step 7. Perform simulation

Once the setup of the simulation and the boundary conditions are fixed, the simulation can be run. Although appearing as the most trivial step, there are important issues to be considered and accounted for, which are affecting the results of the simulation.

In many cases, models describing general physical problems, one of which are phase transformations, and which cannot be solved analytically are formulated in terms of evolution equations describing the rate of change of state parameters as a function of the system constitution at this particular moment. The evolution of state parameters can be obtained by numerical integration in time, which represents the essential part in this step of design of simulation. This needs closer inspection.

In order to integrate a partial differential equation numerically, several parameters affecting numerical stability of the integration procedure can heavily affect the outcome of the simulation procedure. The most important of these is the integration step width, the time integration interval, which can be preset and kept constant in a general way only in the most rare cases. Generally, a *maximum time step* is determined dynamically, depending on the present state of the system, and passed to an adaptive integration step size algorithm. Parameters influencing the integration step width are, among others, the maximum change of concentration in a spatial integration point of the concentration profile, the maximum displacement of a phase boundary, the volumetric growth rate of an individual precipitate in a distribution of precipitates, or the change of mean composition of the matrix due to nucleation and growth of precipitates in a mean-field model.

The available and accessible numerical parameters controlling the integration procedure and their exact nature is dependent on the particular software or algorithm used. In any case, the effect of these parameters on the simulation results must be carefully considered and inspected. Numerical integration parameters should be chosen such that the simulation results do not, or only minimally, depend on the choice of value. In general, a compromise between numerical stability and accuracy (small integration steps) and convergence velocity (large integration steps) should be sought, the bias being laid on the particular simulation purpose.

Luckily, the available computer codes (see Chapter 9) provide automatic adaptive time integration procedures and the user is rarely concerned with this kind of problem. Be aware of it, however, if you write your own code!

Step 8. Interpret results

This step in design of simulation is many times underestimated, as a simulation experimenter tends to put major effort into the design and setup of the simulation. Just like a real experimenter who wants his/her experiments to deliver "curves that can be interpreted" high amount of energy is directed into the mere problem of "getting the simulation to run." Once, the computational procedure has successfully completed, which means that the computation completes without an obvious message of unrecoverable error, the simulation experimenter is enthusiastic and, at the same time, at high risk of taking the results as they are, forgetting that every evolution curve, every number delivered by the computation, must be carefully interpreted before useful conclusion can be arrived at.

With the availability of the simulation results, the simulation experimenter is in charge of checking plausibility of the simulation data, of verifying whether the simulation has delivered meaningful and significant results, of finding out which simulation outcome is due to "real" physical processes (these will be the important results!) and which detail might only be an artifact of the numerical solution procedure, or the way the results are displayed.

As a first important step in the interpretation of the simulation results, the simulation experimenter is advised to display the dependent variables of interest in such a way that the basic features of the quantity of interest are clearly visible in the output graphs. Use log scale when appropriate, and do not use it when inappropriate. As an example, Figure 8.1 displays the equilibrium phase fraction diagram of an alloy with majority (matrix) and minority (precipitate) phases in linear and logarithmic scale. The minority phases are actually the phases of interest and, when inspected in the graph with linear scale, the evolution of the corresponding phase fractions is practically hidden. When displayed in logarithmic scale, both, the phase fractions of majority and minority phases are clearly visible and accessible for inspection.

Another example, where important results might be misinterpreted or hidden in unfavorable presentation of results, is directly related to precipitation simulations. Figure 8.2 shows the evolution of the precipitate phase fraction versus time in arbitrary units for an isothermal precipitation reaction. When plotting the dependent variable in linear scale, only a steep increase is observed in the very beginning of the reaction and a flat line until the simulation end time is reached. In contrast, when using logarithmic scale, the expected sigmoidal shape is visible and the progress of the reaction is fully illustrated.

Even more misleading can be situations, when the simulations are performed apparently correctly but the results are looked at only in the context of an incomplete set of dependent and independent parameters. Certain features of a curve could be misinterpreted if they are not put into proper context and compared to other important associated input and output parameters. A typical example would be the interpretation of the mean radius evolution in the course of a precipitation reaction during continuous cooling based only on the graph shown in Figure 8.3. The plot shows the mean radius evolution versus time. Between 2 and 3 arbitrary time units (au), the radius first increases sharply and then shows a gradual decrease between 3 and 6 au Looking only at these results suggests that the precipitates first rapidly grow to a size of 4.5 au and then gradually dissolve again until their size converges to approximately 2 au Then, the size of the precipitates seems to be constant in the further evolution of the precipitation reaction.

The interpretation of the same curve can be entirely different when looking at the mean radius evolution in the context of other dependent variables, such as number density or phase fraction. Figure 8.4 shows the same simulation results in the integral picture of temperature, phase fraction, mean radius, and number density. Apparently, during cooling, the first precipitates nucleate with a radius being in the order of 4 to 5 au and at a time around 2 to 3 au. The phase fraction at this time is practically zero and the number density is rather low. As the precipitates start growing, their size increases. However, the cooling process continues and supersaturation increases. As the driving force for precipitation increases, too, the critical radius decreases and the following precipitates nucleate with smaller radius and higher number density, that is, with higher nucleation rate. Consequently, the mean radius of the precipitate population decreases because more and more precipitates with smaller radius nucleate. This is clearly visible in the

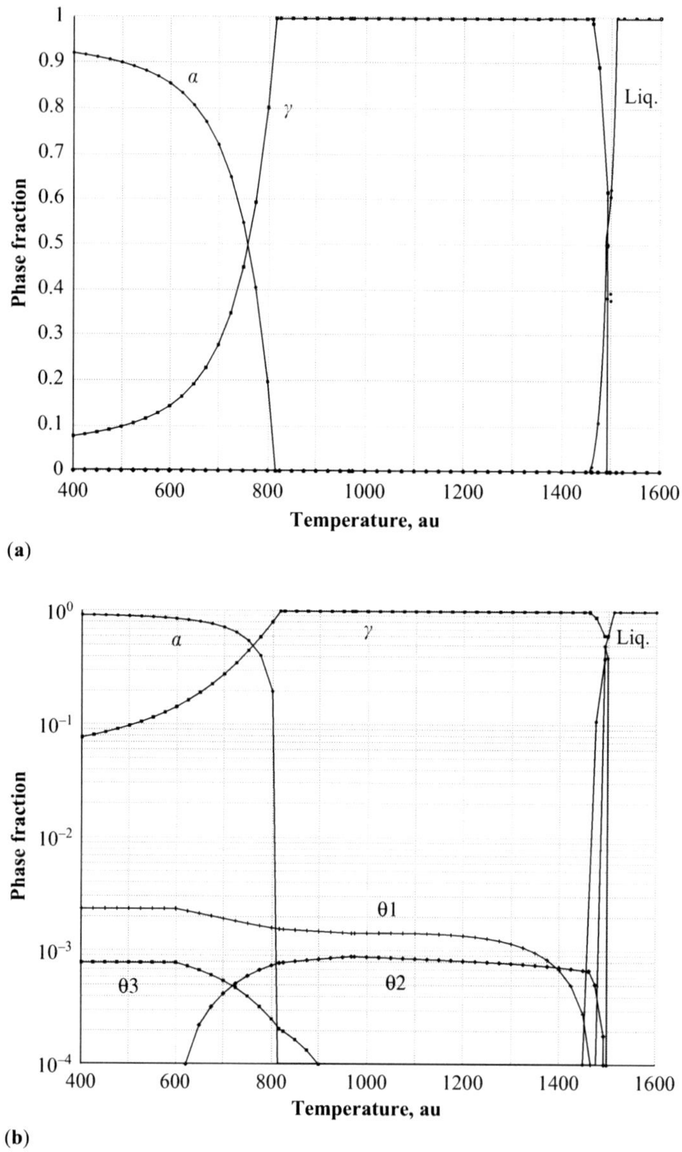

Figure 8.1. Phase fraction diagram of an arbitrary alloy system shown in linear (a) and logarithmic (b) *y*-axis scale. In linear scale, the minority phases are practically invisible, in log scale they are clearly observed.

number density graph as well as the increasing phase fraction in the phase fraction plot. At a time of approximately 6 au, the reaction comes to a stop due to seizing diffusion at lower temperature.

Together with the appropriate setup design and a selection of the proper degree of abstraction/detail of the simulation approach, the interpretation step is an utmost essential procedure. Only if the evolution of each relevant precipitation parameter fits into the global picture that the simulation experimenter has made himself of the simulated process, the simulation results can

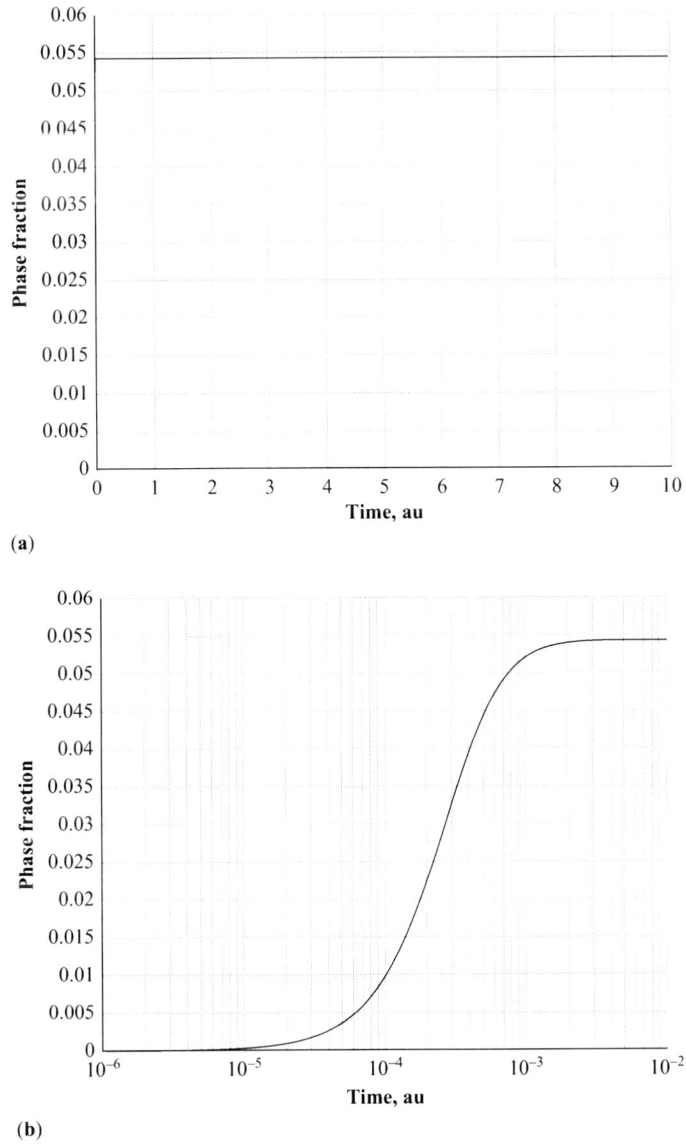

Figure 8.2. Evolution of phase fraction during precipitation in linear (a) and logarithmic (b) time scale (*x*-axis). In linear scale, the phase fraction appears as an almost continuous straight line. In log scale, the sigmoidal shape of the phase fraction evolution is clearly observed.

be of real practical use. The next step, the sensitivity analysis of input parameters, is closely related to this aspect.

Step 9. Investigate sensitivity of input parameters

The interpretation demonstrated in step 8 provides the basis for the next step of analyzing the sensitivity of input parameters that enter the simulation. Once the interaction of simulation parameters

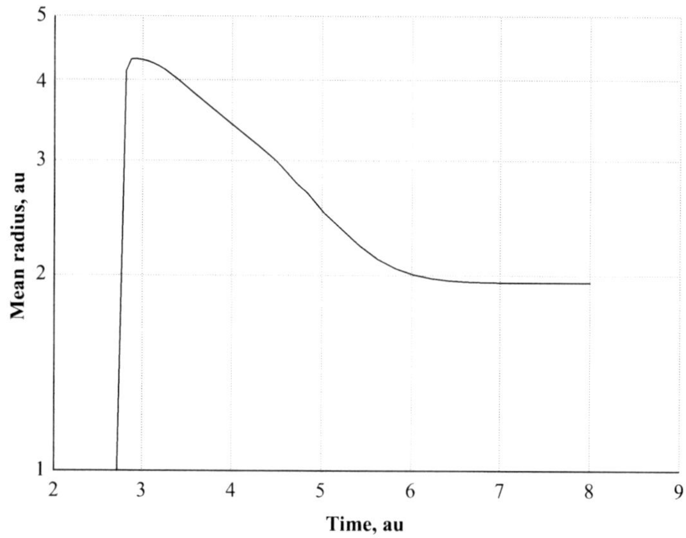

Figure 8.3. Typical mean radius evolution of a precipitate population during continuous cooling from above solution temperature.

and the progress of the simulation is understood, it is important to check whether any of the input parameters

- shows "over-sensitive" or "over-critical" impact on the simulation results, that is, small variations in the input parameters lead to huge variations in the results;
- does not show the expected impact on the results, that is, the changes observed in the results go into an unexpected direction or the results do not change at all, even if the input parameter should be an essential quantity.

The first issue in this context is related to situations where any small input parameter variation changes the simulation results "over-proportionally." For instance, if a relatively small change of simulation temperature either stimulates or suppresses precipitation, check whether your simulation conditions involve a temperature that is located close to a phase boundary. In this case, the simulation might deliver plausible results but in the interpretation of the results the simulation experimenter could be unaware of this situation and his/her suggestion for a practical heat-treatment temperature to his colleagues in the heat-treatment shop could be critical. When designing practical heat-treatment procedures from computational thermokinetics simulations, the simulation experimenter has to be aware of the practical significance of his/her conclusions. Optimum heat-treatment temperatures or optimum dissolution times from simulation are certainly valuable. But, can you trust them blindly? A factor two off in the predicted dissolution time of a precipitate can be considered as an excellent and exciting result in predictive heat-treatment simulation. Practically, this deviation could cost a lot of money if the dissolution times are slightly too short and complete dissolution of precipitates is desired.

The second issue represents yet another important step in the design of simulations concept. A careful and systematic check of the effect of input parameters on the simulation results delivers information on the correctness and efficiency of the global simulation setup and on the

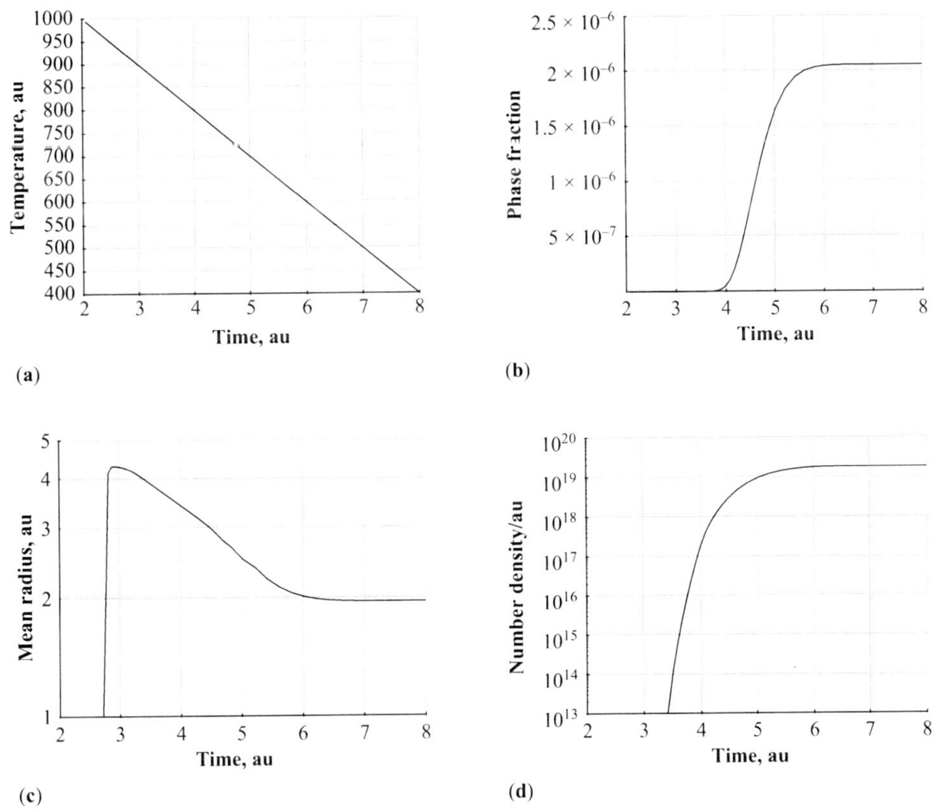

Figure 8.4. Typical evolution of precipitation parameters during continuous cooling from above solution temperature. These parameters are suitable for gathering an integral picture of the precipitation process.

correct definition/implementation of individual process models in the software. As an example, perform the same simulation with a modification in the input for the grain size. Can you see changes in the results? If you have defined grain boundary precipitates in your system, this parameter *should* have an influence. Can you see the impact of this particular variation in the number density or mean radius variables of the grain boundary precipitates? If not, carefully check your simulation setup. Either the setup is erroneous or the choice of model is inappropriate because it is not capable of dealing with this type of situation.

Step 10. Verify simulation on experimental results

Finally, your simulation has to pass the benchmark of experimental verification.[3] When successful, this step in design of simulation is the concluding step: If you find that your simulation fits the experimental results to your satisfaction, your job is done. Congratulations.

[3] If you utilize the technique of interfacial energy fitting to reproduce your experimental results, you can stop reading now. It is *always* possible to get accordance between experiment and simulation if you tune the interfacial energy.

This is a rare case, however, and the last step in design of simulation often brings you back to one of the previous steps. Most of the time, you will observe discrepancies between simulation and experiment. The manifold reasons for this could be found in one of the following subjects:

- *Geometric boundary conditions*: The kinetics of diffusion-controlled transformations are widely controlled by the diffusion distances that the diffusing species have to overcome in the course of the reaction. In simulations involving the moving boundary problem, the reaction comes to completion if the equilibrium volume fractions of transformation and parent phases are established, which means that the phase boundary has migrated to a position that corresponds to this ratio. In the actual simulation, the effective distance that the phase boundary has to travel until completion is dependent on the cell extensions that are chosen in the simulation setup. Does your domain extension represent the experimentally observed geometry? Is your choice of planar, cylindrical, or spherical geometry appropriate?
- *Equilibrium properties of the system*: The fundamental properties of a thermodynamic system are manifested in its equilibrium constitution. If you carry out a kinetic simulation, in the limit of infinitely long time, the system should approach the equilibrium state. If your simulation converges to a different state, check the underlying thermodynamic parameters. If the equilibrium determined by your thermodynamic parameters is incorrect, your kinetic simulation will be as well (garbage in = garbage out).
- *Consistency of precipitation parameters*: If you perform precipitation kinetics simulations and, for example the radius of precipitates after simulation is different from the experimentally observed, make sure that the number density of precipitates is compatible with the measured precipitate mean radius. If your nucleation model predicts a too high number density of particles, the maximum radius of precipitates will be too small due to the constraint of mass balance. Increase or decrease the number density by adaptation of input parameters (grain size, dislocation density, etc.), and you will hopefully find accordance and consistency of these parameters.
- *Plausibility of input parameters*: Recheck the values of your input parameters if your results are unsatisfying (grain size, dislocation density, diffusion data, heat treatment schedule, etc.). Have you used SI-units? Does the software expect SI-units? A typical source of error would be to enter the dislocation density in cm^{-2} instead of m^{-2}.
- *Availability of input parameters*: Are all input parameters properly defined, which are an integral part of your model and submodels? For instance: If you utilize a model for subgrain size evolution (recovery), did you properly define the subgrain boundary energy? Otherwise, you will never observe effects such as subgrain coarsening due to recovery.
- *Consideration of physical mechanisms*: Did you account for all relevant operative physical and chemical processes? For instance, if you simulate precipitation in the presence of mechanical deformation, did you consider that the dislocation density will increase and that pipe-diffusion might become important in the kinetic evolution of the system? Did you account for possible intermediate metastable phases? Quenched-in vacancies?

- *Capabilities of the model*: Is the simulation approach that you are using capable of representing the experimental situation? Is the particular feature of the transformation, which you are interested in, reflected in the model?

This list is incomplete, but it might give the simulation experimenter an impression of the many things that must be considered properly when aiming at sound and predictive simulations of phase transformation processes in solid-state matter. It is a difficult task, per se, but if successful, it is also extremely rewarding. Go for it, and nothing less.

CHAPTER 9

Software for Precipitation Kinetics Simulation

In recent years, the variety of generally available software for simulation of precipitation and phase transformation processes has significantly increased in number. While, in the 1990s, the only major computer program for multicomponent phase transformation modeling was the software package DICTRA, between 2000 and 2010 several software codes addressing this research topic have appeared.

An interesting feature in practically all advanced precipitation kinetics codes is the use of *computational thermodynamics* as the basis for evaluation of driving forces for precipitation, chemical potentials, and multicomponent diffusional mobility data. In most cases, the CALPHAD approach is utilized where Gibbs energies of individual phases are stored in huge databases as a function of chemical composition, temperature, and sometimes pressure.[1] As CALPHAD databases are nowadays available for many types of alloying systems, the combination of computational thermodynamics with general computer codes for kinetic simulations has opened the path for utilizing *computational thermokinetics* in all alloying systems where the necessary thermodynamic and mobility information is at hand.

Precipitation kinetics modeling can be performed at various levels of sophistication. The most often-used approach in this context is certainly the KJMA model, which only requires knowledge of some system-specific parameters (See Section 5.1) and a standard pocket calculator. Within the KJMA framework, the evolution of a phase transformation is evaluated on the basis of the overall transformation progress. KJMA-type models are part of innumerable PhD theses and can be found in a multitude of research papers in scientific literature. The KJMA equation has even been extended to the simultaneous precipitation of several interacting phases. See, for instance, the model by Robson and Bhadeshia [149] describing concurrent precipitation in heat-resistant power plant steel, which is described in Section 5.1.3.

For many applications, the KJMA approach is not sufficiently detailed and introspective, particularly when it comes to more complex heat treatments or the evolution of multimodal precipitate size distributions. In these cases, the classical approach promoted by Kampmann

[1] See Section 1.3.3 for more details on the CALPHAD approach.

and Wagner [152] allows solving this problem numerically, while still using standard desktop computers. This model is nowadays well known in literature as the *numerical Kampmann–Wagner* (NKW) model (see Section 5.3 for details). It is considerably more flexible and general compared to the KJMA approach and it enjoys wide popularity in the precipitation modeling community.

At the time of writing this book, the computer codes available for general precipitation kinetics and phase transformation simulations are[2]

- DICTRA (1994)
- PrecipiCalc (2004)
- *MatCalc* (2004)
- PanPrecipitation (2010)
- TC-Prisma (2011).

Each is described below in the form of a brief overview[3] where the basic features of the software, the philosophy behind the software, its historic background, and some typical examples of application[4] are presented.

9.1 DICTRA—Diffusion-Controlled Transformation

This article has been courteously provided by Gerhard Inden, formerly Max-Planck-Institute for Steel Research, Düsseldorf, Germany.[5]

9.1.1 General Information

In 1987, a 5-year research project COSMOS[6] was started at the Max-Planck-Institute, Düsseldorf, for Steel Research. The project was financially supported by the Volkswagen Foundation and the Federal State of North Rhine-Westphalia. The project was divided into two parts: (*a*) a theoretical part focusing on the computer simulation of diffusion-controlled transformations in multicomponent systems (directed by Professor G. Inden), and (*b*) an experimental part with the aim of providing reliable microstructural information as function of time and temperature serving as a benchmark for the simulations (directed by Professor H.P. Hougardy). The vision of the project was to develop software on the basis of fundamental thermodynamic concepts.

---

[2] Year of first release in parenthesis.

[3] The software descriptions have been compiled with the help of and on the basis of information provided by the software producers. For this reason, the text might be biased by the developer's subjective perception of his software product.

[4] The application examples describing the features of the different software have been suggested by the authors of the corresponding sections and do not necessarily reflect the author's personal preference.

[5] The statements, considerations, and judgements provided in this section are the sole opinion of the author of this section. They do not necessarily represent the opinion of the author of this book.

[6] COSMOS: **CO**mputer **S**upported **MO**deling of **S**teels.

At that time, the CALPHAD approach for describing the thermodynamic properties of multicomponent systems was already well developed. Based on this approach, thermodynamic databases were available and the software Thermo-Calc was the leading tool for computer calculation of phase equilibria in multicomponent systems. It was thus a great fortune, when Professor M. Hillert and Professor J. Ågren from KTH-Stockholm decided to join the project from the beginning. The group in Düsseldorf focused on geometric aspects of the microstructure, for example the cell concept, while the group in Stockholm focused on the treatment of diffusion in multicomponent systems and on the incorporation of thermodynamics using Thermo-Calc. After five years of intensive cooperation, the first version of the DICTRA[7] program was accomplished. In order to ensure a continuous development and its commercial availability, the decision was taken in 1993 to transfer the ownership of the achievements to the Foundation for Applied Thermodynamics in Stockholm, which did already manage the further development and commercial distribution of Thermo-Calc.[8] The basic features of DICTRA are summarized in Table 9.1.

Table 9.1. Software information table for DICTRA.

Software	Diffusion Controlled Transformations
Acronym	DICTRA
Developer	Thermo-Calc Software AB, Stockholm
Licensing	Commercial
Release date	1994
Website	http://www.thermocalc.se/DICTRA.htm
Contact email	info@thermocalc.se
Kinetic model	Moving phase boundary
Nucleation	No
Interfacial energies	Experimentally derived
Evolution equations from	Multicomponent diffusion/local equilibrium
Thermodynamics/mobility	CALPHAD-type databases, unencrypted or encrypted
Programming language	Fortran
Platforms	Windows, Linux
GUI/command-line/API	No/yes/yes
Scripting	Yes
Programming	No
Exporting	Post-script, hpgl plotter, various printer formats

[7] DICTRA: **DI**ffusion **C**ontrolled **TRA**nsfomations.

[8] Thermo-Calc Software, Head Office: Norra Stationsgatan 93, Plan 5, SE-113 64 Stockholm, Sweden. http://www.thermocalc.com

9.1.2 Basic Concepts

Phase transformations are treated in DICTRA as a moving boundary problem [213–215]. Space is subdivided into regions with different phases and compositions. The velocity, v, is obtained from the mass balance condition

$$J_k^{\beta} - J_k^{\alpha} = v\left(\frac{X_k^{\beta}}{v^{\beta}} - \frac{X_k^{\alpha}}{v^{\alpha}}\right), \tag{9.1}$$

where J_k^{φ} is the flux of component k in phase $\varphi = \alpha, \beta$ at the interface, v is the interface velocity, X_k^{α} and X_k^{β} are the mole fractions of component k at the interface, v^{φ} is the molar volume of phases α, β. The unknowns of the problem are the $2(n-1)$ mole fractions X_k^{α} and X_k^{β}, or equivalently the chemical potentials μ_k^{α} and μ_k^{β}, and the velocity v, in total $n + (n-1)$ variables. The mass balance condition (9.1) provides n equations. The remaining $(n - 1)$ equations are obtained from the boundary conditions at the moving interface.

The program DICTRA is interfaced with Thermo-Calc, which handles all thermodynamic calculations. In the present context, it is important to point out that Thermo-Calc provides not only the integral thermodynamic functions, but also the derivatives up to the second order in analytical form. This makes the determination of, for example, thermodynamic potentials, activities and so forth, very effective and accurate. This plays an important role in the treatment of higher order systems.

9.1.2.1 Sharp Interface

The concept of a sharp interface assumes that two phases, for example α and β, are in direct contact with each other and the interface represents a purely geometrical boundary dividing the spatial coordinates into those of α and β. Diffusing elements can cross the boundary without needing any extra driving force. Consequently, the chemical potentials of the diffusing elements must take the same value on either side of the boundary:

$$\mu_k^{\alpha} = \mu_k^{\beta}, \quad \text{for } k = 1, \ldots n-1 \tag{9.2}$$

This is an equilibrium condition, which holds at the boundary and is thus called *local equilibrium*. The numerical values of the chemical potentials are determined by the mass balance condition. From this, the interfacial compositions, x_k^{α} and x_k^{β}, can be deduced.

9.1.2.2 Local Equilibrium

The role of local equilibrium can be illustrated with a ternary system as shown in Figure 9.1. The system Fe-Mn-C was chosen with C as a fast diffuser. Given the alloy composition in the two-phase field $\alpha + \gamma$, the operating tie-line is defined by the velocity of the interface obtained from the flux balance with respect to every diffusing element. In order for α to grow, C must diffuse into the γ-matrix. Therefore, the activity at the interface of γ must be higher than in the matrix.

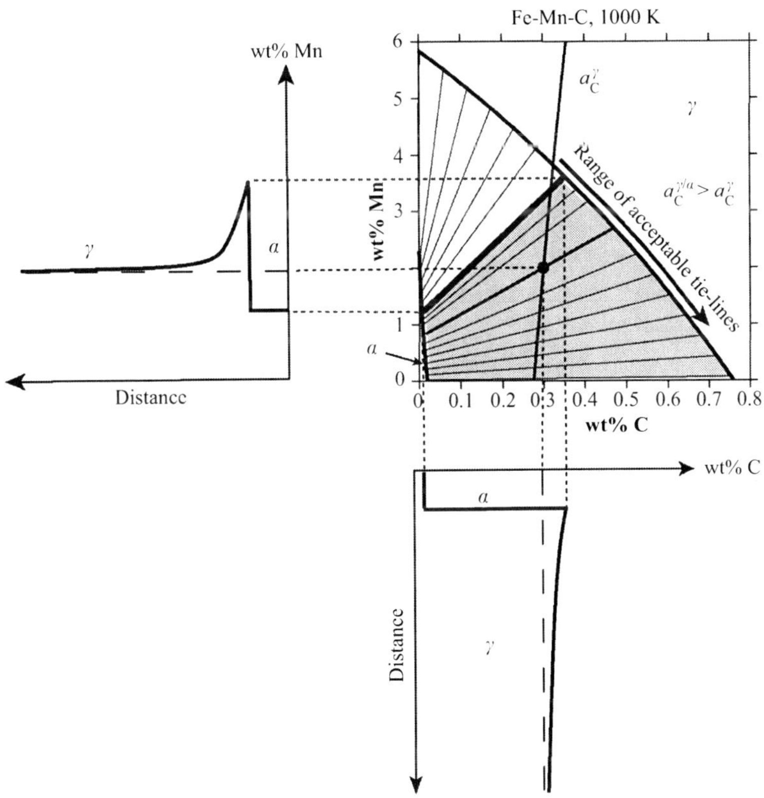

Figure 9.1. Operating tie-line in the beginning of precipitation of ferrite α from austenite γ (bold line).

This defines the range of acceptable tie-lines. Due to the fast diffusion of C, the initial operating tie-line (bold line in Figure 9.1) must be very close to the intersection of the C-isoactivity line of the austenite matrix with the phase boundary. This defines a sufficiently small driving force for C-diffusion to make sure that the velocity obtained from the C-flux can be the same as that from the Mn-flux. In the DICTRA program, the operating tie-line is determined iteratively. The reaction proceeds with this boundary condition until the C-enrichment in γ has decreased the driving force for C-diffusion to an extent that Mn-diffusion becomes competitive. Then, the operating tie-line shifts toward the global equilibrium tie-line (solid black line in Figure 9.1).

Depending on the matrix composition, the local equilibrium condition may lead to either a slow or a fast reaction regime. The reaction is slow if diffusion of, both, substitutional and interstitial elements is required (LE). This is the case shown in Figure 9.1. The fast reaction occurs when the precipitating phase can inherit the substitutional composition of the matrix, so that only the interstitial elements have to diffuse. This is local equilibrium without partitioning between precipitate and matrix (LENP[9]). The details of these reactions are outlined in [216–219].

[9]LENP: Local-Equilibrium Negligible Partitioning.

9.1.2.3 Diffusion

From the beginning of the DICTRA development it was clear that diffusion had to be treated with true thermodynamic forces (see e.g., ref. [56]). The flux of an extensive quantity is proportional to the driving force, which is given by the gradient of its conjugate intensive quantity. In the case of particle flow, the extensive quantity is the number, N_i, of particles i, its conjugate intensive quantity is the chemical potential, μ_i. The particle flux is proportional to the driving force and to the concentration $c_i = N_i/V = x_i/v^\varphi$ with $J_i = -M_i c_i \nabla \mu_i$. The proportionality factor is called mobility, M_i, of particle i. The flux of a component i can thus be written as

$$J_i = -M_i c_i \frac{\partial \mu_i}{\partial x} = -M_i c_i \left(\sum_{j=1}^{n} \frac{\partial \mu_i}{\partial c_j} \frac{\partial c_j}{\partial x} \right) = -\sum_{j=1}^{n} D_{ij} \frac{\partial c_j}{\partial x}. \tag{9.3}$$

The advantage of using chemical potential gradients is twofold: (*a*) only ($n-1$) mobility parameters are needed instead of $(n-1)^2$ diffusion coefficients; in contrast to this, the diffusivity matrix is strongly composition and temperature dependent due to the thermodynamic couplings; (*b*) the mobilities do not exhibit the strong composition and temperature dependence of the diffusivities as the thermodynamic properties of the system are included in the driving force, not in the kinetic parameters. The mobilities depend on the vacancy concentration and on the activated process of atomic exchange with a vacancy. They are thus mainly dependent on the crystal structure and the atomic bonding. Furthermore, the so-called up-hill diffusion against a concentration gradient is automatically included in a treatment based on true thermodynamic forces.

The development of DICTRA had to include the simultaneous development of a mobility database. In the meantime, several updated databases are available: for example MOB2[10] and MOBFE1[11] for Fe-base alloys, MOBNI2[12] for Ni-base alloys.

9.1.2.4 Microstructure

Microstructural aspects may be taken into account by appropriate geometrical arrangements. In order to keep the simulation time at reasonable values, DICTRA works with only one space variable. Consequently, only planar, spherical, and cylindrical geometries can be used (see Figure 9.2). The equivalent of a grain is a cell in DICTRA. The planar cell is a good approximation to describe the thickening of a layer on a grain boundary. The growth of a single particle can be described by a spherical cell with the particle at the center. In the case of small grain size, the thickening of a grain boundary layer may be described with a spherical cell with a layer growing from the periphery of the cell. Within a cell, regions with different phases may be introduced.

At the cell boundary, various boundary conditions can be defined, for example:

- *One cell*: closed system, that is zero flux.
- *Several cells*: local equilibrium at the cell boundaries, that is the activity of the components takes the same value at all cell boundaries. The numerical value is determined by the condition of mass conservation in the total system, that is for every diffusing component, the sum of the fluxes between the cells is zero.

[10] MOB2: Thermo-Calc Alloys Mobility Database, 75 elements included to be used with thermodynamic databases TCFE5 or SSOL.

[11] MOBFE1: Thermo-Calc Fe and steel mobility database, 22 elements included to be used with TCFE6.

[12] MOBNI2: Thermo-Calc Ni-alloys mobility database, 17 elements included to be used with TCNI5.

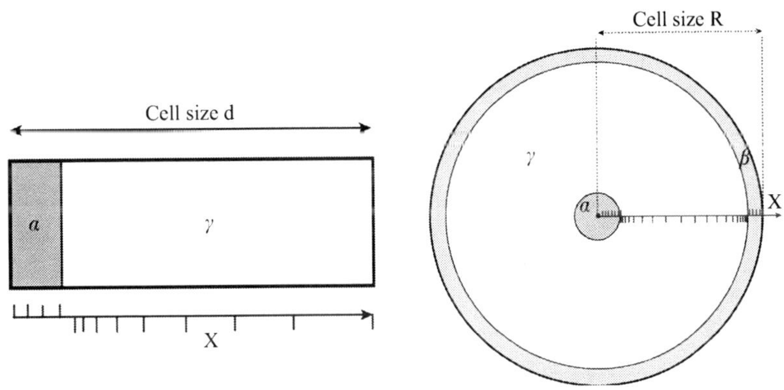

Figure 9.2. Planar and spherical cell. The space discretization may be selected as a linear grid with constant spacing or as a grid with an increasing or decreasing spacing according to a geometric series in order to get a higher grid point density in the vicinity of moving interfaces.

- Activity flux function taking into account a surface reaction in the case of particle exchange with a gas (e.g., carburization or nitridation).
- *Mixed zero flux and activity*: the flux of selected components is set to zero, while the activity of others is set to a prescribed value.

Further conditions are available for treating special reactions like the growth of an external oxide scale, which demands an expanding system.

9.1.2.5 Nucleation and Surface Energy

In DICTRA it is possible to introduce a surface energy, σ, of precipitates. This is an important term in the early stages of growth as well as in the treatment of coarsening. In these cases the surface energy term is a function of the volume of the particle. It is also possible to introduce this term as a function of interface velocity. This allows treating limited interface mobility.

The nucleation process itself is not included in the DICTRA program. However, phases may be defined as "inactive" phases, which become stable when a given driving force for precipitation, ΔG_m^*, is reached. This is of importance when the precipitation of phases is simulated during cooling, for example from a homogeneous single-phase state. It is also possible that under isothermal conditions the competition between coexisting phases leads to a sequential nucleation.

In the case of spherical geometry, the surface energy term is represented by an additive pressure term of the thermodynamic description of the precipitating phase as $\Delta G_m = -\Delta P \cdot v^{\varphi}$, with $\Delta P = -2\sigma/\rho$. The critical radius of the nucleus is obtained as $r^* = -2\sigma v^{\varphi}/\Delta G_m^*$. This critical size can then be entered when the phase starts to appear.

The surface energy term may also lead to strong effects during the early stages of growth. It can lead to the observation that fast growth of a phase according to LENP cannot be started until a certain size has been reached.

9.1.3 DICTRA Precipitation Simulation

The software offers an interactive user interface with an extensive collection of commands. This allows to formulate the problem very efficiently and to interact during simulations. The results are then made available with a post-processor, which is very similar to that of Thermo-Calc.

The way to define a simulation shall be illustrated with an example of simultaneous precipitation of three phases, $M_{23}C_6$, M_7C_3 and cementite, and from ferrite in Fe-Cr-C alloys at $T = 1053$ K (see [220]). Two alloys will be considered: Fe-12%Cr-0.12%C and Fe-7%Cr-0.12%C. The phase diagram in Figure 9.3 shows that in the alloy with 12%Cr, the carbide $M_{23}C_6$ will be the stable precipitate, while M_7C_3 will be the stable only in the alloy with 7%Cr. Figure 9.4 is a blow-up of Figure 9.3 with the metastable boundaries $\alpha + M_{23}C_6$, $\alpha + M_7C_3$, and $\alpha + M_3C$. Both alloys are within the three two-phase fields. Consequently, all three carbides have to be considered as precipitating phases.

Furthermore, the alloy with 7%Cr falls below the isoactivity line of C corresponding to cementite with the same ratio, u_{Cr}, of substitutional elements as the ferritic matrix. Consequently, this alloy may precipitate cementite according to the fast LENP regime.

9.1.3.1 Interactive Formulation of a Problem in DICTRA

The way a problem is defined in DICTRA is shown in the following tables. The commands to be entered by the user are in bold letters. A user's guide and an example collection are available at the Thermo-Calc home page.[13]

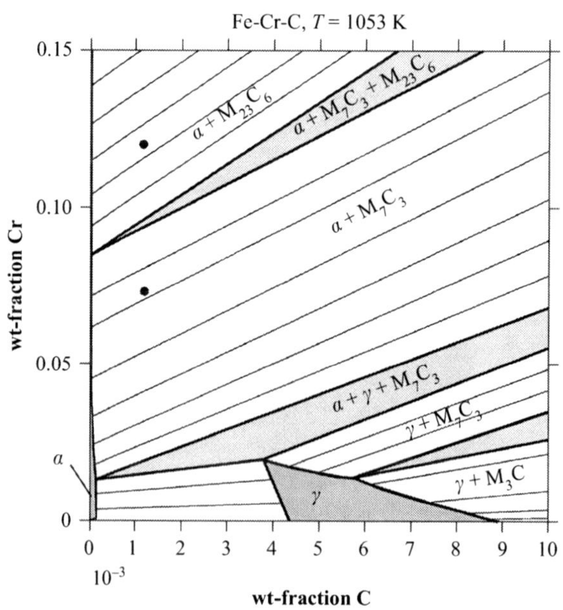

Figure 9.3. Isothermal section of the phase diagram Fe-Cr-C at 1053 K. The dots mark the alloys Fe-12%Cr-0.12%C and Fe-7%Cr-0.12%C.

[13] http://www.thermocalc.com/Library.htm

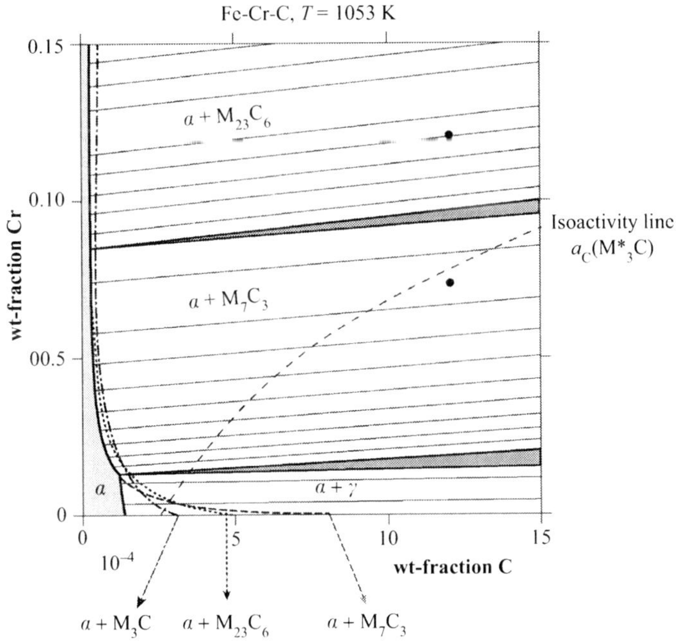

Figure 9.4. Blow-up of the isothermal section in Figure 9.3 showing the metastable phase boundaries of the two-phase regions α + $M_{23}C_6$, α + M_7C_3, and α + M_3C . $M*_3C$ represents cementite with the same ratio $u_{Cr} = X_{Cr}/(X_{Fe} + X_{Cr})$ as in the matrix. The isoactivity line of C in cementite M^*_3C is shown by the dotted line.

First step: Collection of thermodynamic and kinetic data	
SYS: **go database** THERMODYNAMIC DATABASE module Current database: TCS Steels/Fe-Alloys Database v6.2	At first, the thermodynamic database is opened to collect the data of the system.
TDB_TCFE6: **define-system** ELEMENTS: **fe cr c** TDB_TCFE6: **reject phases** *	Here, the steel database TCFE6 is used.
GAS:G LIQUID:L BCC_A2 FCC_A1 HCP_A3 DIAMOND_FCC_A4 GRAPHITE CEMENTITE M23C6 M7C3 M5C2 M3C2 KSI_CARBIDE A1_KAPPA KAPPA FE4N_LP1 FECN_CHI SIGMA CHI_A12 LAVES_PHASE_C14 CR3SI REJECTED TDB_TCFE6: **restore phases bcc m23c6 m7c3 cementite** BCC_A2 M23C6 M7C3 CEMENTITE RESTORED TDB_TCFE6: **get**	The system must be specified by definition of the elements and of the phases to be considered. Therefore, all phases are first suspended, and subsequently those phases are restored, which are to be considered.

(*continued*)

First step: Collection of thermodynamic and kinetic data (*continued*)	
TDB_TCFE6: **append mob2** Current database: TCS Alloys Mobility Database v2.4 APP: **define-system** ELEMENTS: **fe cr c** APP: **reject phases *** APP: **restore phases bcc m23c6 m7c3 cementite** APP :**get** APP: **go dictra-monitor** DIC>	The next step is to get the mobility data for the system. Therefore, the mobility database is appended. In order to specify the system, the same commands as before are given.

Second step: Definition of global conditions holding for the whole system	
DIC> **set-condition** GLOBAL OR BOUNDARY CONDITION /GLOBAL/: **global** VARIABLE : **t** LOW TIME LIMIT /0/: **0** T(TIME,X)= **1053;** HIGH TIME LIMIT /*/: * ANY MORE RANGES /N/: **No**	Temperature *t* is a global condition, which holds for the whole system. It is possible to enter a function of time and position. In the present instance an isothermal condition is selected. It is possible to define various temperature profiles as a function of time. Therefore, the starting and ending time for a temperature/time regime must be specified.

Third step: Definition of the starting microstructural condition in the various cells. Cell one	
DIC> **enter-region carbide** DIC> **enter-region ferrite** ATTACH TO REGION NAMED/CARBIDE/: **carbide** ATTACHED TO THE RIGHT OF CARBIDE /YES/: **yes** DIC> **enter-grid** REGION NAME : /CARBIDE/: **carbide** WIDTH OF REGION /1/: **5e-9** TYPE /LINEAR/: **geometric** NUMBER OF POINTS /50/: **5** VALUE OF R IN THE GEOMETRICAL SERIE: **0.95** DIC> **enter-grid** REGION NAME : /FERRITE/: **ferrite** WIDTH OF REGION /1/: **10e-6** TYPE /LINEAR/: **geometric** NUMBER OF POINTS /50/: **60** VALUE OF R IN THE GEOMETRICAL SERIE: **1.05** DIC> **enter-phase** ACTIVE OR INACTIVE PHASE /ACTIVE/: **active**	Definition of the microstructural properties of cell number one: Within the cell, various regions can be specified. In cell one the growth of $M_{23}C_6$ in a ferritic matrix shall be treated. Therefore two regions must be introduced: region carbide and region ferrite. The radius of the carbide nucleus is selected 5 nm, the cell radius is given the value 10 μm.

(*continued*)

<table>
<tr><th colspan="2">Third step: Definition of the starting microstructural condition in the various cells. Cell one (continued)</th></tr>
<tr><td>ACTIVE OR INACTIVE PHASE /ACTIVE/: active
REGION NAME : /CARBIDE/: carbide
PHASE TYPE /MATRIX/: matrix
PHASE NAME: /NONE/: m23c6
DIC> enter-phase
ACTIVE OR INACTIVE PHASE /ACTIVE/: active
REGION NAME : /FERRITE/: ferrite
PHASE TYPE /MATRIX/: matrix
PHASE NAME: /NONE/: bcc
DIC> enter-composition
REGION NAME : /CARBIDE/: carbide
PHASE NAME: /M23C6/: m23c6
DEPENDENT COMPONENT ? /FE/: Fe
COMPOSITION TYPE /MOLE_FRACTION/: mole-fraction
PROFILE FOR /CR/: Cr
TYPE /LINEAR/: linear
VALUE OF FIRST POINT : 0.50
VALUE OF LAST POINT : /.44/: 0.50</td><td>The spatial grid is selected as a geometrical series in order to get a higher density of grid points near the moving interface. An exponent greater than 1 leads to an increase in spacing with increasing distance.

The type of phase existing in the regions must be specified by the designations of the phases in the database.

Active means that the phase is present, in contrast to an inactive phase, which is not stable.

Matrix defines a regular phase, in contrast to the options lamellar or spheroidal.</td></tr>
<tr><td>DIC> enter-composition
REGION NAME : /FERRITE/: ferrite
PHASE NAME: /BCC_A2/: bcc
DEPENDENT COMPONENT ? /FE/: Fe
COMPOSITION TYPE /MOLE_FRACTION/: weight-fraction
PROFILE FOR /C/: C
TYPE /LINEAR/: linear
VALUE OF FIRST POINT : 0.0012
VALUE OF LAST POINT : /.0012/: 0.0012
PROFILE FOR /CR/: Cr
TYPE /LINEAR/: linear
VALUE OF FIRST POINT : 0.07
VALUE OF LAST POINT : /.07/: 0.07</td><td>The composition of the phases has to be specified. The Cr-content of the carbide corresponds to the critical nucleus composition. The C-content is fixed by stoichiometry.

It is possible to enter a composition profile as a function distance. It is also possible to read a composition/distance profile from a data-file.

Here a constant value is entered.</td></tr>
</table>

Cell two	Cell three
create-new-cell enter-region carbide enter-region ferrite carbide yes enter-grid carbide 5e-9 geo 5 0.95 enter-grid ferrite 10e-6 60 1.05 enter-phase active carbide matrix M7C3 enter-phase active ferrite matrix bcc enter-comp carbide M7C3 Fe m-f Cr lin 0.57 0.57 enter-comp ferrite bcc Fe w-f c lin .0012 .0012 cr lin 0.07 0.07	create-new-cell enter-region carbide enter-region ferrite carbide yes enter-grid carbide 5e-9 geo 5 0.95 enter-grid ferrite 10e-6 60 1.05 enter-phase active carbide matrix M7C3 enter-phase active ferrite matrix bcc enter-comp carbide M7C3 Fe m-f Cr lin 0.50 0.50 enter-comp ferrite bcc Fe w-f c lin .0012 .0012 cr lin 0.07 0.07

In order to treat the simultaneous growth of the three carbides, three cells have to be introduced. Spherical geometry will be selected and a cell size with radius 10 μm. The initial radius of the nuclei will be taken as 5 nm.

In the last table, only the commands that are to be entered by the user are given. Putting together all these commands yields a command file. The use of command files simplifies the starting procedure and the input of modifications.

9.1.3.2 Results of the Simulation

9.1.3.2.1 Fe-12Cr-0.12C

The results of the simulation are shown in Figure 9.5. The precipitation reaction starts very slowly, so that it takes 10^3 s until the volume fraction visibly increases. Then, all three carbides grow. In this particular instance, $M_{23}C_6$ and cementite Fe_3C grow almost at the same rate until the maximum volume fraction for metastable Fe_3C is reached. The subsequent decrease of Fe_3C

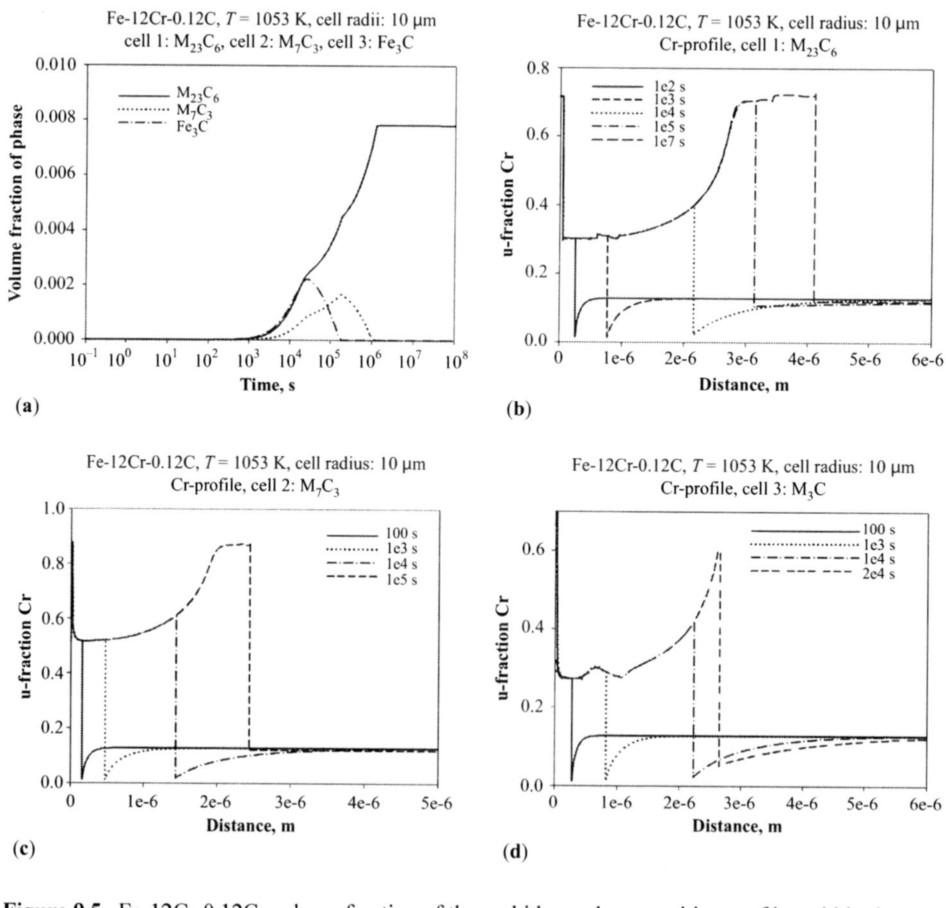

Figure 9.5. Fe-12Cr-0.12C, volume fraction of the carbides and composition profiles within the three cells.

increases the growth of both $M_{23}C_6$ and M_7C_3. After about 2×10^5 s, the volume fraction of M_7C_3 decreases and $M_{23}C_6$ increases to its equilibrium fraction.

The composition profiles corresponding to different time steps show that a considerable partitioning of Cr takes place during growth of all three carbides. This leads to the slow reaction. Furthermore, the growing carbides show a strong variation of the interface composition, which leads to a composition profile inside the carbides. Due to the slow diffusion within these carbides, it is expected that this profile will be maintained.

9.1.3.2.2 Fe-7Cr-0.12C

The results of the simulation are shown in Figure 9.6. The volume fraction of cementite increases rapidly already after 1 s. Figure 9.6(d) shows that this rapid growth is due to the reaction regime LENP, that is cementite grows with the same Cr content as offered by the matrix. This regime holds for 30 s. Then the transition to the LE regime starts with a strong increase in Cr at the interface. This leads to a stasis in the volume fraction versus time curve, Figure 9.6(a). After 100 s the volume fraction of cementite starts increasing again and reaches the maximum after about 2×10^4 s. Up to this time of about 5 h of heat treatment, the microstructure is dominated

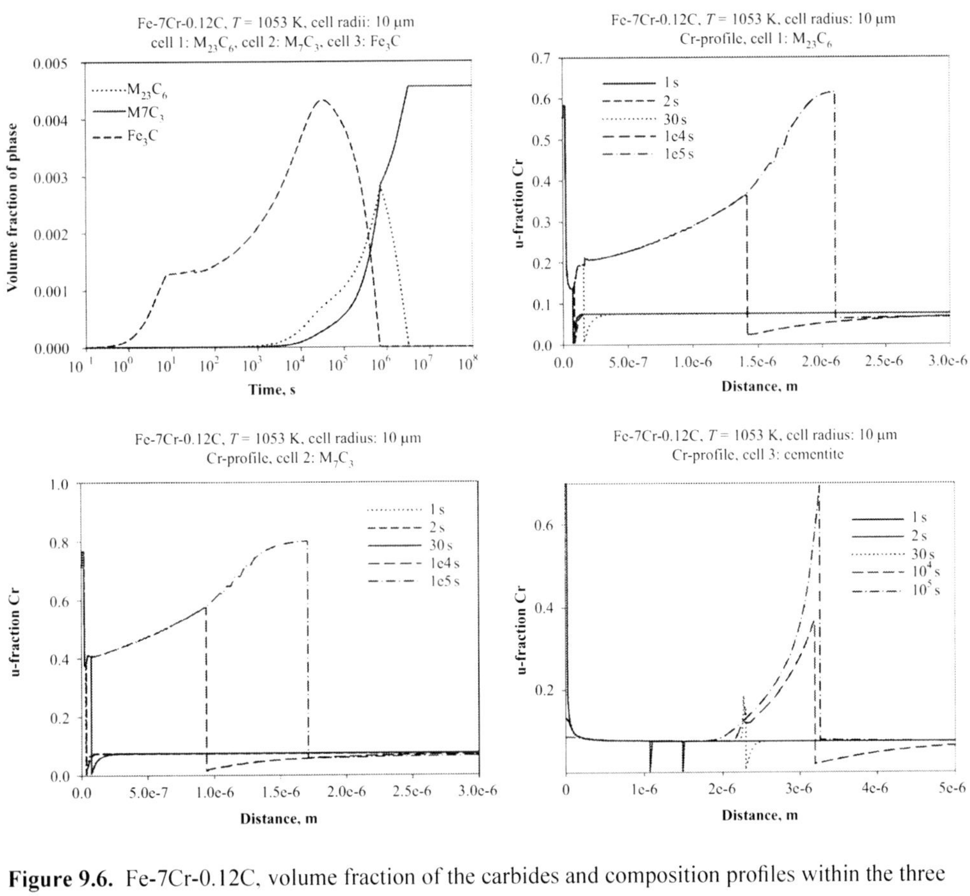

Figure 9.6. Fe-7Cr-0.12C, volume fraction of the carbides and composition profiles within the three cells.

by the presence of cementite precipitates. After about 10^3 s, the metastable $M_{23}C_6$ carbide grows considerably. The stable carbide M_7C_3 starts to exceed negligible amounts only after 10^4 s.

9.1.3.2.3 Gas–Metal Reactions

DICTRA allows various boundary conditions at the cell boundary. This allows treating gas–metal reactions, including expanding systems like the growth of an external oxide scale. The effect of surface reactions, for example in a gas mixture like CO-H_2-H_2O, can be taken into account by an activity-flux function. As an example, the carburization of Ni-30Cr at 850°C is illustrated [221]. In this instance, at the cell boundary, the C-activity with reference to the gas state at the temperature of carburization is fixed to $a_C = 1$.

The C-diffusion from the cell boundary leads to a C-gradient and a precipitation of carbides. In order to treat such a situation, the concept of a "dispersed" system is implemented in DICTRA [215]. The grid points define volume segments within which the global equilibrium is calculated defining the amount of each phase in the segment. In the case of carburization, diffusion takes place in the matrix only. The presence of carbides reduces the effective diffusion distance. This is taken into account by a labyrinth factor. This factor depends on the fraction of precipitates, their morphology, and size distribution. At every time step, first, a diffusion calculation is performed leading to the updated composition profile. In this calculation, the fraction and composition of the precipitates remain unchanged. Then the equilibrium calculation follows. In this calculation the composition within the segment includes the matrix and the precipitates. This model thus assumes that equilibrium is reached within every time step.

The results of the DICTRA calculation are shown in Figure 9.7. In this instance, the labyrinth factor was taken proportional to the square of volume fraction of precipitates.

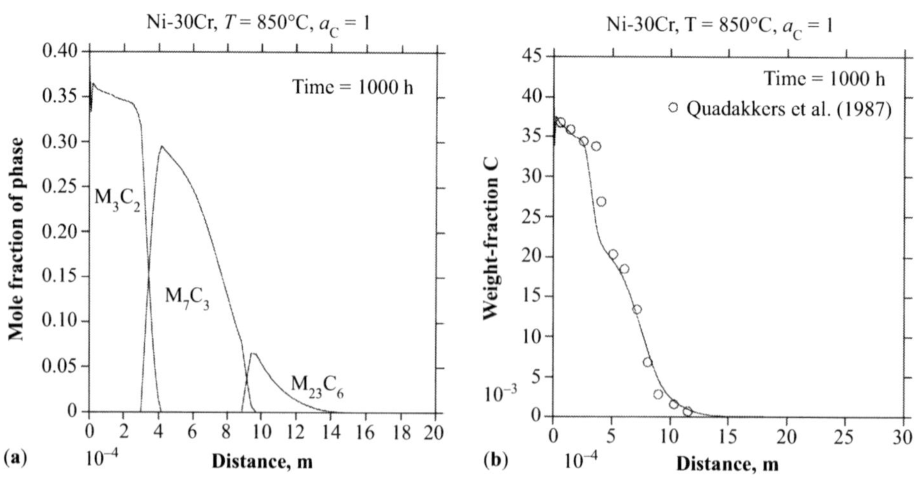

Figure 9.7. Volume fraction of carbide precipitates (a) and total C-content (b) as a function of penetration distance from the cell surface after carburization at 850°C for 1000 h. The experimental data are from reference [222].

9.1.3.2.4 Surface Tension

The effect of surface energy on the growth regime of precipitates shall be illustrated with the precipitation of $M_{23}C_6$ in ferritic Fe-2%Cr-0.2%C at 780°C. This was recently discussed in ref. [223]. Spherical geometry is assumed and a surface energy of $\sigma = 0.4\ \mathrm{J/m^2}$. The molar volume of the carbide is $v^{M_{23}C_6} = 6\times10^{-6}\ \mathrm{m^3/mol}$ and, according to Thermo-Calc, the driving force for nucleation is $\Delta G^* = 4868$ J/mol. With these data, the critical radius of the nucleus is $r^* = 2\sigma v^{M_{23}C_6}/\Delta G^* \cong 1\,\mathrm{nm}$.

Figure 9.8 shows the isothermal section of the phase diagram at $T = 780$°C. The effect of the surface energy added to the Gibbs energy of the nucleus decreases the supersaturation of the carbide relative to the matrix. This leads to a displacement of the phase boundary $\alpha/M_{23}C_6$.

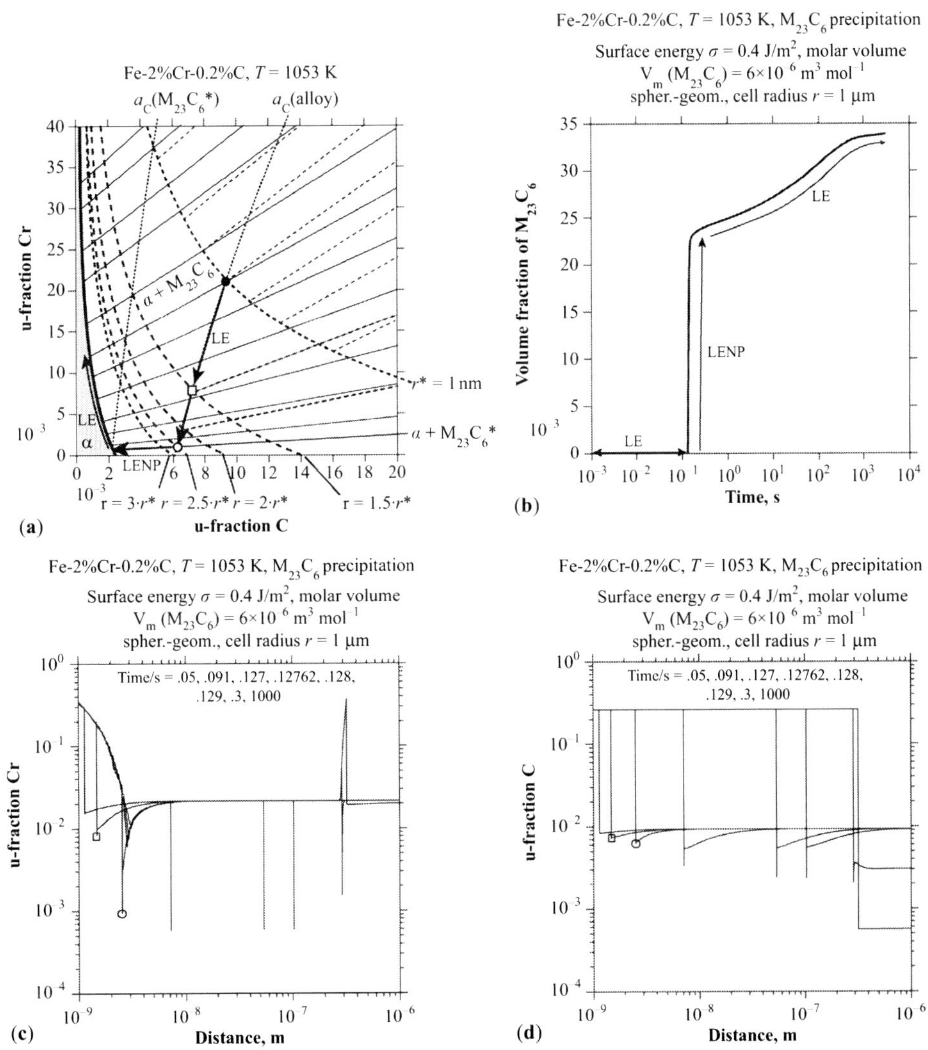

Figure 9.8. Growth of $M_{23}C_6$ in an alloy Fe-2%Cr-0.2%C at $T = 780$°C. The matrix is ferrite, the carbide nucleus is a sphere with radius 1 nm, and the surface energy of the interface $\alpha/M_{23}C_6$ is assumed as $\sigma = 0.4\ \mathrm{J/m^2}$: (a) phase diagram, (b) volume fraction of the carbide as a function of time, (c) Cr-composition profiles at various time steps, (d) C-composition profiles at various time steps.

The critical nucleus is in equilibrium with the matrix. Therefore, for the critical nucleus, the phase boundary passes through the point representing the matrix. The corresponding tie-line defines the composition of the nucleus. During the growth of the particle, the phase boundary moves and the boundary condition in ferrite follows the isoactivity line of C. In Figure 9.8(c) and (d), the Cr- and C-profiles are shown for various time steps. The Cr- and C-composition on the ferrite side of the $a/M_{23}C_6$ interface is marked by symbols for the radii, $r = 1.5r*$ and $r = 2.5r*$. These compositions are depicted in Figure 9.8(a) where they are marked by the same symbols showing the evolution along the C-isoactivity line. During this regime, Cr has to be redistributed because of the carbide composition changes as given by the variation of the tie-lines. This regime lasts up to about 0.13 s. Up to this time, only a negligible volume fraction is formed (see Figure 9.8[b]). Then, the LENP tie-line corresponding to the carbide $M_{23}C_6^*$ is reached. $M_{23}C_6^*$ represents the carbide with the same u-fraction of Cr as the matrix. Growth starts to occur without redistribution of Cr according to the LENP regime with a very high velocity. This can be deduced from Figure 9.8(b). The LENP reaction regime stops when the required strong C-flux can no longer be supplied due to the fact that the C-diffusion field has reached the cell boundary (see Figure 9.8[d]). Then follows again a reaction according to LE with redistribution of Cr.

Figure 9.8(b) shows the evolution of the fraction of carbide as a function of time. The growth of the carbide in the first regime cannot be seen in a linear scaling, as, in spherical geometry, the attained volume fraction is too small.

This example shows that the simulation with DICTRA provides a deep insight into the complex mechanisms of phase transformation.

9.1.4 Further Modules

9.1.4.1 Para-Equilibrium Model

This model assumes that the transformation of a phase γ into α is controlled only by the diffusion of fast diffusing interstitial elements, for example [219,224]. The substitutional elements are treated as immobile components. Consequently, the u-fractions of substitutional components are the same in both phases. Assuming para-equilibrium in a multicomponent system, Fe-X-C, the equilibrium is defined by the conditions $\mu_C^\alpha = \mu_C^\gamma$ and $u_{Fe}\mu_{Fe}^\alpha + \sum_X u_X \mu_X^\alpha = u_{Fe}\mu_{Fe}^\gamma + \sum_X u_X \mu_X^\gamma$. The first condition defines the local equilibrium for C, the second imposes the constraint that this equilibrium has to be within the section of constant fraction of substitutional components. Because of this constraint, a reaction according to para-equilibrium cannot lead to the final equilibrium state. The para-equilibrium diagram is in fact a binary diagram M-C, where M stands for the fixed mixture of substitutional elements.

In contrast to LENP, where the operating tie-line defines identical u-fractions of the substitutional elements in both phases, the condition of para-equilibrium requires that the average diffusion velocity of the substitutional elements is well below the velocity of the moving interface, see for example ref. [225]. This condition is likely to be fulfilled at low temperatures, but it may also be fulfilled at higher temperatures depending on the temperature variation of the mobilities.

The simulation will be illustrated with the transformation of austenite into ferrite in Fe-1.15%Si-0.51%C at 749°C. For this alloy, experimental data were obtained within the COSMOS project using three-dimensional metallographic analysis by serial sectioning. In Figure 9.9, the isothermal section of the Fe-Cr-C phase diagram is shown. The phase diagram

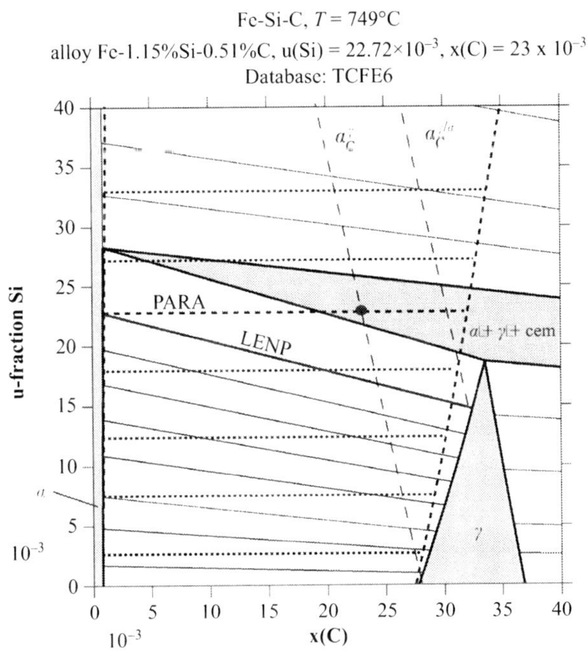

Figure 9.9. Transformation from ferrite into austenite in Fe-1.15%Cr-0.51%C. Isothermal section of the Fe-Cr-C phase diagram. The dotted lines show the para-equilibrium diagram. Both the LENP and the para-equilibrium reaction are possible.

according to para-equilibrium is superimposed. Both reaction regimes are thermodynamically possible: the alloy composition falls into the two-phase field of the para-equilibrium diagram, and, for the LENP reaction, the C-activity at the γ/α-interface is higher than the C-activity of the γ-matrix. C will thus diffuse from the γ-interface into the austenitic matrix, as required for the growth of ferrite.

The comparison with the experimental data presented in Figure 9.10 shows that the LENP regime is adopted in the transformation. It is a general observation that the LENP regime is adopted whenever it is thermodynamically possible. The para-equilibrium regime can be observed in cases where LENP is thermodynamically not possible (see refs. [219,225]). This is in accordance with the principle of maximum entropy production rate.[14]

9.1.4.2 Pearlite Module

In DICTRA, a lamellar type of coupled phase transformation can be simulated as an example for pearlite formation. In this case, both surface diffusion at the moving interface between the lamellae and volume diffusion ahead of the reaction front can be taken into account. The lamellar spacing is a result of the simulation.

[14] See Section 3.5.1.

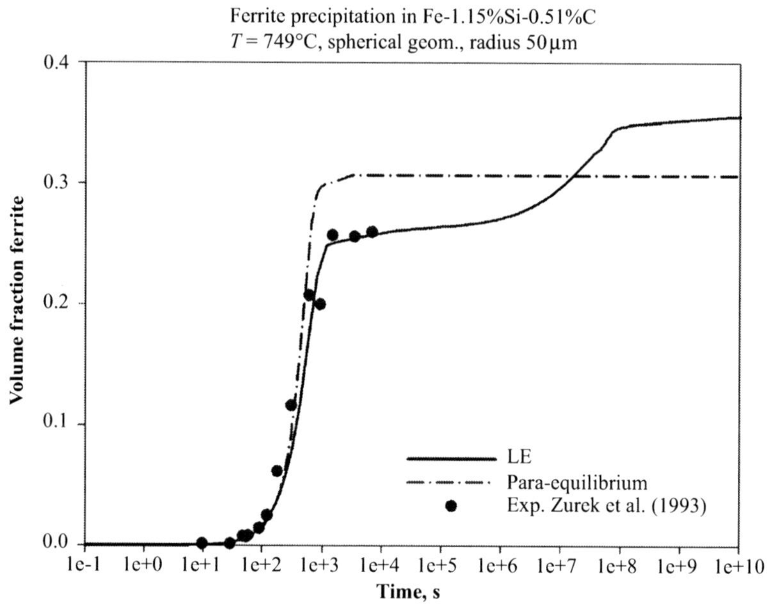

Figure 9.10. Volume fraction of ferrite, calculated according to LENP and para-equilibrium. The experimental data are in agreement with the LENP reaction. The experimental data are from reference [226].

The module is based on the Hillert–Zener model [227–229]. Details of the model are outlined in ref. [229].

9.2 *PrecipiCalc*—Software for 3D Multiphase Precipitation Evolution

This article has been courteously provided by Herng-Jeng Jou of QuesTek Innovations LLC, Evanston, Illinois, USA.[15]

9.2.1 General Information

PrecipiCalc is an efficient high-fidelity 3D multiphase precipitation evolution simulation software grounded in computational CALPHAD-based thermodynamics and multicomponent diffusional databases and nucleation and growth models based on the numerical Kampmann–Wagner approach with mean-field distributed spherical precipitates. Its development was strongly driven by industrial engineering needs to accelerate the design, development, and qualification of high performance alloys.

Since its first development and implementation by QuesTek Innovations LLC at the end of 2000, through several US government funded research programs *PrecipiCalc's* capability has

[15]The statements, considerations, and judgements provided in this section are the sole opinion of the author of this section. They do not necessarily represent the opinion of the author of this book.

been extensively extended, calibrated, and validated for various precipitate phases in several multicomponent alloy systems. Furthermore, *PrecipiCalc* has also been successfully used by QuesTek to assist the computational design and qualification of high performance materials including Fe-, Al-, Ni-, Ti-, and Cu-based alloys. The software implementation of *PrecipiCalc* is efficient and input is text-based to provide flexibility and integration with other software to achieve location-specific and design minimum property prediction. The core technology of *PrecipiCalc* is now being integrated by Thermo-Calc Software AB into their user friendly TC-PRISMA software.

QuesTek has been utilizing the computational *Materials by Design*® technology [230,231] to successfully design, invent, and rapidly bring to market never-before-conceived materials. *Materials by Design* technology was pioneered by Professor Gregory B. Olson of Northwestern University in the mid-1980s. Before *PrecipiCalc* was available, CALPHAD-based parametric precipitation models using driving forces and multicomponent coarsening rate constants were (and still are) commonly used within the computational design of alloy compositions and processing conditions [232–235] to achieve suitable precipitation microstructure and desirable mechanical properties. However, parametric models lack the capability to capture time-dependent precipitate microstructure evolution following the manufacturing processes, such as, for example, complex industrial heat treatment cycles. Hence, there was a strong motivation to develop *PrecipiCalc* software to enhance QuesTek's *Materials by Design* technology and to enable the optimal design of high performance materials.

Another strong motivation of *PrecipiCalc* development was driven by the Accelerated Insertion of Materials (AIM) initiative supported by US Air Force and Defense Advanced Research Projects Agency (DARPA). The introduction of new or existing materials into newly engineered products can involve extensive testing in order to qualify the materials; and this is often costly and time-consuming for mission-critical components in aerospace application. The key objective of the AIM initiative utilized computational materials models and an integrated product design and component qualification framework to accelerate the insertion of materials through focused experimental testing, rapid manufacturing process optimization, and minimum property prediction by linking macroscopic process models to microstructural evolution. During the AIM program, QuesTek worked with major aeroturbine engine manufacturers and university participants to simulate location-specific precipitation microstructure [236] and tensile strength prediction of Ni-base disk superalloys. In addition, design minimum yield strength prediction was successfully demonstrated [237]. After this demonstration for existing Ni-base superalloys under the AIM program, QuesTek has utilized this technology and *PrecipiCalc* simulations to successfully accelerate the application of new computationally designed alloys [234,238].

To actually achieve the goals stated above, the primary features of *PrecipiCalc* models and implementation were driven by the following requirements. Note, *PrecipiCalc* models and a logical calibration protocol have been thoroughly documented in open literature [236,239].

- *Prediction of precipitate particle size distribution evolution during industrial complex nonisothermal heat treatment.* Prediction of γ' microstructure in Ni-base disk superalloys during processing was the earliest application of *PrecipiCalc*. As the γ' precipitation kinetics in these alloys is not quench-suppressible by common industrial processes, capturing the nonisothermal effects, including multistage nucleation, transient dissolution, and growth were of prime importance [236].

- *Multicomponent composition effects.* Industrial alloys usually contain many alloying elements to achieve desirable property combinations. To capture the effects of multicomponent interaction with processing interaction, CALPHAD-based thermodynamics and diffusivity forms the basis of *PrecipiCalc* models.
- *Integration flexibility and requirement.* Precipitate microstructure prediction is just one part of an integrated process–structure–property–performance engineering toolset. It is essential for *PrecipiCalc* software to have the flexibility and capability to integrate with various process modeling tools (often FEM-based) and subsequent property models. Hence, *PrecipiCalc's* implementation maintains a text-based user interface. For users seeking a user-friendly interface, Thermo-Calc Software AB's TC-PRISMA,[16] a product codeveloped [240] by QuesTek and Thermo-Calc Software AB, might be considered.
- *Multiscale multimodal precipitate microstructure.* Some precipitates (such as γ' in Ni-base disk alloys) can develop multimodal distributions with particle sizes ranging from nanoscale up to micron scale developed during heat treatment. Simulation of multiscale 3D microstructure evolution can be extremely computationally expensive and impractical for precipitation models such as phase field and boundary integral models.
- *Model flexibility and extensibility.* Often, special precipitation mechanisms are important for the particular problem at hand. It is strongly desirable to allow easy model extensions for such situations. *PrecipiCalc* has already been extended to address the following special mechanisms: in situ inoculant formation and subsequent grain size prediction in casting, elastic misfit coherency energy, precipitation evolution with coherency loss, size-dependent interfacial energy, precipitation under para-equilibrium bcc-Fe and cementite conditions in secondary hardening steels, fast diffusion route, coprecipitation, collector plate grain boundary precipitate growth models, heterogeneous nucleation on various types of defects, and so forth.
- *Computational efficiency.* The *Materials by Design* and AIM technology often require a large number of precipitation simulations of different compositions, for different locations and process thermal history. *PrecipiCalc* uses a semianalytical method of characteristics [241,242] with an adaptive geometric gridding scheme that achieves excellent efficiency with minimum numerical diffusion for population balance equations to numerically solve the time evolution of particle size distribution based on the underlying multicomponent nucleation and growth models.

Driven by the key requirements for *PrecipiCalc* models and software, a numerical Kampmann and Wagner approach with multicomponent extensions [236] was used as a foundation. The key model assumptions used in *PrecipiCalc* (in order to achieve analytical multicomponent particle growth equation) are discussed below:

- *Mean field assumption*—assuming representative far field compositions around all particles. For precipitate microstructure with a single modal distribution, this is a reasonable assumption. For multimodal distributions with large differences in particle size and number density, the mean field approximation overestimates the interaction between particles

[16] See Section 9.5.

and thus overall coarsening rate. In addition, the mean-field assumption inherently implies a spatially uniform distribution of particles. This can cause inaccuracy due to neglect of nonuniform distribution effects. To overcome the limitation of the mean-field approximation, QuesTek worked with Professor Peter Voorhees in extending multipole expansion methods [243,244] under the US Navy ONR and DARPA joint funded Digital 3D Microstructure initiative. This precipitation simulation code designated *PrecipiCalc-3D* allows explicit treatment of particle spatial arrangements.

- *Classical nucleation theory*—classical nucleation theory (including incubation) provides an adequate description of the output of the nucleation process but neglects details during nucleation, which can be better described by cluster dynamics methods. Even with this shortcoming, classical nucleation theory still accurately captures the particle size in high volume fraction γ' of Ni-based superalloys during quenching experiments [239].
- *Compositions at particle–matrix interface* [236]—*PrecipiCalc* models assume the interfacial compositions to follow equilibrium (tie-line) compositions with Gibbs–Thomson correction, maximum driving force (parallel tangent) compositions with Gibbs–Thomson correction, or compositions with flux balance consideration. However, solute trapping and partial para-equilibrium mechanisms, particularly at low temperatures, can cause deviation from parallel (maximum driving force) or common tangent (tie-line equilibrium) compositions in *PrecipiCalc* models. An example is para-equilibrium cementite formation [244] in steels, where their initial formation is only driven by carbon diffusion while other substitutional solute atoms are all trapped. At current state, *PrecipiCalc* models are not suitable for precipitation with solute trapping mechanisms.
- *3D spherical particles*—from experience, for rod or plate particles with low aspect ratio (<5), the spherical morphology approximation is reasonable for low volume fraction as the spacing between particles is large, so morphology effects to diffusion field are small.

The main features of the *PrecipiCalc* software are summarized in Table 9.2. Details on the software implementation are given subsequently.

9.2.2 Software Implementation

Figure 9.11 shows a schematic diagram of the *PrecipiCalc* software implementation and the flow of its operation. *PrecipiCalc* software contains two executable programs, shown as the two green boxes, one for computation and another for post-processing. The computation program saves results into a binary HDF [245] file, which is then used by post-processing program to generate visualization plots or data files. The post-processor can generate text-based data file of PLOTMTV [246] and GRACE [247] (both work under UNIX/X11 and Windows/Cygwin) for rapid results visualization, and can also output data files to be used by MS Excel and other spreadsheet programs. The operation of both the *PrecipiCalc* computation and post-processing programs can be driven by the end user directly on console or by other software, as shown in the schematic diagram in Figure 9.11. An integration of *PrecipiCalc* with component-level FEM code and tensile yield strength model using iSIGHT in a distributed platform/network was demonstrated under the DARPA AIM program [236], enabling the prediction of spatially distributed microstructure and properties as well as manufacturing variation to forecast minimum properties [237].

Table 9.2. Software information table for *PrecipiCalc*.

Software	***PrecipiCalc***
Acronym	***PrecipiCalc***
Developer	H.-J. Jou QuesTek Innovations LLC (in collaboration with Professors Peter Voorhees and Gregory Olson at Northwestern University) Evanston, Illinois, USA
Licensing	commercial/academic
Release date	2004
Web site	http://www.questek.com/precipicalc.html
Contact email	software@questek.com
Kinetic model	Numerical Kampmann–Wagner (NKW)
Nucleation	Classical nucleation theory (CNT) with multicomponent and nonisothermal extension
Interfacial energies	Experimentally derived
Evolution equations from	Mean field, 3D steady-state multicomponent diffusion
Thermodynamics/ mobility	CALPHAD-type databases, unencrypted or encrypted
Programming language	C++
Platforms	Windows, Linux
GUI/command-line/ API	No/yes/no
Scripting	Yes
Programming	No
Exporting	Ascii, csv, PLOTMTV (postscript), GRACE (PDF, png, etc.)

Both computation and post-processing programs use the same text-based input file, which contains following information:

- *Material compositions*—the overall compositions of simulation systems, including matrix and precipitate phases. However, they are not necessarily the same as overall alloy compositions. For computational efficiency and stability, it is often recommended to input only the relevant compositions for the phases of interest. For example, primary inclusions formed at high temperature typically do not evolve during solid-state heat treatment. Hence, it is desirable to remove them, following a mass balance principle, for the simulations of strengthening precipitates during heat treatment.

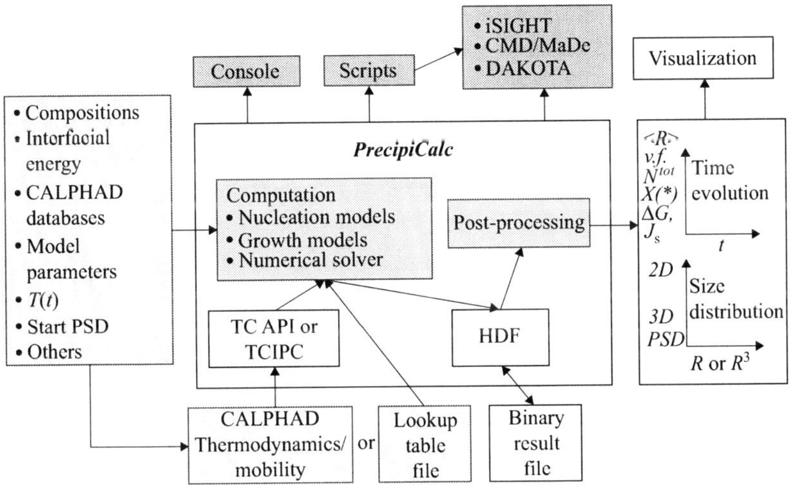

Figure 9.11. Scheme of software implementation for *PrecipiCalc*.

- *Location and names of CALPHAD thermodynamics and mobility databases*—In the case without mobility databases, users can input pre-factor and activity energy describing elemental diffusivity. Alternatively, a special version of *PrecipiCalc* can use the preprocessed lookup table file to bypass CALPHAD engine for speeding up simulations.
- *Phase information*—definition of matrix and precipitate phases should be consistent with the naming convention in the CALPHAD databases used.
- *Matrix–particle interfacial energy*—precipitation simulation, particularly nucleation behavior, can be fairly sensitive to the interfacial energy. Experimental calibration is often needed to achieve the necessary accuracy.
- *Temperature profile*—if desirable, user can provide a separate file (text-based) to describe the time and temperature history.
- *Output parameters*—controlling the results storing times during the evolution. The computation results are saved into an HDF-format binary file, with the file name specified in the input file.

Sometime the following input can also be included for special purposes:

- *Starting particle size distribution (PSD)*—user can put in starting particle size distribution if necessary. This is not needed for supersaturated start condition.
- *Numerical parameters*—there is no need to alter the default numerical parameters for most simulations. If needed, users can modify many parameters to control spatial resolution in particle size distribution and temporal resolution.
- Additional model parameters to fine tune the simulation.

9.2.3 Example of *PrecipiCalc* Simulations

To demonstrate the basic *PrecipiCalc* operation, the following is an example of γ' precipitate evolution in ME3 Ni-base superalloy during the rapid cooling of SSDTA (Single Sensor

Differential Thermal Analysis) experiment, documented in ref. [239]. This nonisothermal simulation is chosen, instead of a simpler isothermal one, to illustrate the *PrecipiCalc* capability of predicting interesting transient multimodal particle size distribution and microstructure. The purpose of the SSDTA experiment and simulation is to determine the nucleation onset temperature for calibrating high quality γ/γ' interfacial energy. Conventionally, isothermal coarsening experiments have been used for interfacial energy calibration [248] with results that would carry the uncertainty of diffusivity value. Nucleation behavior, on the other hand, is highly driven by bulk thermodynamics and interfacial energy with minimum impact from diffusivity. The following is the *PrecipiCalc* input file "ME3-1Ar.inp" for the ME3 simulation with 41°C/s average cooling rate documented in Figure 5 of ref. [239]. The text-based input file follows the "key = value" format in each row, while the order of rows is not important.

```
# File "ME3-1Ar.inp"
# Comment lines starting with "#"

# Output binary HDF file
RESULT_FILE=ME3-1Ar.out

# Output control, output every 1°C and also on input T(t) points
OUTPUT_TIME_FORMAT=2
OUTPUT_TPNODE=1
OUTPUT_DTEMP=1

# Temperature profile T(t) from a separate file
T_FILE_NAME=ME3-1Ar.dat

# Compositions of ME3 with MC/MB2 removed at 1179.6C equilibrium
ELEMENT_UNIT=0
ELEMENT1=NI, 0
ELEMENT2=AL, 7.30378e-002
ELEMENT3=CR, 1.42667e-001
ELEMENT4=CO, 2.03599e-001
ELEMENT5=TI, 4.26280e-002
ELEMENT6=MO, 2.29464e-002
ELEMENT7=W,  6.29101e-003
ELEMENT8=NB, 5.23041e-003
ELEMENT9=TA, 6.81528e-003

# DICTRA Command, thermodynamics and mobility databases
# DICTRA Command not needed for PrecipiCalc with integrated TC-API
DICTRA_CMD = dictra23
DICTRA_TDB = Ni-data7.TDB
DICTRA_MOB = NIST-NiMob.tdb

# Phase Definition, Matrix
M_PHASE_NAME=FCC
M_LP_MODEL=FCC

# Phase Definition, Precipitate
NUM_P_PHASES=1
P_INFO_START=1
P_PHASE_NAME=GAMMA_PRIME
```

```
P_MASS_BALANCE_MODE=1
P_LP_MODEL=GAMMA_PRIME
P_SE_C4=0.0275
P_COMP_MODE=1
P_INFO_END=1

# Using default numerical control, except this one for high dT/dt
DTEMP_MAX=0.1

# help Thermo-Calc/DICTRA with energy minimization convergence
TC_S_S_V=y(gamma,Ni#1)=0.85 y(gamma,Al#1)=0.001 y(gamma,Ti#1)=1e-4
TC_S_S_V=y(gamma,Ta#1)=5e-4 Y(gamma,co#1)=0.15
TC_S_S_V=Y(GAMMA,NI#2)=0.01 Y(GAMMA,CR#2)=0.05 Y(GAMMA,CO#2)=0
TC_S_S_V=Y(GAMMA,AL#2)=0.45 Y(GAMMA,TI#2)=0.45
TC_S_S_V=Y(GAMMA,MO#2)=0.01 Y(GAMMA,TA#2)=0.06

# Print info on screen at least every 3 second
PRINT_TIME_INTERVAL=3
```

Some comments about this input file:

- Compositions of nine components for *PrecipiCalc* simulation came from γ compositions at 1179.6°C equilibrium, without the stable MC and MB_2 phases at that temperature. The resulting concentrations of C, B, and Zr are low, and thus ignored here. 1179.6°C is the holding temperature before the rapid cool during SSDTA experiment. It is assumed that the equilibrium is achieved at the holding temperature and MC/MB_2 do not evolve or convert into other phases during subsequent rapid cool.
- TC_S_S_V keys and values are used to provide guidelines to assist Thermo-Calc/DICTRA engine's free energy minimization.
- DTEMP_MAX is the only numerical parameter used here to resolve lower temperature nucleation behavior better in rapid cooling simulation. The default value of 1 (10 times larger than 0.1 used in the file) can introduce some numerical noise in nucleation behavior although impact to the γ' microstructure evolution results is not noticeable.

Aforementioned input file also specifies an external temperature profile data file "ME3-1Ar.dat," came from SSDTA experiment with latent heat contribution removed. This file has two columns, time (in seconds) and temperature (in°C), as partially shown below.

```
0        1179.590759
0.092    1176.459199
0.192    1173.066359
0.292    1169.68499
0.392    1166.315075
0.492    1162.956596
0.592    1159.609534
:
:
:
```

```
50.292    467.33
51.292    463.36
52.292    459.27
53.292    455.24
54.292    451.04
55.292    447.34
56.292    443.42
61.292    424.86
66.292    409.22
71.292    395.58
76.292    382.88
81.292    370.54
86.292    359.71
96.292    339.88
106.292   321.24
116.292   303.45
```

The following screen dump captures the computation code, "mpc," and its screen output. The computation code "mpc" takes the input file name as the only input argument, and directly goes into simulation after dumping consistency-checking information.

```
$ mpc ME3-1Ar.inp

************ PrecipiCalc (MPC-IPC) ************
             Version 0.9.2
MULTICOMPONENT, MULTIPARTICLE PRECIPITATION
CALCULATION SOFTWARE,    (C)2001-2007 AS AN
UNPUBLISHED WORK, QUESTEK INNOVATIONS  LLC.
ALL  RIGHTS RESERVED.  USE OF THIS SOFTWARE
WITHOUT LICENSE IS STRICTLY PROHIBITED.

COMPILED FOR Linux 2.6.18-194.3.1.el5 ON beta.questek.loc
AT Thu Dec 22 15:13:00 2011 BY hjjou

Info: processing user input file...

Info: validating key-names...
Info: key-name validation complete

Info: input file processing is done

Info: starting precipitate [1] key-value validation ...
Info: precipitate [1] key-value validation ok.

Info: starting global key-value validation ...
Info: global key-value validation ok

Info: checking dependences in user settings ...
Info: dependences in user settings are ok
```

```
Please confirm following DICTRA information before launching ...
    - Element list :       NI AL CR CO TI MO W NB TA
    - Matrix phase :       FCC
    - Precip phases :      GAMMA_PRIME
    - Dictra command :     dictra23
    - TC database :        Ni-data7.TDB
    - Mobility database : NIST-NiMob.tdb

   Info: program is initializing DICTRA based on input ...


   Info: checking if user preset PSD for any precipitates...
   Info: done checking preset PSD.
            start computing...
   Info: start computation ...
   <-- #1/1042
    Global Info:    dt controlled by (dT)max [9]/[10]
     time= 2.9378e-02  Temp= 1.1787e+03  <dt>= 3.2643e-03
    P[1]
     <R> = 0.0000e+00  vf= 0.0000e+00  totN= 0.0000e+00  bin#= -1
     dG  =-9.2820e+01  Js= 0.0000e+00  tau = 0.0000e+00  Js0 =
0.0000e+00
     dt controlled by None(no particles or just out) [0]/[0]
   -->
   <-- #2/1042
    Global Info:    dt controlled by (dT)max [10]/[11]
     time= 5.8757e-02  Temp= 1.1776e+03  <dt>= 2.9378e-03
    P[1]
     <R> = 0.0000e+00  vf= 0.0000e+00  totN= 0.0000e+00  bin#= -1
     dG  =-8.6924e+01  Js= 0.0000e+00  tau = 0.0000e+00  Js0 =
0.0000e+00
     dt controlled by None(no particles or just out) [0]/[0]
   -->


   :
   :

   <-- #1041/1042
    Global Info:    dt controlled by (dT)max [10]/[11]
     time= 1.1585e+02  Temp= 3.0424e+02  <dt>= 5.6211e-02
    P[1]
     <R> = 7.9875e-10  vf= 4.9453e-01  totN= 1.2060e+25  bin#= 236
     dG  = 4.2642e+03  Js= 5.5917e+17  tau = 7.2251e+00  Js0 =
5.9813e+17
     dt controlled by (dVF/dt)max-Nuc [11]/[11]
```

```
-->
<-- #1042/1042
 Global Info:    dt controlled by (dT)max [7]/[8]
  time= 1.1629e+02  Temp= 3.0354e+02  <dt>= 6.3439e-02
 P[1]
  <R> = 7.9875e-10  vf= 4.9453e-01  totN= 1.2060e+25  bin#= 236
  dG  = 4.2759e+03  Js= 5.2768e+17  tau = 7.2251e+00  Js0 =
5.6430e+17
  dt controlled by (dVF/dt)max-Nuc [8]/[8]
-->

Info: this calculation took 28.483333 minutes

Computation is done. Info: normal quit...
Thanks for using PrecipiCalc!!!
```

As there are 1042 output locations along the rapid cooling, only the beginning and ending screen dumps are shown. The simulation takes about 28.5 minutes to complete on a modern desktop computer. With the default DTEMP_MAX=1, the frequency to call Thermo-Calc/DICTRA for thermodynamics and diffusivity quantities update is reduced, and this simulation takes about 3.5 minutes to finish. If thermodynamics and diffusivity data is preprocessed into a lookup table file, the simulation will not call Thermo-Calc/DICTRA and it takes only 1 second to complete on the same machine. This suggests the numerical algorithm in *PrecipiCalc* is efficient, and the majority of computational time is used by Thermo-Calc/DICTRA in this simulation example.

Though the screen dump displays some key information regarding the simulation, the results of interest would need the assistance of post-processing program. The following screen dump shows the initial screen of the post-processing code "mpcpost," which again takes the input file "ME3-1Ar.inp" as command line argument. After showing summary information of the simulation, the post-processor presents a menu on console and waits for user's input for further guideline.

```
$ mpcpost ME3-1Ar.inp

******** PrecipiCalc Post-Processor *******
             Version 0.9.0
MULTICOMPONENT, MULTIPARTICLE PRECIPITATION
CALCULATION SOFTWARE,    (C)2001-2003 AS AN
UNPUBLISHED WORK, QUESTEK INNOVATIONS  LLC.
ALL  RIGHTS RESERVED.  USE OF THIS SOFTWARE
WITHOUT LICENSE IS STRICTLY PROHIBITED.
COMPILED FOR Linux 2.6.18-194.3.1.el5 ON beta
AT Thu Nov 25 17:10:11 2010 BY hjjou


Press <Enter> to continue...
```

```
<<--Summary of Result File "ME3-1Ar.out"-->>
Calculation Info:
 project_name:              PROJECT01
 number_of_nodes:           1042
 working_directory:         NAS01 005/SSDTA/ME3
 T_profile_filename:        ME3-1Ar.dat
 input_filename:            ME3-1Ar.inp
 TDB_filename:              Ni-data7.TDB
 MOB_filename:              NIST-NiMob.tdb
 number_of_phases           1
 number_of_elements         9
 element_list:              NI AL CR CO TI MO W NB TA
 overall_composition:       0.496785 0.0730378 0.142667
0.203599 0.042628 0.0229464 0.00629101 0.00523041
0.00681528
 time_vector:               start from 0.0293783 to
116.292(sec)
 trackflag_vector:                 1
 matrix_phase_name:         FCC
 partical_phase_name:       GAMMA_PRIME
Program Info:
 OS_type:                   Linux 2.6.18-194.3.1.el5
 host_name:                 beta.questek.loc
 compile_time:              Thu Dec 22 16:34:16 2011
 compile_by:                hjjou
 compute_time:              Sat Jan 14 08:32:54 2012

Press <Enter> to continue...

Top-Level Menu Commands:
 U     : view/change general User setting [...]
 I     : view result file Information
 P     : view/generate PSD plot/file
 H     : make particle frequency Histogram data file
 S     : view/generate Summary plot/file
 T     : sTrength calculation [...]
 F     : Find node #'s within a range of time
 D     : finD information by radius range
 L     : Load another result file
 W     : Write original input parameters to a file
 Q     : Quit the program

Please enter your command:
```

As there are many post-processing features, we demonstrate the results of two commonly used features. The first feature is "S: view/generate Summary plot/file" that will produce plots and data files with predicted temporal γ' microstructure and other model quantities.

Using the default GRACE plotting format, the post-processor automatically generates and displays two plots. Figure 9.12 shows the first 50 seconds during quench, with additional

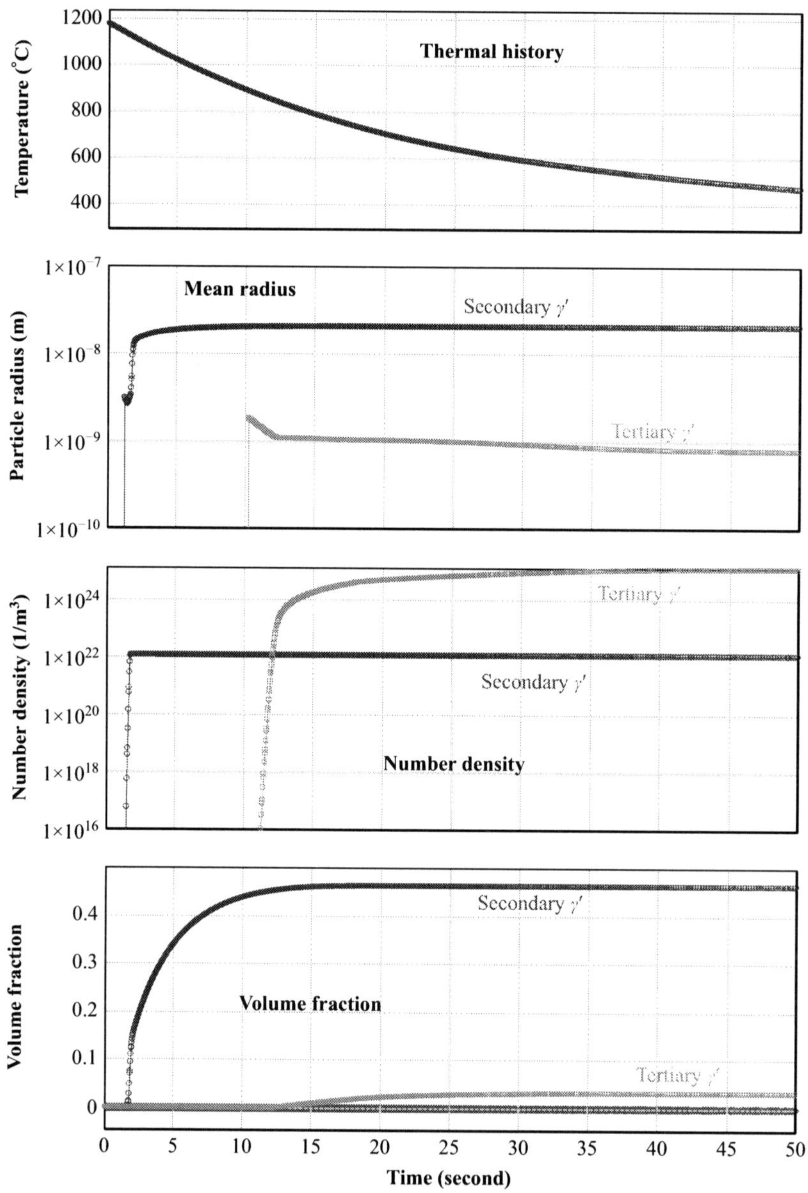

Figure 9.12. Output produced by the *PrecipiCalc* post-processor with the "S: view/generate Summary plot/file" option.

annotation. The first plot contains temperature (showing rapid cooling rates) and calculated γ' microstructure evolution with time. As nonisothermal simulation results in multimodal distributions, the post-processor automatically separates them and presents their mean size, number density, and volume fraction in separate curves. Note that in this supersolvus treated Ni-base superalloy, there is no primary γ' according to conventional naming scheme. Only secondary and tertiary γ' formed during quench are shown.

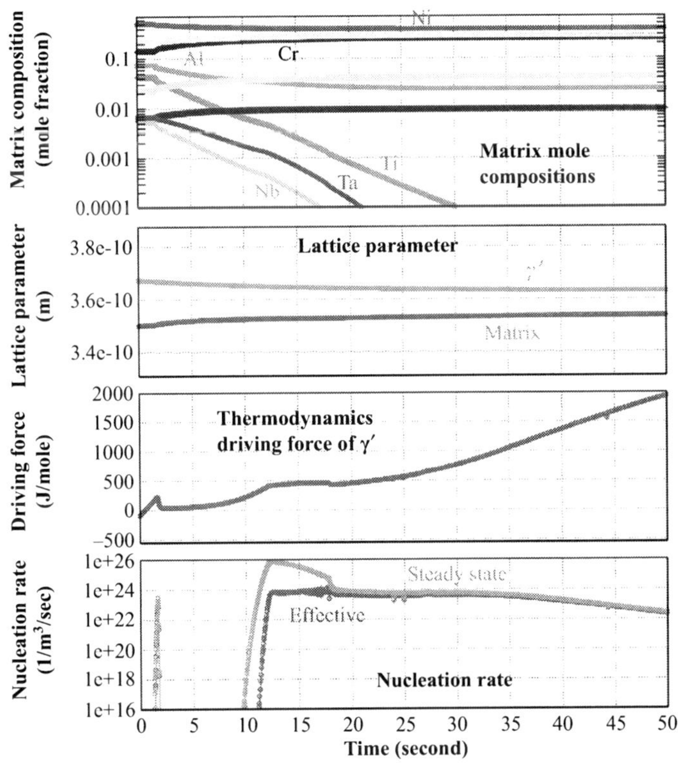

Figure 9.13. Detailed output produced by the *PrecipiCalc* post-processor for various precipitation parameters.

The second summary plot, as shown in Figure 9.13, illustrates the temporal evolution of predicted matrix compositions, lattice parameters, thermodynamic driving force of γ', and nucleation rate. Note that there are two nucleation events driven by the effective nucleation rate (curve for secondary γ'). One occurs at about 2 seconds, forming large low number density γ' and resulting in high volume fraction secondary γ'. This first nucleation burst is fairly brief and is shutdown by significant reduction of thermodynamic driving force, caused by large volume fraction and fast growth of secondary γ' at high temperature.

As temperature drops, growth of secondary γ' slows down, enabling accumulation of γ' driving force to create the second nucleation burst at about 11 seconds. The resulting tertiary γ' forms at a much higher driving force (thus smaller size) than does secondary γ'. Due to slow kinetics at continuously dropping temperature, growth of all γ' particles is insignificant after 25 seconds (600°C) and allows thermodynamics driving force to continue to rise. This continues the nucleation of finer and finer particles at negligible volume fraction. As discussed in ref. [239], combining predicted γ' volume fraction and matrix composition, molar enthalpy evolution and its derivative can be found to compare with experimental nucleation onset temperature.

Another post-processing feature, "P: view/generate PSD plot/file," generates the particle size distributions at selected output times. As an example, the GRACE plot in Figure 9.14(a) (with additional annotation) illustrates the distributed γ' number density at time = 3, 5, and 8 seconds when there is only secondary γ' particles.

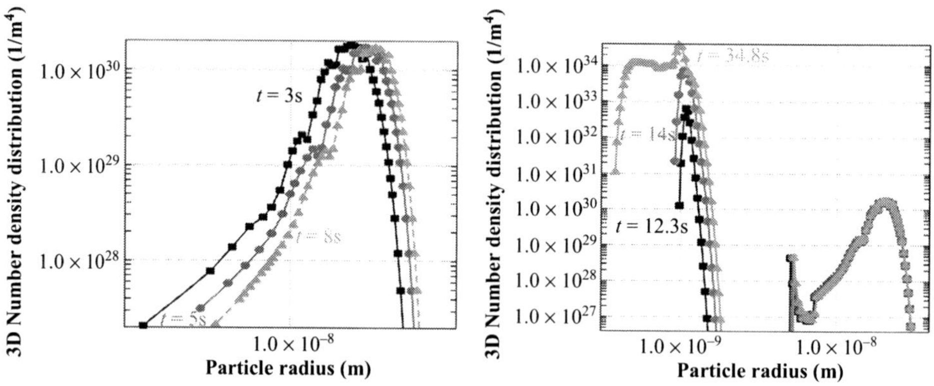

Figure 9.14. Multimodal size distributions of particles generated by *PrecipiCalc*.

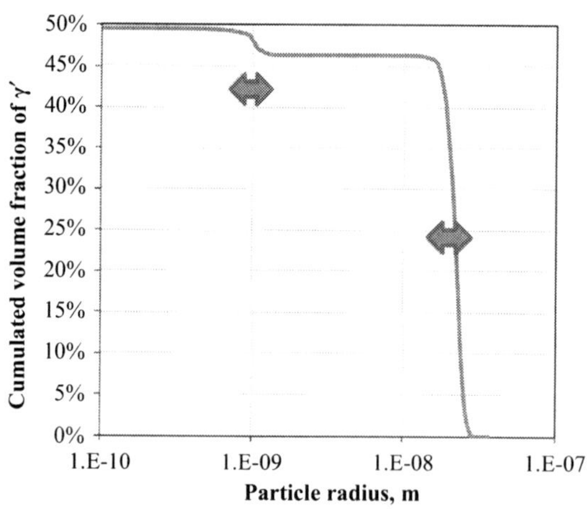

Figure 9.15. Cumulated volume fraction of γ' during heat treatment.

Figure 9.14(b) is the distributed γ' density at three different times after tertiary γ' is formed. Continuing growth and nucleation for the tertiary γ' during rapid cooling can be seen, while the secondary γ' does not change much. The predicted subnanoscale particles (or clusters) at $t = 34.8$ s, though has a high number density, do not have significant volume fraction and impact to material property.

This can be seen clearly by further processing the number density distribution plot into a cumulated γ' volume fraction plot at $t = 34.8$ s as shown in Figure 9.15. The two arrows indicate the major concentration of γ' volume fraction; the subnanoscale distribution (<0.8 nm) and the distribution tail spike for secondary γ' (between 5 and 7 nm) due to brief nucleation do not have significant volume fraction.

The predicted γ' microstructure after SSDTA experiment compares reasonably well to the experimental characterization as documented in Figure 7 and Table 10 of ref. [239]. Ref. [239] also documented other simulations for slower furnace cooling with a cooling rate of 0.56°C/s where *PrecipiCalc* again predicts the coarser γ' size very well. The predicted γ' size distribution driven by transient growth in one of the furnace cool samples also compares very well with experimental SEM results, and the distribution is much narrower than that predicted by coarsening theory.

9.2.4 Summary

The example shown earlier illustrates the implementation and basic operation of the *PrecipiCalc* software. In addition, some of its capability in transient multimodal precipitation microstructure is demonstrated.

PrecipiCalc has been applied to more complex industrial heat treatment involving a combination of isothermal, heat up, and cooling. As documented in ref. [236] for modeling IN100 alloy through industrial heat treatment, many transient γ' microstructure features during heat up processes were captured. For further examples of simulating multiple precipitate phases with *PrecipiCalc*, users can refer to ref. [249]. Application of *PrecipiCalc* in steel and aluminum alloys can be found in ref. [238]. There are numerous *PrecipiCalc* applications to other alloy systems conducted by the engineers and researchers at QuesTek and Northwestern University, including precipitation evolution in heat affected zone of friction stir welded 7050 aluminum alloy [250] and γ' evolution in austenitic steels [251].

9.3 *MatCalc*—The Materials Calculator

9.3.1 General Information

The development of the software package *MatCalc* started with a PhD thesis project (E. Kozeschnik) at the Graz University of Technology, Austria, in 1993. The topic of the thesis was the computer modeling of precipitation in interstitial-free and bake-hardening steel. The development of the *MatCalc* code was made necessary by the absence of general software for precipitation kinetics simulation at the time when the PhD project was started. In the first years of code generation, a CALPHAD-type thermodynamic engine was developed. On top of that, a precipitation kinetics module based on the numerical Kampmann–Wagner model with binary nucleation and a parabolic growth law was built, as well as a moving boundary module using Finite Elements [252] and local thermodynamic equilibrium. The latter two modules have been removed and replaced by alternative approaches later.

In practical application of the early software to complex precipitation and phase transformation problems, it became clear that the classical models available at this time were not fully suitable for accomplishing the ambitious goal of computationally efficient and physically-based *predictive* precipitation kinetics simulations in technologically interesting multinary systems. In a collaborative project of the Graz University of Technology, the Materials Center Leoben Forschung GmbH, and the University of Leoben in Austria, a sequence of strategic projects were initiated with the goal of developing appropriate models for realization of this task.

As a major result of this research, in 2004, the fundamental relations behind the *MatCalc* growth model were published [44,253]. The concept followed in this work brought two entirely new aspects into the field of precipitation modeling: The thermodynamic extremal principle of maximum entropy production (see Section 3.5.1) and the mean-field representation of the precipitation problem (see Section 3.5). With these two approaches, the *evolution of radius* and *chemical composition* of each precipitate could be described by a linear system of equations. Thus, even complex precipitation problems could be treated in a computationally efficient way.

After setting up the basic functionality of the precipitation kinetics framework, further development of *MatCalc* was focused on the development of a theory for a *quantitative prediction of interfacial energies*. This quantity traditionally represents a major undetermined parameter inside nucleation theory, although at the same time being the most essential. Based on a generalization of the nearest-neighbor broken-bond model developed in the 1930s, a methodology has been developed and incorporated into the *MatCalc* code that allows for a determination of interfacial energies based on an effective structural factor and the enthalpy of solution [99]. The effects of interfacial curvature [123] and entropic contributions of diffuse interfaces [133] were incorporated into the model shortly afterward. The corresponding theory is described in Chapter 4.

In recent years, emphasis has been on the development of models describing the interaction of precipitation with simultaneous thermal and thermomechanical treatment, as well as the treatment of simultaneous precipitation and long-range diffusion. Some recent examples of theory development involve a physical model for structural vacancy evolution (Section 7.4.2) or precipitate growth on grain boundaries (Section 6.1).

Character, philosophy, and look-and-feel of the *MatCalc* software are well described by its vision of being a *scientific tool box for enjoyable computational materials engineering*. The models that are incorporated in the software provide a comprehensive computational framework for precipitation and phase transformation modeling, which contains the most relevant physical mechanisms governing the process of solid-state precipitation. As far as possible, no undetermined parameters are used in the modeling approaches, including the interfacial energy, which is computed from the thermodynamic information of the CALPHAD databases as a temperature- and composition-dependent quantity.

In the way the software is designed, the user ought to apply it as a playground for testing and verifying his/her particular understanding and conception of the microstructural processes that occur during the phase transformation and precipitation reactions of interest. For successful operation of the software, the user should at least have a vague picture of the physical processes occurring in the system. The software then provides the necessary tools for testing and verifying these hypotheses. Although the software is per se not designed for minimal user input, but for physical modeling, most of the relevant input parameters are assigned representative default values that allow for a straightforward simulation with a reasonably small amount of user input. In many cases, simulations with most input parameters left on their default values already provide realistic first-order predictions of the evolution of the precipitation process.

Table 9.3 compiles the basic technical information of the *MatCalc* package, as well as contact information and main software features.

9.3.2 The Kinetic Model

The conceptual grounds of the *MatCalc* precipitation kinetics module are founded in (*a*) a comprehensive multicomponent nucleation model and (*b*) in computationally efficient evolution equations for further precipitate development derived from the thermodynamic extremal principle in the framework of a mean-field approach. The former approach is described in detail in Sections 2.3, 2.4, and 2.5 of this book, the latter approach in Section 3.5. The time integration

Table 9.3. Software information table for *MatCalc*.

Software	*MatCalc*—The Materials Calculator
Acronym	***MatCalc***
Developer	E. Kozeschnik et al. (see web site) Institute of Materials Science and Technology Vienna University of Technology Vienna, Austria
Licensing	commercial/academic/free (limited)
Release date	2004
Web site	http://matcalc.at
Contact email	info@matcalc.at
Kinetic model	Numerical Kampmann–Wagner (NKW)
Nucleation	Multicomponent classical nucleation theory (CNT)
Interfacial energies	Generalized nearest-neighbor broken-bond (GBB)
Evolution equations from	Thermodynamic extremal principle (SFFK-model)
Thermodynamics/mobility	CALPHAD-type databases, unencrypted
Programming language	C++, Qt (for GUI)
Platforms	Windows, Linux, Mac
GUI/command-line/API	yes/yes/yes
Scripting	Yes
Programming	Yes
Exporting	Ascii, png, svg, pdf

of the evolution equations is performed within the numerical Kampmann–Wagner approach where the precipitate distribution is treated in terms of discrete size classes for precipitates of identical radius and chemical composition (see Section 5.3).

In the time integration procedure of the evolution equations, the entire simulation schedule is discretized into (sufficiently) small isothermal time increments. Within these time intervals, the nucleation rate equation is evaluated for each of the precipitates assigned to the precipitation reaction and new size classes are added to the list of existing classes if applicable. Then, for each existing class, the growth equations are evaluated and the radius and chemical composition of the size classes are updated. If the radius of a particular size class falls below a certain threshold value, the size class is removed from the list of classes. If the maximum number of size classes is exceeded by nucleation of new classes, *MatCalc* internally rearranges the size class distribution to match the maximum settings. A schematic of the time integration procedure in *MatCalc* is shown in Figure 9.16.

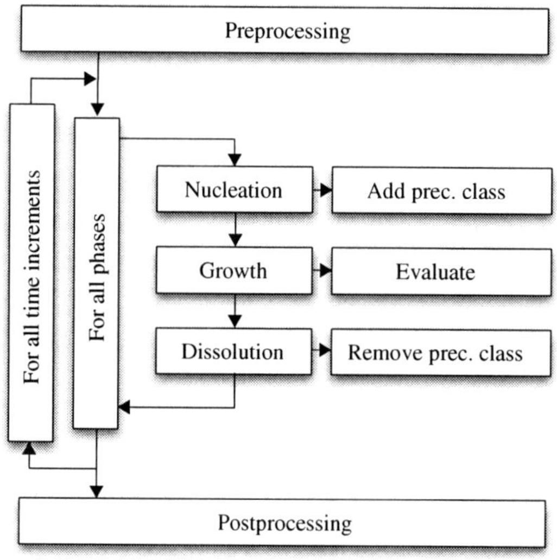

Figure 9.16. Time integration of the evolution equations as implemented in *MatCalc*.

In the actual precipitation simulation, the user first defines the thermodynamic system, nominal chemical composition, and the thermomechanical[17] treatment. The $T(t)$ and $\varepsilon(t)$ schedules can be conveniently entered in the form of *heat treatments*, which are shown later in Figure 9.24. The definition of the thermodynamic system includes the list of possible precipitates and precipitation domains, the latter defining the "containers" where precipitates can form. Each of these objects is associated with a thermodynamic phase as defined in the CALPHAD database. It is easily possible to set up multiple precipitation domains within one simulation, thus reflecting the changes in matrix phase (austenite, ferrite, martensite, etc.) as observed, for example, during heat treatment of steel. In Figure 9.17, the relation between precipitation domains and precipitate phases is schematically shown.

The precipitates are commonly assigned to an active precipitation domain which means that all thermophysical properties of the precipitate are evaluated with respect to the thermodynamic matrix phase associated with the precipitation domain. The precipitation domain contains all relevant information defining the properties of the matrix, such as dislocation density, grain size, and so forth, and many more.

The precipitate itself contains all relevant precipitate-related properties, such as volumetric misfit, nucleation properties, and more. The precipitate size distributions are organized in the form of size classes, where *MatCalc* keeps record of a list where groups of precipitates are lumped together with identical radius and chemical composition. A representative size distribution list for one precipitate could look like the one shown at the bottom of Figure 9.17. Please

[17]The *MatCalc* software has several features that incorporate the impact of mechanical deformation into the simulation procedure. These include, among others, the consideration of pipe diffusion along dislocations, dislocation evolution by generation, annihilation, and creep, dynamic grain size evolution, and the generation and annihilation of excess structural vacancies.

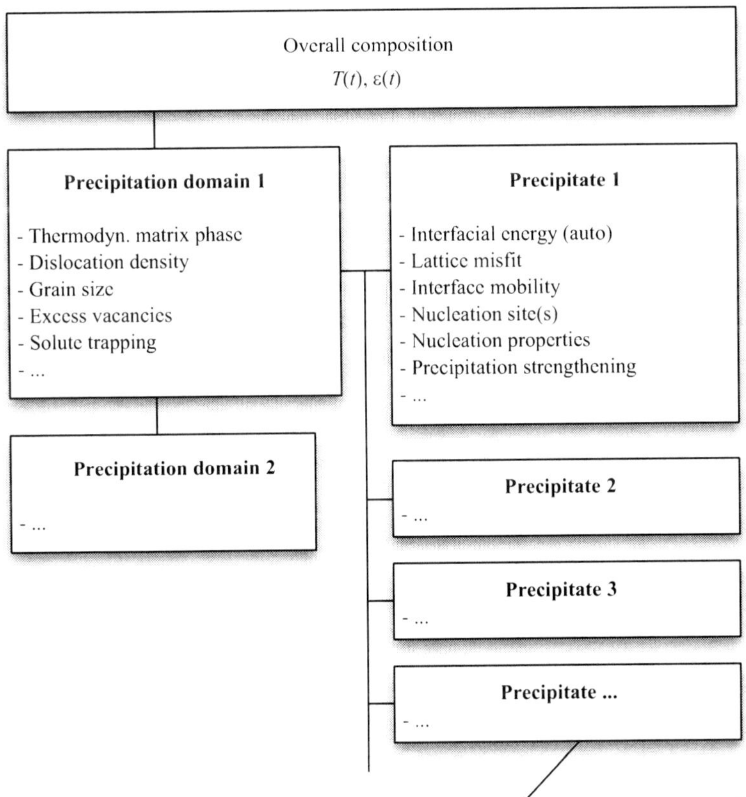

N_k	ρ_k	$c_{k,C}$	$c_{k,Cr}$	$c_{k,Fe}$	$c_{k,...}$
1e12	1e-9	0.25	0.61	0.04	...
2e13	2e-9	0.25	0.62	0.02	...
5e14	3e-9	0.25	0.64	0.02	...
2e15	4e-9	0.25	0.68	0.01	...
...	...	...	...	...	...

Figure 9.17. Data structure and relation between precipitation objects in *MatCalc*.

note that each precipitate size class is also related to a particular chemical composition, which is controlled by the evolution equations described in Section 3.5.2.

The time integration of the evolution equations is performed within the Euler forward integration scheme. Several numerical parameters can be defined, which control the behavior of the adaptive step size procedure and, thus, control the accuracy and performance of the algorithm. In most cases, the default values bring about a convenient balance between these two quantities and the users need not change these settings for successful simulation. Further details of the integration procedure are outlined in ref. [253].

9.3.3 *MatCalc* Precipitation Simulation in the GUI Version

On the example of AlN precipitation in steel, the handling of the *MatCalc* software in the Graphical User Interface (GUI) is briefly demonstrated in this section. After starting the program and opening a new workspace, the basic setup of the simulation is performed in the database dialog with the selection of thermodynamic and mobility databases, and elements and phases involved in the calculation. In the present example, this selection comprises the "*mc_fe.tdb*" thermodynamic database, the "*mc_fe.ddb*" diffusion database, the elements Fe, Al and N,[18] as well as the phases BCC_A2, FCC_A1 and ALN (Figure 9.18).

In a next step, the precipitation domains are created, representing the "containers" where precipitates can form. A precipitation domain stores information on the thermodynamic matrix phases, which are FCC_A1 for the austenite and BCC_A2 for the ferrite domains, as well as microstructural information about dislocation density, grain size, evolution of grain size, subgrain size and dislocation density, treatment of excess vacancies, trapping of vacancies and interstitials on solute atoms, dislocations and grain boundaries, precipitation strengthening, as well as mechanical properties such as Young's modulus, Poisson's ratio, and average grain boundary energies (Figure 9.19).

After creating the precipitation domains, the precipitate phases are created and set up in the phase status dialog. Here, all properties of equilibrium phases as well as precipitates are defined. Figure 9.20 displays the corresponding dialog for the thermodynamic matrix phase of the austenite precipitation domain. The most relevant settings in this tab of the dialog are the phase flags defining the status of the phase in the calculations, as well as the major constituents

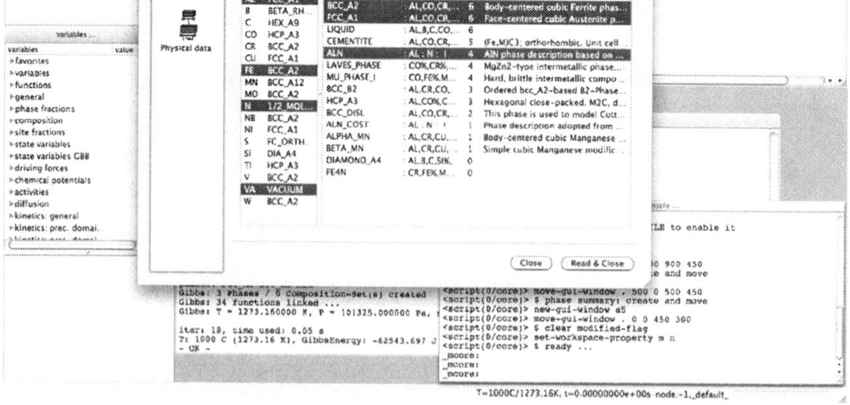

Figure 9.18. *MatCalc* database dialog: Setup of the thermodynamic system.

[18]The element VA represents vacant interstitial lattice sites and is selected automatically by the software.

that define the internal composition of the equilibrium phases. Figure 9.21 shows the general settings for the precipitate phases, which are the maximum number of size classes in the precipitation simulation, the precipitate shape (oblate, spherical, or prolate), and the interfacial energy. The shape of the precipitate enters the evaluation procedures for precipitation strengthening as

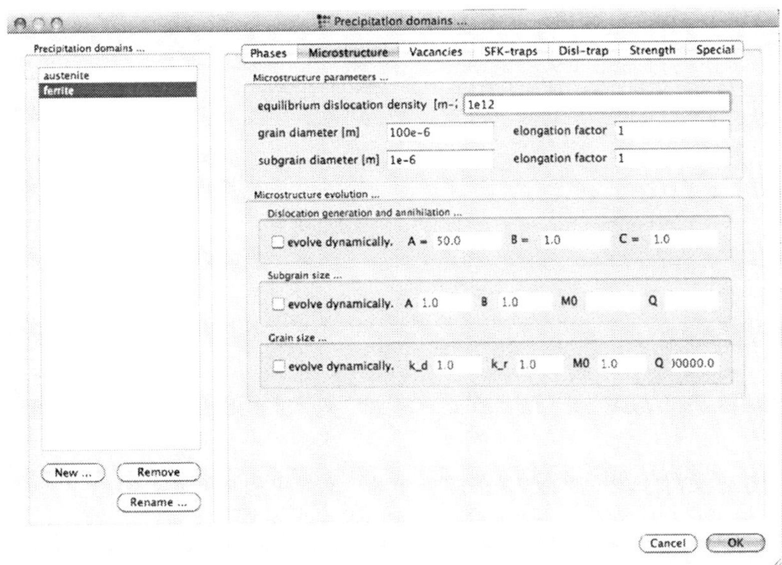

Figure 9.19. *MatCalc* precipitation domain dialog: Definition of the properties of the precipitate "container."

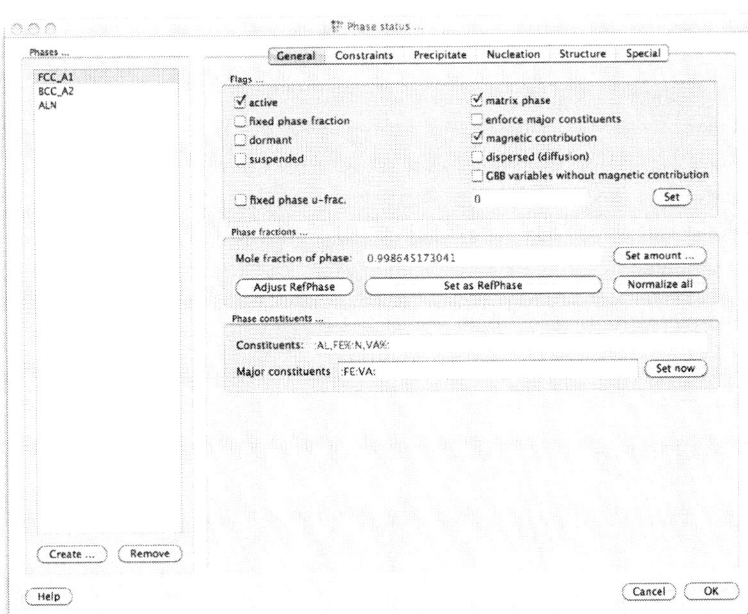

Figure 9.20. *MatCalc* phase status dialog: General settings for equilibrium phases.

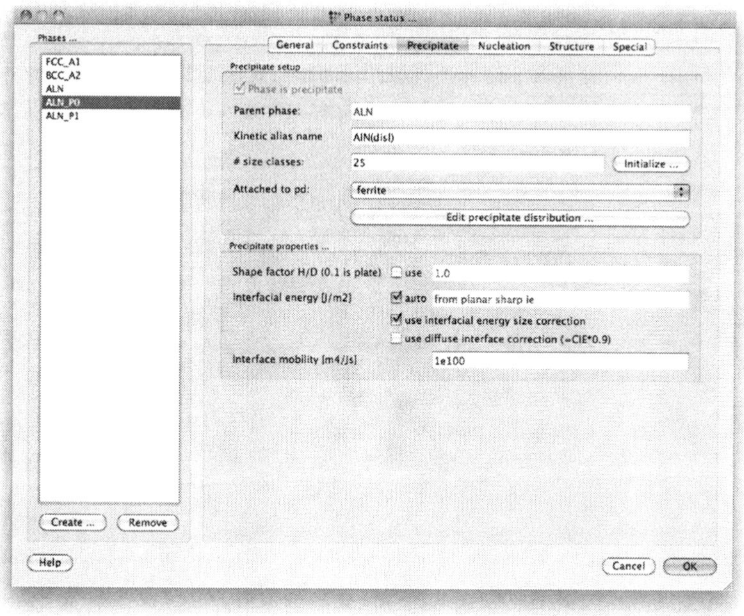

Figure 9.21. *MatCalc* phase status dialog: General settings for precipitate phases.

well as precipitate growth. The interfacial energy is commonly evaluated automatically from the generalized nearest-neighbor broken-bond model as described in Chapter 4. The settings in this dialog tab are most often left on their default values.

The nucleation properties of each phase are defined in the "nucleation" tab of the phase status dialog. In the present example, nucleation of AlN is allowed on dislocations as well as on grain boundaries. For each of these two nucleation sites, a separate precipitate population is created, as visible in the two AlN-phases with suffices "P0" and "P1." The settings for the nucleation properties of the two populations are shown in Figures 9.22 and 9.23. The differences in diffusion geometry at dislocations and at grain boundaries have been described in Section 6.1.

The settings in the "nucleation" tab of the phase status dialog are essential for defining the correct type and treatment of the precipitate. Again, most parameters can be left on their default values, except for the definition of the potential nucleation site(s). In the present example, dislocations and grain boundaries are selected for the two precipitate populations. For the dislocation precipitate, a special feature is selected that instructs *MatCalc* to take into account the elastic energy caused by the volume misfit between AlN and fcc Fe-matrix in the evaluation of the nucleation barrier, G^*.

The last information that is needed by *MatCalc* in the precipitation kinetics simulation is the time–temperature profile, the so-called *heat treatment* (Figure 9.24). A heat treatment is a list of linear time–temperature segments that define the temperature and deformation profile. In the present example, the heat treatment is started above the solution temperature of the AlN precipitates. After cooling to 1000°C and holding for 5 minutes, the material is quenched to room temperature. At 700°C, the transformation from austenite to ferrite is assumed to occur, as visible in the setting for the active precipitation domain of the heat treatment dialog. At this point, all precipitates are moved from the austenite domain to the ferrite domain. Finally, the sample is heated up again to 650°C for 30 minutes for additional heat treatment.

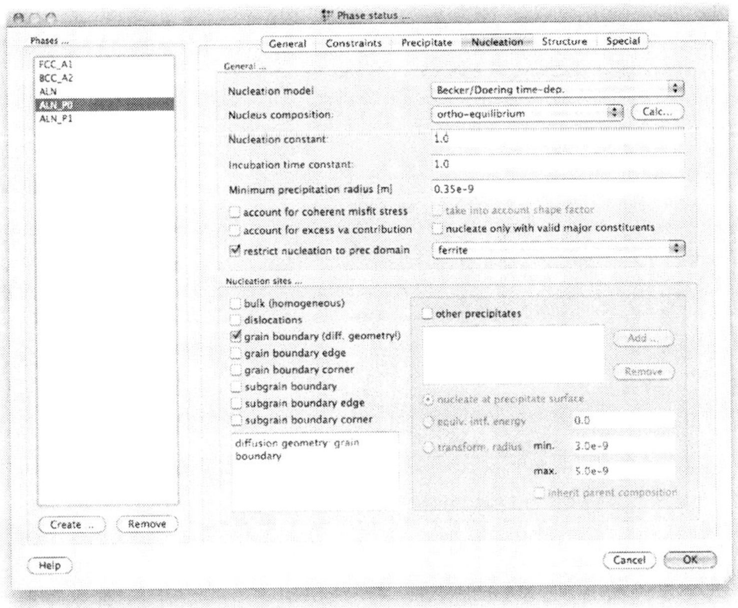

Figure 9.22. *MatCalc* phase status dialog: Nucleation settings for AlN grain boundary precipitate ALN_P0.

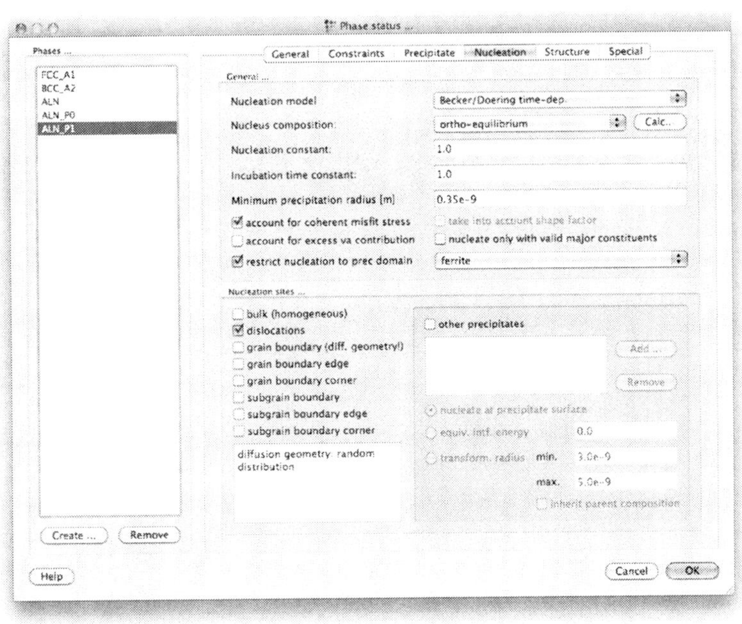

Figure 9.23. *MatCalc* phase status dialog: Nucleation settings for AlN precipitate population nucleating on dislocations, ALN_P1.

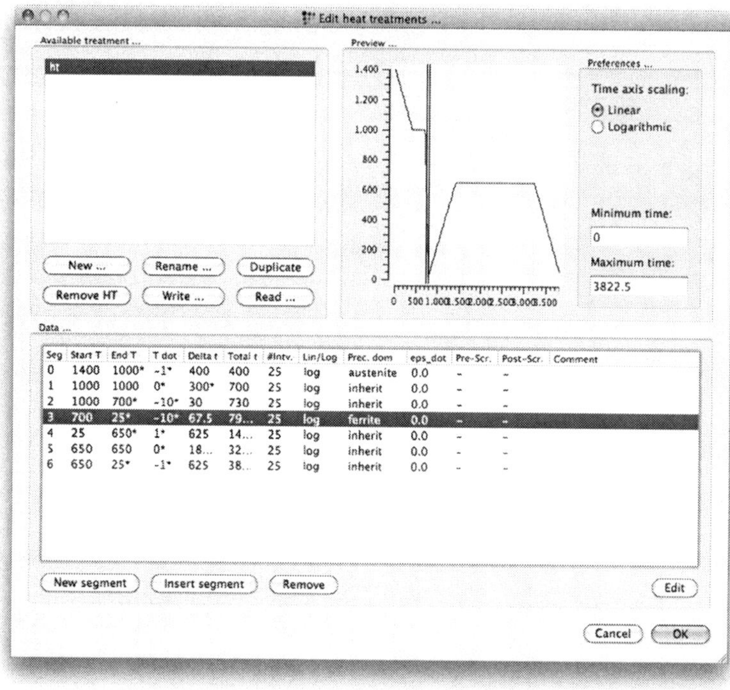

Figure 9.24. Heat treatment definition in *MatCalc*. Each heat treatment is a list of linear temperature segments, with additional information on the active precipitation domain, post- or presegment scripts or deformation rates.

Figure 9.25 shows a screen capture of the results of this example calculation. The first plot in the left top corner displays the applied temperature profile, the other graphs show the evolution of phase fraction, number density, and mean precipitate radius. The dashed lines are for the dislocation precipitates, the solid lines for the precipitate population on the grain boundaries.

9.3.4 *MatCalc* Precipitation Simulation Using Scripting

As an alternative to controlling *MatCalc* simulations from the GUI, the software can be used in command-line mode via the *MatCalc* console. All actions and settings that can be performed in the graphical user environment are also accessible on command-line level. Sequences of commands can be compiled into script files and run automatically by the software. For flexible programming, conditions (if/else/endif) and loops (for/endfor; while/endwhile) are available. Below, some parts of a typical *MatCalc* script are reproduced.[19]

[19] In the script menu of the *MatCalc* software, various template scripts for different simulation problems are provided with the software that can be utilized by the user for his/her own simulation task.

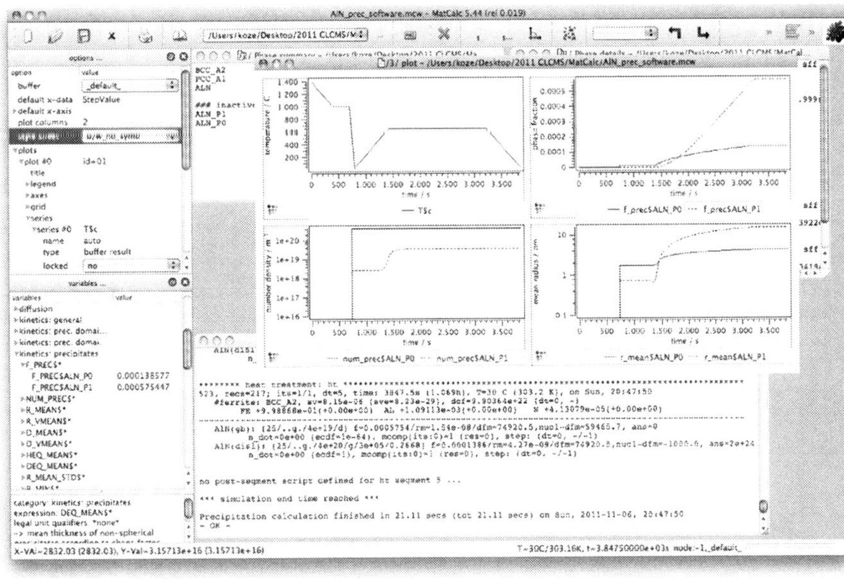

Figure 9.25. Heat treatment definition in *MatCalc*. Each heat treatment is a list of linear temperature segments, with additional information on the active precipitation domain, post- or presegment scripts or deformation rates.

The first part of a script commonly contains a description of what the script does. It verifies the correct *MatCalc* version and opens and loads the thermodynamic and mobility databases. The commands performing this task might read like this:[20]

```
$$*************************************************************
******************** SETUP INFORMATION *********************
*************************************************************$$

$ make sure we work in the correct module
use-module core              $ select core module for
kinetic simulation

$ open new workspace
new-workspace

$ give some information about the script
@echo n
set-workspace-info +Selected elements:
set-workspace-info +C, Mn, Al, Cr, N, Fe
set-workspace-info +
set-workspace-info +Matrix phase: Ferrite (BCC_A2)
```

[20]Note that "$" marks a comment and two "$$" signs embrace a comment block.

```
set-workspace-info +Precipitate phase: AlN
set-workspace-info +
echo y

$$************************************************************
******************** SYSTEM SETUP ************************
************************************************************$$
$ verify correct MatCalc version (is accessible as inter-
nal variable)
if (matcalc_version<5440008)
   send-dialog-string "MatCalc version must be 5.44.0008
or higher to run this script. Stopping."
   stop_run_script           $ stop script
endif


$ show console window and notify user to perform selection
move-gui-window c show              $ bring console
window to front
send-dialog-string "please enter simulation temperature
in console window ..."


set-variable-value sim_temp 1000    $ default value for
temperature
input-variable-value sim_temp "Please enter simulation
temperature"
echo y


$$************************************************************
***************** GLOBAL VARIABLES ************************
************************************************************$$


set-variable-value npc 50           $ number of
precipitate classes


$$************************************************************
***** DATABASES, CHEMICAL COMPOSITION, SELECTED PHASES *****
************************************************************$$


open-thermo-database mc_fe.tdb      $ open thermodynamic
database
select-elements Al C Cr Fe Mn N     $ select elements
select-phase bcc_a2 aln             $ select phases
read-thermodyn-database             $ read thermodynamic
database
```

```
$ define composition ...
enter-composition wp al=0.07 cr=0.039 c=0.025 mn=0.2
n=0.01
read-mobility-database mc_fe.ddb     $ read diffusion
data
```

The following code block is used to define precipitation domains and precipitates.

```
$$**********************************************************
******* PRECIPITATION DOMAINS, PRECIPITATES ***************
*********************************************************$$

$----------------- PRECIPITATION DOMAINS ----------------$

create-precipitation-domain ferrite     $ create precipi-
tation domain
set-precipitation-parameter ferrite x bcc_a2    $ matrix
phase
set-precipitation-parameter ferrite t d e 1e12  $ dislo-
cation density
set-precipitation-parameter ferrite t g 7.5e-6  $ ferrite
grain size
$ temperature-dependent Young's mod.
set-precipitation-parameter ferrite s m
(193000-73,333*T$C)*1e6
$ diffusion ratio gb and bulk / dislocations and bulk
set-precipitation-parameter ferrite s g s y (10^(12-
0.01*T$C))
set-precipitation-parameter ferrite s d s y (10^(8-
0.005*T$C))

$----------------- PRECIPITATE PHASES ----------------$

create-new-phase aln p AlN(gb)                  $ new prec.
phase at gb
set-precipitation-parameter aln_p0 c npc       $ # of size
classes
set-precipitation-parameter aln_p0 d ferrite  $ precipi-
tation domain
set-precipitation-parameter aln_p0 n s g       $ nucle-
ation sites at gb
```

```
create-new-phase aln p AlN(disl)                    $ new prec.
phase @disl
set-precipitation-parameter aln_p1 c npc           $ # of size
classes
set-precipitation-parameter aln_p1 d ferrite       $ precipi-
tation domain
set-precipitation-parameter aln_p1 n s d           $ nucle-
ation sites
set-precipitation-parameter aln_p1 n f y           $ volume
misfit in nucl.
set-precipitation-parameter aln_p1 t m n 0.27 $ use volu-
metric misfit
```

Usually, in the next code block, definitions for heat treatments, setup of output windows for plots,and so forth, follow. This is not reproduced here, however.

Finally, in the last example code block, the precipitation simulation is set up and started. Before this, some initial equilibrium is calculated to initialize variables.

```
$$*******************************************************
************** START PRECIPITATE SIMULATION *************
*******************************************************$$


set-temperature-celsius 1400           $ define some
initial T
set-automatic-startvalues              $ initialize equil.
calc.
calc-equilibrium                       $ calculate equilib-
rium


$ speed up simulations, modify numerical parameters
set-simulation-parameter c n f 1.0  $ maximum radius
growth


$ define simulation parameters
set-simulation-parameter t h ht 10  $ use heat treatment
set-simulation-parameter s r        $ reset kinetic calc

$------------- START PRECIPITATE SIMULATION -------------
start-precipitate-simulation        $ let's go

$$*******************************************************
*************RECIPITATE SIMULATION FINISHED*************
*******************************************************$$
```

9.3.5 Using *MatCalc* with External Software

In addition to the GUI and command-line versions of *MatCalc*, libraries are available for use in third-party software, such as MatLAB and Mathematica. The *MatCalc* libraries can also be linked directly to external C++ or Delphi code and the functions called via the *MatCalc* API.

9.3.6 Software-Relevant Literature and Web Sources

9.3.6.1 Modeling

- Warczok, P., J. Zenisek, and E. Kozeschnik. 2012. "Atomistic and Continuums Modeling of Cluster Migration and Coagulation in Precipitation Reactions." *Computational Materials Science* 60: 59–65.
- Fischer, F. D., J. Svoboda, F. Appel, and E. Kozeschnik. 2011. "Modeling of Excess Vacancy Annihilation at Different Types of Sinks." *Acta Materialia* 59: 3463–72.
- Svoboda, J., E. Gamsjäger, F. D. Fischer, Y. Liu, and E. Kozeschnik. 2011. "Diffusion Processes in a Migrating Interface: The Thick-Interface Model." *Acta Materialia* 59: 4775–86.
- Holzer, I., and E. Kozeschnik. 2010. "Computer Simulation of the Yield Strength Evolution in Cu-Precipitation Strengthened Ferritic Steel." *Materials Sciences Engineering A* 527: 3546–51.
- Kozeschnik, E., J. Svoboda, R. Radis, and F. D. Fischer. 2010. "Mean-Field Model for the Growth and Coarsening of Stoichiometric Precipitates at Grain Boundaries." *Modelling and Simulation in Materials Science and Engineering* 18: 015011 (19pp).
- Sonderegger, B., and E. Kozeschnik. 2010. "Interfacial Energy of Diffuse Phase Boundaries in the Generalized Broken-Bond Approach." *Metallurgical and Materials Transactions* 41A: 3262–9.
- Sonderegger, B., and E. Kozeschnik. 2009. "Size Dependence of the Interfacial Energy in the Generalized Nearest-Neighbor Broken-Bond Approach." *Scripta Materialia* 60: 635–8.
- Sonderegger, B., and E. Kozeschnik. 2009. "Generalized Nearest Neighbor – Broken Bond Analysis of Randomly Oriented Coherent Interfaces in Multi-Component fcc and bcc Structures." *Metallurgical and Materials Transactions* 40A: 499–510.
- Kozeschnik, E. 2008. "Thermodynamic Prediction of the Equilibrium Chemical Composition of Critical Nuclei: bcc Cu Precipitation in Alpha-Fe." *Scripta Materialia* 59: 1018–21.
- Kozeschnik, E., J. Svoboda, and F. D. Fischer. 2006. "Shape Factors in Modeling of Precipitation." *Materials Science Engineering A* 441: 68–72.
- Kozeschnik, E., W. Rindler, and B. Buchmayr. 2007. "Scheil-Gulliver Simulation with Partial Redistribution of Fast Diffusers and Simultaneous Solid-Solid Phase Transformations." *International Journal of Materials Research* 98 (9): 826–31.
- Kozeschnik, E., J. Svoboda, and F. D. Fischer. 2005. "Modified Evolution Equations for the Precipitation Kinetics of Complex Phases in Multi-Component Systems." *CALPHAD* 28 (4): 379–82.
- Svoboda, J., F. D. Fischer, P. Fratzl, and E. Kozeschnik. 2004. "Modelling of Kinetics in Multi-Component Multi-Phase Systems with Spherical Precipitates I: Theory." *Materials Science Engineering A* 385 (1–2): 166–74.

- Svoboda, J., F. D. Fischer, P. Fratzl, and E. Kozeschnik. 2004. "Modelling of Kinetics in Multi-Component Multi-Phase Systems with Spherical Precipitates II: Numerical Solution and Application." *Materials Science Engineering A* 385 (1–2): 157–65.

9.3.6.2 Application

- Povoden-Karadeniz, E., and E. Kozeschnik. 2012. "Simulation of Precipitation Kinetics and Precipitation Strengthening of B2-precipitates in Martensitic PH 13–8 Mo Steel." *ISIJ International* 52: 610–15.
- Radis, R., Ch. Schlacher, E. Kozeschnik, P. Mayr, N. Enzinger, H. Schröttner, and Ch. Sommitsch. 2012. "Loss of Ductility Caused by AlN Precipitation in Hadfield Steel." *Metallurgical and Materials Transactions* 43A: 1132–9.
- Falahati, A., P. Lang, and E. Kozeschnik. 2012. "Precipitation in Al-alloy 6016 – The Role of Excess Vacancies." *Materials Science Forum* 706-709: 317–22.
- Zamberger, S., and E. Kozeschnik. 2012. "Carbo-Nitride Precipitation in Tempered Martensite – Computer Simulation and Experiment." *Materials Science Forum* 706–709: 1586–91.
- Warczok, P., D. Reith, M. Schober, H. Leitner, R. Podloucky, and E. Kozeschnik. 2011. "Investigation of Cu Precipitation in bcc-Fe – Comparison of Numerical Analysis with Experiment." *International Journal of Material Research* 102: 709–16.
- Falahati, A., E. Povoden-Karadeniz, P. Lang, P. Warczok, and E. Kozeschnik. 2010. "Thermo-Kinetic Computer Simulation of Differential Scanning Calorimetry Curves of AlMgSi Alloys." *International Journal of Materials Research* 101: 1089–96.
- Radis, R., and E. Kozeschnik. 2010. "Concurrent Precipitation of AlN and VN in Microalloyed Steel." *Steel Research International* 81: 681–5.
- Pudar, M., S. Zamberger, K. Spiradek-Hahn, R. Radis, and E. Kozeschnik. 2010. "Computational Analysis of Precipitation During Continuous Casting of Microalloyed Steel." *Steel Research International* 81: 372–80.
- Shim, Jae-Hyeok, E. Kozeschnik, Woo-Sang Jung, Seung-Cheol Lee, Dong-Ik Kim, Jin-Yoo Suh, Young-Su Lee, and Young Whan Cho. 2010. "Numerical Simulation of Long-Term Precipitate Evolution in Austenitic Heat-Resistant Steels." *CALPHAD* 34: 105–12.
- Radis, R., and E. Kozeschnik. 2010. "Kinetics of AlN Precipitation in Microalloyed Steel." *Modelling and Simulation in Materials Science Engineering* 18: 055003 (16pp).
- Radis, R., M. Schaffer, M. Albu, G. Kothleitner, P. Pölt, and E. Kozeschnik. 2009. "Multimodal Size Distributions of γ' Precipitates During Continuous Cooling of UDIMET 720 Li." *Acta Materialia* 57: 5739–47.
- Kozeschnik, E. 2008. "Thermodynamic Prediction of the Equilibrium Chemical Composition of Critical Nuclei: bcc Cu Precipitation in Alpha-Fe." *S cripta Materialia* 59: 1018–21.
- Kozeschnik, E., and H. K. D. H. Bhadeshia. 2008. "Influence of Silicon on Cementite Precipitation in Steels." *Journal of Materials Science Technology* 24 (3): 343–7.
- Kozeschnik, E., P. Warbichler, I. Letofsky-Papst, S. Brett, and B. Buchmayr. 2002. "Dissimilar 2.25Cr-9Cr and 2Cr-0.5CrMoV Steel Welds: II. Identification of Precipitates." Science and Technology of Welding & Joining 7 (2): 69–76.

• Kozeschnik, E., P. Pölt, S. Brett, and B. Buchmayr. 2002. "Dissimilar 2.25Cr-9Cr and 2Cr-0.5CrMoV Steel Welds: I. Characterisation of the Weld Zone and Numerical Simulation." *Science and Technology of Welding & Joining* 7 (2): 63–8.

9.3.6.3 Examples

• http://www.matcalc.at -> documentation -> examples.

9.4 PanPrecipitation—An Integrated Computational tool for Precipitation Simulation of MultiComponent Alloys

This article has been courteously provided by Weisheng Cao, S.-L. Chen and F. Zhang of CompuTherm, LLC, 437 S. Yellowstone Dr., Madison WI 53719, USA.[21]

9.4.1 Introduction

Precipitation hardening or age hardening provides one of the most widely used mechanisms for strengthening of structural materials. A simulation tool that can be used to understand the microstructural evolution and the correlated hardening responses during precipitation is highly in demand. However, precipitation is a very complex process that involves simultaneous occurrence of nucleation, growth, and coarsening. Accurate modeling of the precipitation process, therefore, requires a synchronous consideration of all these contributions. Moreover, the needed instant phase equilibrium information and chemical driving force requires the simulation to be coupled directly with thermodynamic calculation.

In this regard, an integrated computational tool, PanPrecipitation [253,254], is being developed to study precipitation kinetics and precipitation hardening behavior. It is fully coupled with the automatic thermodynamic calculation engine, PanEngine [254,255], so that reliable thermodynamic properties and atomic mobility data necessary for precipitation simulation can be obtained instantly during the simulation. Another unique feature of PanPrecipitation is that it provides multiple built-in precipitation models at multilevels. Therefore, the user can choose an appropriate model for the precipitation simulation. Furthermore, the generic data structure also facilitates its integration with other models.

Figure 9.26 schematically shows the overall picture of PanPrecipitation, which includes two parts: the problem of interest to be studied and the integrated modeling framework aiming to tackle the problem. The problem of our interests is how alloy chemistry and heat-treatment condition affect the microstructure and the related mechanical properties, while the computational tool consists of three separate modules that, working with the thermodynamic/mobility and precipitation databases, respectively deals with the thermodynamic/mobility calculation based on the CALPHAD method, the simulation of microstructure using the built-in kinetic models, and the prediction of mechanical properties according to the proposed age hardening

[21] The statements, considerations, and judgements provided in this section are the sole opinion of the author of this section. They do not necessarily represent the opinion of the author of this book.

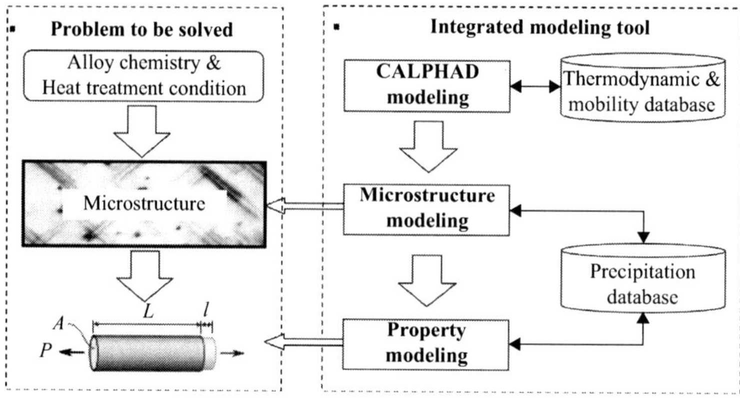

Figure 9.26. Overall picture of the materials problem to be studied and the proposed modeling tool.

models. As such, the developed computational tool can simulate the thermodynamic/kinetic properties and the evolution of microstructure as well as the corresponding mechanical property responses under different heat-treatment processes.

The basic features and capabilities of the PanPrecipitation software are compiled in Table 9.4.

9.4.2 Kinetic Models

In order to consider the simultaneous nucleation, growth, and coarsening, two advanced kinetic models were adopted. These include the Fast-Acting model based on Langer and Schwartz theory [150] for simulation of the evolution of particle number density and mean size, and the more advanced NKW (numerical Kampmann–Wagner) model [152] for predication of the full evolution of the particle size distribution (PSD) in addition to average quantities. The NKW model is based on Kampmann and Wagner's work as implemented in a numerical framework [152] and extended to handle both homogeneous and heterogeneous nucleation dealing with various morphologies for the simulation of coprecipitation kinetics of multicomponent alloys under arbitrary heat-treatment conditions. The following gives a brief introduction to the NKW model along with its submodels for nucleation, growth, and coarsening. More detailed descriptions can be found in refs. [115,152,254,256].

Specifically in the NKW model the continuous particle size distribution is divided into a large number of size classes. The program takes a simulation step at every sample time hit. To maintain both accuracy and efficiency between two adjacent simulation steps, a fifth-order Runge–Kutta scheme is used to generate an adaptive step size based on the continuity equation. At each simulation step, the number of new particles is first calculated using classical nucleation theory and then these new particles are allocated to an appropriate size class. The transient nucleation rate, J, is given by[22]

[22]For a discussion of this equation, see Section 2.3.4, equation (2.90), on time-dependent nucleation.

Table 9.4. Software information table for PanPrecipitation.

Software	PanPrecipitation (a kinetic module of Pandat)
Acronym	PanPrecipitation
Developer	Weisheng Cao and Shuanglin ChenCompuTherm, LLC 437 S. Yellowstone Dr., Madison WI 53719 USA.
Licensing	commercial/academic/free (limited)
Release date	2010
Web site	www.computherm.com
Contact email	info@computherm.com
Kinetic model	Numerical Kampmann–Wagner model (NKW), Fast-Acting model (Langer–Schwartz), JMAK
Nucleation	Multicomponent Classical Nucleation Theory (CNT)
Interfacial energies	From property database
Evolution equations from	Thermodynamic extremal principle (SFFK, Section 3.5), Chen et al.-model (Section 3.3.4) and Morral et al.-model [91]
Thermodynamics/mobility	CALPHAD-type databases, unencrypted
Programming language	C++ for shared library, C# for GUI
Platforms	Windows, Linux, Mac
GUI/command-line/API	yes/yes_no/yes_no
Scripting	Yes
Programming	Yes
Exporting	Xml, Ascii, excel, emf, png, svg, pdf, jpeg, bmp, gif

$$J = NZ\beta^* \exp\left(-\frac{G^*}{k_B T}\right)\exp\left(-\frac{\tau}{t}\right). \tag{9.4}$$

The pre-exponential terms in equation (9.4) are: N_v, the nucleation site density, Z, the Zeldovich factor, and β^*, the atomic attachment rate. Time is indicated by t, the incubation time for nucleation by τ, the Boltzmann constant by R_g, and the temperature by T. The nucleation barrier is defined as

$$G^* = \frac{4\pi\gamma}{3}(r^*)^2, \tag{9.5}$$

where r^* is the radius of the critical nucleus[23] and

$$r^* = -\frac{2\gamma}{\Delta G_{vol}}. \tag{9.6}$$

[23] See Section 2.3.2.

ΔG_{vol} is the volume free energy change on nucleation and γ is the specific energy of the matrix–particle interface.

Next, the growth of existing particles for each size class is computed by assuming diffusion-controlled growth where the Gibbs–Thomson size effect is also taken into account. In previous studies for binary alloys or alloys with stoichiometric compositions, the growth model proposed by Morral and Purdy [91] was adopted and modified to handle the growth/dissolution of various precipitate phases with different morphologies. For more complex alloys with significant homogenous range, the other two models were proposed: the SFFK model [44,71] derived from the principle of maximum entropy production[24] [61,63] and the growth model proposed by Chen, Stepsson, and Ågren [57].[25] Both models have been implemented in the current framework.

Finally, the particle size distribution and the volume fraction of the particles are updated and the matrix composition is recalculated according to the mass conservation equation at the end of each simulation step. The updated values are then used in the next time step.

It should be noted that coarsening in the NKW model arises naturally and becomes dominant when the supersaturation is sufficiently low. In the Fast-Acting model, however, the coarsening is considered explicitly. Accordingly, a model based on the LSW theory [74,75] is employed,[26] which characterizes the curvature-driven particle growth. Its accuracy and efficiency are also ensured by controlling the adaptive step size with a fifth-order Runge–Kutta scheme. In addition to the more advanced NKW and Fast-Acting models, the simpler JMAK model [144–147,150,152] is also implemented[27] and modified for prediction of the overall phase transformation kinetics.

9.4.3 Software Design and Data Structure

PanPrecipitation is built as a shared library and integrated into PANDAT [254,256] as a specific module that extends the capability of PANDAT for kinetic simulation and property prediction while taking full advantage of the automatic thermodynamic calculation engine (PanEngine) and the user-friendly PANDAT Graphical User Interface (PanGUI). The integrated computational tool, complying with the architecture of PANDAT [254,255], logically consists of three separate layers: the user interface layer (PanGUI), the calculation layer (PanPrecipitation and PanEngine), and the data layer (thermodynamic & mobility and precipitation databases). PanGUI is to translate and pass the commands and simulation conditions to PanPrecipitation and then display the returned results as texts, tables, or graphs. PanPrecipitation processes the commands/simulation conditions and reads the model parameters stored in the precipitation database. PanEngine, combined with thermodynamic/mobility databases, provides atomic mobility data and thermodynamic properties (such as chemical driving force and phase equilibrium information) that are necessary for the simulation of microstructure (volume fraction, number density, average particle size, PSD, etc.) in PanPrecipitation. This information can thus

[24] See Section 3.5.1.
[25] See Section 3.3.4.
[26] See Section 3.6.1.
[27] See Section 5.1.

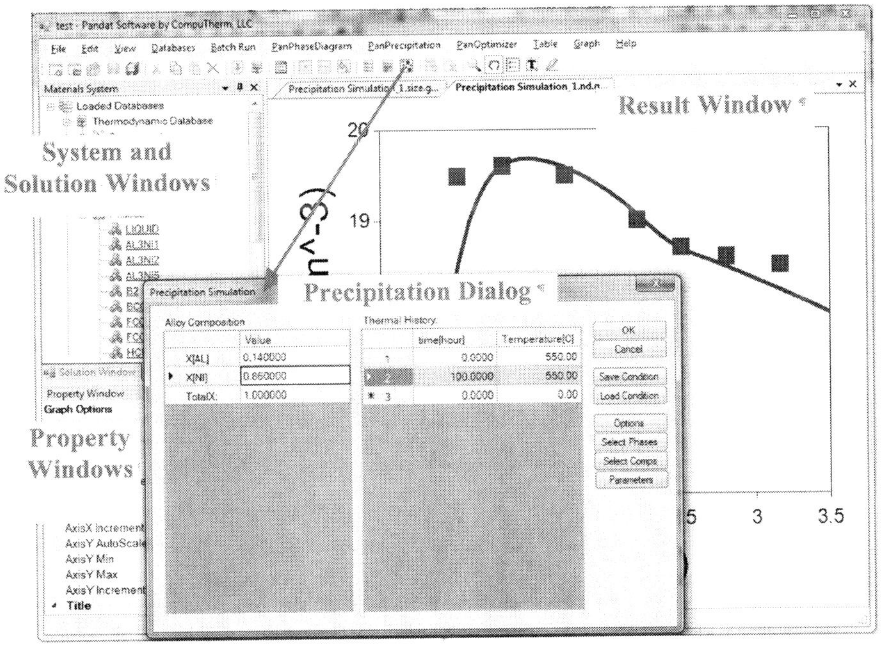

Figure 9.27. PANDAT Graphical User Interface (PanGUI) with precipitation simulation dialog.

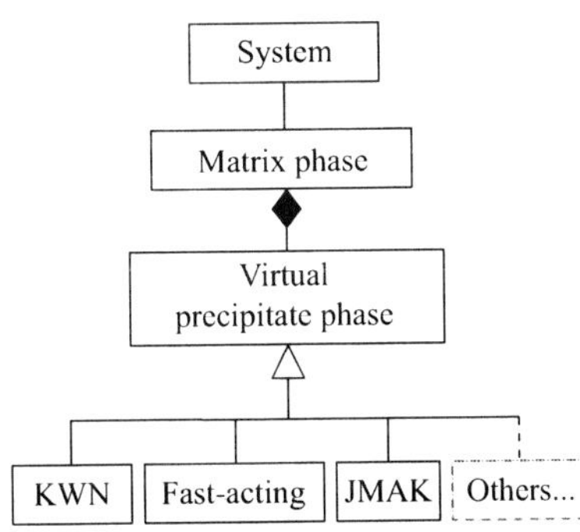

Figure 9.28. Data structure of system information in PanPrecipitation.

be used for predication of mechanical properties. After the simulation is complete, results will be returned to PanGUI for user's viewing.

PanPrecipitation is designed following the object-oriented (OO) concept and written in C++ with generic data structures, balancing performance, maintainability, and extensibility. The basic data structure for storing precipitation information in the system of interest is schematically shown in Figure 9.28.

In general, a system contains a matrix phase and a number of precipitate phases. The precipitation behavior of different precipitate phases can be described by different kinetic models. In addition to the three built-in models—the Numerical Kampmann–Wagner (NKW) [152] model, the Fast-Acting model [150], and the modified JMAK model—this generic data structure also enables its integration with user-defined kinetic models.

Based on the above data structure, input parameters for the matrix and its precipitate phases are organized in "Extensible Markup Language" (XML) format, which is a standard markup language and well known for its extensibility. In accordance with the XML syntax, a set of well-formed tags are specially designed to define the kinetic and property models for each precipitate phase along with the corresponding model parameters such as interfacial energy, molar volume, and nucleation site parameter, which is the so-called "Precipitation Database" and should be validated with the available experimental data. An example precipitate database for a Ni-14Al at% alloy can be defined as:

```
<Alloy name="NI-14Al">
    <MatrixPhase name="FCC_A1">
        <ParameterTable name="Parameters for gamma">
            <Parameter name="Molar_Volume" value="7.1E-6"
description="Molar Volume" />
        </ParameterTable >
        <PrecipitatePhase name="L12_FCC" model="Fast_Acting"
morphology="Sphere"
nucleation="Modified_Homo" growth="SFFK">
            <ParameterTable name="Parameters for Gamma_
prime">
                <Parameter name="Molar_Volume"
value="7.1E-6" description="Molar Volume" />
                <Parameter name="Interfacial_Energy" val-
ue="0.025"
description="Interfacial Energy" />
                <Parameter name="Atomic_Spacing"
value="3.621E-10"
description="Atomic Spacing" />
            </ParameterTable >
        </PrecipitatePhase >
    </MatrixPhase >
</Alloy >
```

9.4.4 Examples

To date, PanPrecipitation has been used for modeling of the microstructural evolution in a number of alloy systems. Three examples are given here, which will cover the simulations of:

1. the full evolution of precipitation behavior in a model Ni–14 at%Al alloy with the NKW model [254];
2. the coarsening behavior of Rene88 DT using the Fast-Acting model [255];
3. the precipitation hardening behavior in a series of commercial Al-Mg-Si alloys [255].

9.4.4.1 Example 1: Precipitation Behavior of a Model Ni-14 at% Al Alloy

As a first validation, PanPrecipitation was used to simulate the kinetics of precipitation in a Ni–14 at% Al alloy isothermally annealed at 550°C. In this case, the disordered fcc (Ni) solid solution, which is usually referred as the γ phase, is stable in the temperature range from the solidus to the solvus. Below the solvus temperature, γ'-Ni_3Al precipitates form the solid solution during the aging process. This example has been studied previously by Kampmann and Wagner [152]. In order to perform a direct comparison and obtain the full evolution of PSD in addition to average quantities, the NKW model was used in the simulation with the PanNickel 7.0 database providing the relevant thermodynamic model parameters and mobility data. The input parameters for kinetic models—the interfacial energy, the nucleation site density—were optimized in terms of the measured number density and average particle size [105].

Figure 9.29 plots the NKW-predicted evolution of particle radius, supersaturation, and number density as a function of aging time, superimposed with the experimental data. As can be seen, the predictions and the measurements are in good agreement. It has been noted earlier that the precipitation process typically involves the simultaneous occurrence of nucleation, growth, and coarsening. This simulation indeed reveals a significant overlap of nucleation, growth, and coarsening processes due to the small value of (γ/γ') interphase boundary energy and the observed large number of particles that interfere with each other and thus lead to the coarsening of larger particles at the expense of smaller ones in the very early stage of precipitation. This phenomenon is very different from the precipitation behavior of another well-studied model alloy Cu-1.02 at% Co aged at 600°C. The previous studies [115,255] indicate a plateau period for such an alloy, which happens after the nucleation is effectively complete and the coarsening has yet to begin.

The NKW-predicted γ' size distributions for the Ni–14 at% Al after aging at 550°C for 21, 217, and 3720 minutes are shown in Figure 9.30. These distributions are plotted in terms of the particle frequency, which is computed by normalizing the total number of particles to be one. At the early stages of the precipitation, the nucleation process is dominant and the number density of γ' particles increases rapidly. Therefore, the PSD is characterized by a sharp peak around the critical nucleus size as shown in Figure 9.30 for $t = 21$ minutes. At later stages, the transformation falls into the coarsening regime and the PSD is then flattened as indicated in the case of $t = 3720$ minutes. It is interesting to point out that the PSD, at an intermediate stage of $t = 217$ minutes, displays a double peak, which is consistent with the previous findings from both experimental observation [104] and model prediction [115].

9.4.4.2 Example 2: Coarsening of Rene88DT

In the second example, the coarsening behavior of the commercial alloy Rene88DT (Ni-1.99Al-12.99Co-15.67Cr-4.00Mo-3.88W-3.72Ti-0.70Nb in wt%) was studied with PanPrecipitation

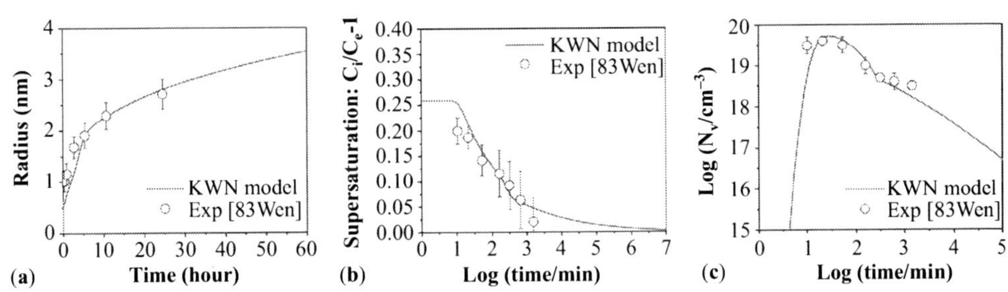

Figure 9.29. NKW-predicted evolution of average γ' particle size, supersaturation, and number den sity compared with the experimental data [105].

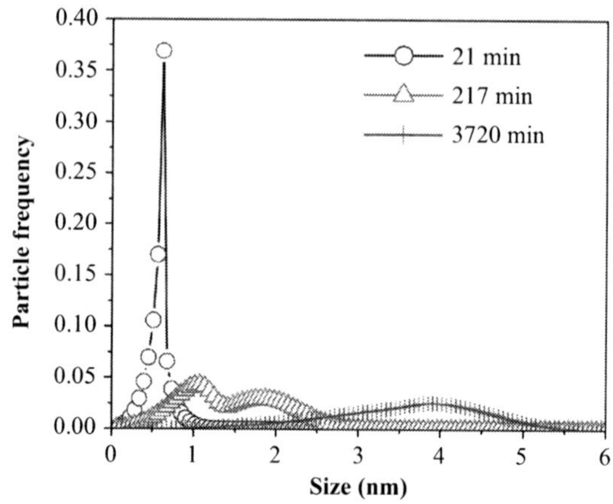

Figure 9.30. NKW-predicted evolution of average γ' particle size, supersaturation, and number density compared with the experimental data [105].

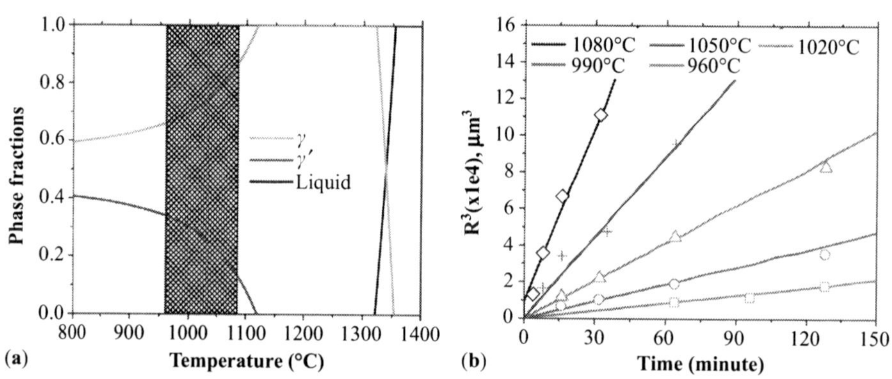

Figure 9.31. (a) Phase fractions of liquid, g, and γ' as a function of temperature for Alloy Rene88DT showing the temperature window where the experimental data are available; (b) The predicted temporal evolution of mean γ' size in Alloy Rene88DT aged at different temperatures 960°C, 990°C, 1020°C, 1050°C, and 1080°C compared with the experimental data [258].

[255]. Figure 9.31(a) shows the calculated phase fractions as a function of temperature together with the temperature window where the two phases $\gamma + \gamma'$ are stable and the experimental data are available.

Figure 9.31(b) plots the cube of the predicated mean γ' size against time at five different temperatures: 1080°C, 1050°C, 1020°C, 990°C, and 960°C, which were obtained using the Fast-Acting model. In good agreement with the measured values by Mao [258], the coarsening rate increases as the temperature increases. At each temperature, it is found that the γ' coarsening kinetics is consistent with the LSW theory even when the holding time is short, that is, R^3 is proportional to time.

9.4.4.3 Example 3: Precipitation Hardening Behavior of Al-Mg-Si Alloys

Aluminum alloys have been heavily used in automotive and aerospace industries due to their excellent strength to density ratio. The Al-Mg-Si (AA6xxx series) alloys are considered one of the best candidates because they provide a medium to high strength while having rather good formability and weldability. Alloys of this series, having Mg and Si as the major elements, are usually precipitation hardened by aging process. In this simulation, the NKW model was used for simulation of microstructure evolution. As a first approximation, the precipitate phases are treated as one single phase with the composition Mg_5Si_6 and the particles are assumed to be spherical.

Figure 9.32 shows the model predicted hardness of the alloy AA6082 in the cases of artificial aging and subsequent reheating at various temperatures. The experimental data by Anderson [259] are also plotted for comparison. The comparison suggests that the proposed modeling approach is able to reasonably reproduce the hardening or softening responses to the artificial aging and the reheating processes.

As a further validation, the precipitation behavior of a group of AA6005 alloys was also investigated. These alloys have been studied by Myhr et al. [260] through both experimentation and theoretical modeling. The degree of agreement between the simulations and the experimental data is summarized in Figure 9.33. In general, a fairly good agreement has been achieved for all combinations of materials and heat treatment conditions being investigated.

9.4.5 Discussion

It should be particularly pointed out that due to the complexity of the precipitation process, the kinetic models discussed in this study are by no means comprehensive. Many times, a case-by-case study is needed to understand a particular material that has undergone a particular process. In this regard, the current modeling framework was designed in a way that the built-in kinetic models allow for modification and new kinetic models can also be integrated for special cases.

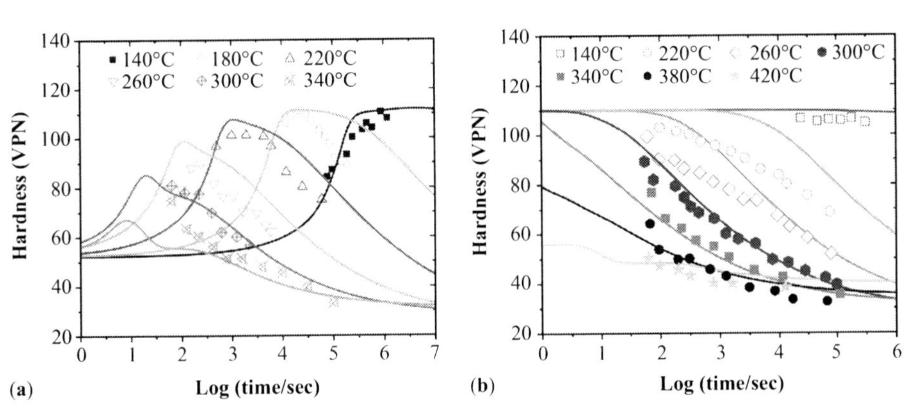

Figure 9.32. Simulated and measured aging curves of an AA6082 alloy (a) during artificial aging at various temperatures; (b) during reheating at various temperatures from the peak-aged condition. Experimental data are taken from Anderson [259].

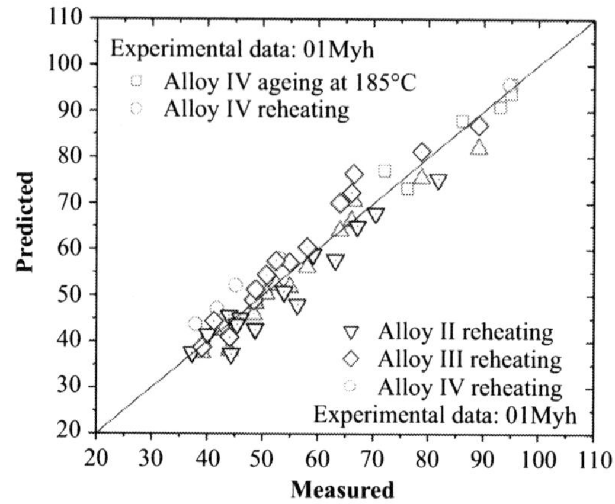

Figure 9.33. Comparison between predicted and measured hardness for a range of AA6xxx alloys under different heat treatment conditions. Experimental data are taken from Myhr et al. [260].

Meanwhile, the necessary thermodynamic and mobility inputs can be obtained without any difficulty. Such an example is given by considering the modeling of heterogeneous nucleation taking place in Al-Mg-Si alloy. The equation for heterogeneous nucleation barrier

$$G^* = \frac{A}{\left(\Delta G_{\mathrm{vol}}\right)^2}, \tag{9.7}$$

with A being a parameter related to the potency of the heterogeneous nucleation site in the parent material, which was used to replace the original one derived from the classical homogeneous nucleation theory, while the chemical driving force is provided by the thermodynamic calculation based on the CALPHAD method. Furthermore, at a high level, the proposed framework can be easily integrated with a third party program for more applications. For instance, by coupling PanPrecipitation with ProCAST, a comprehensive modeling work is currently undertaken for the prediction of the final microstructure and mechanical properties of aluminum casting alloys after heat treatment [261]. After specifying the alloy chemical composition, solidification process, and heat-treatment parameters, the model would predict the microstructure and potential defects through various stages of the component lifecycle: casting, solid-solution heat treatment, and artificial aging.

The importance of the related thermodynamic/mobility and precipitation databases should be specifically emphasized in order to make reliable simulations. The precipitation kinetics is sensitive to the input parameters particularly to the value of the interfacial energy, which is usually difficult to be measured directly by experiment. One must perform extensive validations so that the chosen parameters could reasonably describe the precipitation behavior of the key alloys with abundant experimental data. The precipitation databases containing all these parameters are still under development. Even though the CALPHAD calculation, combined with the thermodynamic/mobility database, shows rather good predictive capability in obtaining the necessary thermodynamic properties and diffusivity data, it could still be risky when the calculation is outside the assessed composition ranges.

9.5 TC-PRISMA

9.5.1 General Information

TC-Prisma is a computational tool for simulation of diffusion-controlled precipitation processes in multicomponent multiparticle systems. TC-Prisma is built on the *Thermo-Calc* and *DICTRA* softwares (see Section 9.1) that are utilized for evaluation of chemical driving forces, operating tie-lines under local-equilibrium condition at the moving phase boundaries, as well as diffusive atomic mobilities. These two software packages, as well as TC-Prisma itself, are developed by the Swedish company *Thermo-Calc Software AB*. Parts of the code are adopted from the *PrecipiCalc* (see Section 9.2) software developed by Questek Innovations LLC.[28]

In version 1 of the TC-Prisma software, simulations of concurrent nucleation and growth can be performed under isothermal conditions and for a maximum of five precipitation phases. In addition to the thermodynamic and diffusional mobility data available with the Thermo-Calc and DICTRA packages, physical properties, such as interfacial energy values for precipitation phases, are provided exemplarily for Fe-alloys in the form of a *property database*. In most practical cases, the user is requested to provide suitable interfacial energy data manually from either literature or from fitting to appropriate experiments. An overview of the software features is summarized in Table 9.5.

Table 9.5. Software information table for TC-Prisma.[29]

Software	TC Prisma (integrated with Thermo-Calc[30] and DICTRA[31])
Acronym	TC Prisma
Developer	Thermo-Calc Software AB, Stockholm
Licensing	Commercial
Release date	2011
Web site	http://www.thermocalc.se/TC-PRISMA.htm
Contact email	info@thermocalc.se
Kinetic model	Numerical Kampmann–Wagner model (NKW), Langer–Schwartz model (LS)
Nucleation	Multicomponent Classical Nucleation Theory (CNT)
Interfacial energies	Manually, or from property database

(Continued)

[28] See corresponding comment in Section 9.2.1, page 368.

[29] Unfortunately, the distributor/developer of TC-Prisma (Thermo-Calc Software AB) did not provide in-depth information on their software for presentation in the present book. The information given here is, therefore, based on the TC-Prisma User Manual from 11/2011 as well as the Thermo-Calc web site, only. The demonstration simulation has been performed by the author of this book using TC-Prisma, version 1.

[30] *Thermo-Calc*: computational thermodynamics software, http://www.thermocalc.com

[31] See Section 9.1 for details.

Table 9.5. Software information table for TC-Prisma (*continued*)

Evolution equations from	Chen et al.-model (see Section 3.3.4)
Thermodynamics/mobility	CALPHAD-type databases, unencrypted or encrypted
Programming language	Java (GUI), kernel: unknown
Platforms	Windows
GUI/command-line/API	yes/no/no
Scripting	Yes (macro)
Programming	No
Exporting	TXT, XLS, PNG, JPG, GIF, BMP

9.5.2 Kinetic Model

The simulation algorithm of TC-Prisma is based on the Numerical Kampmann–Wagner (NKW) model (see Section 5.3) adopting the original concept of the Langer–Schwartz model (see Section 5.2). In the numerical implementation, the temporal evolution of the mean radius and particle density over the whole course of the precipitation reaction is evaluated by solving a set of rate equations derived with certain assumptions for the nucleation and growth rates of the discrete set of precipitate size classes. Details on these models are described in the above-mentioned sections of this book.

The growth equations implemented in TC-Prisma are based on the model developed by Chen et al. [57], which is described in Section 3.3.4. The transient nucleation rate expression is evaluated according to equation (2.90), Section 2.3.4, using the corresponding submodels summarized in Table 2.6 of Section 2.5.1. For the incubation time, the approximation of Feder et al. [25] is used, equation (2.68).

9.5.3 Performing TC-Prisma Simulations "From Scratch"

Simulations in TC-Prisma can be run either "from scratch" or using script files. When starting new simulations, it is recommended that the user either defines the simulation setup from the GUI dialogs or that he/she uses an existing script and modifies the file according to the desired simulation conditions. Both methods are briefly demonstrated below.

When starting a simulation "from scratch," the user goes through the following steps indicated on the example of cementite precipitation in bcc ferrite.

9.5.3.1 Define System

The startup window of TC-Prisma allows the user to enter the thermodynamic system properties either based on a selection of appropriate thermodynamic and diffusion databases (radio button "Calphad Database Calculation") or by entering the corresponding data (thermodynamic solvus, diffusivity, interfacial energy, etc.) manually (radio button "UserDefined Binary Calculation"). The corresponding dialogs are shown in Figures 9.34, 9.35, and 9.36.

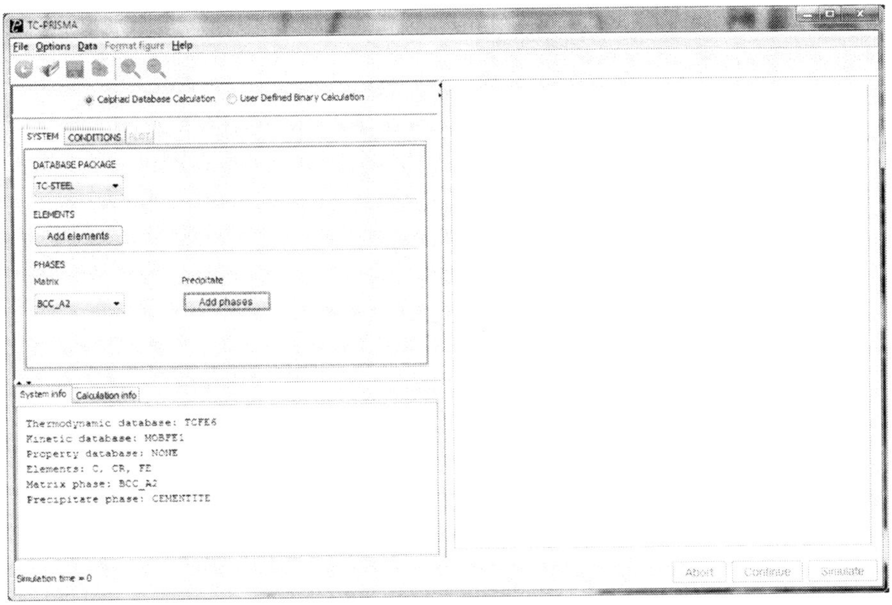

Figure 9.34. Welcome screen of the TC-Prisma software for starting a new simulation in "Calphad Database Calculation" mode.

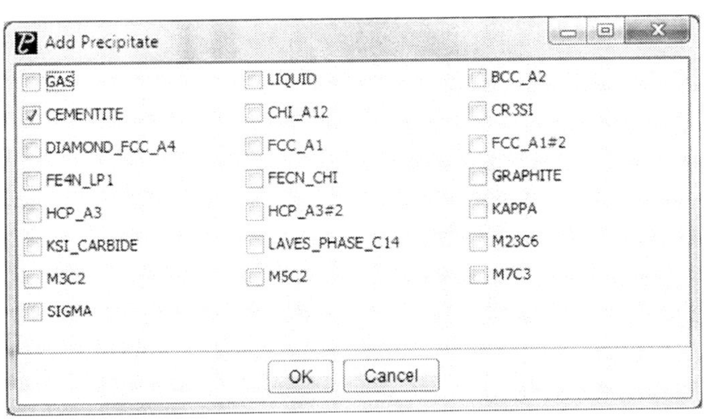

Figure 9.35. Selecting precipitate phases in TC-Prisma using the TC-STEEL database.

In the "system" tab, the user can select a predefined database package from the drop-list of available packages or define a new set of packages. Then, the elements considered in the simulation can be selected, as well as the thermodynamic matrix phase and up to five precipitate phases.

9.5.3.2 Define Simulation Conditions

After activating the "Conditions" tab, the user enters information such as the materials chemistry, temperature, simulation end time, as well as the nucleation site properties. The corresponding screen dumps are shown in Figures 9.37 and 9.38.

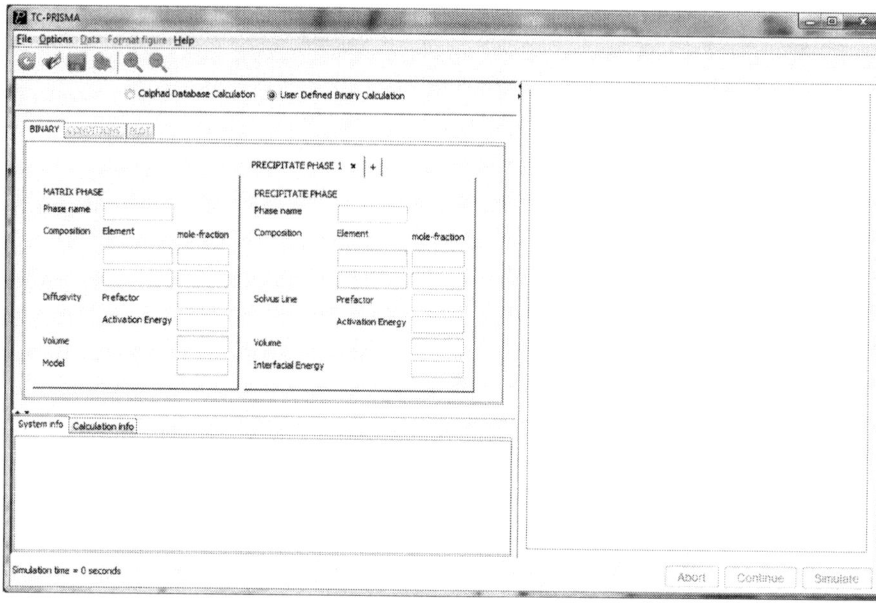

Figure 9.36. Welcome screen of the TC-Prisma software for starting a new simulation in "User Defined Binary Calculation" mode.

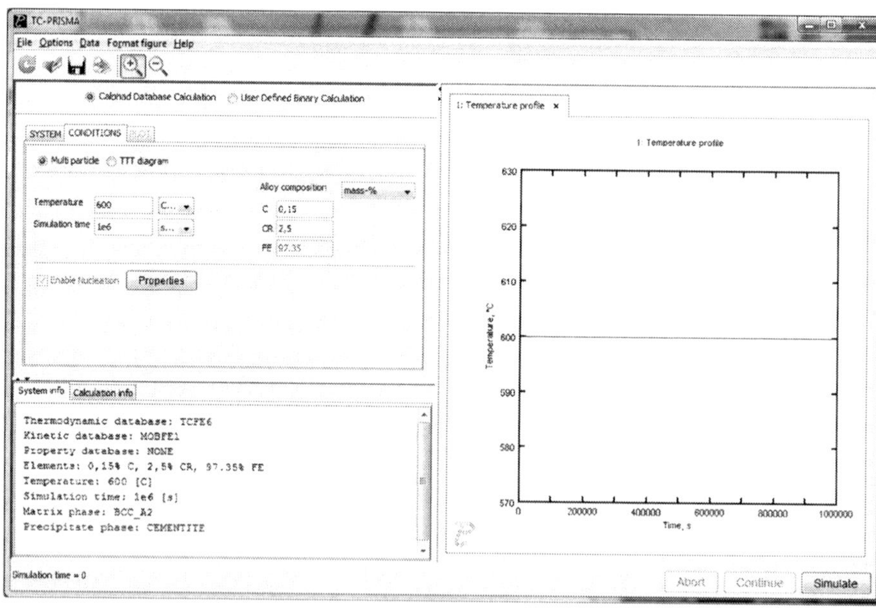

Figure 9.37. Defining the TC-Prisma simulation conditions.

Based on the user selection, the number of available nucleation sites is evaluated for bulk (homogeneous) nucleation or heterogeneous nucleation at grain boundaries, grain corners and edges, as well as dislocations. The actual site densities are evaluated in analogy to the methods outlined in Section 2.4.2.

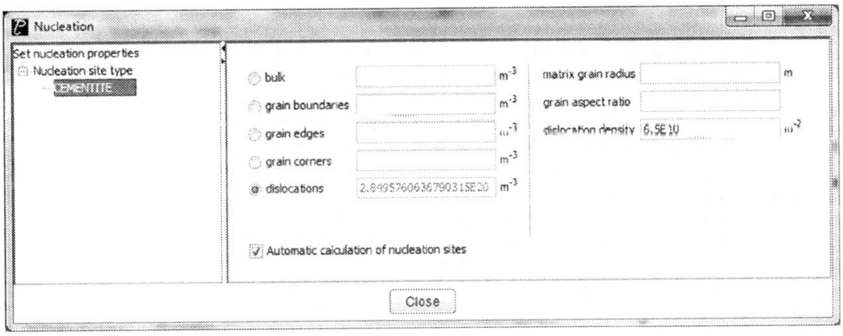

Figure 9.38. Defining nucleation sites in TC-Prisma.

Several special parameters can be entered in a separate dialog accessible from the "data" menu, which are the interfacial energy, phase boundary mobility (to allow deviations from local equilibrium due to interface friction), phase energy addition (to modify phase diagram properties), and mobility enhancement or molar volume.

9.5.3.3 Start Simulation

After pressing the "Simulate" button, the TC-Prisma calculation is initiated and the simulation progress is displayed in the "System info" and "Calculation info" tab windows.

9.5.3.4 Plot Results

When the simulation is completed, the user can display the simulation results in the "Plot" tab after selecting the desired output quantity in the corresponding drop list (see Figure 9.39).

9.5.4 Performing Simulations Using Scripts

After setting up the simulations using the GUI features of TC-Prisma, the setup can be saved (exported) into TC-Prisma macro files. For the example shown in the previous section, TC-Prisma produces the following output:

```
#TC-PRISMA Script File

#Wed Aug 01 18:07:34 CEST 2012

######################################
#Units. Temperature, Time and Amount
######################################
#Temperaure unit. CELSIUS, KELVIN, FARHEIT
TEMPERATURE_UNIT=CELSIUS
```

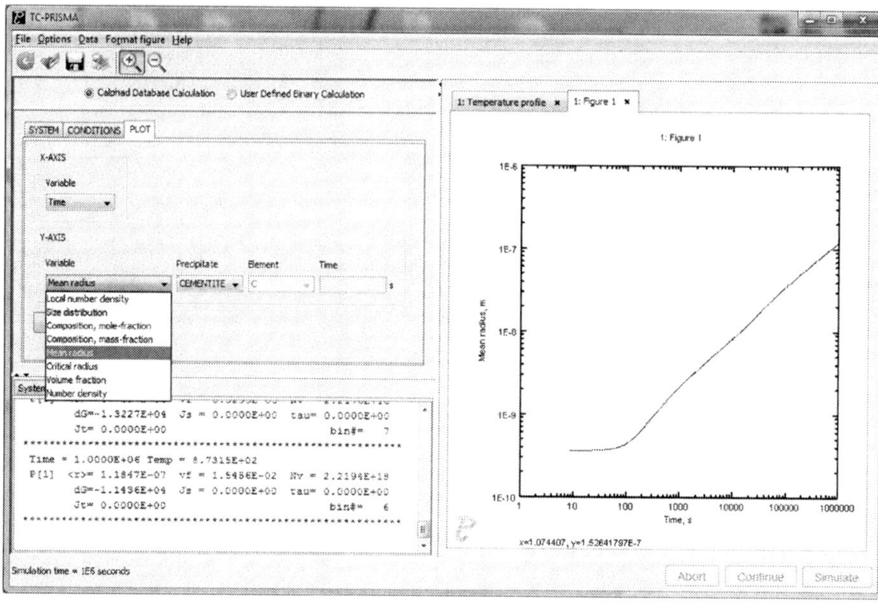

Figure 9.39. Plotting the results of a TC-Prisma simulation.

```
#Amount unit.
AMOUNT_UNIT=MASS_PERCENT

#Time Unit
TIME_UNIT=SECONDS


#####################################
#Calculation setup. Database package
#####################################
#Database Package (e.g. TCSTEEL)
DATABASE_PACKAGE=TC-STEEL

#Growth Rate model: Advanced or Simplified
GROWTH_RATE_MODEL=SIMPLIFIED

#a comment
DISPLAY_RESULT=true

#Calculation Base type. Normal or Binary
CALCULATION_BASE_TYPE=NORMAL

#Grid Distribution for the integration method.  Geometric
or Random
GRID_DISTRIBUTION=GEOMETRIC
```

```
#Type of calculation: single, multi or ttt
CALCULATION_DISPLAY_TYPE=MULTI


#####################################
#Nucleation Data
#####################################
#Automatic Calculation of Nucleation Sites
AUTOMATIC_CALCULATION_OF_NUCLEATION_SITES=true

#Matrix Grain Radius
GRAIN_RADIUS=1.0E-4

#Grain Aspect Ratio
ASPECT_RATIO=1.0

#Dislocation Density
DISLOCATION_DENSITY=6.5E10

#Nucleation Sites
NUCLEATION_SITES=2.8694433948090053E20

#Shape
SHAPE=sphere

#Nucleation Model
CALCULATION_METHOD=default

#Interfacial Energy parameter. From database or other
INTERFACIAL_ENERGY_PARAMETER=DATABASE;

#Nucleation type. Bulk, boundary, corner, edge, disloca-
tion
NUCLEATION_TYPE=DISLOCATION


#####################################
#System Data
#####################################
#Selected components (Semicolon separated list)
CHECKED_COMPONENTS=C;CR;FE;

#Selected composition (Semicolon separated list)
COMPONENT_VALUES=0,15;2,5;97.35;

#Matrix phase
MATRIX_PHASE=BCC_A2
```

```
#Precipitate Phases (Semicolon separated list)
PRECIPITATE_PHASES=CEMENTITE;


#####################################
#General Conditions
#####################################
#Simulation Time
SIMULATION_TIME=1e6


#####################################
#TTT-Diagram specific Conditions
#####################################
#TTT Minimum Temperature
TTT_TEMP_MIN=

#TTT Maximum Temperature
TTT_TEMP_MAX=

#TTT Temperature Step Size
TTT_TEMP_STEP=

#TTT Phase Fraction
TTT_PHASE_FRACTION=


#####################################
#Multi-particle specific Conditions
#####################################
#Temperature
TEMPERATURE=600


#####################################
#Other stuff
#####################################
#a comment
CHECKED_PHASES=CEMENTITE;

#a comment
INTERFACIAL_ENERGY=0.0;

#a comment
MOBILITY_FROM_DATABASE=true;

#a comment
MOBILITY=0.0;
```

```
#a comment
PHASE_ENERGY_ADDITION=0.0;

#a comment
PREFACTOR=0.0

#a comment
ACTIVATION_ENERGY=0.0

#a comment
VOLUME_FROM_DATABASE=true

#a comment
MATRIX_VOLUME=0.0

#a comment
SECONDARY_MATRIX_VOLUME=0.0

#a comment
PRECIPITATE_VOLUMES=0.0;

#a comment
NUCLEATION_ENABLED=true

#####################################
#Numerical Parameters
#####################################
#Maximum timestep fraction
MAX_TIMESTEP=0.1

#Maximum nuclei size
MAX_NUCLEI_SIZE=10000.0

#Minimum particle density for a size class
MIN_PARTICLE_DENSITY_SIZE_CLASS=1.0E10

#Maximum particle density for a size class
MAX_PARTICLE_DENSITY_SIZE_CLASS=1.0E19

#Number of grid points over one order of magnitude in R
R_POINTS_PER_MAGNITUDE=150.0

#Maximum number of grid points over one order of magni-
tude in R
MAX_R_POINTS_PER_MAGNITUDE=200.0

#Minimum number of grid points over one order of magni-
tude in R
MIN_R_POINTS_PER_MAGNITUDE=100.0
```

```
#Maximum relative volume fraction of subcritical parti-
cles allowed to dissolve in one time step
MAX_VOL_FRAC_DISSOLVE_TSTEP=0.01

#Maximum relative radius change
MAX_REL_R_CHANGE=0.01

#Relative radius change for avoiding class collision
REL_R_CHANGE_CLASS_COLLISION=0.5

#Maximum overall volume change
MAX_OVERALL_VOL_CHANGE=0.0010

#Maximum relative change of nucleation rate in logarith-
mic scale
MAX_REL_CHANGE_NUC_RATE_LOG=0.5

#Maximum relative change of critical radius
MAX_REL_CHANGE_CRITICAL_RADIUS=0.1

#Minimum radius for a nucleus to be considered as a par-
ticle
MIN_RADIUS_NUCLEUS_AS_PARTICLE=5.0E-10
```

These settings can be read back into TC-Prisma to reproduce the simulation setup.

9.6 Comparison of Software Codes

The software codes, that are presently available for simulation of phase transformation and precipitation kinetics, are widely different in terms of theoretical concept and software realization. The user, uncertain of which software offers the highest potential for solving his/her particular problem, has the choice between various approaches and program implementations.[32] The following discussion could be useful in any kind of decision process concerning the question, "which software is most suitable and useful for a given simulation objective."

A major distinction between the various software programs can be made in terms of the total number of transformation products that can be treated in the simulation and the degree of detail that is accounted for. The most fundamental concept of all simulation models is realized in the DICTRA code, which solves the moving phase boundary problem simultaneously with the transient multicomponent diffusion equations. The phase boundary movement and, thus, the phase transformation process are evaluated without extra degrees of freedom in the solution process in addition to the chemical potentials and diffusional mobilities of the diffusing species. The solutions delivered by DICTRA might be considered as *rigorous*, which means that only

[32] Some aspects that might help the user in selecting the appropriate code are also given in Chapter 8 in the discussion of the *Design of Simulation* concept.

minimum necessary simplifications[33] are utilized to solve the phase transformation problem and that the simulation thus represents a situation that is closest to the real conditions. However, this advantage carries along with it the drawback that the solution procedure is computationally expensive and can only be applied to a limited number of individual precipitates. Treatment of size distributions is out of the scope of this type of model.

Larger ensembles of particles can be treated with codes that utilize the Langer–Schwartz or numerical Kampmann–Wagner type algorithms. These approaches are suitable for tracking the evolution of entire size distributions of precipitates, however, at the cost of "rigorousity" in the treatment of individual precipitate nucleation and growth processes. The models implemented in softwares, such as *PrecipiCalc*, *MatCalc*, PanPrecipitation, or TC-Prisma, involve approximations of the real situation, which, make them less "exact." However, these simplifications bring sufficient numerical benefit to make the approaches *applicable* to simulations on larger scale, that is, to simulations of entire precipitate distributions. Each of these software products utilizes specific models for the description of the growth and shrinkage of individual particles, and they use different, yet comparable, algorithms for the evolution of the size distributions. These are described in the corresponding sections in this chapter.

When dealing with precipitation modeling, a major step in the evolution of precipitate distributions is the nucleation process. In this aspect, the different software products can be distinguished into three groups: software that does not deal with nucleation at all, software that treats nucleation on a phenomenological basis, and software that tries to deal with this process rigorously. The DICTRA package is vastly focused on the fundamental solution of the moving boundary problem and does not take nucleation into account at all. In DICTRA, the simulation starts with a supercritical particle, the evolution of which is then described in terms of growth and dissolution. *PrecipiCalc*, PanPrecipitation, and TC-Prisma use classical nucleation theory to describe the nucleation process of the precipitates, however, these programs consider the interfacial energy as a quantity, which is not known a priori and which must be "identified and adjusted" from appropriate experiment. This means that the choice of value for the interface energy is made such that the considered experimental data can be well reproduced. Typical values for interfacial energies are provided in so-called *property databases*, which compile interfacial energy data that have been fitted previously on representative experiments.

The software *MatCalc* follows a different approach. Here, nucleation is considered as a process that is fully determined by theoretical models. In *MatCalc*, the interfacial energy is not treated as a fitting parameter but computed from the generalized nearest-neighbor broken-bond model (GBB), fully taking into account the chemical composition, temperature, and precipitate size dependence of this quantity. Also, the impact of heterogeneous nucleation sites is accounted for on a rigorous basis. As all input parameters in the *MatCalc* nucleation and growth approach are taken from independent thermodynamic and diffusion databases, the results delivered by this software might be considered as being of *predictive* nature.

A criterion that is sometimes put on software products is the availability of a GUI. Software, such as DICTRA and PrecipiCalc, are used in command/script mode only, whereas PanPrecipitation, *MatCalc*, and TC-Prisma also offer GUIs. While the mere availability of a nice GUI should not be a primary criterion in the selection of a software product, the possibilities of post-processing and statistical evaluation of simulation results could be.

[33]The major assumption made in DICTRA is the "local equilibrium hypothesis" at the phase boundary.

Finally, it is important to consider the required degree of expertise of the end user when selecting the appropriate software for the given user/problem combination. An experienced user with sound theoretical background in physical metallurgy and computational materials engineering is probably well advised to choose the particular type of model and software that gives him/her the most flexibility in setting up the simulation problem. The more knowledge and expertise one can put into the simulation setup, the more precise will be the answer and the more *introspection* and understanding of the process can be gained.

For inexperienced users, this obvious advantage might be "counter-productive," however, because the flexibility and variety in possibilities for simulation setup could be confusing and he/she runs into the danger of misinterpretation of the simulation results. In these cases, the user might be better off with softwares that only allow for a limited amount of user interference and parameter setting. This offers the advantage that the simulation will always provide simple, clear, and unambiguous answers. In cases, the answers might be too simple, though,[34] and the simulation results should be considered with proper care.

[34] *Make everything as simple as possible, but not simpler* (Albert Einstein).

Appendix

Interfacial Energies of Typical Precipitate–Matrix Systems

The following tables contain calculated interfacial energy values as obtained from the GBB model described in Section 4.3. The calculations have been performed with the software *MatCalc,*[1] version 5.51, the corresponding thermodynamic databases are listed in the tables.

It is important to note that the interfacial energy is a strong function of chemical composition of the precipitate, volumetric misfit, interfacial orientation, and so forth, and it is not easily possible to list all possible combination of cases. The values given here, however, represent the upper limit of the interfacial energy value as obtained with the GBB approach for the sharp interface assumption. The evaluation is based on the fully supersaturated state, which means that the phase fraction of precipitate is constrained to zero. The composition of the precipitates is set according to the parallel tangent construction outlined in Section 1.4, which ensures that the precipitate experiences maximum chemical driving force.

It is also necessary to point out that the values in the following tables cannot be directly used in evaluation of the critical nucleation barrier, as nucleation involves rather small spherical particles where the influence of interface curvature must be taken into account. Typically, the size correction discussed in Section 4.4 reduces the effective interfacial energy value during nucleation to 60 to 90% of the planar sharp-interface value. The size correction function is taken into account in evaluation of the nucleus composition for minimum nucleation barrier. The role of entropy in diffuse interfaces is also not accounted for in these values and should be considered by the interested reader.

[1] See Section 9.3 details.

Table A.1. Typical interfacial energies for precipitates in bcc Fe-base alloys[2] obtained with the GBB model. For some precipitates, para (p) composition values and compositions with minimum nucleation barrier (*) are shown in italic. All values for fully supersaturated state.

Nominal comp. at%	**Precipitate comp. at%**	**Precipitate**	**Temp. °C**	**Intf. energy mJ/m²**
Fe – 0.1C	75.0Fe – 25.0C	Fe_3C	25–400	175–160
”	75.0Fe – 25.0C	”	400–800	160–200
Fe – 2C	75.0Fe – 25.0C	Fe_3C	25–400	175–160
”	75.0Fe – 25.0C	”	400–800	160–200
Fe – 1C – 2Mn	75.0Mn – 25.0C	$(Fe,Mn)_3C$	25	395
”	21.2Fe – 53.8Mn – 25.0C	”	400	345
”	66.8Fe – 8.2Mn – 25.0C	”	800	220
”	*73.5Fe – 1.5Mn – 25.0C (p)*	”	*25–400*	*175–160*
”	*73.5Fe – 1.5Mn – 25.0C (p)*	”	*400–800*	*160–195*
”	*74.7Fe – 0.3Mn – 25.0C (*)*	”	*25*	*175*
”	*74.0Fe – 1.0Mn – 25.0C (*)*	”	*400*	*160*
”	*72.0Fe –3.0Mn – 25.0C (*)*	”	*800*	*200*
Fe – 1C – 2Si	75.0Fe – 25C	Fe_3C	25–400	180–160
”	75.0Fe – 25C	”	400–800	160–200
”	*73.5Fe – 1.5Si – 25C (p)*	$(Fe,Si)_3C$	*200*	*155*
”	*73.5Fe – 1.5Si – 25C (p)*	”	*600*	*147*
”	*71.9Fe – 3.1Si – 25C (*)*	”	*200*	*125*
”	*75.0Fe – 25C (*)*	”	*600*	*175*
Fe – 1C – 2Cr	75.0Cr – 25C	$(Fe,Cr)_3C$	25	435
”	71.3Cr – 25C – 0.4Fe	”	400	420
”	22.7Cr – 25C – 52.3Fe	”	800	250
”	*1.5Cr – 25C – 73.5Fe (p)*	”	*25–400*	*165–150*
”	*1.5Cr – 25C – 73.5Fe (p)*	”	*400–800*	*150–180*
”	*0.3Cr – 25C – 74.7Fe (*)*	”	*25*	*160*
”	*2.0Cr – 25C – 73.0Fe (*)*	”	*400*	*150*
”	*8.7Cr – 25C – 66.3Fe (*)*	”	*800*	*205*
”	69.7Cr – 30C – 0.3C	$(Fe,Cr)_7C_3$	25	465
”	68.4Cr – 30C – 1.6Fe	”	400	455
”	41.0Cr – 30C – 29.0Fe	”	800	345
”	*1.4Cr – 30C – 68.6Fe (p)*	”	*25–400*	*140–145*

[2] Thermodynamic data from *MatCalc* database, mc_fe.tdb', version 1.009.

Table A.1. (*Continued*)

Nominal comp. at%	Precipitate comp. at%	Precipitate	Temp. °C	Intf. energy mJ/m^2
”	*1.4Cr – 30C – 68.6Fe (p)*	”	*400–800*	*145–170*
”	*30C – 70.0Fe (*)*	”	*25*	*135*
”	*3.9Cr – 30C – 66.1Fe (*)*	”	*400*	*160*
”	*1.9Cr – 30C – 50.7Fe (*)*	”	*800*	*255*
”	79.3Cr – 20.7C	$(Fe,Cr)_{23}C_6$	25	410
”	65.8Cr – 20.7C – 13.5Fe	”	400	360
”	25.7Cr – 20.7C – 53.6Fe	”	800	230
”	*1.6Cr – 20.7C – 77.7Fe (p)*	”	*25–400*	*105–95*
”	*1.6Cr – 20.7C – 77.7Fe (p)*	”	*400–800*	*95–125*
”	*20.7C – 79.3Fe (*)*	”	*25*	*97*
”	*0.8Cr – 20.7C – 78.5Fe (*)*	”	*400*	*90*
”	*8.4Cr – 20.7C – 70.9Fe (*)*	”	*800*	*160*
Fe – 0.2N – 2Cr	66.6Cr –33.3N	Cr_2N	25–400	545–540
”	65.7Cr – 33.3N – 1.0Fe	”	400–770	540–545
”	63.7Cr – 33.3N – 3.0Fe	”	770–800	545–520
”	*8.1Cr – 33.3N – 58.6Fe (*)*	”	*25*	*135*
”	*29.8Cr – 33.3N – 36.9Fe (*)*	”	*400*	*275*
”	*50.2Cr – 33.0N – 16.8Fe (*)*	”	*800*	*425*
Fe – 2C – 0.02N – 1Cr	66.7Cr – 23.8N – 9.5C	$Cr_2(C,N)$	300	480
”	66.6Cr – 14N – 19.4C	”	500	465
”	65.6Cr – 9.3N – 24.1C	”	700	475
”	*26.8Cr – 49.6Fe – 8.8N – 26.8C (*)*	*(Cr,Fe)2(C,N)*	*300*	*120*
”	*48.8Cr – 17.9Fe – 8.2N – 25.0C (*)*	”	*500*	*335*
”	*60.0Cr – 66.7Fe – 8.0N – 25.4C (*)*	”	*700*	*430*
Fe – 0.1N – 5Cr – 1V	63.4V – 33.3N – 3.3Cr	$(V,Cr)_2N$	300	700
”	51V – 33.3N – 15.7Cr	”	800	630
Fe – 0.1Al – 0.02N	50.0Al – 50.0N	AlN	300–700	640–750
Fe – 0.1Nb – 0.5C	50.5Nb – 49.5C	NbC	300–770	850–920
”	50.6Nb – 49.4C	”	770–800	920–890

(*Continued*)

Table A.1. Typical interfacial energies for precipitates in bcc Fe-base alloys[2] obtained with the GBB model. For some precipitates, para (p) composition values and compositions with minimum nucleation barrier (*) are shown in italic. All values for fully supersaturated state (*Continued*).

Nominal comp. at%	Precipitate comp. at%	Precipitate	Temp. °C	Intf. energy mJ/m^2
”	*25.3Nb – 25.1Fe – 49.6C (*)*	*(Nb,Fe)C*	*400*	*470*
”	*31.0Nb – 19.2Fe – 49.8C (*)*	”	*500*	*560*
”	*36.7Nb – 13.4Fe – 50C (*)*	”	*600*	*660*
”	*42.0Nb – 8.0Fe – 50C (*)*	”	*700*	*760*
Fe – 0.1Nb – 0.5N	50Nb – 50N	NbN	300–750	865–930
	50.2Nb – 49.8N	”	750–800	930–890
Fe – 0.1Nb – 1C – 0.3N	50Nb – 44.3N – 5.7C	Nb(C,N)	300	860
”	50Nb – 41.1N – 8.9C	”	500	870
”	50Nb – 38.0N – 11.9C	”	700	900
”	*26.3Nb – 24.4Fe – 40.5C – 8.8N (*)*	*(Nb,Fe)(C,N)*	*500*	*490*
”	*31.7Nb – 18.6Fe – 37.8C – 11.9N (*)*	”	*600*	*580*
”	*36.8Nb – 13.3Fe – 34.6C – 15.3N (*)*	”	*700*	*680*
Fe – 0.1V – 0.5C	50.6V – 49.4C	VC	300	630
”	51.5V – 48.5C	”	700	605
”	*33.1V – 21.5Fe – 45.4C (*)*	*(V,Fe)C*	*500*	*430*
”	*45.5V – 6.0Fe – 48.5C (*)*	”	*600*	*560*
”	*49.2V – 2.2Fe – 48.6C (*)*	”	*700*	*585*
Fe – 0.1V – 0.5N	50.1V – 49.9N	VN	300–700	750–725
”	*27.2V – 22.9Fe – 50.0N (*)*	*(V,Fe)N*	*400*	*510*
”	*30.7V – 19.4Fe – 49.9N (*)*	”	*500*	*545*
”	*33.9V – 16.3Fe – 49.8N (*)*	”	*600*	*575*
”	*36.9V – 13.3Fe – 49.8N (*)*	”	*700*	*600*
Fe – 0.1Ti – 0.5C	50.5Ti – 49.5C	TiC	300	820
”	51.3Ti – 48.7C	”	700	870
”	*37.6Ti – 15.4Fe – 47.2C (*)*	*(Ti,Fe)C*	*400*	*640*
”	*42.2Ti – 10.6Fe – 47.3C (*)*	”	*500*	*715*
”	*45.3Ti – 7.5Fe – 47.3C (*)*	”	*600*	*765*
”	*47.3Ti – 5.4Fe – 47.4C (*)*	”	*700*	*815*

Table A.1. (*Continued*).

Nominal comp. at%	Precipitate comp. at%	Precipitate	Temp. °C	Intf. energy mJ/m²
Fe – 0.1Ti – 0.5N	50Ti – 50N	TiN	300–700	1085–1125
”	*23.4Ti – 26.6Fe – 50.0N (*)*	*(Ti,Fe)N*	*400*	*570*
”	*27.2Ti – 22.8Fe – 50.0N (*)*	”	*500*	*645*
”	*31.2Ti – 18.8Fe – 50.0N (*)*	”	*600*	*725*
”	*35.3Ti – 14.7Fe – 50.0N (*)*	”	*700*	*820*
Fe – 10Cr – 0.1V – 0.5N	33.3V – 33.3N – 30.4Cr	Z-phase	300–700	665–660
”	50.3V – 49.7N	(V,Cr)N	300	680
”	49.7V – 47.8N – 2.5Cr	”	700	640
Fe – 10Cr – 0.1Nb – 0.5N	33.3Nb – 33.3N – 26.1Cr	Z-phase	300–700	725–775
”	50Nb – 50N	(Nb,Cr)N	300–700	795–845
Fe – 1C – 1Mo	66.7Mo – 33.3C	Mo_2C	300	560
”	65.6Mo – 1.1Fe – 33.3C	”	700	570
”	*31.9Mo – 40.4Fe – 27.7C (*)*	*(Mo,Fe)2C*	*400*	*180*
”	*37.6Mo – 30.5Fe – 31.9C (*)*	”	*500*	*270*
”	*42.3Mo – 24.8Fe – 32.8C (*)*	”	*600*	*325*
”	*46.7Mo – 20.2Fe – 33.1C (*)*	”	*700*	*380*
”	57.1Mo – 28.6Fe – 14.3C	*(Mo,Fe)6C*	300	440
”	50.4Mo – 35.3Fe – 14.3C	”	700	390
”	*29.0Mo – 56.8Fe – 14.3C (*)*	”	*400*	*70*
”	*30.0Mo – 55.8Fe – 14.3C (*)*	”	*500*	*90*
”	*31.6Mo – 54.1Fe – 14.3C (*)*	”	*600*	*125*
”	*33.9Mo – 51.8Fe – 14.3C (*)*	”	*700*	*170*
Fe – 1C – 1W	51.9W – 33.8Fe – 14.3C	W_6C	300	395
”	43W – 42.7Fe – 14.3C	”	700	440
Fe – 2Mo	66.6Fe – 33.4Mo	Laves	300	105
”	67.1Fe – 32.9Mo	”	700	135
Fe – 2Mo – 1Si	51Fe – 33.3Mo – 15.7Si	Laves	300	385
”	56.8Fe – 33.3Mo – 9.9Si	”	700	370
”	*74.8Fe – 25.9Mo – 0.3Si (*)*	”	*400*	*40*
”	*72.4Fe – 26.9Mo – 0.7Si (*)*	”	*500*	*75*
”	*68.9Fe – 29.3Mo – 1.7Si (*)*	”	*600*	*135*
”	*64.6Fe – 31.8Mo – 3.6Si (*)*	”	*700*	*230*

(*Continued*)

Table A.1. Typical interfacial energies for precipitates in bcc Fe-base alloys[2] obtained with the GBB model. For some precipitates, para (p) composition values and compositions with minimum nucleation barrier (*) are shown in italic. All values for fully supersaturated state (*Continued*).

Nominal comp. at%	**Precipitate comp. at%**	**Precipitate**	**Temp. °C**	**Intf. energy mJ/m²**
Fe – 2W	66.4Fe – 33.6W	Laves	300–700	120–200
Fe – 33Cr	65.7Cr – 34.3Fe	Sigma	300	25
”	55Cr – 45Fe	”	500	15
”	48.1Cr – 51.9Fe	”	630	25
”	45.2Cr – 54.8Fe	”	700	15
Fe – 0.01N – 0.01B	50B – 50N	BN	300–700	950
Fe – 0.1C – 0.01B	79.3Fe – 12.5C – 8.2B	$Fe_{23}(CB)_6$	300	105
”	79.3Fe – 10.9C – 9.8B	”	800	135
Fe –1Cr – 0.01B	50.0Cr – 50.0B	CrB	300	710
”	49.4Cr – 0.6Fe – 50.0B	”	500	710
”	42.8Cr – 7.2Fe – 50.0B	”	550	655
”	34.4Cr – 15.6Fe – 50.0B	”	800	600
”	*27.8Cr – 22.2Fe – 50.0B (*)*	”	*300*	*515*
”	*33.3Cr – 16.7Fe – 50.0B (*)*	”	*500*	*575*
”	*35.6Cr – 14.4Fe – 50.0B (*)*	”	*700*	*610*
”	16.8Fe – 49.9Cr – 33.3B	Fe_2B	300	625
”	27.5Fe – 39.2Cr – 33.3B	”	650	550
”	48.2Fe – 18.5Cr – 33.3B	”	700	330
”	52.2Fe – 14.5Cr – 33.3B	”	800	310
”	*56.8Fe – 9.9Cr – 33.3B (*)*	”	*300*	*300*
”	*50.6Fe – 16.1Cr – 33.3B (*)*	”	*500*	*305*
”	*54.0Fe – 12.7Cr – 33.3B (*)*	”	*700*	*295*
Fe – 3Cu	100Cu	bcc-Cu	300–700	470–665
”	*60.2Cu – 49.8Fe (*)*	*bcc-(Cu,Fe)*	*400*	*170*
”	*81.2Cu – 18.8Fe (*)*	”	*500*	*350*
”	*94.5Cu – 5.5Fe (*)*	”	*600*	*515*
”	*98.0Cu – 2.0Fe (*)*	”	*700*	*630*
”	100Cu	fcc-Cu	300–700	520–700

Table A.1. (*Continued*).

Nominal comp. at%	Precipitate comp. at%	Precipitate	Temp. °C	Intf. energy mJ/m^2
Fe – 1Mn – 0.1S	50Mn – 50S	MnS	300–700	1125–1400
”	*5.0Mn – 45.0Fe – 50S (*)*	*(Mn,Fe)S*	*500*	*470*
”	*10.5Mn – 39.5Fe – 50S (*)*	”	*600*	*600*
”	*17.8Mn – 32.2Fe – 50S (*)*	”	*700*	*800*
Fe – 1Cu – 0.1S	47Cu – 40.6S – 12.4Fe	Cu_2S	300	510
”	39.1Cu – 39.9S – 21Fe	”	700	580
”	*21.8Cu – 42.8S – 35.5Fe (*)*	*(Cu,Fe)2S*	*500*	*435*
”	*22.5Cu – 41.9S – 35.6Fe (*)*	”	*600*	*460*
”	*23.6Cu – 41.4S – 35.0Fe (*)*	”	*700*	*500*
Fe – 0.1Ti – 1C – 0.1S	50Ti – 25C – 25S	$Ti_4C_2S_2$	300–700	515–605
Fe – 6Al – 6Ni	50Ni – 34.9Al – 15.1Fe	NiAl(B_2)	300	235
”	48.2Ni – 31Al – 20.8Fe	”	650	335
”	27.5Ni – 21.8Al – 50.7Fe	”	800	110
”	*13.8Ni – 12.3Al – 73.9Fe (*)*	”	*400*	*20***
”	*16.1Ni – 15.9Al – 67.9Fe (*)*	”	*500*	*26***
”	*13.7Ni – 11.4Al – 74.9Fe (*)*	”	*600*	*23***
”	*18.9Ni – 15.3Al – 65.8Fe (*)*	”	*700*	*58***
”	*22.5Ni – 18.1Al – 59.4Fe (*)*	”	*800*	*74***
Fe – 25Co – 15Mo	45.1Mo – 34.0Fe – 20.9Co	μ-phase	300–700	150–260
Fe – 3Si – 0.1N	42.9Si – 57.1N	Si_3N_4	300–700	425–445
Fe – 3Mn – 3Si – 0.1N	25.0Si – 25.0Mn – 50.0N	$MnSiN_2$	300	490
”	”	”	550	513
”	”	”	700	560
”	”	”	800	510

** The critical nucleation energy landscape for the ordered BCC_B2 phase is evaluated for an assumed volumetric misfit of 3% (200 J/mol). This is necessary because G^* does not show a pronounced G^* minimum, otherwise, and, in the numerical search, the optimum nucleus composition becomes identical to the matrix composition.

Table A.2. Typical interfacial energies for precipitates in fcc Fe-base alloys[3] obtained with the GBB model. For some precipitates, para (p) composition values and compositions with minimum nucleation barrier (*) are shown in italic. All values for fully supersaturated state.

Nominal comp. at%	**Precipitate comp. at%**	**Precipitate**	**Temp. °C**	**Intf. energy mJ/m²**
Fe – 8C	75.0Fe – 25C	Fe_3C	800–1200	77–83
Fe – 8C – 2Mn	72.0Fe – 3.0Mn – 25C	$(Fe,Mn)_3C$	800–1200	77–83
Fe – 8C – 2Si	75.0Fe – 25C	Fe_3C	200	90
"	75.0Fe – 25C	"	800	79
"	75.0Fe – 25C	"	1200	86
"	*73.3Fe – 1.6Si – 25C (p)*	*$(Fe,Si)_3C$*	*200*	*62*
"	*73.3Fe – 1.6Si – 25C (p)*	"	*600*	*52*
"	*64.8Fe – 0.1Si – 25C (*)*	"	*200*	*88*
"	*75.0Fe – 25C (*)*	"	*600*	*80*
Fe – 8C – 2Cr	41.1Fe – 33.9Cr – 25C	$(Fe,Cr)_3C$	400	132
"	59.9Fe – 15.1Cr – 25C	"	600	95
"	65.3Fe – 9.7Cr – 25C	"	800	86
"	67.7Fe – 7.3Cr – 25C	"	1000	84
"	69.0Fe – 6.0Cr – 25C	"	1200	86
"	*73.3Fe – 1.6Cr – 25C (p)*	"	*400*	*77*
"	*73.3Fe – 1.6Cr – 25C (p)*	"	*800*	*74*
"	*71.3Fe – 3.7Cr – 25C (*)*	"	*400*	*80*
"	*69.5Fe – 5.5Cr – 25C (*)*	"	*800*	*79*
"	12.9Fe – 57.1Cr – 30C	$(Fe,Cr)_7C_3$	400	210
"	44.1Fe – 25.9Cr – 30C	"	800	120
"	56.8Fe – 13.2Cr – 30C	"	1200	81
"	*68.5Fe – 1.5Cr – 30C (p)*	"	*400*	*50*
"	*63.0Fe – 7.0Cr – 30C (*)*	"	*400*	*66*
"	42.7Fe – 36.6Cr – 20.7C	$(Fe,Cr)_{23}C_6$	400	135
"	66.3Fe – 13.0Cr – 20.7C	"	800	71
"	73.5Fe – 5.8Cr – 20.7C	"	1200	58
"	*77.6Fe – 1.7Cr – 20.7C (p)*	"	*400*	*42*
"	*76.4Fe – 2.9Cr – 20.7C (*)*	"	*400*	*46*

[3] Thermodynamic data from *MatCalc* database, mc_fe.tdb', version 1.028.

Table A.2. (*Continued*).

Nominal comp. at%	Precipitate comp. at%	Precipitate	Temp. °C	Intf. energy mJ/m^2
Fe – 0.2N – 2Cr	66.6Cr –33.3N	$(Cr,Fe)_2N$	400	331
”	62.7Cr – 4.0Fe – 33.3N	”	800	330
”	47.6Cr – 21.5Fe – 30.9N	”	1100	235
”	*45.3Cr – 22.1Fe – 32.6N (*)*	”	*400*	*178*
Fe – 2C – 0.02N – 1Cr	66.7Cr –26.6N – 6.7C	$Cr_2(C,N)$	300	310
”	66.7Cr –21.3N – 12.0C	”	500	315
”	66.7Cr –16.5N – 16.8C	”	700	304
Fe – 0.1N – 5Cr – 1V	65.7V – 33.3N – 1.0Cr	$(V,Cr)_2N$	400	605
”	53.3V – 33.3N – 9.0Cr – 4.3Fe	”	800	575
”	*56.3V – 33.3N – 7.8Cr – 2.5Fe (*)*	”	*400*	*560*
Fe – 0.1Al – 0.02N	50Al – 50N	AlN	600	450
”	”	”	1400	550
Fe – 0.1Nb – 0.5C	51.5Nb – 48.5C	NbC	400–1200	670
”	*34.5Nb – 15.5Fe – 50.0C (*)*	*(Nb,Fe)C*	*400*	*415*
”	*48.7Nb – 1.3Fe – 50.0C (*)*	”	*800*	*630*
”	*50.1Nb – 49.9C (*)*	”	*1200*	*660*
Fe – 0.1Nb – 0.5N	50Nb – 50N	NbN	800–1400	710–760
”	*29.1Nb – 21.3Fe – 49.7N (*)*	*(Nb,Fe)N*	*400*	*350*
”	*44.5Nb – 5.6Fe – 49.9N (*)*	”	*800*	*620*
Fe – 0.1Nb – 1C – 0.3N	50.0Nb – 48.1N – 1.7C	Nb(C,N)	400	650
”	50.7Nb – 41.3N – 8.0C	”	1200	730
”	*29.3Nb –21.2Fe –1.5C – 35.0N (*)*	*(Nb,Fe)(C,N)*	*400*	*340*
”	*44.5Nb –5.7Fe –19.8C – 30.1N (*)*	”	*800*	*590*
”	*50.2Nb –0.3Fe –21.6C –27.9N (*)*	”	*1200*	*700*
Fe – 0.1V – 0.5C	53.5V – 46.5C	VC	400–1200	465–445
”	*44.5V – 7.8Fe – 47.7C (*)*	*(V,Fe)C*	*400*	*380*
”	*53.4V – 0.3Fe – 46.3C (*)*	”	*800*	*460*

(*Continued*)

Table A.2. Typical interfacial energies for precipitates in fcc Fe-base alloys[3] obtained with the GBB model. For some precipitates, para (p) composition values and compositions with minimum nucleation barrier (*) are shown in italic. All values for fully supersaturated state (*Continued*).

Nominal comp. at%	Precipitate comp. at%	Precipitate	Temp. °C	Intf. energy mJ/m^2
Fe – 0.1V – 0.5N	50.0V – 49.9N	VN	400–1200	570–640
"	*28.3V – 22.0Fe – 49.8N (*)*	*(V,Fe)N*	*400*	*325*
"	*40.7V – 9.6Fe – 49.7N (*)*	"	*800*	*510*
"	*46.7V – 4.0Fe – 49.3N (*)*	"	*1200*	*620*
Fe – 0.1Ti – 0.5C	51.3Ti – 48.7C	TiC	400–770	640
"	48.4Ti – 8.3Fe – 43.3C	"	1200	565
"	*30.6Ti – 25.4Fe – 44.0C (*)*	*(Ti,Fe)C*	*400*	*350*
"	*43.8Ti – 10.9Fe – 43.8C (*)*	"	*800*	*530*
"	*48.4Ti – 6.2Fe – 45.4C (*)*	"	*1200*	*590*
Fe – 0.1Ti – 0.5N	50Ti – 50N	TiN	600–1400	910–1050
"	*21.8Ti – 28.6Fe – 49.6N (*)*	*(Ti,Fe)N*	*400*	*335*
"	*34.0Ti – 16.1Fe – 50.0N (*)*	"	*800*	*605*
"	*43.3Ti – 6.7Fe – 50.0N (*)*	"	*1100*	*825*
"	*48.7Ti – 1.4Fe – 50.0N (*)*	"	*1400*	*980*
Fe – 10Cr – 0.1V – 0.5N	34.0V – 32.0N – 31.0Cr – 3.0Fe	Z-phase	400–1200	470–540
"	52.5V – 47.5N	VN	400	460
"	48.7V – 48.6N – 2.6Cr	"	800	480
"	33.2V – 48.5N – 17.8Cr	"	1200	450
"	*5.9V – 49.9N – 44.2Cr (*)*	*(V,Cr)N*	*400*	*225*
"	*20.1V – 49.7N – 30.1Cr (*)*	"	*700*	*325*
"	*33.0V – 49.1N – 17.6Cr (*)*	"	*1000*	*420*
Fe – 10Cr – 0.1Nb – 0.5N	33.3Nb – 33.3N – 26.1Cr – 7.3Fe	Z-phase	400–1200	530–600
"	51Nb – 49N	NbN	400–1200	560–630
"	*7.2Nb – 49.9N – 43.0Cr (*)*	*(Nb,Cr)N*	*400*	*250*
"	*23.5Nb – 49.8N – 26.6Cr (*)*	"	*700*	*395*
"	*37.7Nb – 49.5N – 12.8Cr (*)*	"	*1000*	*525*
Fe – 2C – 2Mo	66.7Mo – 33.3C	Mo_2C	400	460
"	65.3Mo – 1.4Fe – 33.3C	"	800	445
"	50.0Mo – 18.0Fe – 32.1C	"	1200	275
"	*42.2Mo – 25.5Fe – 32.3C (*)*	*(Mo,Fe)2C*	*400*	*205*

Table A.2. (*Continued*).

Nominal comp. at%	Precipitate comp. at%	Precipitate	Temp. °C	Intf. energy mJ/m²
”	*49.2Mo – 17.8Fe – 33.0C (*)*	”	*600*	*275*
”	*56.2Mo – 10.6Fe – 33.2C (*)*	”	*800*	*350*
”	57.1Mo – 28.6Fe – 14.3C	Mo_6C	400	410
”	49.6Mo – 36.1Fe – 14.3C	”	800	190
”	34.4Mo – 51.3Fe – 14.3C	”	1200	140
”	*30.4Mo – 55.3Fe – 14.3C (*)*	*(Mo,Fe)6C*	*400*	*80*
”	*33.7Mo – 52.0Fe – 14.3C (*)*	”	*600*	*130*
”	*35.9Mo – 49.8Fe – 14.3C (*)*	”	*800*	*165*
Fe – 2C – 2W	49.3W – 36.4Fe – 14.3C	W_6C	400	315
”	39.4W – 46.3Fe – 14.3C	”	1200	280
Fe – 2Mo	66.9Fe – 33.1Mo	Laves	400	190
”	67.6Fe – 32.4Mo	”	800	170
”	78.6Fe – 21.4Mo	”	1200	75
Fe – 2Mo – 1Si	52.2Fe – 33.3Mo – 14.5Si	Laves	400	460
”	58.5Fe – 33.3Mo – 8.2Si	”	800	375
”	78.2Fe – 21.8Mo – 0.1Si	”	1200	80
”	*65.9Fe – 33.8Mo – 0.3Si (*)*	*Laves*	*400*	*195*
”	*62.6Fe – 32.7Mo – 4.7Si (*)*	”	*800*	*295*
Fe – 33Cr	61.6Cr – 38.4Fe	Sigma	400	50
”	50.6Cr – 49.4Fe	”	800	35
”	44.4Cr – 55.6Fe	”	1200	25
Fe – 0.01N – 0.01B	50B – 50N	BN	400–1200	730–880
Fe – 0.1C – 0.01B	79.3Fe – 7.8C – 12.9B	$Fe_{23}(CB)_6$	400	105
”	79.3Fe – 3.7C – 17.0B	”	800	115
Fe –1Cr – 0.01B	49.8Cr – 0.2Fe – 50.0B	CrB	400	695
”	35.9Cr – 14.1Fe – 50.0B	”	700	600
”	31.0Cr – 19.0Fe – 50.0B	”	1000	560
”	*26.7Cr – 23.3Fe – 50.0B (*)*	”	*400*	*520*
”	*30.6Cr – 19.4Fe – 50.0B (*)*	”	*600*	*555*
”	*32.2Cr – 17.8Fe – 50.0B (*)*	”	*800*	*570*
”	19.4Fe – 47.3Cr – 33.3B	Fe_2B	400	605
”	26.0Fe – 40.7Cr – 33.3B	”	600	555

(*Continued*)

Table A.2. Typical interfacial energies for precipitates in fcc Fe-base alloys[3] obtained with the GBB model. For some precipitates, para (p) composition values and compositions with minimum nucleation barrier (*) are shown in italic. All values for fully supersaturated state (*Continued*).

Nominal comp. at%	Precipitate comp. at%	Precipitate	Temp. °C	Intf. energy mJ/m^2
"	47.3Fe – 19.4Cr – 33.3B	"	650	350
"	55.0Fe – 11.7Cr – 33.3B	"	1000	305
"	*11.1Cr – 55.6Fe – 33.3B (*)*	"	*400*	*345*
"	*11.0Cr – 55.7Fe – 33.3B (*)*	"	*600*	*310*
"	*10.6Cr – 56.1Fe – 33.3B (*)*	"	*800*	*300*
Fe – 3Cu	100.0Cu	fcc-Cu	400	395
"	99.9Cu – 0.1Fe	"	700	395
"	97.2Cu – 2.8Fe	"	1000	365
"	*47.5Cu – 52.5Fe (*)*	*fcc-(Cu,Fe)*	*400*	*60*
"	*74.6Cu – 25.4Fe (*)*	"	*600*	*185*
"	*92.1Cu – 7.9Fe (*)*	"	*800*	*315*
Fe – 1Mn – 0.1S	50Mn – 50S	MnS	400	1090
"	50Mn – 50S	"	800	1150
"	50Mn – 50S	"	1200	1200
"	*2.2Mn – 47.8Fe – 50S (*)*	*(Mn,Fe)S*	*400*	*460*
"	*18.8Mn – 31.2Fe – 50S (*)*	"	*800*	*715*
"	*38.8Mn – 12.2Fe – 50S (*)*	"	*1200*	*1025*
Fe – 1Cu – 0.1S	41.3Cu – 40.8S – 17.8Fe	Cu_2S	400	480
"	34.3Cu – 40.4S – 25.3Fe	"	800	460
"	31.0Cu – 39.9S – 29.1Fe	"	1200	450
"	*23.2Cu – 46.1S – 30.7Fe (*)*	*(Cu,Fe)2S*	*400*	*420*
"	*26.1Cu – 41.0S – 32.9Fe (*)*	"	*800*	*435*
"	*28.5Cu – 40.0S – 31.6Fe (*)*	"	*1200*	*440*
Fe – 0.1Ti – 1C – 0.1S	50.0Ti –50.0S	TiS	400–1200	260
"	50.0Ti – 25.0C – 25.0S	$Ti_4C_2S_2$	400–1200	425
"	51.5Ti – 48.5C	TiC	400	638
"	52.0Ti – 47.8C – 0.2Fe	"	850	642
"	48.3Ti – 43.7C – 8.0Fe	"	1200	565
Fe – 3Si – 0.1N	42.9Si – 57.1N	Si_3N_4	400–800	210–260
Fe – 3Mn – 3Si – 0.1N	25.0Si – 25.0Mn – 50.0N	$MnSiN_2$	400–1200	260–350

Table A.2. (*Continued*).

Nominal comp. at%	Precipitate comp. at%	Precipitate	Temp. °C	Intf. energy mJ/m^2
Fe – 18Cr – 8Ni	56.1Cr – 42.2Fe – 1.7Ni	σ	400	51
”	46.7Cr – 50.7Fe – 2.6Ni	”	600	35
Fe – 19Cr – 12Ni – 2.5Mn – 1.5Mo – 0.25C – 0.6N [4]	63.6Cr – 3.5Mo – 23.2N – 9.7C	Cr_2N	400	160
”	61.4Cr – 5.3Mo – 16.7N – 16.5C	”	600	175
”	63.1Cr – 3.5Mo – 0.3Fe – 24.4N – 8.7C	”	800	155
”	62.9Cr – 3.2Mo – 1.2Fe – 27.3N – 5.1C – 0.2Mn – 0.1Ni	”	1000	145
”	*53.5Cr – 13.9Fe – 2.5Mo – 1.7Mn – 1.1Ni –26.6N – 0.7C (*)*	”	*400*	*28*
”	*58.2Cr – 7.6Fe – 2.8Mo – 0.8Mn – 0.6Ni –26.6N – 1.6C (*)*	”	*600*	*76*
”	*61.8Cr – 2.7Fe – 3.1Mo – 0.3Mn – 0.2Ni –28.4N – 3.5C (*)*	”	*800*	*125*
”	41.9Fe – 36.2Cr – 16.7Mo – 5.2Ni	X (chi)	400	66
”	49.1Fe – 28.5Cr – 14.4Mo – 8.0Ni	”	600	43
”	53.7Fe – 25.8Cr – 11.8Mo – 8.7Ni	”	800	32
”	56.3Fe – 24.4Cr – 10.2Mo – 9.1Ni	”	1000	27
”	*57.8Fe – 23.5Cr – 9.2Mo – 9.5Ni (*)*	”	*400*	*24*
”	*55.8Fe – 24.8Cr – 10.5Mo – 8.9Ni (*)*	”	*600*	*28*
”	*53.7Fe – 25.8Cr – 11.8Mo – 8.7Ni (*)*	”	*800*	*32*
”	42.5Fe – 23.8Cr – 29.6Mo – 4.1Ni	Laves	400	135
”	47.1Fe – 19.0Cr – 28.7Mo – 5.2Ni	”	600	145

(*Continued*)

[4] AISI 316LN (Fe – 18Cr – 13Ni – 2.5Mo – 2.5Mn – 0.05C – 0.15N, in wt%).

Table A.2. Typical interfacial energies for precipitates in fcc Fe-base alloys[3] obtained with the GBB model. For some precipitates, para (p) composition values and compositions with minimum nucleation barrier (*) are shown in italic. All values for fully supersaturated state (*Continued*).

Nominal comp. at%	Precipitate comp. at%	Precipitate	Temp. °C	Intf. energy mJ/m^2
"	48.1Fe – 17.6Cr – 27.9Mo – 6.4Ni	"	800	140
"	48.8Fe – 16.6Cr – 27.1Mo – 7.5Ni	"	1000	130
"	*49.6Fe – 15.3Cr – 25.6Mo – 9.4Ni (*)*	"	*400*	*115*
"	*49.0Fe – 16.4Cr – 27.0Mo – 7.6Ni (*)*	"	*600*	*130*
"	*48.5Fe – 17.0Cr – 27.5Mo – 7.0Ni (*)*	"	*800*	*135*
"	28.1Fe – 13.6Cr – 31.7Mo – 26.6Ni	σ (sigma)	400	145
"	30.9Fe – 13.8Cr – 28.8Mo – 26.4Ni	"	600	135
"	34.3Fe – 14.2Cr – 25.8Mo – 25.5Ni – 0.2Mn	"	800	115
"	40.2Fe – 15.9Cr – 21.2Mo – 22.1Ni – 0.6Mn	"	1000	80
"	*55.9Fe – 26.4Cr – 7.7Mo – 8.5Ni – 1.5Mn (*)*	"	*400*	*14*
"	*52.6Fe – 27.3Cr – 9.3Mo – 9.4Ni – 1.4Mn (*)*	"	*600*	*19*
"	*43.1Fe – 17.5Cr – 18.8Mo – 19.8Ni – 0.8Mn (*)*	"	*800*	*60*
"	66.9Cr – 20.7C – 10.3Mo – 2.0Fe	$M_{23}C_6$	400	280
"	64.0Cr – 20.7C – 10.1Mo – 5.0Fe – 0.2Ni	"	600	270
"	60.5Cr – 20.7C – 8.9Mo – 9.3Fe – 0.5Ni – 0.1Mn	"	800	255
"	56.4Cr – 20.7C – 5.9Mo – 15.4Fe – 1.1Ni – 0.5Mn	"	1000	220
"	*44.7Cr – 20.7C – 1.7Mo – 28.9Fe – 2.8Ni – 1.2Mn (*)*	"	*400*	*155*
"	*53.3Cr – 20.7C – 4.1Mo – 19.6Fe – 1.5Ni – 0.8Mn (*)*	"	*600*	*195*
"	*56.0Cr – 20.7C – 5.6Mo – 16.1Fe – 1.1Ni – 0.5Mn (*)*	"	*800*	*215*

Table A.3. Typical interfacial energies for precipitates in Ni-base alloys[5] obtained with the GBB model. For some precipitates, compositions with minimum nucleation barrier (*) are given exemplarily. All values for fully supersaturated state.

Nominal comp. at%	**Precipitate comp. at%, (bal. = Fe)**	**Precipitate**	**Temp. °C**	**Intf. energy mJ/m²**
Ni – 15Al	75.0Ni – 25.0Al	Ni_3Al (γ')	400	50
"	75.1Ni – 24.9Al	"	800	48
"	77.5Ni – 22.5Al	"	1200	20
"	*77.6Ni – 22.4Al (*)*	"	*400*	*19*
"	*76.4Ni – 23.6Al (*)*	"	*600*	*33*
"	*75.8Ni – 24.2Al (*)*	"	*800*	*40*
Ni – 15Ti	75.0Ni – 25.0Ti	Ni_3Ti (γ')	400	138
"	75.1Ni – 24.9Ti	"	800	137
"	76.1Ni – 23.9Ti	"	1200	127
"	*79.7Ni – 20.3Ti (*)*	"	*400*	*92*
"	*76.5Ni – 23.5Ti (*)*	"	*600*	*123*
"	*75.8Ni – 24.2Ti (*)*	"	*800*	*130*
Ni – 10Al – 10Ti	75.0Ni – 9.0Al – 16.0Ti	Ni_3(Ti,Al) (γ')	400	121
"	75.0Ni – 11.3Al – 13.7Ti	"	800	115
"	75.1Ni – 13.1Al – 11.8Ti	"	1200	107
"	*79.4Ni – 14.7Al – 5.9Ti (*)*	"	*400*	*35*
"	*78.5Ni – 14.7Al – 6.8Ti (*)*	"	*600*	*47*
"	*77.1Ni – 14.6Al – 8.4Ti (*)*	"	*800*	*71*
Ni – 15Nb	75.0Ni – 25.0Nb	Ni_3Nb (γ')	400	139
"	75.1Ni – 24.9Nb	"	800	137
"	76.8Ni – 23.2Nb	"	1200	120
"	*78.4Ni – 21.6Nb (*)*	"	*400*	*104*
"	*75.8Ni – 24.2Nb (*)*	"	*600*	*130*
"	*75.4Ni – 24.6Nb (*)*	"	*800*	*134*
"	75.0Ni – 25.0Nb	Ni_3Nb (γ'')	400	141
"	75.0Ni – 25.0Nb	"	800	141
"	75.7Ni – 24.3Nb	"	1200	133
"	*79.3Ni – 20.7Nb (*)*	"	*400*	*95*

(*Continued*)

[5] Thermodynamic data from *MatCalc* database, mc_ni.tdb', version 1.026.

Table A.3. Typical interfacial energies for precipitates in Ni-base alloys[5] obtained with the GBB model. For some precipitates, compositions with minimum nucleation barrier (*) are given exemplarily. All values for fully supersaturated state (*Continued*).

Nominal comp. at%	Precipitate comp. at%, (bal. = Fe)	Precipitate	Temp. °C	Intf. energy mJ/m^2
"	*76.2Ni – 23.8Nb (*)*	"	*600*	*128*
"	*75.6Ni – 24.4Nb (*)*	"	*800*	*135*
"	75.0Ni – 25.0Nb	Ni_3Nb (δ)	400	156
"	75.1Ni – 24.9Nb	"	800	155
"	75.9Ni – 24.1Nb	"	1200	147
"	*83.1Ni – 16.9Nb (*)*	"	*400*	*70*
"	*81.0Ni – 19.0Nb (*)*	"	*600*	*94*
"	*78.2Ni – 21.8Nb (*)*	"	*800*	*123*
Ni – 10Al – 5Nb – 5Ti	75.0Ni – 10.2Al – 14.1Ti – 0.7Nb	γ'	400	113
"	75.0Ni – 12.7Al – 11.4Ti – 0.9Nb	"	800	106
"	75.3Ni – 14.6Al – 9.2Ti – 0.9Nb	"	1200	95
"	*78.4Ni – 15.6Al – 5.2Ti – 0.8Nb (*)*	"	*400*	*41*
"	*77.3Ni – 15.6Al – 6.3Ti – 0.8Nb (*)*	"	*600*	*58*
"	*76.0Ni – 15.2Al – 7.9Ti – 0.9Nb*	"	*800*	*81*
"	75.0Ni – 12.8Ti – 12.2Nb	γ''	400	132
"	75.1Ni – 0.4Al – 10.9Ti – 13.7Nb	"	800	128
"	*75.6Ni – 4.5Al – 5.4Ti – 14.5Nb (*)*	"	*400*	*101*
"	*75.1Ni – 0.8Al – 10.1Ti – 14.0Nb (*)*	"	*600*	*125*
"	75.0Ni –25.0Nb	δ	400	138
"	75.1Ni – 1.2Al – 23.7Nb	"	800	134
"	*76.8Ni – 4.8Al – 18.4Nb (*)*	"	*400*	*100*
"	*75.6Ni – 2.6Al – 21.8Nb (*)*	"	*600*	*123*
Ni – 15Cr –1C	70.0Cr – 30.0C	M_7C_3	400	190
"	68.2Cr – 1.8Ni – 30.0C	"	800	175
"	60.7Cr – 9.3Ni – 30.0C	"	1200	140
"	*46.3Cr – 23.7Ni – 30.0C (*)*	"	*400*	*92*

Table A.3. (*Continued*).

Nominal comp. at%	Precipitate comp. at%, (bal. = Fe)	Precipitate	Temp. °C	Intf. energy mJ/m^2
"	*53.9Cr – 16.1Ni – 30.0C (*)*	"	*600*	*119*
"	*60.9Cr – 9.1Ni – 30.0C (*)*	"	*800*	*144*
"	76.0Cr – 3.3Ni – 20.7C	$M_{23}C_6$	400	153
"	57.0Cr – 22.3Ni – 20.7C	"	800	98
"	38.2Cr – 41.1Ni – 20.7C	"	1200	53
"	*40.3Cr – 39.0Ni – 20.7C (*)*	"	*400*	*58*
"	*48.9Cr – 30.4Ni – 20.7C (*)*	"	*600*	*78*
"	*53.9Cr – 25.4Ni – 20.7C (*)*	"	*800*	*90*
Ni – 5Mo – 1C	10.3Mo – 69.0Ni – 20.7C	$M_{23}C_6$	400	247
"	10.3Mo – 69.0Ni – 20.7C	"	1200	245
"	*10.3Mo – 69.0Ni – 20.7C (*)*	"	*400*	*247*
"	*8.5Mo – 70.8Ni – 20.7C (*)*	"	*800*	*206*
"	31.1Mo – 54.6Ni – 14.3C	M_6C	400	153
"	30.7Mo – 55.0Ni – 14.3C	"	1200	150
"	*31.1Mo – 54.6Ni – 14.3C (*)*	"	*400*	*153*
"	*30.7Mo – 55.0Ni – 14.3C (*)*	"	*800*	*151*
Ni – 12Cr – 5Co – 11Al – 4Ta – 1.3W[6]	72.6Ni – 18.5Ta – 5.0Al – 1.8Co – 1.6W – 0.5Cr	γ'	400	130
"	69.0Ni – 11.7Ta – 11.5Al – 3.9Co – 1.6W – 2.3Cr	"	800	99
"	67.5Ni – 6.5Ta – 14.9Al – 4.4Co – 1.4W – 5.3Cr	"	1200	64
"	*68.3Ni – 2.8Ta – 14.8Al – 4.7Co – 1.0W – 8.4Cr (*)*	"	*400*	*17*
"	*68.0Ni – 3.3Ta – 15.0Al – 4.6Co – 1.1W – 8.0Cr (*)*	"	*600*	*25*
"	*67.8Ni – 4.3Ta – 15.1Al – 4.5Co – 1.2W – 7.1Cr (*)*	"	*800*	*39*
"	*67.7Ni – 5.0Ta – 15.2Al – 4.4Co – 1.2W – 6.5Cr (*)*	"	*1000*	*48*

(*Continued*)

[6] PWA 1480 (Ni – 10Cr – 5Co – 5Al – 12Ta – 4W, wt%).

Table A.3. Typical interfacial energies for precipitates in Ni-base alloys[5] obtained with the GBB model. For some precipitates, compositions with minimum nucleation barrier (*) are given exemplarily. All values for fully supersaturated state (*Continued*).

Nominal comp. at%	Precipitate comp. at%, (bal. = Fe)	Precipitate	Temp. °C	Intf. energy mJ/m²
Ni – 15Cr – 10Co – 6Al – 6Ti – 2.5Mo – 0.8C[7]	72.8Ni – 19.1Ti – 5.9Al – 2.2Co	γ'	400	157
”	70.9Ni – 15.3Ti – 9.5Al – 3.6Co – 0.5Cr – 0.2Mo	”	800	138
”	68.4Ni – 11.5Ti – 12.1Al – 5.3Co – 2.3Cr – 0.4Mo	”	1200	107
”	*66.4Ni – 5.7Ti – 12.4Al – 7.9Co – 6.8Cr – 0.8Mo (*)*	”	*400*	*38*
”	*66.8Ni – 7.1Ti – 12.6Al – 7.3Co – 5.6Cr – 0.6Mo (*)*	”	*600*	*56*
”	*67.5Ni – 9.4Ti – 12.6Al – 6.3Co – 3.7Cr – 0.5Mo (*)*	”	*800*	*84*
”	*68.1Ni – 10.6Ti – 12.4Al – 5.6Co – 2.8Cr – 0.5Mo (*)*	”	*1000*	*98*
”	51.4Ti – 48.6C	TiC	400	475
”	50.8Ti – 48.1C – 1.1Cr	”	800	480
”	48.6Ti – 47.4C – 3.4Cr – 0.6Mo	”	1200	478
”	*0.8Ti – 49.3C – 25.5Cr – 24.6Mo (*)*	*(Cr,Mo)C*[8]	*400*	*143*
”	*2.7Ti – 48.2C – 24.4Cr – 24.7Mo (*)*	”	*600*	*162*
”	*5.6Ti – 46.5C – 23.1Cr – 23.5Mo – 1.2Co – 0.1Ni (*)*	”	*800*	*183*
”	*8.7Ti – 45.0C – 21.9Cr – 23.0Mo – 1.2Co – 0.2Ni (*)*	”	*1000*	*206*
”	67.6Cr – 20.7C – 10.3Mo – 1.4Ni	$M_{23}C_6$	400	303
”	61.5Cr – 20.7C – 10.1Mo – 7.5Ni – 0.2Co	”	800	326
”	54.5Cr – 20.7C – 9.5Mo – 14.4Ni – 0.8Co	”	1200	303

[7] RENE 80 (Ni – 14Cr – 10Co – 3Al – 5Ti – 4Mo – 0.17C, wt%).

[8] The Ti-carbide composition evaluated for minimum nucleation barrier in the fully supersaturated state is unstable with respect to the (Cr,Mo)-carbide composition.

Table A.3. (*Continued*).

Nominal comp. at%	Precipitate comp. at%, (bal. = Fe)	Precipitate	Temp. °C	Intf. energy mJ/m^2
"	*34.2Cr – 20.7C – 1.1Mo – 39.7Ni – 4.3Co (*)*	"	*400*	*96*
"	*38.8Cr – 20.7C – 1.7Mo – 34.6Ni – 4.2Co (*)*	"	*600*	*128*
"	*42.4Cr – 20.7C – 2.6Mo – 30.2Ni – 4.1Co (*)*	"	*800*	*160*
"	*45.3Cr – 20.7C – 3.8Mo – 26.5Ni – 3.7Co (*)*	"	*1000*	*193*
"	69.8Cr – 30.0C – 0.2Mo	M_7C_3	400	297
"	68.9Cr – 30.0C – 0.7Mo – 0.4Co	"	800	291
"	66.9Cr – 30.0C – 1.4Mo – 0.8Co – 0.9Ni	"	1200	282
"	*35.1Cr – 30.0C – 4.1Mo – 2.5Co – 28.2Ni (*)*	"	*400*	*108*
"	*41.2Cr – 30.0C – 3.9Mo – 2.5Co – 22.4Ni (*)*	"	*600*	*141*
"	*47.1Cr – 30.0C – 3.6Mo – 2.4Co – 16.9Ni (*)*	"	*800*	*173*
"	*52.1Cr – 30.0C – 3.3Mo – 2.2Co – 12.4Ni (*)*	"	*1000*	*202*
Ni – 21Cr – 19Fe – 3Nb – 2Mo– 1Ti – 0.2C[9]	74.3Ni – 17.4Nb – 7.6Ti – 0.4Fe – 0.2Cr – 0.1Mo	γ'	400	196
"	71.8Ni – 17.1Nb – 6.4Ti – 2.6Fe – 1.5Cr – 0.6Mo	"	700	179
"	67.2Ni – 15.2Nb – 4.0Ti – 6.4Fe – 5.9Cr – 1.4Mo	"	1000	134
"	*64.3Ni – 11.4Nb – 2.0Ti – 9.3Fe – 11.1Cr – 1.9Mo (*)*	"	*400*	*77*
"	*66.1Ni – 14.2Nb – 3.3Ti – 7.4Fe – 7.4Cr – 1.6Mo (*)*	"	*600*	*118*
"	*67.6Ni – 15.4Nb – 4.1Ti – 6.1Fe – 5.4Cr – 1.3Mo (*)*	"	*800*	*138*

(*Continued*)

[9] Inconel 718 (Ni – 19Cr – 3Mo – 18Fe – 1Ti – 5Nb – 0.4C, wt%).

Table A.3. Typical interfacial energies for precipitates in Ni-base alloys[5] obtained with the GBB model. For some precipitates, compositions with minimum nucleation barrier (*) are given exemplarily. All values for fully supersaturated state (*Continued*).

Nominal comp. at%	**Precipitate comp. at%, (bal. = Fe)**	**Precipitate**	**Temp. °C**	**Intf. energy mJ/m²**
”	74.6Ni – 18.4Nb – 6.6Ti – 0.4Fe	γ''	400	204
”	73.3Ni – 19.7Nb – 4.9Ti – 1.7Fe – 0.3Cr	”	700	196
”	69.2Ni – 15.8Nb – 3.0Ti – 5.9Fe – 5.5Cr – 0.6Mo	”	1000	199
”	*59.5Ni – 7.9Nb – 1.1Ti – 15.8Fe – 14.2Cr – 1.5Mo (*)*	”	*400*	*38*
”	*64.6Ni – 11.5Nb – 1.8Ti – 10.7Fe – 10.3Cr – 1.1Mo (*)*	”	*600*	*78*
”	*68.0Ni – 14.6Nb – 2.6Ti – 7.2Fe – 6.9Cr – 0.7Mo (*)*	”	*800*	*116*
”	31.6Ti – 19.1Nb – 49.3C	TiC	400	488
”	29.2Ti – 21.6Nb – 49.2C	”	700	486
”	27.1Ti – 23.7Nb – 49.0C – 0.2Cr	”	1000	485
”	25.2Ti – 25.2Nb – 48.8C – 0.6Cr – 0.2Mo	”	1300	481
”	*13.6Ti – 20.3Nb – 49.9C – 16.2Cr (*)*	*(Ti,Nb,Cr)C*	*400*	*362*
”	*17.2Ti – 24.9Nb – 49.9C – 8.0Cr (*)*	”	*600*	*419*
”	*19.7Ti – 26.6Nb – 49.7C – 4.0Cr (*)*	”	*800*	*448*
”	66.9Cr – 20.7C – 10.3Mo – 1.7Ni – 0.3Fe	$M_{23}C_6$	400	307
”	61.5Cr – 20.7C – 10.2Mo – 6.3Ni – 1.3Fe	”	700	292
”	55.3Cr – 20.7C – 9.7Mo – 11.4Ni – 3.0Fe	”	1000	271
”	*42.2Cr – 20.7C – 3.5Mo – 22.3Ni – 11.3Fe (*)*	”	*400*	*150*
”	*47.0Cr – 20.7C – 5.1Mo – 18.2Ni – 9.0Fe (*)*	”	*600*	*189*
”	*50.3Cr – 20.7C – 7.7Mo – 15.4Ni – 5.8Fe (*)*	”	*800*	*233*

Table A.3. (*Continued*).

Nominal comp. at%	**Precipitate comp. at%, (bal. = Fe)**	**Precipitate**	**Temp. °C**	**Intf. energy mJ/m²**
Ni – 27Cr – 19Co – 1.0Al – 2.5Ti – 1Nb[10]	24.1Ni – 24.8Nb – 50.5Co – 0.3Cr – 0.1Ti	γ'	400	378[11]
”	59.0Ni – 12.3Nb – 16.0Co – 0.3Cr – 12.2Ti – 0.2Al	”	600	227
”	63.1Ni – 6.4Nb – 12.7Co – 0.6Cr – 15.9Ti – 1.7Al	”	800	194
”	62.0Ni – 3.8Nb – 14.9Co – 1.5Cr – 15.1Ti – 2.7Al	”	1000	165
”	*58.8Ni – 4.3Nb – 18.9Co – 11.7Cr – 4.0Ti – 2.3Al (*)*	”	*400*	*51*
”	*61.5Ni –3.4Nb – 16.4Co – 5.0Cr – 10.4Ti – 3.3Al (*)*	”	*600*	*108*
”	*61.4Ni –3.3Nb – 16.1Co – 2.8Cr – 13.3Ti – 3.1Al (*)*	”	*800*	*138*
”	71.5Cr – 26.7Co – 1.9Ni	σ	400	147
”	68.2Cr – 25.7Co – 6.1Ni	”	600	128
”	64.2Cr – 24.4Co – 11.4Ni	”	800	105
”	*55.8Cr – 22.1Co – 22.1Ni (*)*	”	*400*	*63*
”	*59.8Cr – 23.2Co – 17.0Ni (*)*	”	*600*	*83*
”	*61.8Cr – 23.8Co – 14.4Ni (*)*	”	*800*	*93*
Ni – 19Cr – 14Co –5Al – 6Ti[12]	71.2Ni – 21.3Ti – 3.8Co – 3.7Al	γ'	400	170
”	68.8Ni – 17.6Ti – 5.8Co – 7.2Al – 0.6Cr	”	800	150
”	65.9Ni – 13.5Ti – 8.6Co – 9.6Al – 2.5Cr	”	1200	114
”	*63.3Ni – 7.3Ti – 11.8Co – 10.2Al – 7.4Cr (*)*	”	*400*	*42*
”	*64.2Ni – 9.2Ti – 10.9Co – 10.2Al – 5.5Cr (*)*	”	*600*	*65*
”	*64.9Ni – 11.6Ti – 9.7Co – 10.1Al – 3.7Cr (*)*	”	*800*	*94*

(*Continued*)

[10] Inconel 740 (Ni – 25Cr – 20Co – 0.5Al – 2Ti – 2Nb, wt%).
[11] At the lowest temperature, the Co-rich γ'-phase is more stable than the Ni-rich variant.
[12] UDIMET 720 (Ni – 18Cr – 15Co – 2.5Al – 5Ti, in wt%).

Table A.3. Typical interfacial energies for precipitates in Ni-base alloys[5] obtained with the GBB model. For some precipitates, compositions with minimum nucleation barrier (*) are given exemplarily. All values for fully supersaturated state (*Continued*).

Nominal comp. at%	Precipitate comp. at%, (bal. = Fe)	Precipitate	Temp. °C	Intf. energy mJ/m²
Ni – 22Cr – 1Co – 3Al – 3Ti – 0.25C[13]	74.9Ni – 18.7Ti – 6.2Al – 0.2Co	γ'	400	153
”	74.9Ni – 18.7Ti – 6.2Al – 0.2Co	”	700	141
”	74.2Ni – 12.3Ti – 10.6Al – 2.5Cr – 0.4Co	”	1000	117
”	*74.3Ni – 5.3Ti – 9.7Al – 10.0Cr – 0.8Co (*)*	”	*400*	*47*
”	*74.1Ni – 9.1Ti – 10.8Al – 5.4Cr – 0.5Co (*)*	”	*600*	*88*
”	*74.1Ni – 11.4Ti – 10.8Al – 3.2Cr – 0.5Co (*)*	”	*800*	*110*
”	50.8Ti – 48.7C – 0.6Cr	TiC	400	417
”	48.9Ti – 48.3C – 2.9Cr	”	700	414
”	46.2Ti – 47.9C – 5.9Cr	”	1000	408
”	43.3Ti – 47.3C – 9.3Cr	”	1300	401
”	*20.9Ti – 48.7C – 30.4Cr (*)*	”	*400*	*272*
”	*28.0Ti – 47.8C – 24.2Cr (*)*	”	*600*	*315*
”	*32.9Ti – 47.2C – 19.3Cr – 0.5Ni – 0.1Ni (*)*	”	*800*	*337*
”	77.7Cr – 20.7C – 1.6Co	$M_{23}C_6$	400	258
”	77.0Cr – 20.7C – 1.3Co – 1.0Ni	”	700	251
”	73.8Cr – 20.7C – 1.0Co – 4.5Ni	”	1000	234
”	*44.1Cr – 20.7C – 0.4Co – 34.8Ni (*)*	”	*400*	*94*
”	*50.5Cr – 20.7C – 0.4Co – 28.4Ni (*)*	”	*600*	*122*
”	*56.3Cr – 20.7C – 0.5Co – 22.5Ni (*)*	”	*800*	*148*

[13] Nimonic 80A (Ni – 20Cr – 1Co – 1.5Al – 2.5Ti – 0.05C, in wt%).

Table A.4. Typical interfacial energies for precipitates in Al-base alloys[14] obtained with the GBB model. All values for fully supersaturated state.

Nominal comp. at%	Precipitate comp. at%, (bal. = Fe)	Precipitate	Temp. °C	Intf. energy mJ/m^2
Al – 3Cu	33.1Cu – 66.9Al	Al_2Cu (Θ)	25–400	88
”	31.6Cu – 68.4Al	”	400–600	87–74
”	33.3Cu – 66.7Al	Al_2Cu (Θ′)	25–600	91
”	25.0Cu – 75.0Al	Al_3Cu (Θ″/GP)	25–600	38
Al – 3Cu – 1Mg	25.0Cu – 25.0Mg – 50.0Al	Al_2CuMg (S)	25–600	122
Al – 3Cu – 1Mg – 1Si	9.5Cu – 38.1Mg – 33.3Si– 19.0Al	$Al_4Cu_2Mg_8Si_7$ (Q_high)	25–600	275
Al – 3Cu – 1Mn	8.0Cu – 12.0Mn – 80.0Al	$Al_{20}Cu_2Mn_3$ (T)	25–600	54
Al – 3Cu – 1Fe	20.0Cu – 10.0Fe – 70.0Al	Al_7Cu_2Fe (S)	25–600	111
Al – 1Mn	14.3Mn – 85.7Al	Al_6Mn (G)	25–600	46
”	7.7Mn – 92.3Al	$Al_{12}Mn$	25–600	30
Al – 1Cr	12.5Cr – 87.5Al	$Al_{13}Cr_2$ (Θ)	25–600	82
Al – 1Cr – 1Si	26.6Cr – 27.7Si – 45.6Al	$(Al,Si)_{11}Cr_4$ (α)	25	195
”	26.6Cr – 15.6Si – 57.8Al	”	600	160
”	19.0Cr – 19.0Si – 62.0Al	$Al_{13}Cr_4Si_4$ (β)	25–600	190
”	19.0Cr – 19.0Si – 62.0Al	$Al_{58}Cr_{31}Si_{11}$ (α)	25–600	75
Al – 1Fe	24.1Fe – 75.9Al	$Al_{13}Fe_4$ (β)	25–600	57
Al – 1Fe – 1Si	15.2Fe – 16.3Si – 68.5Al	$Al_9Fe_2Si_2$ (β/T6)	25	195
”	15.2Fe – 13.5Si – 71.3Al	”	600	193
”	19.0Fe – 14.9Si – 66.1Al	Al_8Fe_2Si (α'/T5)	25	185
”	19.0Fe – 11.0Si – 70.0Al	”	600	178
Al – 1Fe – 1Mn – 1Si	17.4Fe – 0.0Mn – 12.6Si – 70.0Al	$Al_{19-20}(Fe,Mn)_5Si_2$ (α)	25	235
”	16.5Fe – 0.9Mn – 7.9Si – 74.7Al	”	600	225
”	0.5Fe – 22.5Mn – 19.2Si – 57.7Al	$Al_4(Fe,Mn)_6Si_4$ (β)	25	155
”	19.5Fe – 3.5Mn – 13.3Si – 63.6Al	”	600	110
Al – 1Mg – 1Si	66.7Mg – 33.3Si	Mg_2Si (β)	25–600	375
”	64.3Mg – 35.7Si	$Mg_{1.8}Si$ (β')	25–600	335
”	45.5Mg – 51.4Si – 3.1Al	Mg_5Si_6 (β'')	25–600	255

(*Continued*)

[14] Thermodynamic data from *MatCalc* database, mc_al.tdb', version 1.011.

Table A.4. Typical interfacial energies for precipitates in Al-base alloys[14] obtained with the GBB model. All values for fully supersaturated state (*Continued*).

Nominal comp. at%	**Precipitate comp. at%, (bal. = Fe)**	**Precipitate**	**Temp. °C**	**Intf. energy mJ/m²**
”	28.4Mg – 25.0Si – 46.6Al	GP (-Matsuda)	25	90
”	”	”	300	69
”	100.0Si	Si (Diamond)	25–600	395
Al – 3Mg – 5Zn	66.7Zn – 33.3Mg	$MgZn_2$ (η)	25	190
”	63.8Zn – 33.3Mg – 2.9Al	”	300	180
”	50.3Zn – 30.1Mg – 19.6Al	”	600	125
”	11.8Zn – 9.8Mg – 78.4Al	AlMgZn (GP)	25–600	28
Al – 3Mg – 5Zn – 2Cu	52.2Zn – 33.3Mg – 14.5Cu	$MgZn_2$ (σ)	25	198
”	50.6Zn – 33.3Mg – 16.0Cu – 0.01Al	”	300	198
”	48.2Zn – 33.3Mg – 16.4Cu – 2.1Al	”	600	188

References

1. Wagner, C. 1952. *Thermodynamics of Alloys.* Addison-Wesley Press.
2. Falk, G., and W. Ruppel. 1976. *Energie und Entropie.* Berlin: Springer Verlag.
3. Callen, H. B. 1985. *Thermodynamics and an Introduction to Thermostatics.* New York: John Wiley & Sons.
4. Hack, K. ed. 1996. *The SGTE Casebook – Thermodynamics at Work.* London: The Institute of Materials.
5. Malanowski, S., and A. Anderko. 1996. *Modelling Phase Equilibria.* Canada: John Wiley & Sons.
6. Hillert, M. 2008. *Phase Equilibria, Phase Diagrams and Phase Transformations – Their Thermodynamic Basis.* Cambridge: Cambridge University Press.
7. Bragg, W. L., and E. J. Williams. 1934. "The Effect of Thermal Agitation on Atomic Arrangement in Alloys." *Proceedings of the Royal Society of London A* 145 (855): 699–730. doi: http://dx.doi.org/10.1098/rspa.1934.0132.
8. Redlich, O., and A. T. Kister. 1948. "Algebraic Representation of Thermodynamic Properties and the Classification of Solutions." *Industrial and Engineering Chemistry* 40 (2): 345–8. doi: http://dx.doi.org/10.1021/ie50458a036.
9. Saunders, N., and A. P. Miodownik. 1998. *CALPHAD (calculation of phase diagrams): A Comprehensive Guide.* Pergamon Materials Series, Pergamon Press.
10. Lukas, H. L., S. G. Fries, and B. Sundman. 2007. *Computational Thermodynamics: The CALPHAD Method.* Cambridge University Press.
11. Janssens, Koenraad G. F., Dierk Raabe, Ernst Kozeschnik, Mark A. Miodownik, and Britta Nestler. 2007. *Computational Materials Engineering – An Introduction to Computational Microstructure Evolution of Polycrystalline Materials.* Elsevier Publishing.
12. Perez, M. 2005. "*Gibbs–Thomson effects in phase transformations.*" *Scripta Materialia* 52 (8): 709–12. doi: http://dx.doi.org/10.1016/j.scriptamat.2004.12.026.
13. Eshelby, J. D. 1957. "The Determination of the Elastic Field of an Ellipsoidal Inclusion, and Related Problems." *Proceedings of the Royal Society of London A* 241: 376–96. doi: http://dx.doi.org/10.1098/rspa.1957.0133.
14. Glicksman, M. E. 2000. *Diffusion in Solids.* John Wiley & Sons.
15. Thomson, W. 1870. "On the Equilibrium of Vapour at a Curved Surface of Liquid." *Proceedings of the Royal Society of Edinburgh* 7: 63–8.
16. Gibbs, J. W. 1961. *The Scientific Papers.* New York: Dover Publications Inc.
17. von Smoluchowski, M. 1908. "Molekular-kinetische Theorie der Opaleszenz von Gasen im kritischen Zustande, sowie einiger Erscheinungen." *Annalen der Physik* 25: 205–26. doi: http://dx.doi.org/10.1002/andp.19083300203.
18. Einstein, A. 1910. "Theorie der Opaleszenz von homogenen Flüssigkeiten und Flüssigkeitsgemischen in der Nähe des kritischen Zustandes." *Annalen der Physik* 33: 1275–98. doi: http://dx.doi.org/10.1002/andp.19103381612.
19. Volmer, M., and A. Weber. 1926. "Keimbildung in übersättigten Gebilden." *Zeitschrift für Physikalische Chemie* 119: 277–301.

20. Becker, R., and W. Döring. 1935. "Kinetische Behandlung der Keimbildung in übersättigten Dämpfen." *Annalen der Physik* 24: 719–52. doi: http://dx.doi.org/10.1002/andp.19354160806.
21. Farkas, L. 1927. "Keimbildungsgeschwindigkeit in übersättigten Dämpfen." *Zeitschrift für Physikalische Chemie* 125: 236–42.
22. Volmer, M., and H. Flood. 1934. "Tröpfchenbildung in Dämpfen." *Zeitschrift für Physikalische Chemie* 170: 273–85.
23. Zeldovich, J. B. 1943. "On the Theory of New Phase Formation Cavitation." *Acta Physicochimica URSS* 18: 1–22.
24. Wakeshima, H. 1954. "Time Lag in Self-Nucleation." *Journal of Chemical Physics* 22: 1614–15. doi: http://dx.doi.org/10.1063/1.1740475.
25. Feder, J., K. C. Russell, J. Lothe, and G. M. Pound. 1966. "Homogeneous Nucleation and Growth of Droplets in Vapor." *Advances in Physics* 15: 111–78. doi: http://dx.doi.org/10.1080/00018736600101264.
26. Mitchell, R. J., M. Preuss, M. C. Hardy, and S. Tin. 2006. "Influence of Composition and Cooling Rate on Constrained and Unconstrained Lattice Parameters in Advanced Polycrystalline Nickel–Base Superalloys." *Materials Science and Engineering A* 423: 282–91. doi: http://dx.doi.org/10.1016/j.msea.2006.02.039.
27. Turnbull, D., and J. C. Fisher. 1949. "Rate of Nucleation in Condensed Systems." *Journal of Chemical Physics* 17: 71–3. doi: http://dx.doi.org/10.1063/1.1747055.
28. Russell, K. C. 1980. "Nucleation in Solids: The Induction and Steady State Effects." *Advances in Colloid and Interface Science* 13: 205–318. doi: http://dx.doi.org/10.1016/0001-8686(80)80003-0.
29. Frenkel, J. 1939. "A General Theory of Heterophase Fluctuations and Pre-Transition Phenomena." *Journal of Chemical Physics URSS* 1: 315–24. doi: http://dx.doi.org/10.1063/1.1750484.
30. Turnbull, D. 1948. "Transient Nucleation." *Transactions of the American Institute of Mining, Metallurgical and Petroleum Engineers* 175: 774–83.
31. Soisson, F., and G. Martin. 2000. "Monte Carlo Simulations of the Decomposition of Metastable Solid Solutions: Transient and Steady-State Nucleation Kinetics." *Physical Review B* 62: 203–14. doi: http://dx.doi.org/10.1103/PhysRevB.62.203.
32. Russell, K. C. 1969. "The Role of Excess Vacancies in Precipitation." *Scripta Metallurgica* 3: 313–16. doi: http://dx.doi.org/10.1016/0036-9748(69)90294-4.
33. Thomson, W. 1887. "On the Division of Space with Minimum Partitional Area." *Philosophical Magazine* 24: 503–14. doi: http://dx.doi.org/10.1080/14786448708628135.
34. Rajek, J. 2005. PhD thesis, Institute for Materials Science, Welding and Forming, Graz University of Technology, Austria.
35. Clemm, P. J., and J. C. Fisher. 1955. "The Influence of Grain Boundaries on the Nucleation of Secondary Phases." *Acta Metallurgica* 3: 70–3. doi: http://dx.doi.org/10.1016/0001-6160(55)90014-6.
36. Kozeschnik, E., and E. Gamsjäger. 2006. "High-Speed Quenching Dilatometer Investigation of the Austenite to Ferrite Transformation in Fe-C." *Metallurgical and Materials Transactions* 37A: 1791–7. doi: http://dx.doi.org/10.1007/s11661-006-0121-z.
37. Read, W. T., and W. Shockley. 1950. "Dislocation Models of Crystal Grain Boundaries." *Physical Review* 78: 275–89. doi: http://dx.doi.org/10.1103/PhysRev.78.275.
38. Murr, L. E. 1975. *Interfacial Phenomena in Metals and Alloys* (p. 131). Addison-Wesley.
39. Holm, E. A., D. L. Olmsted, and S.M. Foiles. 2010. "Comparing Grain Boundary Energies in Face-Centered Cubic Metals: Al, Au, Cu and Ni." *Scripta Materialia* 63: 905–8.
40. Roth, T. A. 1975. "The Surface and Grain Boundary Energies of Iron, Cobalt and Nickel." Materials Science Engineering 18: 183–92.
41. Shibuta, Y., S. Takamoto, and T. Suzuki. 2009. "Dependence of the Grain Boundary Energy on the Alloy Composition in the bcc Iron–Chromium Alloy: A Molecular Dynamics Study." *Computational Materals Science* 44: 1025–9. doi: http://dx.doi.org/10.1016/j.commatsci.2008.07.013.

42. Ochs, T., and C. Elsässer. 2000. "Symmetrical Tilt Grain Boundaries in bcc Transition Metals: Comparison of Semiempirical with Ab-Initio Total-Energy Calculations." *Philosophical Magazine A* 80: 2405–23. doi: http://dx.doi.org/10.1080/01418610008216481.
43. Rohrer, G. S., E. A. Holm, A. D. Rollett, S. M. Foiles, Jia Li, and D. L. Olmsted. 2010. "Comparing Calculated and Measured Grain Boundary Energies in Nickel." *Acta Materialia* 58: 5063–9. doi: http://dx.doi.org/10.1016/j.actamat.2010.05.042.
44. Svoboda, J., F. D. Fischer, P. Fratzl, and E. Kozeschnik. 2004. "Modelling of Kinetics in Multi-Component Multi-Phase Systems with Spherical Precipitates I. – Theory" *Materials Science Engineering A* 385: 166–74. doi: http://dx.doi.org/10.1016/S0921-5093(04)00820-2.
45. Hobstetter, J. N. 1949. "Stable Transformation Nuclei in Solids." *Transactions of AIME* 180: 121–30.
46. Kozeschnik, E. 2008. "Thermodynamic Prediction of the Equilibrium Chemical Composition of Critical Nuclei: Bcc Cu Precipitation in α-Fe." *Scripta Materialia* 59: 1018–21. doi: http://dx.doi.org/10.1016/j.scriptamat.2008.07.008.
47. Kozeschnik, E., J. Svoboda, and F. D. Fischer. 2005. "On the Role of Chemical Composition in Multi-Component Nucleation." in Proceedings of the International Conference Solid-Solid Phase Transformations in Inorganic Materials, PTM 2005, Phoenix, AZ, USA, pp. 301–10.
48. Zener, C. M. 1949. "Theory of Growth of Spherical Precipitates from Solid Solution." *Journal of Applied Physics* 20: 950–3. doi: http://dx.doi.org/10.1063/1.1698258.
49. Rivera-Diaz-del-Castillo, P. E. J., and H. K. D. H. Bhadeshia. 2001. "Theory for Growth of Spherical Precipitates with Capillarity Effects." *Journal of Materials Science and Technology* 17: 30–2. doi: http://dx.doi.org/10.1179/026708301101509089.
50. Coates, D. E. 1972. "Diffusion Controlled Precipitate Growth in Ternary Systems I." *Metallurgical and Materials Transactions B* 3: 1203–12. doi: 10.1007/BF02642453.
51. Coates, D. E. 1973. "Diffusion Controlled Precipitate Growth in Ternary Systems II." *Metallurgical and Materials Transactions* 4: 1077–86. doi: http://dx.doi.org/10.1007/BF02645611.
52. M. Hillert, "Paraequilibrium", internal report, Swedish Institute for Metal Research, Stockholm, 1953.
53. M. Hillert, J. Ågren, "On the definitions of paraequilibrium and orthoequilibrium", Scripta Mater. 50 (2004) 697–699.
54. Purdy, G. R., D. H. Weichert, and J. S. Kirkaldy. 1964. "Growth of Proeutectoid Ferrite in Ternary Iron-Carbon-Manganese Austenite." *Transactions of the Metallurgical Society of AIME* 230: 1025–34.
55. J.R. Bradley and H.I. Aaronson, "Growth-kinetics of grain-boundary ferrite allotriomorphs in Fe-C-X alloys", Metal. Trans. 12A (1981) 1729–1741
56. Schneider, A., and G. Inden. 2005. "Simulation of the Kinetics of Precipitation Reactions in Ferritic Steels." *Acta Materialia* 53: 519–31. doi: http://dx.doi.org/10.1016/j.actamat.2004.10.008.
57. Chen, Qing, J. Stepsson, and J. Ågren. 2007. "Analytical Treatment of Diffusion During Precipitate Growth in Multicomponent Systems." *Acta Materialia* 56: 1890–6. doi: http://dx.doi.org/10.1016/j.actamat.2007.12.037.
58. Thornton, K., J. Ågren, and P. W. Vorhees. 2003. "Modelling the Evolution of Phase Boundaries in Solids at the Meso- and Nano-Scales." *Acta Materialia* 51: 5675–710. doi: http://dx.doi.org/10.1016/j.actamat.2003.08.008.
59. Christian, J. W. 2002. *The Theory of Transformations in Metals and Alloys, Part I and II*. Oxford: Pergamon Elsevier Science.
60. Sietsma, J., and S. van der Zwaag. 2004. "A Concise Model for Mixed-Mode Phase Transformations in the Solid State." *Acta Materialia* 52: 4143–52. doi: http://dx.doi.org/10.1016/j.actamat.2004.05.027.
61. Svoboda, J., and I. Turek. 1991. "On Diffusion-Controlled Evolution of Closed Solid-State Thermodynamic Systems at Constant Temperature and Pressure." *Philosophical Magazine B* 64: 749–59. doi: http://dx.doi.org/10.1080/13642819108207635.
62. Leech, J. W. 1960. *Classical Mechanics*. New York: Wiley.

63. Onsager, L. 1931. "Reciprocal Relations in Irreversible Processes. I." *Physical Review* 37: 405–26. doi: http://dx.doi.org/10.1103/PhysRev.37.405.
64. Svoboda, J., I. Turek, and F. D. Fischer. 2005. "Application of the Thermodynamic Extremal Principle to Modeling of Thermodynamic Processes in Material Sciences." *Philosophical Magazine* 85: 3699–707. doi: http://dx.doi.org/10.1080/14786430500267760.
65. Svoboda, J., I. Turek, and V. Sklenicka. 1990. "Unified Thermodynamic Treatment of Cavity Nucleation and Growth in High-Temperature Creep." *Acta Metallurgica et Materialia* 38: 573–8. doi: http://dx.doi.org/10.1016/0956-7151(90)90211-X.
66. Gill, S. P. A., M. G. Cornforth, and A. C. F. Cocks. 2001. "Modelling Microstructure Evolution in Engineering Materials." *International Journal of Plasticity* 17: 669–90. doi: http://dx.doi.org/10.1016/S0749-6419(00)00066-8.
67. Svoboda, J., and P. Lukas. 2000. "Creep Deformation Modelling of Superalloy Single Crystals." *Acta Materialia* 48: 2519–28. doi: http://dx.doi.org/10.1016/S1359-6454(00)00078-1.
68. Fischer, F. D., J. Svoboda, and P. Fratzl. 2003. "A Thermodynamic Approach to Grain Growth and Coarsening." *Philosophical Magazine A* 83: 1075–93. doi: http://dx.doi.org/10.1080/0141861031000068966.
69. Svoboda, J., E. Gamsjäger, F. D. Fischer, and P. Fratzl. 2004. "Application of the Thermodynamic Extremal Principle to the Diffusional Phase Transformations." *Acta Materialia* 52: 959–67. doi: http://dx.doi.org/10.1016/j.actamat.2003.10.030.
70. Svoboda, J., E. Gamsjäger, F. D. Fischer, Y. Liu, and E. Kozeschnik. 2011. "Diffusion Processes in a Migrating Interface: The Thick-Interface Model." *Acta Materialia* 59: 4775–86. doi: http://dx.doi.org/10.1016/j.actamat.2011.04.020.
71. Kozeschnik, E., J. Svoboda, and F. D. Fischer. 2004. "Modified Evolution Equations for the Precipitation Kinetics of Complex Phases in Multi-Component Systems." *CALPHAD* 28: 379–82. doi: http://dx.doi.org/10.1016/j.calphad.2004.11.003.
72. Ostwald, W. 1896. *Lehrbuch der Allgemeinen Chemie*, volume 2, part 1. Leipzig, Germany.
73. Ostwald, W. 1900. "On the Assumed Isomerism of Red and Yellow Mercury Oxide and the Surface-Tension of Solid Bodies." *Zeitschrift für Physikalische Chemie* 37: 495–503.
74. Wagner, C. 1961. "Theorie der Alterung von Niederschlägen durch Umlösen (Ostwald-Reifung)." *Zeitschrift für Elektrochemie* 65: 581–91.
75. Lifshitz, I. M., and V. V. Slyozov. 1961. "The Kinetics of Precipitation from Supersaturated Solid Solutions." *Journal of Physics and Chemistry of Solids* 19: 35–50. doi: http://dx.doi.org/10.1016/0022-3697(61)90054-3.
76. Voorhees, P. W. 1985. "The Theory of Ostwald Ripening." *Journal of Statistical Physics* 38: 231–52. doi: http://dx.doi.org/10.1007/BF01017860.
77. Ardell, A. J. 1972. "Effect of Volume Fraction on Particle Coarsening – Theoretical Considerations." *Acta Metallurgica* 20: 61–71. doi: http://dx.doi.org/10.1016/0001-6160(72)90114-9.
78. Brailsford, A. D., and P. Wynblatt. 1979. "The Dependence of Ostwald Ripening Kinetics on Particle Volume Fraction." *Acta Metallurgica* 27: 489–97. doi: http://dx.doi.org/10.1016/0001-6160(79)90041-5.
79. Davies, C. K. L., P. Nash, and R. N. Stevens. 1980. "The Effect of Volume Fraction of Precipitate on Ostwald Ripening." *Acta Metallurgica* 28: 179–89. doi: http://dx.doi.org/10.1016/0001-6160(80)90067-X.
80. Tsumaraya, K., and Y. Miyata. 1983. "Coarsening Models Incorporating Both Diffusion Geometry and Volume Fraction of Particles." *Acta Metallurgica* 31: 437–52. doi: http://dx.doi.org/10.1016/0001-6160(83)90221-3.
81. Marqusee, J. A., and J. Ross. 1984. "Theory of Ostwald Ripening: Competitive Growth and Its Dependence on Volume Fraction." *Journal of Chemical Physics* 80: 536–43. doi: http://dx.doi.org/10.1063/1.446427.

82. Tokuyama, M., and K. Kawasaki. 1984. "Statistical-Mechanical Theory of Coarsening of Spherical Droplets." *Physica A* 123: 386–411. doi: http://dx.doi.org/10.1016/0378-4371(84)90162-6.
83. Voorhees, P. W., and M. E. Glicksman. 1984. "Solution to the Multi-Particle Diffusion Problem with Applications to Ostwald Ripening – I. Theory." *Acta Metallurgica* 32: 2001–11. doi: http://dx.doi.org/10.1016/0001-6160(84)90180-9.
84. Voorhees, P. W., and M. E. Glicksman. 1984. "Solution to the Multi-Particle Diffusion Problem with Applications to Ostwald Ripening – II. Computer Simulations." *Acta Metallurgica* 32: 2013–30. doi: http://dx.doi.org/10.1016/0001-6160(84)90181-0.
85. Enomoto, Y., M. Tokuyama, and K. Kawasaki. 1986. "Finite Volume Fraction Effects on Ostwald Ripening." *Acta Metallurgica* 34: 2119–28. doi: http://dx.doi.org/10.1016/0001-6160(86)90157-4.
86. Marder, M. 1987. "Correlations and Ostwald Ripening." *Physical Reviews A* 36: 858–74. doi: http://dx.doi.org/10.1103/PhysRevA.36.858.
87. Marsh, S. P., and M. E. Glicksman. 1996. "Kinetics of Phase Coarsening in Dense Systems." *Acta Materialia* 44: 3761–71. doi: http://dx.doi.org/10.1016/1359-6454(95)00451-3.
88. Calderon, H. A., P. W. Voorhees, J. L. Murray, and G. Kostorz. 1994. "Ostwald Ripening in Concentrated Alloys." *Acta Metallurgica et Materialia* 42: 991–1000. doi: http://dx.doi.org/10.1016/0956-7151(94)90293-3.
89. Wang, K. G., M. E. Glicksman, and K. Rajan. 2005. "Length Scales in Phase Coarsening: Theory, Simulation, and Experiment." *Computational Materials Science* 34: 235–53. doi: http://dx.doi.org/10.1016/j.commatsci.2004.11.005.
90. Umantsev, A., and G. B. Olson. 1993. "Ostwald Ripening in Multicomponent Alloys." *Scripta Metallurgica et Materialia* 29: 1135–40. doi: http://dx.doi.org/10.1016/0956-716X(93)90191-T.
91. Morral, J. E., and G. R. Purdy. 1994. "Particle Coarsening in Binary and Multicomponent Alloys." *Scripta Metallurgica et Materialia* 30: 905–8. doi: http://dx.doi.org/10.1016/0956-716X(94)90413-8.
92. Ågren, J., M. T. Clavaguera-Mora, J. Golczewski, G. Inden, H. Kumar, and Ch. Sigli. 2000. "Applications of Computational Thermodynamics." *CALPHAD* 24: 41–54. doi: http://dx.doi.org/10.1016/S0364-5916(00)00014-6.
93. Kuehmann, C. J., and P. W. Voorhees. 1996. "Ostwald Ripening in Ternary Alloys." *Metallurgical and Materials Transactions* 27A: 937–42. doi: http://dx.doi.org/10.1007/BF02649761.
94. Hoyt, J. J. 1999. "Coarsening in Multiphase Multicomponent Systems – I. The Mean Field Limit." *Acta Materialia* 47: 345–51. doi: http://dx.doi.org/10.1016/S1359-6454(98)00316-4.
95. Rivera-Diaz-del-Castillo, P. E. J. 2002. "Precipitate Coarsening in Multicomponent Systems." *Scripta Materialia* 47: 113–17. doi: http://dx.doi.org/10.1016/S1359-6462(02)00103-3.
96. Ardell, A. J. 1995. "Interfacial Free-Energies and Solute Diffusivities from Data on Ostwald Ripening." *Interface Science* 3: 119–25. doi: http://dx.doi.org/10.1007/BF00207013.
97. Becker, R. 1938. "Die Keimbildung bei der Ausscheidung in Metallischen Mischkristallen." *Annalen der Physik* 32: 128–40. doi: http://dx.doi.org/10.1002/andp.19384240115.
98. Turnbull, D. 1955. *Impurities and Imperfections* (pp. 121–44). American Society of Metals.
99. Sonderegger, B., and E. Kozeschnik. 2009. "Generalized Nearest Neighbor – Broken Bond Analysis of Randomly Oriented Coherent Interfaces in Multi-Component fcc and bcc Structures." *Metallurgical and Materials Transactions A* 40: 499–510. doi: 10.1007/s11661-008-9752-6.
100. Borchers, Ch., and R. Bormann. 2005. "Determination of Low-Temperature Interfacial Energies from a Pair Interaction Model." *Acta Materialia* 53: 3695–701. doi: http://dx.doi.org/10.1016/j.actamat.2005.04.022.
101. Li, X., N. Saunders, and A. P. Miodownik. 2002. "The Coarsening Kinetics of γ′ Precipitates in Nickel-Based Alloys." *Metallurgical and Materials Transactions A* 33: 3367–73. doi: 10.1007/s11661-002-0325-9.
102. Lennard-Jones, J. E. 1931. "Cohesion." *Proceedings of the Physical Society* 43: 461–82. doi: http://dx.doi.org/10.1088/0959-5309/43/5/301.

103. Ardell, A. J. 1968. "An Application of Theory of Particle Coarsening – Gamma' Precipitate in Ni-Al Alloys." *Acta Metallurgica* 16: 511–16. doi: http://dx.doi.org/10.1016/0001-6160(68)90125-9.
104. Xiao, S. Q., and P. Haasen. 1991. "HREM Investigation of Homogeneous Decomposition in a Ni-12 at. %Al Alloy." *Acta Metallurgica et Materialia* 39: 651–9. doi: http://dx.doi.org/10.1016/0956-7151(91)90133-L.
105. Wendt, H., and P. Haasen. 1983. "Nucleation and Growth of γ'-Precipitates in Ni-14 at. % Al." *Acta Metallurgica et Materialia* 31: 1649–59. doi: http://dx.doi.org/10.1016/0001-6160(83)90163-3.
106. Hirata, T., and D. H. Kirkwood. 1977. "Prediction and Measurement of Precipitate Number Densities in a Nickel-6.05 wt-Percent-Aluminium Alloy." *Acta Metallurgica* 25: 1425–34. doi: http://dx.doi.org/10.1016/0001-6160(77)90073-6.
107. Price, D. L., and B. R. Cooper. 1996. "Full-Potential LMTO Calculation of Ni/Ni_3Al Interface Energies." *Materials Research Society Symposium Proceedings* 408: 463–8. doi: http://dx.doi.org/10.1557/PROC-408-463.
108. Staron, P., and R. Kampmann. 2000. "Early-Stage Secomposition Kinetics in Ni-Al Alloys – I. Small and Wide-Angle Neutron Scattering Investigation on Ni-13 at.% Al and Clusterdynamic Modelling." *Acta Materialia* 48: 701–12. doi: http://dx.doi.org/10.1016/S1359-6454(99)00398-5.
109. Mishin, Y. 2004. "Atomistic Modeling of the Gamma and Gamma'-Phases of the Ni-Al System." *Acta Materialia* 52: 1451–67. doi: http://dx.doi.org/10.1016/j.actamat.2003.11.026.
110. Farkas, D., M. F. de Campos, R. M. de Souza, and H. Goldenstein. 1994. "Atomistic Structure of the Coherent Ni/Ni_3Al Interface." *Scripta Metallurgica et Materialia* 30: 367–71. doi: http://dx.doi.org/10.1016/0956-716X(94)90390-5.
111. Chen, J. K., D. Farkas, and W. T. Reynolds, Jr. 1997. "Atomistic Simulation of an fcc/bcc Interface in Ni-Cr Alloys." *Acta Materialia* 45: 4415–21. doi: http://dx.doi.org/10.1016/S1359-6454(97)00159-6.
112. Monzen, R., K. Takada, and K. Matsuda 2003. "Coarsening Kinetics of Cu Particles in an Fe-1.5% Cu Alloy." *Zeitschrift für Metallkunde* 94: 1241–6.
113. Servi, I. S., and D. Turnbull. 1966. "Thermodynamics and Kinetics of Precipitation in Copper-Cobalt System." *Acta Metallirgica* 14: 161–9. doi: http://dx.doi.org/10.1016/0001-6160(66)90297-5.
114. Shiflet, G. J., Y. W. Lee, H. I. Aaronson, and K. C. Russell. 1981. "A Reassessment of the Comparison of Classical Homogeneous Nucleation Theory and Experiment in Cu-Co Alloys." *Scripta Metallurgica* 15: 719–22. doi: http://dx.doi.org/10.1016/0036-9748(81)90006-5.
115. Robson, J. D. 2004. "Modelling the Evolution of Particle Size Distribution During Nucleation, Growth and Coarsening." *Materials Science and Technology* 20: 441–8. doi: http://dx.doi.org/10.1179/026708304225016725.
116. Stowell, M. J. 2002. "Precipitate Nucleation: Does Capillarity Theory Work?" *Materials Science and Technology* 18: 139–44. doi: http://dx.doi.org/10.1179/026708301125000348.
117. Oriani, R. A. 1964. "Ostwald Ripening of Precipitates in Solid Matrices." *Acta Metallurgica* 12: 1399–409. doi: http://dx.doi.org/10.1016/0001-6160(64)90128-2.
118. Yang, S. C., and P. Nash. 1988. "Application of Thermodynamic Factors in Analysis of Particle Coarsening Kinetics in Fe-Cu and Co-Cu." *Materials Science and Technology* 4: 860–6. doi: http://dx.doi.org/10.1179/026708388790221223.
119. Calderon, H., and M. E. Fine. 1984. "Coarsening Kinetics of Coherent NiAl-Type Precipitates in Fe-Ni-Al and Fe-Ni-Al-Mo Alloys." *Materials Science Engineering* 63: 197–208. doi: http://dx.doi.org/10.1016/0025-5416(84)90121-6.
120. White. R. J., and S. B. Fisher. 1978. "Precipitate and Growth Kinetics of Gamma' in NIMONIC PE16." *Materials Science Engineering* 33: 149–57. doi: http://dx.doi.org/10.1016/0025-5416(78)90167-2.
121. Ricks, R. A., A. J. Porter, and R. C. Ecob. 1983. "The Growth of Gamma'-Precipitates in Nickel-Base Super-Alloys." *Acta Metallurgica* 31: 45–53. doi: http://dx.doi.org/10.1016/0001-6160(83)90062-7.
122. Wang, Y., Z.-K. Liu, L.-Q. Chen, and C. Wolverton. 2007. "First-Principles Calculations of β''-Mg_5Si_6/α-Al Interfaces." *Acta Materialia* 55: 5934–47. doi: http://dx.doi.org/10.1016/j.actamat.2007.06.045.

123. Sonderegger, B., and E. Kozeschnik. 2009. "Size Dependence of the Interfacial Energy in the Generalized Nearest-Neighbor Broken-Bond Approach." *Scripta Materialia* 60: 635–8. doi: http://dx.doi.org/10.1016/j.scriptamat.2008.12.025.
124. Tolman, R. C. 1949. "The Effect of Droplet Size on Surface Tension." *Journal of Chemical Physics* 17 (3): 333-337. doi: http://dx.doi.org/10.1063/1.1747247.
125. Kashchiev, D. 2000. *Nucleation: Basic Theory with Applications* (pp. 79–82). Burlington, MA: Butterworth-Heinemann.
126. Schmelzer, J., I. Gutzow, and J. Schmelzer, Jr. 1996. "Curvature-Dependent Surface Tension and Nucleation Theory." *Journal of Colloid and Interface Science* 178 (2): 657–65. doi: http://dx.doi.org/10.1006/jcis.1996.0163.
127. Rasmussen, D. H. 1982. "Energetics of Homogeneous Nucleation—Approach to a Physical Spinodal." *Journal of Crystal Growth* 56 (1): 45–55. doi: http://dx.doi.org/10.1016/0022-0248(82)90011-2.
128. W. Vogelsberger and G. Marx, „Curvature dependence of surface-tension of small drops", Z. Phys. Chem. 257 (1976) 580–586.
129. Cahn, J. W., and J. E. Hilliard. 1958. "Free Energy of a Nonuniform System. 1. Interfacial Free Energy." *Journal of Chemical Physics* 28: 258–67. doi: http://dx.doi.org/10.1063/1.1744102.
130. Cahn, J. W. 1959. "Free Energy of a Nonuniform System. 2. Thermodynamic Basis." *Journal of Chemical Physics* 30: 1121–4. doi: http://dx.doi.org/10.1063/1.1730145.
131. Cahn, J. W., and J. E. Hilliard. 1959. "Free Energy of a Nonuniform System. 1. Interfacial Free Energy." *Journal of Chemical Physics* 31: 688–99. doi: http://dx.doi.org/10.1063/1.1730447.
132. Steinbach, I. 2009. "Phase Field Models in Materials Science." *Modelling and Simulation in Materials Science and Technology* 17: 73001. doi: 10.1088/0965-0393/17/7/073001.
133. Sonderegger, B., and E. Kozeschnik. 2010. "Interfacial Energy of Diffuse Phase Boundaries in the Generalized Broken-Bond Approach." *Metallurgical and Materials Transactions A* 41: 3262–9. doi: 10.1007/s11661-010-0370-8.
134. Kikuchi, R. 1951. "A Theory of Cooperative Phenomena." *Physical Review* 81: 988–1003. doi: http://dx.doi.org/10.1103/PhysRev.81.988.
135. Kikuchi, R., and K. Masuda-Jindo. 2002. "Cluster Variation Method in the Computational Materials Science." *CALPHAD* 26: 33–54. doi: http://dx.doi.org/10.1016/S0364-5916(02)00023-8.
136. Oates, W. A., F. Zhang, S. L. Chen, and Y. A. Chang. 1999. "Improved Cluster-Site Approximation for the Entropy of Mixing in Multicomponent Solid Solutions." *Physical Review B* 59: 11221–5.
137. Lee, Y. W., and H. I. Aaronson. 1980. "Anistropy of Coherent Interphase Boundary Energy." *Acta Metallurgica* 28 (4): 539–48. doi: http://dx.doi.org/10.1016/0001-6160(80)90143-1.
138. Cao, W., J. Zhu, F. Zhang, W. A. Oates, M. Asta, and Y. A. Chang. 2006. "Application of the Cluster/Site Approximation to the Calculation of Coherent Interphase Boundary Energies." *Acta Materialia* 54 (2): 377–83. doi: http://dx.doi.org/10.1016/j.actamat.2005.09.010.
139. Asta, M. D. 1996. *Theory and Application of the Cluster and Path Probability Methods* (pp. 247–54). New York: Plenum.
140. X. Li, N. Saunders, and A.P. Miodownik, "The coarsening kinetics of gamma' particles in nickel-based alloys" Metall. Mater. Trans. 33A (2002) 3367–3373.
141. Yin, Y. F., and R. G. Faulkner. 2003. "Simulations of Precipitation in Ferritic Steels." *Materials Sciences Technology* 19: 91–8. doi: http://dx.doi.org/10.1179/026708303225008518.
142. Xu, K., and B. G. Thomas. 2012. "Particle-Size-Grouping Model of Precipitation Kinetics in Microalloyed Steels." Metallurgical and. Materials Transactions 43A: 1079–96. doi: http://dx.doi.org/10.1007/s11661-011-0938-y.
143. Kolmogorov, A. N. 1937. "On Statistical Theory of metal Crystallization." In *Izvestia Academy of Science, USSR, Ser. Math* 3: 355–60.
144. Johnson, W. A., and R. F. Mehl. 1939. "Reaction Kinetics in Process of Nucleation and Growth." *Transactions of the American Institute of Mining and Metallurgical Engineering* 135: 416–58.

145. Avrami, M. 1939. "Kinetics of Phase Change. I General Theory." *Journal of Chemical Physics* 7: 1103–12. doi: http://dx.doi.org/10.1063/1.1750380.
146. Avrami, M. 1940. "Kinetics of Phase Change. II Transformation-Time Relations for Random Distribution of Nuclei." *Journal of Chemical Physics* 8: 212–24. doi: http://dx.doi.org/10.1063/1.1750631.
147. Avrami, M. 1941. "Granulation, Phase Change, and Microstructure Kinetics of Phase Change. III." *Journal of Chemical Physics* 9: 177–84. doi: http://dx.doi.org/10.1063/1.1750872.
148. Fanfoni, M., and M. Tomellini. 1998. "The Johnson-Mehl-Avrami-Kolmogorov Model: A Brief Review." *Il Nuovo cimento* 20D: 1171–82. doi: http://dx.doi.org/10.1007/BF03185527.
149. Robson, J. D., and H. K. D. H. Bhadeshia. 1997. "Modeling Precipitation Sequences in Power Plant Steels Part 1 – Kinetic Theory." *Materials Science Technology* 13: 631–9. doi: http://dx.doi.org/10.1179/026708397790285944.
150. Robson, J. D., and H. K. D. H. Bhadeshia. 1996. "Kinetics of Precipitation in Power Plant Steels." *CALPHAD* 20: 447–60. doi: http://dx.doi.org/10.1016/S0364-5916(97)00007-2.
151. Langer, J. S., and A. J. Schwartz. 1980. "Kinetics of Nucleation in Near-Critical Fluids." *Physical Review A* 21: 948–58. doi: http://dx.doi.org/10.1103/PhysRevA.21.948.
152. Binder, K., and D. Stauffer. 1976. "Statistical Theory of Nucleation, Condensation and Coagulation." *Advances in Physics* 25: 343–96. doi: http://dx.doi.org/10.1080/00018737600101402.
153. Kampmann, R., and R. Wagner. 1984. "Kinetics of Precipitation in Metastable Binary Alloys–Theory and Application to Cu-1.9at%Ti and Ni-14at%Al". In *Decomposition of Alloys: The Early Stages* edited by P. Haasen et al. (pp. 91–103). Oxford: Pergamon Press.
154. Radis, R., M. Schaffer, M. Albu, G. Kothleitner, P. Pölt, and E. Kozeschnik. 2009. "Multimodal Size Distributions of γ′ Precipitates During Continuous Cooling of UDIMET 720 Li." *Acta Materialia* 57: 5739–47. doi: http://dx.doi.org/10.1016/j.actamat.2009.08.002.
155. Robson, J. D. 2004. "Modelling the Overlap of Nucleation, Growth and Coarsening During Precipitation." *Acta Materialia* 52: 4669–76. doi: http://dx.doi.org/10.1016/j.actamat.2004.06.024.
156. Lange, W. F., M. Enomoto, and H. I. Aaronson. 1989. "Precipitate Nucleation Kinetics at Grain-Boundaries." *International Materials Review* 34: 125–50.
157. Aaron, H. B., and H. I. Aaronson. 1968. "Growth of Grain Boundary Precipitates in Al-4 Percent Cu by Interfacial Diffusion." *Acta Metallurgica* 16: 789–98. doi: http://dx.doi.org/10.1016/0001-6160(68)90097-7.
158. Brailsford, A. D., and H. B. Aaron. 1969. "Growth of Grain-Boundary Precipitates." *Journal of Applied Physics* 40: 1702–10. doi: http://dx.doi.org/10.1063/1.1657835.
159. Kozeschnik, E., J. Svoboda, R. Radis, and F. D. Fischer. 2010. "Mean-Field Model for the Growth and Coarsening of Stoichiometric Precipitates at Grain Boundaries." Modelling and Simulation in Materials Sciences and Engineering 18: 015011 ff. doi: http://dx.doi.org/10.1088/0965-0393/18/1/015011.
160. Wilson, F. G., and T. Gladman. 1988. "Aluminium Nitride in Steel." *International Materials Review* 33: 221–86. doi: http://dx.doi.org/10.1179/095066088790324102.
161. Leslie, W. C., R. L. Rickett, C. L. Dotson, and C. S. Walton. 1954. "Solution and Precipitation of Aluminium Nitride in Relation to the Structure of Low Carbon Steel." *Transactions of the ASM* 46: 1470–99.
162. Radis, R., and E. Kozeschnik. 2010. "Kinetics of AlN Precipitation in Microalloyed Steel." *Modelling and Simulation in Materials Sciences Engineering* 18: 015011 ff. doi: http://dx.doi.org/10.1088/0965-0393/18/5/055003.
163. Ardell, A. J. 1972. "On the Coarsening of Grain Boundary Precipitates." *Acta Metallurgica* 20: 601–9. doi: http://dx.doi.org/10.1016/0001-6160(72)90015-6.
164. Vengrenovitch, R. D. 1982. "On the Ostwald Ripening Theory." *Acta Metallurgica* 30: 1079–86. doi: http://dx.doi.org/10.1016/0001-6160(82)90004-9.
165. Hoyt, J. J. 1991. "On the Coarsening of Precipitates Located on Grain Boundaries and Dislocations." *Acta Metallurigca et Materialia* 39: 2091–8. doi: http://dx.doi.org/10.1016/0956-7151(91)90179-5.

166. Kozeschnik, E., J. Svoboda, and F. D. Fischer. 2006. "Shape Factors in Modeling of Precipitation." *Materials Sciences Engineering A* 441: 68–72. doi: http://dx.doi.org/10.1016/j.msea.2006.08.088.
167. Warczok, P., J. Zenisek, and E. Kozeschnik. 2012. "Atomistic and Continuums Modeling of Cluster Migration and Coagulation in Precipitation Reactions." *Computational Materials Science* 60: 59–65. doi: http://dx.doi.org/10.1016/j.commatsci.2012.02.033.
168. Binder, K., and D. Stauffer. 1974. "Theory for the Slowing Down of the Relaxation and Spinodal Decomposition of Binary Mixtures." *Physical Review Letters* 33: 1006–9. doi: http://dx.doi.org/10.1103/PhysRevLett.33.1006.
169. Binder, K. 1977. "Theory for the Dynamics of 'Clusters' II. Critical Diffusion in Bindary Systems and the Kinetics of Phase Separation." *Physical Reviews B* 15: 4425–47. doi: http://dx.doi.org/10.1103/PhysRevB.15.4425.
170. Fratzl, P., and O. Penrose. 1997. "Competing Mechanisms for Precipitate Coarsening in Phase Separation with Vacancy Dynamics." *Physical Reviews B* 55: R6101–4. doi: http://dx.doi.org/10.1103/PhysRevB.55.R6101.
171. Soisson, F., and Chu-Chun Fu. 2007. "Cu-Precipitation Kinetics in a-Fe from Atomistic Simulations: Vacancy Trapping Effects and Cu-Cluster Mobility." *Physical Reviews B* 76: 214102. doi: http://dx.doi.org/10.1103/PhysRevB.76.214102.
172. von Smoluchowski, M. 1917. "Versuch Einer Mathematischen Theorie der Koagulationskinetik Kolloider Lösungen." *Zeitschrift Physikalische Chemie* 92: 129–68.
173. Lindborg, U., and K. Torssell. 1968. "A Collision Model for the Growth and Separation of Deoxidation Products." *Transactions of the Metallurgical Society of AIME* 242: 94–102.
174. Saffman, P. G., and J. S. Turner. 1956. "On the Collision of Drops in Turbulent Clouds." *Journal of Fluid Mechanics* 1: 16-30. doi: http://dx.doi.org/10.1017/S0022112056000020.
175. Friedlander, S. K., and C. S. Wang. 1966. "The Self-Preserving Particle Size Distribution for Coagulation by Brownian Motion." *Journal of Colloid and Interface Science* 22: 126-132. doi: http://dx.doi.org/10.1016/0021-9797(66)90073-7.
176. Levic, V. 1962. *Physicochemical Hydrodynamics*, 2nd edition. Prentice-Hall
177. Warczok, P., Y. Shan, M. Schober, H. Leitner, and E. Kozeschnik. 2011. "Analysis of Clustering Characteristics During Early Stages of Cu Precipitation in bcc-Fe." *Solid State Phenomena* 172–174: 309–14. doi: http://dx.doi.org/10.4028/www.scientific.net/SSP.172-174.309.
178. Bongartz, K., W. J. Quadakkers, R. Schulten, and H. Nickel. 1989. "A Mathematical Model Describing Carburization in Multielement Alloy Systems." *Metallurgical Transactions* 20A: 1021–8. doi: http://dx.doi.org/10.1007/BF02650138.
179. Buchmayr, B., and J. S. Kirkaldy. 1990. "Preferential Concentration and Depletion of Carbides after Annealing of Mismatched Chromium Alloy Weldments." In *Fundamentals and Application of Ternary Diffusion*, edited by G. R. Purdy, 164–72. Pergamon Press.
180. Engström, A., L. Höglund, and J. Ågren. 1994. "Computer Simulation of Diffusion in Multiphase Systems." *Metallurgical and Materials Transactions* 25A: 303–8. doi: http://dx.doi.org/10.1007/BF02652288.
181. Helander, T., and J. Ågren. 1997. "Computer Simulation of Multicomponent Diffusion in Joints of Dissimilar Steels." *Metallurgical and Materials Transactions* 28A: 1127–34. doi: http://dx.doi.org/10.1007/s11661-997-0133-3.
182. Kozeschnik, E., P. Pölt, S. Brett, and B. Buchmayr. 2002. "Dissimilar 2.25Cr-9Cr and 2Cr-0.5CrMoV Steel Welds, I. Characterisation of the Weld Zone and Numerical Simulation." *Science and Technology of Welding & Joining* 7: 63–8. doi: http://dx.doi.org/10.1179/136217102225003005.
183. Kozeschnik, E., P. Warbichler, I. Letofsky-Papst, S. Brett, and B. Buchmayr. 2002. "Dissimilar 2.25Cr-9Cr and 2Cr-0.5CrMoV Steel Welds, II. Identification of Precipitates." *Science and Technology of Welding & Joining* 7: 69–76. doi: http://dx.doi.org/10.1179/136217102225003014.
184. Shewmon, P. G. 1986. *Diffusion in Solids*. Warrendale, PA: The Minerals, Metals & Materials Society.

185. Kirkaldy, J. S., and D. J. Young. 1987. *Diffusion in the Condensed State.* London: The Institute of Metals.
186. Crank, J. 1975. *Mathematics of Diffusion.* Oxford University Press.
187. Philibert, J. 1986. *Diffusion et Transport de Matire dans les Solides.* Editions de Physique.
188. Kirkendall, E., L. Thomassen, and C. Upthegrove. 1939. *Rates of Diffusion of Copper and Zinc in Alpha Brass,* volume 133, pp. 186–203. Transactions of the AIME.
189. Kirkendall, E. O. 1942. *Diffusion of Zinc in Alpha Brass*, volume 147, pp. 104–10. Transactions of the AIME.
190. Smigelskas, A. D., and E. O. Kirkendall. 1947. *Zinc Diffusion in Alpha Brass*, volume 171, pp. 130–42. Transactions of the AIME.
191. Nakajima, H. 1949. "The Discovery and Acceptance of the Kirkendall Effect: The Result of a Short Research Career." *JOM* 49: 15–19. doi: http://dx.doi.org/10.1007/BF02914706.
192. Matsumoto, M., and T. Nishimura. 1998. "Mersenne Twister: A 623-Dimensionally Equidistributed Uniform Pseudo-Random Number Generator." *ACM Transactions on Modeling and Computer Simulation* 8: 3–30. doi: http://dx.doi.org/10.1145/272991.272995.
193. Gupta, Devendra, ed. 2005. *Diffusion Processes in Advanced Technological Materials.* Springer Verlag.
194. Einstein, A. 1905. "Über die von der Molekularkinetischen Theorie der Wärme Geforderte Bewegung von in Ruhenden Flüssigkeiten Suspendierten Teilchen." *Annelen der Physik* 322 (8): 549–60. doi: http://dx.doi.org/10.1002/andp.19053220806.
195. von Smoluchowski, M. 1906. "Zur Kinetischen Theorie der BrownscheFn Molekularbewegung und der Suspensionen." *Annalen der Physik* 326 (14): 756–80. doi: http://dx.doi.org/10.1002/andp.19063261405.
196. Graham, Th. 1850. "Über die Bewegung der Gase." Ann. Chem. Pharm. (Liebigs Ann. Chem.) 76: 138–59. doi: http://dx.doi.org/10.1002/jlac.18500760206.
197. Fick, A. E. 1855. "Über Diffusion." Poggendorf's Annalen der Physik 94: 59. doi: http://dx.doi.org/10.1002/andp.18551700105.
198. Cohen, T. H., and M. E. Glicksman. 1995. "Multicomponent Diffusion: Implementation of the Square-Root Diffusivity Method via the Profiler Computer Program." Modelling and Simulation in Materials Sciences and Engineering 3: 585–96. doi: http://dx.doi.org/10.1088/0965-0393/3/5/001.
199. Darken, L. S. 1948. *Diffusion, Mobility and Their Interrelation Through Free Energy in Binary Metallic Systems*, volume 175, pp. 184–201. Transactions of the AIME.
200. Jönsson, B. 1992. "On Ferromagnetic Ordering and Lattice Diffusion – A Simple Model." *Zeitschrift für Metallkunde* 83: 349–55.
201. Helander, T., and J. Ågren. 1999. "A Phenomenological Treatment of Diffusion in Al-Fe and Al-Ni Alloys Having B2-b.c.c Ordered Structure." *Acta Materialia* 47: 1141–52. doi: http://dx.doi.org/10.1016/S1359-6454(99)00010-5.
202. Landolt-Börnstein materials database http://www.springermaterials.com/navigation, access 30.8.2012.
203. Stechauner, G., and E. Kozeschnik. 2012. "Grain boundary and dislocation pipe diffusion in Al, Fe and Ni." Vienna University of Technology, unpublished research.
204. Jönsson, B. 1995. "Assessment of the Mobilities of Cr, Fe and Ni in bcc Cr-Fe-Ni Alloys." *ISIJ International* 35: 1415–21. doi: http://dx.doi.org/10.2355/isijinternational.35.1415.
205. Fischer, F. D., J. Svoboda, F. Appel, and E. Kozeschnik. 2011. "Modeling of Excess Vacancy Annihilation at Different Types of Sinks." *Acta Materialia* 59: 3463–72. doi: http://dx.doi.org/10.1016/j.actamat.2011.02.020.
206. Svoboda, J., F. D. Fischer, and P. Fratzl 2006. "Diffusion and Creep in Multi-Component Alloys with Non-Ideal Sources and Sinks for Vacancies." *Acta Materialia* 54: 3043–53. doi: http://dx.doi.org/10.1016/j.actamat.2006.02.041.
207. Svoboda, J., and F. D. Fischer. 2009. "Vacancy-Driven Stress Relaxation in Layers." *Acta Materialia* 57: 4649–57. doi: http://dx.doi.org/10.1016/j.actamat.2009.06.016.

208. Böhm, H. 1992. *Einführung in die Metallkunde* (p. 71). Mannheim: B.I. Hochschultaschenbuch, Band 196, Bibliographisches Institut.
209. Guinier, A. 1938. "Structure of Age-Hardened Aluminium-Copper Alloys." *Nature* 142: 569–70. doi. http://dx.doi.org/10.1038/142569b0.
210. Preston, G. D. 1938. "Structure of Age-Hardened Aluminium-Copper Alloys." *Nature* 142. 570. doi: http://dx.doi.org/10.1038/142570a0.
211. Hardouin Duparc, O. B. M. 2010. "The Preston of the Guinier-Preston Zones. Guinier." *Metallurgical and Materials Transactions* 41A: 1873–82. doi: http://dx.doi.org/10.1007/s11661-010-0320-5.
212. Pogatscher, S., H. Antrekowitsch, H. Leitner, T. Ebner, and P. J. Uggowitzer. 2011. "Mechanisms Controlling the Artificial Aging of Al–Mg–Si Alloys." *Acta Materialia* 59: 3352–63. doi: http://dx.doi.org/10.1016/j.actamat.2011.02.010.
213. Arieta, F. G., and C. M. Sellars. 1993. "Precipitation Kinetics of Nb(C,N) in Austenite for the Low C High Nb HSLA Steel." In *International Symposium on Low-Carbon Steels for the 90s*, edited by. R. Asfahani and G. Dither. The Minerals, Metals and Materials Society.
214. Crusius, S., G. Inden, U. Knoop, L. Höglund, and J. Ågren. 1992. "On the Numerical Treatment of Moving Boundary Problems." *Zeitschrift für Metallkunde* 9: 673–8.
215. Crusius, S., L. Höglund, U. Knoop, G. Inden, and J. Ågren. 1992. "On the Growth of Ferrite Allotriomorphs in Fe-C Alloys." *Zeitschrift für Metallkunde* 10: 729–37.
216. Borgenstam, A., A. Engström, L. Höglund, and J. Ågren. 2000. "DICTRA, a Tool for Simulation of Diffusional Transformations." *Journal of Phase Equilibria* 21: 269–80.
217. Inden, G., and C. R. Hutchinson. 2003. "Interfacial Conditions at the Moving Interface During Growth of Ferrite from Austenite in Fe-C-X alloys." In *Austenite Formation and Decomposition*, edited by E. B. Damm and M. J. Merwin, pp. 65–79. TMS.
218. Inden, G. 2003. "Computer Modeling of Diffusion Controlled Transformations." In *Thermodynamics, Microstructures and Plasticity*, edited by A. Finel et al., pp. 135–53. Kluwer Academic Publishers.
219. Hillert, M., J. Odqvist, and J. Ågren. 2004. "Interface Conditions During Diffusion Controlled Phase Transformations." *Scripta Materialia* 50: 547–50.
220. Schneider, A., and G. Inden. 2004. "Computer Simulation of Diffusion Controlled Phase Transformations." In *Continuum Scale Simulation of Engineering Materials*, edited by D. Raabe et al., pp. 3–36. Wiley-VCH.
221. Inden, G. 2007. "Diffusion and Phase Transformations in Multi-Component Systems." *Defect and Diffusion Forum* 263: 11–20.
222. Engström, A., L. Höglund, and J. Ågren. 1994. "Computer Simulation of Diffusion in Multiphase Systems." *Metallurgical and Materials Transactions A* 25: 1127–34.
223. Quadakkers, W. J., R. Schulten, K. Bongartz, and H. Nickel. 1987. "Mathematical Modelling for Carbon Diffusion and Carbide Precipitation in NiCr-Based Alloys." *10th International Conference on Metal Corrosion*, pp. 1737–46. Madras/India: Trans Tech Publisher.
224. Chen, W., J. Jeppson, and J. Ågren. 2008. "Analytical Treatment of Diffusion During Precipitate Growth in Multicomponent Systems." *Acta Materialia* 56: 1890–6.
225. Ghosh, G., and G. B. Olson. 2001. "Simulation of Para-Equilibrium Growth in Multicomponent Systems." *Metallurgical and Materials Transactions* 32A: 455–67.
226. Inden, G. 2008. "Kinetics of Phase Transformation in Multi-Component Systems." In *Materials Issues for Generation IV Systems*, edited by V. Ghetta et al., pp. 113–40. Springer Science.
227. Zureck, C. 1993. PhD-Thesis Aachen. VDI-Verlag, Series 5, No. 370.
228. Hillert, M. 1969. "The Role of Interfaces in Phase Transformations." Proceedings of the *Int. Conf. on The Mechanism of Phase Transformations in Crystalline Solids*, Manchester, pp. 231–47. London: The Institute of Metals.
229. Hillert, M. 1972. "On Theories of Growth During Discontinuous Precipitation." *Metallurgical and Materials Transactions* 33: 2729–41.

230. Jönsson, B. 1992. *TRITA-MAC Report 0478 of the Division of Physical Metallurgy.* Stockholm: Royal Institute of Technology.
231. Olson, G. B. 1997. "Computational Design of Hierarchically Structured Materials." *Science* 277: 1237–42.
232. Olson, G. B. 2000. "Designing a New Material World." *Science* 288: 993–8.
233. Saha, A., and G. B. Olson. 2007. "Computer-Aided Design of Transformation Toughened Blast Resistant Naval Hull Steels: Part I." *Journal of Computer Aided Materials Design* 14: 177–200.
234. Campbell, C. E., and G. B. Olson. 2000. "Systems Design of High Performance Stainless Steels I. Conceptual and Computational Design." *Journal of Computer Aided Materials Design* 7: 145–70.
235. Kuehmann, C. J., and G. B. Olson. 2009. "Computational Materials Design and Engineering." *Materials Science and Technology* 25: 472–8.
236. Kuehmann, C. J., B. E. Tufts, and P. Trester. 2008. "Computational Design for Ultra-High-Strength Alloy." *Advanced Materials and Processes* 166: 37–40.
237. Jou, H.-J., P. Voorhees, and G. B. Olson. 2004. "Computer Simulations for the Prediction of Microstructure/Property Variation in Aeroturbine Disks." *Superalloys* 2004: 877–86.
238. Harlow, D. G. 2007. "Probabilistic Property Prediction." *Engineering Fracture Mechanics* 74: 2943–51.
239. Jou, H.-J., A. Misra, and C. Kuehmann. "Precipitation Modeling for Accelerated Alloy Development." in 53rd AIAA/ASME/ASCE/AHS/ASC Structures, Structural Dynamics, and Materials Conference, April 23-26, 2012, Honolulu, Hawaii, AIAA-2012-1488-893.
240. Olson, G. B., H.-J. Jou, J. Jung, J. T. Sebastian, A. Misra, I. Locci, and D. Hull. 2008. "Precipitation Model Validation in 3(rd) Generation Aeroturbine Disc Alloys." *Superalloys* 2008: 923–32.
241. http://www.questek.com/press-releases.html?NewsID=12; and http://www.questek.com/press-releases.html?NewsID=51
242. Kumar, S., and D. Ramkrishna. 1996. "On the Solution of Population Balance Equations by Discretization .1. A fixed Pivot Technique." *Chemical Engineering Science* 51: 1311–32.
243. Bouaswaig, A. E., and S. Engell. 2010. "Comparison of High Resolution Schemes for Solving Population Balances." *Industrial & Engineering Chemistry Research* 49: 5911–24.
244. Akaiwa, N., and P. Voorhees. 1994. "Late-Stage Phase-Separation – Dynamics, Spatial Correlations, and Structure Functions." *Physical Review E* 49: 3860–80.
245. Farjami, S., and P. Voorhees. 2012. Unpublished research.
246. http://www.hdfgroup.org.
247. http://www.phy.ornl.gov/csep/CSEP/CORNELL/TUTORIAL/PLOTMTV/OVERVIEW.html.
248. http://plasma-gate.weizmann.ac.il/Grace.
249. Rastogi, P. K., and A. J. Ardell. 1969. "Coherent Solubilities of Gamma' in Ni-Al Ni-Si and Ni-Ti Alloys." *Acta Metallurgica* 17: 595–602.
250. Jou, H.-J., G. Olson, T. Gabb, A. Garg, and D. Miller. 2012. "Characterization and Computational Modeling of Minor Precipitate Phases in LSHR Alloy." *Superalloys* 2012 (in print).
251. Glamm, R. J. 2011. "Materials Design for Joinable, High Performance Aluminum Alloys." PhD thesis. Northwestern University, USA.
252. Feinberg Z. D. 2012. "Design and Optimization of an Austenitic TRIP Steel for Blast and Fragment Protection." PhD thesis. Northwestern University, USA.
253. Kozeschnik, E. 1999. "Multi-Component Diffusion Simulation based on Finite Elements." *Metallurgical and Materials Transactions* 30A: 2575–82.
254. Kozeschnik, E., J. Svoboda, P. Fratzl, and F. D. Fischer. 2004. "Modelling of Kinetics in Multi-Component Multi-Phase Systems with Spherical Precipitates II. – Numerical Solution and Application." *Materials Science Engineering A* 385: 157–65.
255. Cao, W., S. L. Chen, F. Zhang, K. Wu, Y. Yang, Y. A. Chang, R. Schmid-Fetzer, and W. A. Oates. 2009. "PANDAT Software with PanEngine, PanOptimizer and PanPrecipitation for Multi-Component Phase Diagram Calculation and Materials Property Simulation." *CALPHAD* 33: 328–42.

256. Cao, W., F. Zhang, S.-L. Chen, C. Zhang, and Y. A. Chang. 2011. "An Integrated Computational Tool for Precipitation Simulation." *JOM* 63: 29–34.
257. Chen, S. L., S. Daniel, F. Zhang, Y. A. Chang, X. Y. Yan, F. Y. Xie, R. Schmid-Fetzer, and W. A. Oates. 2002. "The PANDAT Software Package and its Applications." *CALPHAD* 26: 175–88.
258. Robson, J. D., M. J. Jones, and P. B. Prangnell. 2003. "Extension of the N-Model to Predict Competing Homogeneous and Heterogeneous Precipitation in Al-Sc Alloys." *Acta Materialia* 51: 1453–68.
259. Mao, J. 2002. "Gamma Prime Precipitation Modeling and Strength Responses in Powder Metallurgy Superalloys." p. 140. PhD Thesis in Department of Mechanical and Aerospace Engineering. West Virginia University, Morgantown, West Virginia.
260. Anderson, W. A. 1959. *Precipitation from Solid Solution.* Metals Park, OH: American Society for Metals.
261. Myhr, O. R., O. Grong, and S. J. Andersen. 2001. "Modelling of the Age Hardening Behaviour of Al-Mg-Si Alloys." *Acta Materialia* 49: 65–75.
262. Guo, J., W. Cao, S. Scott, et al. 2010. "Prediction of Microstructure and Mechanical Properties in Aluminum Castings after Heat Treatment" In *TMS 2010 Annual Meeting & Exhibition*. Seattle, WA.

INDEX

Check Out Other Momentum Press Material Science and Engineering Titles

Diffuse Scattering and the Fundamental Properties of Materials, *Rozaliya Barabash, Gene E. Ice, and Patrice E. A. Turchi,* Editors
Bonds and Bands in Semiconductors, Second Edition, *J. C. Phillips and Gerald Lucovsky*
11 Books from our Characterization Series that include such subjects as metals and alloys, tribological materials, silicon processing, catalytic materials, and ceramics just to name a few
Plastics Technology Handbook – Volumes 1 and 2, *Don Rosato, Marlene Rosato, and Nick R. Schott,* Editors
Polymer Testing: New Instrumental Methods, *Muralisrinivasan Subramanian*
Solid-State NMR: Basic Principles & Practice, *David C. Apperley, Robert K. Harris, and Paul Hodgkinson*

Announcing Digital Content Crafted by Librarians

Momentum Press offers digital content as authoritative treatments of advanced engineering topics, by leaders in their fields. Hosted on ebrary, MP provides practitioners, researchers, faculty and students in engineering, science and industry with innovative electronic content in sensors and controls engineering, advanced energy engineering, manufacturing, and materials science. **Momentum Press offers library-friendly terms:**

- perpetual access for a one-time fee
- no subscriptions or access fees required
- unlimited concurrent usage permitted
- downloadable PDFs provided
- free MARC records included
- free trials

The **Momentum Press** digital library is very affordable, with no obligation to buy in future years.

For more information, please visit **www.momentumpress.net/library** or to set up a trial in the US, please contact **Adam Chesler**, *adam.chesler@momentumpress.net*.